Automated Reasoning

Automated Reasoning Series

VOLUME 1

Automated Reasoning

Essays in Honor of Woody Bledsoe

Edited by

ROBERT S. BOYER
Computer Sciences Department,
University of Texas at Austin, U.S.A.

SPRINGER SCIENCE+BUSINESS MEDIA, B.V.

ISBN 978-94-010-5542-0 ISBN 978-94-011-3488-0 (eBook)
DOI 10.1007/978-94-011-3488-0

Printed on acid-free paper

Contents

Preface

These essays have been written to honor W. W. Bledsoe, a scientist who has contributed to such diverse fields as mathematics, systems analysis, pattern recognition, biology, artificial intelligence, and automated reasoning. The first essay provides a sketch of his life, emphasizing his scientific contributions. The diversity of the fields to which Bledsoe has contributed is reflected in the range of the other essays, which are original scientific contributions by some of his many friends and colleagues. Bledsoe is a founding father of the field of automated reasoning, and a majority of the essays are on that topic. These essays are collected together here not only to acknowledge Bledsoe's manifold and substantial scientific contributions but also to express our appreciation for the great care and energy that he has devoted to nurturing many of the scientists working in those scientific fields he has helped found.

Robert S. Boyer
Austin

February, 1991

Acknowledgements

Thanks to Larry Wos, editor of the *Journal of Automated Reasoning*, and Derek Middleton and Martin Scrivener, Kluwer Academic editors, for supporting the idea of initiating this collection of essays. Thanks to A. Michael Ballantyne and Michael Spivak, for help with LaTeX, especially in identifying many formatting problems and providing fixes.

Thanks to the following individuals for help in obtaining financial support of the Bledsoe Symposium (planned for November 15 and 16, 1991); S. Kamal Abdali, from the National Science Foundation (NSF Grant CCR-9102868); Alfred Dale, from the Computer Sciences Department of the University of Texas at Austin; Alfred King, from the King Foundation; Vesko Marinov, from Ricoh; Grant Dove, from MCC; Donald Good, from Computational Logic; and Neil Iscoe, from EDS.

We are grateful for research support under National Science Foundation Grant MIP-9017499.

The views and conclusions expressed in this volume are those of the authors and should not be interpreted as necessarily representing the official policies, either expressed or implied, of the National Science Foundation, the United States government, or any of the corporate sponsors of the Bledsoe Symposium.

R. S. B.

Chapter 1

A Biographical Sketch of W. W. Bledsoe

Anne Olivia Boyer
Robert S. Boyer

614 West 32nd Street
Austin, Texas

1.1 Introduction

W. W. Bledsoe is a major figure in the evolution of the new scientific field *artificial intelligence* and one of the founding fathers of the related scientific field *automated reasoning*. Into the following biographical sketch of Bledsoe we weave personal, educational, historical, social, and scientific commentary. At the time we write, Bledsoe is an active contributor to science and education at the University of Texas at Austin. We hope that our fondness for Bledsoe, whom we have known well for twenty-three years, has not clouded our assessment of his many achievements. We are certain that we have failed to treat adequately many aspects of Bledsoe's life prior to our first meeting him in 1966, and sadly fear that lack of space and lack of investigative effort cause us to omit quite a few interesting aspects of his career since then. We hope, however, that this short sketch of Bledsoe will please his friends and perhaps provide some useful information for a future biographer or historian of science.

1.2 Early Days

Woodrow Wilson ("Woody") Bledsoe was born November 12, 1921, in Maysville, Oklahoma. He was named for the then recent president of the United States, whose policies and idealism his parents greatly admired. At the time of Woody's birth, the Bledsoe clan had swelled to a considerable size. His father Thomas brought six children with him when he married his second wife, Eva Matthews, and together they had six more children. Woody was Eva's fourth child. The others were Pickens, Mary Ellen, Sula, Charles, and Tom.

It was not an easy childhood. Thomas Bledsoe had owned a plantation in Alabama, but a hired hand upset some turpentine one evening, a fire broke out, and the farm was lost. After that, it was moving from place to place, sharecropping. Eventually, the family settled in Oklahoma. Like so many other farmers in Oklahoma at that time, Thomas found that trying to wrest a living from the land was exhausting and hardly profitable.

The children, however, were happy living in the country and scarcely aware that they lived in poverty. Most unfortunately for the family, Thomas died in 1934. It was a blow from which he never recovered, says Woody, who was only twelve at the time. The Bledsoes were obliged to move into town (Lexington, Oklahoma), and the children were forced to confront their poverty. Now they knew they were, indeed, very poor.

"We scrambled," says Woody, who mowed lawns and delivered papers and did any honorable work to keep body and soul together. His mother was managing a small dress shop, but there were few women who could afford a new dress in those days, and sales were few. Woody worked for about a dollar a day. At one time he worked for a restaurant that provided employees free pies, and Woody once found himself covered with boils after eating several pecan pies in a single day.

The Depression. Hard times all around. Elsewhere in Oklahoma, people were packing up and fleeing the Dust Bowl. Everything seemed to conspire against even the most hardworking folks. Famous folksinger Woody Guthrie, also from Oklahoma, must have had kids like Woody Bledsoe in mind when he sang his now classic songs about the grit and gumption of the poor people in Oklahoma then. Woody Bledsoe might have found the going rough, but he did not cave in to adversity, and discovered that even in poverty there are joys to be savored in life. He did very well in school, and distinguished himself in mathematics, especially under the guidance of teachers Ellen Sherman and Lois Peyton. He also proved to be extremely adept at tennis, which he continues to play enthusiastically. In his high school years, Woody ran away from home several times, catching rides by hitchhiking or riding freight trains throughout Texas and Oklahoma—a common enough method of transportation during the Depression. In his senior year, to spare his mother some of the burden of supporting so many children, he left home again. He traveled to Calliham in South Texas, arriving unannounced at the door of an acquaintance (Lon Claunch), with whom he spent a year, graduating from Calliham High School in 1939.

The next year held poverty and uncertainty. Woody moved back to Oklahoma and spent most of the next year washing dishes in a restaurant. Hoping to break free from drudgery, working part time to support himself, he entered the University of Oklahoma, but the effort to support himself proved too much even for Woody's apparently boundless energy, and he dropped out to join the army on May 27, 1940.

Life in the army was not much of an improvement, at least in terms of personal fulfillment. As Bledsoe puts it, "Soldiers of the old army were shunned by civilians," and he felt intensely lonely. World War II, however, changed all that, when so many of the young men in the nation received a summons from the government to join the fight. Now it was "okay," even admirable, to be in the armed forces. Bledsoe attended Officers Candidacy School in 1942, was commissioned in the Corps of Engineers, and eventually achieved the rank of captain.

In July, 1943, one of his buddies, Richard Norgaard, in the Corps of Engineers persuaded him to come home with him on leave to Salt Lake City. As it happened, Richard had a pretty sister, Virginia, and as both recall, it was love at first sight. Inspired by her beauty, the usually circumspect Bledsoe was even moved to write a poem in her praise. They corresponded, Bledsoe called her by telephone to propose marriage, and on January 29, 1944, the couple was married in her home.

Eventually, four children were born to them: Margaret, Greg, Pamela, and Lance. Margaret died shortly after birth, in a dysentery epidemic during World War II, while Bledsoe was in Europe. Incredibly, fifty-six of the ninety-three babies born in the hospital during the ten days of Virginia's confinement died because of a shortage of penicillin and skilled medical help due to the war effort. Many years later, in 1985, Greg was killed in an automobile accident. Greg's death was the most severe blow Woody and Virgina have ever sustained. Greg left three children. Pamela, who has a daughter, is a traffic consultant. Lance, who has four children, has followed his father's footsteps into the computer industry.

Woody and Virginia have shared their joys and struggles together for nearly fifty years now, and both attribute much of their spiritual strength and growth to their commitment to the Church of Jesus Christ of the Latter Day Saints (Mormons). Woody was not a Mormon when he married Virginia, but he sat by her side in church many Sundays for eighteen years, read *The Book of Mormon*, and so was eventually converted, on September 21, 1961. He remains very active in the church to this day; he has served as a bishop twice (altogether about four years) and once as a counselor of the stake presidency. (In terms of Roman Catholic organization, the position of Mormon bishop is roughly equivalent to head of a parish and a stake presidency is roughly equivalent to head of a small diocese.)

In August 1944, Bledsoe participated in the war in Europe as a member of Patton's Third Army, in the Army Corps of Engineers. Bledsoe received the Bronze Star medal for his heroic activities in arranging the transportation of troops across the Rhine in March, 1945. At that time, all the Rhine bridges except the one at Remagen had been destroyed by the retreating German army. Patton's Third Army decided to cross the Rhine by boats near Frankfurt rather than suffer the delay of waiting for bridge construction. Therefore the Army Corps of Engineers hauled naval landing craft (designed for beach landings) by truck across Europe to ferry troops across the Rhine. Bledsoe, by then an Army captain, recalls that there was only light enemy fire during the crossing; his main contribution was figuring out how to get the very large landing craft down narrow roads and actually into the water. His first "research" was experimenting with techniques for launching the craft from trucks into the Mosel River; initial experiments were disastrous, but ingenuity prevailed. The simple idea of backing the trucks into the water, floating off the boats, and hauling the trucks back out with tractors turned out to be the key. While hauling trucks down to the Rhine, Bledsoe temporarily confused some allied forces about the location of German artillery by dynamiting away a railroad bridge that created an overhead blockage against the passing of the landing craft. Near the Rhine, Bledsoe took his orders from Patton's headquarters in Kaiserslautern, Germany. (Kaiserslautern is now a major center of work in artificial intelligence, under the direction of Bledsoe's longtime colleague Michael Richter. At the Tenth Conference on Automated Deduction, held in Kaiserslautern in 1990, Bledsoe delivered the banquet address in the nearby Hambacher Schloß, site of one of the earliest German democratic movements.) Bledsoe remained in Europe until October, 1945.

Upon returning to the United States, Bledsoe enrolled as an undergraduate at the University of Utah in Salt Lake City, Utah. After considering electrical engineering and physics, Bledsoe chose to major in mathematics. Perhaps the decisive point in Bledsoe's decision to become a mathematician was his discovery on his own of a proof of the nontrivial theorem that every real-valued continuous function defined on the reals has a Riemann integral over any closed interval. Bledsoe proved this theorem in a course taught by Ferdinand Biesele in the Socratic or Moore style, wherein the students find and present proofs in class rather than listen to lectures. As a mathematics professor, Bledsoe himself would emphasize to his students the importance of finding for themselves proofs of hard theorems. His research achievements have been to a significant extent in the area of automating proof discovery.

1.3 Berkeley

After graduation from Utah, Bledsoe pursued a Ph. D. in mathematics at the University of California at Berkeley under a full fellowship. Berkeley was then (and is still) a dominant force in American mathematics. Among the many renowned faculty members there from whom Bledsoe took classes were the logician A. Tarski and the topologist J. Kelley. Bledsoe did his Ph. D. thesis under the analyst A. P. Morse, about whom we briefly digress.

1.3.1 A. P. Morse

Bledsoe's advisor A. P. Morse believed in the viability and desirability of being absolutely formal in the statement of mathematical theorems. In his book on set theory [58], Morse formally develops in about a hundred pages a flexible syntax akin to ordinary mathematical notation, predicate logic, set theory (including ordinal and cardinal arithmetic), and elementary topology. Remarkably, each theorem is stated in the formal logic. Considering the relatively elementary character of the theorems established in *Principia Mathematica*, we suspect that Morse's book stands as the most mathematically advanced but totally formal treatise on mathematics ever published. Bledsoe's work was heavily influenced by Morse's work in several ways, and thus we elaborate on what we regard as some of Morse's contributions to formal reasoning.

- Morse provides a logic in which mathematics can be conveniently expressed via an extremely flexible and extensible syntax and via a very sturdy and powerful definition principle. (Most formal logics, e.g., that of [61], are intended to be *studied* rather than *used* by people. As a result, their grammars tend to be unnatural because they do not support common notatation. Worse, from the practical point of view, these logics fail to have principles of definition, especially for the definition of concepts introducing new kinds of quantification such as the set builder notation or the definite integral. Rather, definition is regarded as an extralogical, abbreviational, inessential, and extraneous activity.)

- Whenever possible, Morse formulates theorems that are mechanical in spirit, like the equations for ordinary algebraic operations. Morse goes out of his way, for example, to define his concepts in such a way that theorems can be stated as equations, without hypotheses.

Thus it is not surprising that it was a student of Morse (namely Bledsoe) who produced the first powerful prover for set theory, a prover that incorporated a rich conceptual and definitional facility and that derived much of its power from equational set theory theorems.

After he graduated from Berkeley, Bledsoe continued an extensive collaboration and correspondence with A. P. Morse. In 1981, during a meeting of the

American Mathematical Society in San Francisco, Bledsoe organized a dinner held in honor of A. P. Morse; attendees included not only many of Morse's Ph. D. students but also many distinguished members of the mathematics department at the University of California at Berkeley, including A. Tarski and J. Kelley.

1.4 Mathematical Contributions

Bledsoe regards mathematics as his first and greatest academic love, over and above pattern recognition, artificial intelligence, systems analysis, and automated reasoning. Bledsoe has made a number of contributions to the mathematical literature, including these publications in regular mathematics journals: [5, 6, 7, 9, 17, 27, 30, 32].

We now briefly review Bledsoe's mathematical work. Our review is literal in part because the authors of this biography lack the mathematical background to assess the work properly. Why then do we review it at all? We review it in order to emphasize the point that Bledsoe is a "real mathematician." We believe this point deserves emphasis as a lesson to potential contributors to automated reasoning. Most of the most important contributors to automated reasoning had already become serious mathematicians before beginning research on automated reasoning, e.g., Martin Davis, Hao Wang, Bledsoe, Don Knuth, Wu Wen-Tsün, and Larry Wos. We suspect that this is because substantial mathematical ability is required to contribute to this field.

As a graduate student, before finishing his Ph. D. thesis, Bledsoe published two papers [5, 6]. The first of these [5] presents the following charming result:

> In the following ρ metrizes S and ρ' metrizes S'. We agree that a function f is *neighborly at the point* x if and only if for each $\epsilon > 0$ there exists an open sphere α of S such that $\rho(x, y) + \rho'(f(x), f(y)) \leq \epsilon$ whenever $y \in \alpha$. We also agree that a function is *neighborly* if and only if it is neighborly at each point of S. Obviously every continuous function is neighborly.
>
> It is well known that if g is a function on S to S' and if f is such a sequence of continuous functions that $\lim_{n\to\infty} \rho'(f_n(x), g(x)) = 0$ for each x in S, then the points of discontinuity of g form a set of first ρ category. It is the principal purpose of the present note to show that this same conclusion can be drawn when the approximating functions are merely restricted to being neighborly.

Bledsoe's Ph. D. dissertation, written under the supervision of A. P. Morse, is reported in [9]. The topic was "product measures." Concerning the general problem of product measures, Bledsoe remarks in [30]

> It has been known for many years that the product of two regular borel measures on compact hausdorff topological spaces may not be borel in the product topology. The problem of defining a new product measure that extends the classical product measure and carries over this borel property has been approached in different ways ...

Among the approaches is that presented in [9]. In summarizing this work, Bledsoe and Morse remark

> One of our aims is to free topological Fubini theory from the usual restraints of local compactness. In this connection we shall regard Radon measures as very special indeed. Another of our aims is to associate with any two measures μ and ν such a product measure ϕ that in the event μ and ν *happen* to be topological measures of a rather general sort, then our topology-free measure ϕ will actually be a similar measure under which the Borel subsets of the product space are measurable. We shall reach both these goals even though an obvious obstacle in our path to the second is the fact that the open sets of a topological product are not always contained in the σ-field generated by the open rectangles. The product measure we construct in order to overcome this difficulty allows more summable functions and greater freedom of action than the product measures considered heretofore. ...
>
> Suppose μ is an (outer) measure defined for all the subsets of a topological space $\mathcal{S}$. Suppose that $\mu(\mathcal{S}) < \infty$ and that each open set is not only measurable but equal in measure to the upper bound of the measure of closed subsets. Suppose further that from each covering of $\mathcal{S}$ by open sets a countable subfamily can be extracted which covers almost all of $\mathcal{S}$. If ν is another measure like μ, then our associated product measure bears to the topological product space the same relation as μ does to $\mathcal{S}$, and at the same time satisfies the Fubini equality for summable functions. The novel feature of our product measure is that we require to be of measure zero each set whose characteristic function integrates iteratively in both orders to zero.

In some cases, Bledsoe and Morse adopt a language, style, and formality that is quintessentially Morsean in its formality, precision, and succinctness, e.g., the following definition of Mspr:

> Mspr $\mu\nu\mathcal{F}$ = E$\phi \in$ Msr rct rlm μ rlm ν [$\mu \in$ Measure, $\nu \in$ Measure, $\mathcal{F} \in$ bsc ϕ, and, for each $a \in \mathcal{F}$,
>
> $$\int\!\!\int \mathrm{Cr}(x,y)A\mu\, dx\, v\, dy = \phi(A) = \int\!\!\int \mathrm{Cr}(x,y)A\nu\, dy\, \mu\, dx]$$

Bledsoe continued to pursue his investigations on product measures in the 70's. In [30] he and Wilks extend his Ph. D. dissertation work [9] by considering the case in which one of the two measure spaces is borel and the other is constrained by a separation of variables condition.

In [17] Bledsoe, writing again with Morse, generalizes two well-known results from the theory of metric spaces to topological measure theory. The principal result concerns the construction of a measure from a nonnegative set function.

Also, a short note by Bledsoe appears in the Monthly [27] proving that

> Every finite set $\{z_1, z_2, \ldots, z_n\}$ of complex numbers has a subset S such that
>
> $$\left|\sum_{z \in S} z\right| \geq \frac{1}{\pi} \sum_{j=1}^{n} |z_j| .$$

The two well-known mathematicians R. H. Bing and R. D. Mauldin collaborated with Bledsoe in [32] on a collection of set theory theorems with connections to the continuum hypothesis.

1.5 Sandia

After finishing his Ph. D. in mathematics at the University of California at Berkeley in 1953, Bledsoe decided to pursue research rather than take an academic post at Michigan or Virginia. He decided to work in the mathematics department at Sandia Corporation in Albuquerque, New Mexico, which did work almost exclusively on weapons for the Atomic Energy Commission (now the Department of Energy). Bledsoe was attracted both by the desire to focus his mathematical talent on questions in physics and by the desire to be involved in the then rapidly developing field of nuclear weapons. (The first hydrogen bomb was exploded shortly before Bledsoe joined Sandia.) Also influential in his decision to join Sandia was the fact that then, as now, the salary of mathematicians in industry was often an integer multiple of salaries in academia. Albuquerque became the Bledsoes' favorite town. At Sandia, Bledsoe worked in "systems analysis" in the mathematics department. His work there is almost entirely classified. Systems analysis is a term used to express engineering concerned with the interactions of components in a "system" that is itself composed of many other systems, e.g., a telephone system or a modern battlefield. (After Bledsoe left Sandia, an official there told him, in an attempt to entice him to return, that he was the best systems analyst that Sandia had ever had.) Concerning systems analysis, Bledsoe later remarks in [15]:

> In analyzing a system, we often are searching for a set of characteristics which will make the system most optimum in some

> sense. Frequently, this boils down to determining the "best" values of a set of parameters, and we find ourselves doing a *parameter study* in which we vary parametric values according to some policy while we search for an optimal value of a response function. Also, we often try to determine tradeoffs between the parameters.
>
> The system may include some random elements, such as a weather or enemy action, and often the response function we are trying to optimize is itself a random variable.
>
> Many such systems arise in the *war games* studied in weapons systems analysis. There the parameters varied are such things as aircraft velocity, yields of offensive weapons, yields and types of defensive weapons, spacing of vehicles, warning time, etc.

The one unclassified document that Bledsoe produced concerning his systems analysis work at Sandia is entitled "Program for Computing Probabilities of Fallout from a Large-Scale Thermonuclear Attack" [10]. The report presents a computer program for computing the distribution of fallout from nuclear weapons explosions. The program allows for variations in longitude, latitude, wind velocity, megatonage, and for such probabilities as one weapon destroying another. As a reminder of the difficulty of using early computers, we cannot resist mentioning a few archaic operating instructions necessary in those days:

- 1. Put the magnetic drum in a "four interlace" condition.
- 3. Put all weapons cards in READ hopper.
- 8. Start at 40100 (if MS1 is in "ON" position, machine halts with PAK 40104, so hit COMPUTE button).

An extensively commented program listing of about 1400 lines of code is included, with all of the instructions given in octal, and with no evidence of an assembler much less a compiler having been available for coding the computer, an ERA 1103.

By 1957, Bledsoe had become the head of the mathematics department at Sandia and remained so until his departure in 1960. He collaborated there frequently with his good friend Joe Weihe.

While working for Sandia, Bledsoe witnessed several atomic bomb explosions in Nevada. He also traveled to the South Pacific and witnessed a hydrogen bomb explosion (from the island of Eniwetock).

During his work for Sandia, Bledsoe recalls encounters with the extraordinary mathematicians S. Ulam and John von Neumann. Ulam showed him a hand drawing of the first hydrogen bomb design, for which the basic calculations were done on a hand calculator. The calculations could have been refined: the first hydrogen bomb explosion, the "Mike shot," was twice as powerful as expected, almost destroying some observers.

Once, when giving a lecture at Lawrence Livermore Labs on the possibility of using very small atomic "shells" in artillery, a person in the audience asked whether the calculation was "linear" in a certain sense. Bledsoe replied "No," at which point Herb York, the head of Livermore Labs said: "I agree with John von Neumann, I too believe it is linear." This was Bledsoe's introduction to von Neumann. Indeed, it was not "linear," but von Neumann's remark was close to true, something Bledsoe subsequently emphasized.

Toward the end of his stay at Sandia, Bledsoe and a very close friend, the biologist and genius polymath Iben Browning, began work on using computers to do recognition, first of characters and later of faces. Turning to work on what has become known as artificial intelligence was a revolutionary event in Bledsoe's intellectual life. Many years later, in his presidential address to the American Association of Artificial Intelligence in 1985, "I Had a Dream" [42], Bledsoe recalls that after he had his first dream about artificial intelligence:

> When I awoke from this daydream, ... I decided then and there to quit my job and set about spending the rest of my life helping bring this dream to reality.

Bledsoe and Browning's initial contribution to artificial intelligence was presented in the paper "Pattern Recognition and Reading by Machine" [11]. Although the principal topic at hand was the recognition of characters, the entire topic of pattern recognition was clearly imagined. Distinguishing the planned work from previous work by others, Bledsoe and Browning remark:

> All of these approaches prove upon inspection to center upon analysis of the specific characteristics of patterns into parts, followed by a synthesis of the whole from the parts. In these studies, pattern recognition of the whole, that is, Gestalt recognition, was chosen as a more fruitful avenue of approach and as a satisfactory problem for the initial phases of the over-all study.

In simplified form, the Bledsoe-Browning algorithm for recognizing characters is based upon the following idea. We imagine that we wish to recognize single characters that are presented as bit patterns in a fixed rectangular array of bits (i.e., pixels). Initially, we learn some characters. Later we wish to recognize some. When presented with a character to recognize, we will independently compute a "score" for the possibility that the character might be an A, B, or C, etc., and choose the possibility with the highest score. How do we score? Even before we have begun the learning process, we randomly divide the rectangular array into *pairs* of pixels, $p_1, p_2, p_3 \ldots$ During the learning phase, whenever we are asked to learn a given pattern as an instance of a specific letter, say A, we note for each pixel pair p_i the *state* of that pair: (0,0), (0,1), (1,0), or (1,1), depending on the states (on or off) of the individual pixels. Later when we are trying to recognize a character, we score one point towards the possibility that the character is an A for each i such that the p_ith pixel pair has the same state as it did when learning at least one instance of A. Clearly, the algorithm can be generalized for tuples other than pairs, and in fact this method is sometimes known as the "n-tuple method" instead of the "Bledsoe-Browning method."

An analysis of Bledsoe and Browning's method may be found in Ullman's [66]. Ullman presents results of tests he performed showing the Bledsoe-Browning method obtaining recognition rates of about 93% on hand-printed numerals and of about 95% on multifont printed numerals, using n-tuples with $n = 14$, at the cost of about 42 million bits of storage. Related work by Ullman on the Bledsoe and Browning n-tuple method may be found in [64, 65]. The latter reduces fourfold what Ullman terms the "storage gluttony of the Bledsoe and Browning system."

Besides the basic n-tuple method, the paper [11] describes several additional techniques that were employed to increase the accuracy of character recognition. The most successful of these techniques was to finalize the identity of characters using context: "using a vocabulary of words of the length in question, add in their proper order the letter scores of each word in the vocabulary to obtain a total score for each word."

Several themes that were to influence the remainder of Bledsoe's work in artificial intelligence can already be seen here:

- Reliance on partial information for making informed guesses: heuristics.
- Reliance on a variety of partial techniques, working together, to solve a problem, rather than seeking a single "uniform" strategy for solving a problem.
- Asking the question "how can we make a machine do something like what people do?"—e.g., using contextual information.

1.6 Panoramic Research, Inc.

While working at Sandia, Bledsoe, Iben Browning, and Lloyd Lockingen received encouragement from Department of Defense sources, promising financial support to set up a small research firm to pursue topics related to pattern recognition. The three left Sandia in 1960 and established the company Panoramic Research in Palo Alto, California; Panoramic grew to a staff of about 10 people. Bledsoe was president of Panoramic from 1963 to 1965, before his departure for the University of Texas at Austin in January 1966.

Panoramic was a very early company to focus upon research in the general area of artificial intelligence, including pattern recognition, facial recognition, genetics, and neural nets. Frequent visitors to Panoramic included Marvin Minsky, Raj Reddy, John McCarthy, Hans Bremermann, Seymour Papert, and Danny Bobrow, many of whom were founding fathers of artificial intelligence.

While at Panoramic, Bledsoe contributed a short note [13] that describes how the Heisenberg uncertainty principle can be used to show limits on the speed of a single von Neumann computer. For example,

> Thus a computer with a billion bits of storage can have an access time of no less than 10^{-18} seconds, unless materials are used whose density is greater than 20 grams/cm^3.

Perhaps Bledsoe's major research project at Panoramic was on computer facial recognition. Working with Helen Chan, Charles Bisson, and others, Bledsoe produced a system that was remarkably successful at recognizing faces in photographs. The system worked as follows.

A. Database Setup. A human operator examines a photograph and abstracts from it certain facial features, all of which are distances between fixed points on the face, for example the distance from the top to the bottom of an ear. The entry of this data was facilitated by the use of a very early analog-to-digital data entry device known as the Grafacon or Rand Tablet. About forty photographs per hour could be processed in this way by a relatively new operator. Before storing the distances in the database, the distances were mechanically normalized to compensate for head position, e.g., rotation and tilt.

B. Identification. Given a new photograph that we wish to match against some of the photographs already entered into the database, we compute a "pseudo distance" between the new photograph and each of the previously entered photographs. The distance is computed by summing the differences of similar features, first dividing each difference by a standard deviation of measurement for that feature. Photograph pairs that are "close enough" are then presented to a human, who then finally decides which photographs really match.

In experiments involving a database of several thousand different photographs, the computer was able to reduce the number of photographs that needed to be considered by the human to one percent of the photographs in the database. Although one of the authors of this note (RSB) has seen a report on this excellent project, Bledsoe was not permitted to publish the work because of the sensitivity of the photographs used in the experiment! Many reports on this project were produced, including [18, 19, 22, 24].

Even after Bledsoe left Panoramic, he continued to collaborate on the topic of facial recognition with Peter Hart at Stanford Research Institute, as reported in [26]. In that report are described experiments in facial recognition using (a) slight modifications of the technique mentioned above, (b) using modifications of those techniques based on Bayesian decision theory, and (c) using humans to do the entire job of photograph identification. It was concluded that both methods (a) and (b) worked approximately as well as one another and both worked considerably better than (c), but that (b) performed somewhat better on high quality photographs than (a).

While at Panoramic, Bledsoe also made some mathematical analyses of genetic mutation [15, 14] and addressed a National Academy of Sciences meeting on the topic of "convergence rates in pseudo-evolution" in Austin in 1962. He was one of the first to study the evolution of genetic structures by computer. (To this day, Bledsoe is not yet aware of any reasonable computer simulation of genetic mutations that would predict the evolution of entire new species.) Through his work on genetics, Bledsoe also began working with Wilson Stone, a renowned biologist at the University of Texas at Austin.

Another research topic pursued at Panoramic was an algorithm that an aircraft could use to fly low over mountainous terrain so as to avoid detection [20, 21].

After several years of the constant press of fundraising and management incumbent on the president of a small research firm, Bledsoe decided to pursue an academic career. He was strongly encouraged to join the University of Texas by Wilson Stone, who offered him *carte blanche* choice of department and salary.

1.7 The University of Texas

Bledsoe's many contributions to automated reasoning all were made after his move to the University of Texas. We discuss these contributions in a subsequent section.

1.7.1 Teaching

From the beginning of his tenure at Texas, Bledsoe has been a teacher of considerable inspirational influence. In 1966, in the first course at the University of Texas on artificial intelligence, a Lisp compiler for the CDC 6600 was produced by students (including James B. Morris) as a class project. Especially memorable to one of the authors (RSB) was a class that Bledsoe taught in the fall of 1967 in the Moore style, but based on Morse's book on set theory [58]. Among those attending the class were Forrest Baskett (later professor of computer science at Stanford, who conceived the Sun workstation); James B. Morris (later professor of computer science at Purdue); Robert Anderson (later professor of computer science at the University of Houston); Dallas Lankford (later professor of mathematics at the Louisiana Tech University) and Robert S. Boyer (later professor of computer science and mathematics at the University of Texas at Austin). Of these five, four went on to do Ph. D.'s in automated reasoning under Bledsoe.

Mark Moriconi, a former Bledsoe student and now director of the Computer Science Laboratory of SRI International, observes the following about Bledsoe's influence on his students (private communication):

Bledsoe is very focused, a stickler for detail, and always works from concrete examples. His approach to problem solving strongly influenced the way that I work on all kinds of things, both professionally and in everyday life, and I am sure that it has influenced other students as well. Moreover, he instilled high standards for quality in his students, always by example and not by fiat. Without question, Bledsoe has been an effective role model for his students, which after all is one of the most important, but difficult to meet, responsibilities of a professor.

Bledsoe's Ph. D. Students (to date)

John Wade Ulrich	1968, Computer Sciences
Stephen Charles Darden	1969, Computer Sciences
Charles Edward Wilks	1969, Mathematics
James Bertram Morris	1969, Computer Sciences
Robert Brockett Anderson	1970, Mathematics
Robert Stephen Boyer	1971, Mathematics
Dallas Lankford	1972, Mathematics
Vesko Genov Marinov	1973, Computer Sciences
Mark Steven Moriconi	1977, Computer Sciences
John Threecivelous Minor	1979, Computer Sciences
William Mabry Tyson	1981, Computer Sciences
Tie Cheng Wang	1986, Computer Sciences
Larry Marvin Hines	1988, Computer Sciences
Don Simon	1990, Computer Sciences

1.7.2 Visits to Other Universities

While a professor at the University of Texas, Bledsoe has traveled extensively. He spent the 1970-71 academic year at the Artificial Intelligence Laboratory of the Massachusetts Institute of Technology, as a guest of Marvin Minsky. In 1973 Bledsoe visited Bernard Meltzer, J Moore, and the authors at the Metamathematics Unit at Edinburgh University. (While in Edinburgh he toured the Highlands in Moore's taxi and played golf at St. Andrews.) In 1978 Bledsoe spent a semester at Carnegie-Mellon University, where he began a collaboration with Doug Lenat that continued later at MCC.

1.7.3 Administration

Despite having received assurances from the University of Texas that he would not be asked to serve as chairman of the mathematics department, Bledsoe was nevertheless lured into being chairman twice, 1967-69 and 1973-75. The first period of Bledsoe's chairmanship involved considerable administrative tension because the department was sharply divided between the "third floor" group, which included the eighty-year old topologist R. L. Moore, and the rest of the department. (The third floor group was so named because it occupied the entire third floor of the mathematics building. The group had in common a dedication to the Moore method of teaching, which opposed the use of lectures and textbooks.) Eventually the remainder of the mathematics department and Dean John Silber (who was himself fired not much later) determined finally to retire Moore, who had taught at the University since 1920.

1.8 IJCAI, CADE, and AAAI

Bledsoe has been exceptionally active in the building of organizations and conferences for the fields of artificial intelligence and automated reasoning. For example, he was the general chairman of the International Joint Conference on Artificial Intelligence at MIT in 1977 and a member of the IJCAI board of trustees from 1978 to 1983. He hosted the Conference on Automated Deduction in Austin in 1979. He has served numerous times on program committees for IJCAI, AAAI, and CADE.

In 1983, Bledsoe was named president-elect of the American Association of Artificial Intelligence. In 1984, he helped organize the annual AAAI meeting in Austin, where he was installed as president of AAAI. In an address to the 1984 AAAI meeting, outgoing president John McCarthy expressed dismay at the extremely large size of the meeting, over four thousand; in less than twenty years the field had grown from a handful of practitioners to a giant academic and commercial enterprise. Sociologists of academia may note that organizing such major conferences is remarkably frustrating and time consuming. Bledsoe vowed at the end of the conference that he would never again run another conference at the University of Texas because of the bureaucratic ensnarlments he encountered in attempting to provide space for the many computer manufacturers who wished to advertize their wares at the conference, a surprising vow from a man well known for his commitment to positive thinking.

In his 1985 presidential address to the American Association of Artificial Intelligence [42], we find no technical monologue but rather an extremely frank and personal expression of Bledsoe's vision for artificial intelligence and his faith in its future. He opens the address by remarking:

> Twenty-five years ago I had a dream, a *daydream*, if you will. A dream shared with many of you. I dreamed of a special kind of computer, which had eyes and ears and arms and legs, in addition to its "brain."

In the address, Bledsoe is clear that the dream has not yet been fulfilled:

> The twenty-five years have not been totally kind to my dream: Shakey[1] liked shaking more than running and thinking, and was laid aside for a season; language translation sputtered, died, and was resurrected; facial recognition was pushed back on the researchers stack; automatic provers showed signs of growing pains, which disheartened the fainthearted; no machine stepped forward to try the Turing test; robot arms were duplicating block castles instead of playing squash; etc., etc., etc.; many AI researchers lost faith and dropped out.

Yet Bledsoe is adamant that progress is being made:

> First, let me express my annoyance with some of the distracted individuals who criticize AI researchers for not "jumping to infinity" in one leap. Somehow, to them it is OK to work step by step on the dream of obtaining controlled thermonuclear energy or a cure for cancer or a cure for the common cold, but no such step-by-step process is allowed for those trying (partially) to duplicate the intelligent behavior of human beings. To these cynics, a natural language system that converses with us in a limited form of English is not a legitimate step toward passing the Turing test. ...Indeed, almost all of our AI accomplishments have been of the *partial* kind: natural language processors that handle a *subset* of English (or French, etc.); systems that recognize and synthesize *limited* forms of speech; character recognition machines that read only typewritten characters; expert systems that perform a variety of tasks (but not all that a human can); theorem provers that can prove difficult theorems in a *particular* area of mathematics or that can handle the inferencing needed for elementary expert systems, including monotonic reasoning; programs that play expert-level chess; programs that exhibit an elementary level of learning and reasoning by analogy. And the list goes on.

In this presidential address, Bledsoe also describes the characteristics of a good researcher, enumerates the important research topics in AI, and expresses considerable enthusiasm for the prospects of progress in AI.

1.9 MCC

In late 1983, after an intensive consideration of alternative sites, and with considerable financial incentives from the University of Texas, Admiral Bobby Inman (former assistant director of the CIA and former director of the NSA)

[1] A robot developed at SRI in the early 70's.

opened up in Austin the Microelectronics and Computer Technology Corporation. MCC, as it is generally known, was backed by commitments of research funding by over twenty American computer companies to do advanced research. Such unprecedented collaboration was endorsed by a congressional act explicitly waiving anti-trust prohibitions against collaborative efforts on research. The consortium was developed in part as a response to a fear that the Japanese computer industry was on the road to surpassing American dominance in computing.

In early 1984, Bledsoe indicated to Inman a general interest in supporting the goals of MCC, even to the extent of spending some time at MCC. Inman quickly offered Bledsoe the position of vice-president for artificial intelligence, a multi-million dollar per year budget, and the charter to set up a first class AI research center. Bledsoe did set up such a center and directed it for three years.

Shortly after Bledsoe took leave from the University of Texas to work at MCC, the Austin Chamber of Commerce's glossy publication *Austin Magazine* did a story on artificial intelligence. On the cover is a handsome photograph of Bledsoe, with the headline "Machines Who Think—Austin scientists are at the forefront of research to create a new breed of computers that think like humans." In the photograph, Bledsoe is leaning on a computer terminal, whose display reads "I Think, Therefore I Am." A subhead reads "MCC's Woodrow Bledsoe and friend."

During this period, perhaps the most visible artificial intelligence project established at MCC was the CYC project of Doug Lenat. This ambitious project has as its objective the encoding into a computer of all the basic principles and fundamental facts of knowledge necessary for reasoning, perhaps the most ambitious scientific undertaking ever.

Bledsoe left MCC in 1987 to return to the joys of teaching and doing research in automated reasoning.

1.10 Return to the University of Texas

Upon returning to the University of Texas at Austin, Bledsoe was appointed to the "million dollar" Peter O'Donnell, Jr. Chair in Computing Systems, named after a philanthropist extremely generous in endowing major chairs in science and engineering at the University of Texas.

Under the philanthropy of the computer scientist and entrepreneur Ed Fredkin, some prizes in the area of artificial intelligence have been established. Bledsoe chaired and initially formed one such prize committee, now known as the Automatic Theorem Proving Prize Committee of the American Mathematical Society. Under Bledsoe's chairmanship, the committee's first prizes were awarded at a special session in the 1983 Denver meeting of the American Mathematical Society, which Bledsoe organized. The first "milestone" prize of the committee was given to Hao Wang, and the first "current" prize was awarded to L. Wos and S. Winker. A second milestone prize was later awarded to J. A. Robinson. Bledsoe left the committee in 1985. In 1991 Bledsoe was awarded the third "milestone" prize, by a committee then chaired by the Harvard mathematician D. Mumford. In characteristic bashfulness and modesty, Bledsoe's first remark after learning of the award was that it should have been awarded to someone else. The award was presented by John McCarthy at a special session on automated reasoning at the 1991 meeting of the American Mathematical Society in San Francisco. The citation for the prize reads thus:

> The Milestone Award of 1991 goes to Woodrow W. Bledsoe, who has been a central figure in Automatic Theorem Proving, inspiring and guiding this field for over twenty years. His broad view of the subject, using resolution and non-resolution techniques, his deep study of theorem proving in analysis and with inequalities, and his work on interactive theorem provers distinguish him as a major innovator in the field.

Bledsoe received the 1991 Distinguished Service Award of the International Joint Conferences on Artificial Intelligence.

1.11 Automated Reasoning

We now review Bledsoe's contributions to automated reasoning. Bledsoe's receipt of the "milestone" award for contributions to automatic theorem proving is a clear indication than his contributions in automated reasoning have been truly major. What are the respects in which Bledsoe has contributed to automated reasoning?

- Wrote, with J. Morris, one of the first, and one of the best-ever "proof-checking" programs, which was used to check a substantial part of A. P. Morse's [58].
- Invented a remarkably easy-to-use method for doing completeness proofs of resolution strategies and showed the completeness of a new strategy for linear resolution.
- First emphasized the importance of building a prover with many different kinds of heuristic components, reflecting the diversity of reasoning methods employed by mathematicians. Among the components incorporated together in Bledsoe provers have been:
  - Ordinary resolution
  - Natural deduction

  - Special arithmetic decision procedures
  - Rewriting procedures
  - Inductive procedures
  - Special heuristics for analysis
  - Use of example generation techniques
  - Use of analogical reasoning
  - Use of type inference

- Demonstrated that by using such a variety of techniques, proofs for moderately deep theorems in analysis could be automatically discovered.
- Virtually singlehandedly identified the subfield of "non-resolution" theorem proving with several comprehensive survey articles [34, 47].

Extremely characteristic of Bledsoe is his "try it out and see" attitude toward ideas in Artificial Intelligence. To quote further from his AAAI address

> The principle I want to make is this: when you have what looks like a good idea, give it your best shot, waste a little money to get some early feedback. Don't take forever to study the problem, because that is even more expensive (and less exciting).

An excellent summary of Bledsoe's work may be found in [41]. This paper gives an overview of techniques used in Bledsoe's prover, including reduction, induction, resolution, natural deduction, the limit heuristic, variable restrictions and elimination, and shielding. Bledsoe also enumerates theorems in analysis proved entirely automatically by his prover, including:

- The limit of sums and products are the sums and products of the limits.
- The sums, products, and compositions of continuous functions are continuous.
- Differentiable and uniformly continuous functions are continuous.
- If a function f is continuous on a compact set, then it is uniformly continuous on f, and $f[S]$ is compact.
- The Bolzano-Weierstrass Theorem.
- The Intermediate Value Theorem

1.11.1 Mechanical Proof Checking

How did Bledsoe happen to shift from the study of pattern recognition and evolution to automated reasoning? Both while working at Panoramic and after moving to the University of Texas, Bledsoe continued to consult at Sandia. In the summer of 1966, Bledsoe began consulting with Sandia employees E. Gilbert, D. Morrison, and J. Weihe on a mechanical proof-checking program for the Morse logic [58]. Weihe, like Bledsoe, had done his Ph. D. under A. P. Morse, as had T. McMinn, who also consulted on this proof-checking project, and had written the trenchant foreward to [58]. This remarkable proof-checker project is described in [25]. Among the rules of inference of the Morse logic supported by the checker were:

- detachment
- variable instantiation
- schematic (second order) instantiation
- change of bound variable
- universalization

The proof-checker included quite a few additional, derived rules of inference, including tautology checking [25]. Over the course of several years, Bledsoe and Morris developed this checker to the point that they were able to use it to check a substantial percentage of the theorems in Morse's book. This experience convinced him that it was, from the mathematical point of view, not only feasible but even easy to proof check almost any part of mathematics if the user of the proof-checking system were sufficiently patient to provide enough intermediate steps. But Bledsoe became both bored by the tedium of providing steps (especially since at this time the technology was to submit long trays of IBM punch cards and wait hours for a listing of a "run") and excited about the possibility of getting a computer to take much larger inference steps.

1.11.2 Resolution—Yeas and Nays

Like most of the early workers in automated reasoning, Bledsoe became fascinated by the infinite improvement over previous techniques represented by the resolution technique of J. A. Robinson, which had been developed at Argonne National Laboratory in the summer of 1963. Working with his Ph. D. student Robert Anderson, Bledsoe made important mathematical contributions to the theory of resolution, including a new "linear format" and a technique for proving completeness [28]. The completeness technique, known as the excess literal technique, is surprisingly easy to apply. The idea is simply this:

when trying to prove the (ground) completeness of a new resolution strategy, do the induction "on" the "excess literal" measure. The excess literal measure of a set of clauses is the numeric difference between the number of occurrences of literals in the set of clauses and the number of clauses in the set. This method has been used successfully by many people, including one of the authors of this note (RSB) in his Ph. D. thesis under Bledsoe, which introduced the "locking" restriction of resolution and proved its completeness [48, 51]. Other Ph. D. students of Bledsoe who made serious contributions to the resolution literature include Robert Anderson [1, 2], Dallas Lankford [55], James B. Morris [56, 57], and T. C. Wang [46].

One of the first extensions Bledsoe made to his proof-checker was a resolution theorem-prover coded by his student James B. Morris. But despite the fascination that Bledsoe has always felt for the simplicity and power of the resolution approach to theorem-proving, his practical experience with using it to prove difficult theorems was and remains largely a disappointment to him. As he remarks in his [34], "by the early 70's there was emerging a belief that resolution type systems could never really 'hack' it, could not prove really hard mathematical theorems, without some extensive changes in philosophy."

1.11.3 PROVER

In order to develop a theorem prover that was actually capable of proving deep theorems using a feasible amount of computing resources, Bledsoe began the construction and evolution of his most well-known theorem-proving system, which has been known as the "UT Interactive Theorem Prover," or simply as "PROVER." The initial substantial components of this prover, in addition to a resolution component, are described in [29]:

- A splitting routine for proving separable subgoals completely independently.
- A simplification (rewriting, normalizing, or demodulating) routine together with about thirty built-in, set-theoretic rewrite rules.
- An induction routine for setting up base cases and induction steps.
- An equality substitution routine.
- A governing scheme called CYCLE to decide which of the above routines to call in which order.

This prover used resolution as a "black box," but only for short periods of run time. That is, the resolution component could not call the reducer or other routines and did not contribute to finding a proof except by succeeding entirely by itself on very simple subgoals. With this first prover, Bledsoe was able to prove some nontrivial set theory theorems that he found impossible to get solely by resolution, including the theorem that every natural integer is an ordinal.

With a strong background in "analysis," it was natural for Bledsoe to move from set theory to elementary theorems about limits and continuous functions. But to achieve the goal of obtaining a satisfactory number of mechanical proofs in this area, Bledsoe found it necessary to make quite a number of additional changes to his prover.

First, he abandoned resolution altogether, replacing it with a natural deduction theorem-proving program called IMPLY. Unlike the resolution routine that he had been using, IMPLY provided a natural environment for embedding some of his other heuristics and the reducer.

Second, intertwined with IMPLY was not only the rewriter, but a new typing facility, which associated with terms some expressions that were (conservative) characterizations of sets over which the terms could range. As an example, a variable x might be known to range over the interval $[a, b]$. Explicit association of types with certain expressions, together with simple operations for taking set unions and intersections, made it possible to build into the deduction machinery set-theoretic reasoning rules that would be difficult to implement efficiently in a pure resolution setting.

Third, he added a mechanism for solving linear inequalities. Although it is possible to express as clauses appropriate axioms for linear order and polynomials, it appears vastly more efficient to embed such axioms in a decision procedure or even a partial decision procedure rather than to permit such axioms to "slosh around" in a general clause database.

Bledsoe also added to IMPLY the "limit heuristic," a very simple trick for proving certain inequalities. This trick "uses up" a hypothesis inequality by factoring it out of the desired conclusion.

These additions are first described in [31], which was written at M. I. T. Further development of the IMPLY prover is described in [33]. This IMPLY prover of Bledsoe's was influential in the design of the prover of Boyer and Moore [49, 50].

Probably the most significant subsequent improvement to IMPLY is the mechanism reported in [36] for instantiating set variables. Here Bledsoe describes a mechanism for constructing sets with specified properties, and uses it to prove such theorems as "if a set B contains an open neighborhood of each of its points, then B is open" and The Intermediate Value Theorem.

1.11.4 Work With A. Michael Ballantyne

Perhaps the colleague who has worked most on automated reasoning with Bledsoe is Mike Ballantyne. Ballantyne was clearly among the most creative of the many Bledsoe graduate students; Mike also consulted at MCC under Bledsoe. Two Ballantyne-Bledsoe projects are especially worth mentioning.

Perhaps the high point of success with the IMPLY prover is their work on nonstandard analysis [3]. In that paper, some very interesting theorems in intermediate analysis are proved, including the Bolzano-Weierstrass theorem and the theorem that the image of a continuous function on a compact set is compact. Although these theorems were proved by the IMPLY prover, with its natural deduction and rewriting facilities, what is most remarkable about this work is that the axiomatization used for the reals is based upon the non-standard analysis of A. Robinson [60], with its infinitesimals and infinite integers. It is a remarkable fact that proofs of some theorems in intermediate analysis are much simpler in a nonstandard setting than in the usual ϵ and δ approach.

The usefulness of examples in finding proofs has been well known at least since the earliest days of geometry. A very few efforts at using examples in automated reasoning have been successful, notably the early geometry work of Gelernter. In [4], Ballantyne and Bledsoe demonstrate the value of using examples in the area of analysis, especially topology. In that article is described a system, named GRAPHER, that constructs counterexamples as an aid to eliminating false subgoals arising in proofs. The program considers finite families of finite sets together with considerable knowledge of elementary topology. Among many, many decidedly nontrivial inferences, the program found an example of a subspace of a normal space that is not normal.

1.11.5 Work With Larry Hines

Another long time colleague and student of Bledsoe is Larry Hines. The most well-known result of Bledsoe and Hines is the variable elimination method, which is a technique for proving, in a resolution setting, theorems involving inequalities. Although the basic ideas of variable elimination are easily stated [38], a considerable amount of deep thinking has gone into proving that this technique is "complete" [43]. The completeness work was done with Rob Shostak and Ken Kunen.

1.11.6 Program Verification and Natural Language

In 1974 Bledsoe began to contribute to the field of *mechanical program verification.* The IMPLY prover that he developed was incorporated into a verification system for Pascal [52] and was later utilized in the verification system for Gypsy [53, 54, 63], which has been used to verify several significant computing systems. See also [35].

In 1990, Bledsoe's student Don Simon finished a Ph. D. thesis [62] about a program that could proof check number theory proofs directly from a textbook ([67]) typeset in LaTeX.

1.11.7 The Future of Automated Reasoning

In Bledsoe's view, some of the earliest work on automated reasoning, such as that of Newell, Simon, and Shaw [59], started a trend in the direction of employing "people-oriented" reasoning techniques, but this trend was "overshadowed by the emergence of other techniques emphasizing speed, and thus the field took a turn away from people methods." In Bledsoe's view, it is "time to make a significant change in direction" to pursue more automated reasoning studies along the lines of "people methods." According to Bledsoe's [45], the following are crucial elements of mathematical thought that an automated reasoning system should incorporate, all analogous to the way that people reason:

- analogy
- generation and use of examples
- use of counterexamples to prune the search
- conjecturing of lemmas and subgoals
- intelligent fetching of useful lemmas and definitions from a large knowledge base
- an agenda mechanism for controlling the search
- mathematical pattern recognition
- learning
- planning and abstraction
- higher-level and meta reasoning

Since his return to the University of Texas in 1987, Bledsoe has primarily focused on "reasoning by analogy," which he believes is perhaps the most important key to advances in automated reasoning.

References

[1] Robert Anderson (1970): Completeness Results for E-Resolution. Proc. AFIPS 1970 Spring Joint Computer Conf., 36, 653–6.

[2] Robert Anderson (1970): Some Theoretical Aspects of Automatic Theorem Proving. Ph. D. Dissertation, University of Texas at Austin.

[3] A. Michael Ballantyne and W. W. Bledsoe (1977): Automatic Proofs of Theorems in Analysis Using Nonstandard Techniques. Journal of the Association for Computing Machinery 24(3), 353–374.

[4] A. M. Ballantyne and W. W. Bledsoe (1982): On Generating and Using Examples in Proof Discovery. In: J. Hayes and D. Michie, eds.: Machine Intelligence 10. Harwood: Chichester. 3–39.

[5] W. W. Bledsoe (1952): Neighborly Functions. Proceedings of the American Mathematical Society 3(1), 144–115.

[6] W. W. Bledsoe (1952): Some Aspects of Covering Theory. Proceedings of the American Mathematical Society 3(5), 804–812.

[7] W. W. Bledsoe, M. J. Norris, and G. F. Rose (1954): On a Differential Inequality. Proceedings of the American Mathematical Society 5(6), 934–939.

[8] W. W. Bledsoe, et al. (1955): A Systems Study Comparing Various Fuzing Possibilities for the Atlas ICBM. Sandia Corporation Technical Memorandum 155–55–51, Ref. Sym: 1923 (266), August 4, 1955.

[9] W. W. Bledsoe and A. P. Morse (1955): Product Measures. Transactions of the American Mathematical Society 79(1), 173–215.

[10] W. W. Bledsoe (1957): Program for Computing Probabilities of Fallout from a Large-Scale Thermonuclear Attack, SC–4109 (TR) Systems Analysis, Sandia Corporation. Available from the Office of Technical Services, Department of Commerce, Washington, D. C.

[11] W. W. Bledsoe and I. Browning (1959): Pattern Recognition and Reading by Machine. Proceedings of the Eastern Joint Computer Conference, 225–232.

[12] W. W. Bledsoe (1961): Further Results on the N-tuple Pattern Recognition Method. IRE Transactions on Electronic Computers EC–10(1), 96–97.

[13] W. W. Bledsoe (1961): A Basic Limitation on the Speed of Digital Computers. IRE Transactions on Electronic Computers EC–10(3), 530.

[14] W. W. Bledsoe (1961): Lethally Dependent Genes Using Instant Selection. Technical Report PRI 1, Panoramic Research Inc.

[15] W. W. Bledsoe (1961): The Use of Biological Concepts in the Analytical Study of Systems. Technical report PRI 2, Panoramic Research, Inc. A talk presented to the Operations Research Society of America, National Meeting, San Francisco, CA, November 10, 1961.

[16] W. W. Bledsoe and C. L Bisson (1962): Improved Memory-Matrices for the N-tuple Pattern Recognition Method. IRE Transactions on Electronic Computers EC–11, 414–415.

[17] W. W. Bledsoe and A. P. Morse (1963): A Topological Measure Construction. Pacific Journal of Mathematics 13(4), 1067–1084.

[18] W. W. Bledsoe et al. (1964): Facial Recognition Project Report. Technical report PRI 10, Panoramic Research, Inc.

[19] W. W. Bledsoe (1965): The Model Method in Facial Recognition. Technical Report PRI 15, Panoramic Research, Inc. Not publicly available.

[20] W. W. Bledsoe (1965): Two-dimensional Terrain Following a "Back and Fill" Method. Technical report PRI 23, Panoramic Research, Inc.

[21] W. W. Bledsoe (1965): Simulation of Three-Dimensional Navigation. Technical report PRI 24, Panoramic Research, Inc.

[22] W. W. Bledsoe and Helen Chan (1965): A Man-Machine Facial Recognition System—Some Preliminary Results. Technical report PRI 19A, Panoramic Research, Inc. Not publicly available.

[23] W. W. Bledsoe (1966): Some Results on Multicategory Pattern Recognition. Journal of the Association for Computing Machinery 13(2), 304–316.

[24] W. W. Bledsoe (1966): Man-machine Facial Recognition. Technical Report PRI 22, Panoramic Research, Inc.

[25] W. W. Bledsoe and E. J. Gilbert (1967): Automatic Theorem Proof-Checking in Set Theory—A Preliminary Report. Research Report SC–RR–67–525. Sandia Laboratories.

[26] W. W. Bledsoe (1968): Semiautomatic Facial Recognition. Technical Report SRI Project 6693. Stanford Research Institute, Menlo Park, California. Not publicly available.

[27] W. W. Bledsoe (1970): An Inequality about Complex Numbers. American Mathematical Monthly 77(2), 180–182.

[28] R. Anderson and W. W. Bledsoe (1970): A Linear Format for Resolution with Merging and a New Technique for Establishing Completeness. Journal of the Association for Computing Machinery 17, 525–534.

[29] W. W. Bledsoe (1971): Splitting and Reduction Heuristics in Automatic Theorem Proving. Artificial Intelligence 2, 55–77.

[30] W. W. Bledsoe and C. E. Wilks (1972): On Borel Product Measures. Pacific Journal of Mathematics 42(3), 569–579.

[31] W. W. Bledsoe, R. S. Boyer, and W. H. Henneman (1972): Computer Proofs of Limit Theorems. Artificial Intelligence 3, 27–60.

[32] R. H. Bing, W. W. Bledsoe, and R. D. Mauldin (1974): Sets Generated by Rectangles. Pacific Journal of Mathematics 51(1), 27–36.

[33] W. W. Bledsoe and Peter Bruell (1974): A Man-Machine Theorem-Proving System. Artificial Intelligence 5, 51–72.

[34] W. W. Bledsoe (1977): Non-resolution Theorem Proving. Artificial Intelligence 9, 1–35.

[35] W. W. Bledsoe and Mabry Tyson (1977): Typing and Proof by Cases in Program Verification. In: D. Michie and E. W. Elcock, eds.: Machine Intelligence 8. Chichester: Horwood. 30–51.

[36] W. W. Bledsoe (1979): A Maximal Method for Set Variables in Automatic Theorem-proving. In: J. E. Hayes, D. Michie, and L. I. Mikulich, eds.: Machine Intelligence 9. Chichester: Horwood. 55–100.

[37] W. W. Bledsoe and Mabry Tyson (1979): Conflicting Bindings and Generalized Substitutions. Proceedings of the Fourth International Workshop on Automated Deduction.

[38] W. W. Bledsoe and Larry M. Hines (1980): Variable Elimination and Chaining in a Resolution-Based Prover for Inequalities. Proceedings of the 5th Conference on Automated Deduction, Les Arcs, France, (eds. W. Bibel and R. Kowalski). Springer, 70–87.

[39] W. W. Bledsoe (1983): Using Examples to Generate Instantiations of Set Variables. Proceedings of the Eigth International Joint Conference on Artificial Intelligence, 892–901.

[40] W. W. Bledsoe and D. W. Loveland, eds. (1984): Automated Theorem Proving: After 25 years (Contemporary Mathematics Volume 29). Providence, Rhode Island: American Mathematical Society.

[41] W. W. Bledsoe (1984): Some Automatic Proofs in Analysis. In: [40]. 89–118.

[42] W. W. Bledsoe (1986): I Had a Dream: AAAI Presidential Address, 19 August 1985. The AI Magazine 7(1), 57–61.

[43] W. W. Bledsoe, K. Kunen, and R. Shostak (1985): Completeness Results for Inequality Provers. Artificial Intelligence Journal 27, 255–288.

[44] W. W. Bledsoe and L. J. Henschen (1985): What is Automated Theorem Proving? Journal of Automated Reasoning 1, 23–28.

[45] W. W. Bledsoe (1986): Some Thoughts on Proof Discovery. Proceedings of the 1986 Symposium on Logic Programming, Salt Lake City, 2–10. Also Technical Report AI–208–86. Microelectronics and Computer Technology Corporation, Austin, Texas.

[46] W. W. Bledsoe and T. C. Wang (1987): Hierarchical Deduction. Journal of Automated Reasoning 3, 35–77.

[47] W. W. Bledsoe and R. Hodges (1988): A Survey of Automated Deduction. In: H. Shrobe, ed.: Exploring Artificial Intelligence, Survey of Talks from the National Conference on Artificial Intelligence, Seattle, 1987. 483–543.

[48] R. S. Boyer (1971): Locking: A Restriction of Resolution. Ph. D. Dissertation, University of Texas at Austin.

[49] R. S. Boyer and J. S. Moore (1979): A Computational Logic. New York: Academic Press.

[50] R. S. Boyer and J. S. Moore (1988): A Computational Logic Handbook. New York: Academic Press.

[51] C. Chang and R. Lee (1973): Symbolic Logic and Mechanical Theorem-Proving. New York: Academic Press.

[52] D. I. Good, R. L. London, and W. W. Bledsoe (1975): An Interactive Program Verification System, IEEE Transactions on Software Engineering 1(1), 59–67.

[53] D. I. Good, Ann E. Siebert, and Lawrence M. Smith (1982): Message Flow Modulator—Final Report, Institute for Computing Science and Computer Applications, University of Texas at Austin, ICSCA–CMP–34.

[54] D. I. Good (1985): Mechanical Proofs about Computer Programs. In: C. A. R. Hoare and J. C. Shepherdson, eds.: Mathematical Logic and Programming Languages. Prentice-Hall. 55–75.

[55] Dallas Lankford (1972): Equality Atom Term Locking. Ph. D. Dissertation. University of Texas at Austin.

[56] James B. Morris (1969): A Proof Procedure for the First Order Predicate Calculus with Equality. Ph. D. Dissertation. University of Texas at Austin.

[57] James B. Morris (1969): E-Resolution: Extension of Resolution to Include the Equality Relation. Proceedings of the International Joint Conference on Artificial Intelligence, Washington, D. C., 287–294.

[58] A. P. Morse (1965): A Theory of Sets. New York: Academic Press.

[59] A. Newell, J. Shaw, and H. Simon (1957): Empirical Explorations of the Logic Theory Machine: A Case Study in Heuristics. In: E. Feigenbaum and J. Feldman, eds.: Computers and Thought. McGraw-Hill: New York. 109–133.

[60] A. Robinson (1963): Nonstandard Analysis. Amsterdam: North-Holland.

[61] J. R. Shoenfield (1967): Mathematical Logic. Reading: Addison-Wesley.

[62] D. Simon (1990): Checking Number Theory Proofs in Natural Language. Ph. D. Dissertation. University of Texas at Austin.

[63] M. Smith, A. Siebert, B. DiVitto, and D. Good (1981): A Verified Encrypted Packet Interface, Association for Computing Machinery SIGSOFT, 6(3).

[64] J. R. Ullman (1969): Experiments with the N-tuple Method of Pattern Recognition. IEEE Transactions on Computers C–18(12), 1135–1137.

[65] J. R. Ullman (1971): Reduction of the Storage Requirements of Bledsoe and Browning's N-tuple Method of Pattern Recognition. Pattern Recognition 3, 297–306.

[66] J. R. Ullman (1973): Pattern Recognition Techniques. London: Butterworths.

[67] W. J. LeVeque (1962): Elementary Theory of Numbers. Mineola, N. Y.: Dover.

Chapter 2

METEORs: High Performance Theorem Provers using Model Elimination[1]

O. L. Astrachan
D. W. Loveland

Computer Science Department
Duke University
Durham, North Carolina

2.1 Introduction

It has been thought for some time that linear resolution procedures are not the best resolution formats for doing very large unguided searches such as are required for proving significant mathematical theorems. Although very efficient with storage space, the depth-first implementations that take advantage of the linearity seem to preclude use of subsumption and similar devices to reduce redundant search. However, research in logic programming has led to sophisticated architectures giving high-speed performance for Prolog, which is based on a linear input resolution procedure. In particular, the WAM architecture, due to D. H. D. Warren (see [33]) has proven very successful. This

[1]Both authors have received partial support from U. S. Army Research Grant DAAL03-88-K-0082 and NSF Grant CCR-8900383 for the research and writing of this paper.

has led to renewed interest in Model Elimination (ME) ([18, 19, 20]), which is a linear input proof procedure for full first-order logic, similar to but distinct from resolution. Interest has revived because there are situations where the significantly higher inference rate may offset the removal of some redundant computations.

That this offset can be realized is shown here by applying our theorem prover METEOR successfully to two of Woody Bledsoe's challenge problems [5]. To our knowledge (at present) METEOR is the only other prover besides the STR+VE program of Hines-Bledsoe [6] to find automated proofs for both problems. It is appropriate to note here that theorem provers developed by Woody Bledsoe and colleagues set the standards others try to match; that is partly why we honor him with this volume. Our limited success relative to STR+VE is interesting only because we employ a prover presently devoid of specialized mechanisms for designated predicates. Our failure to prove a third challenge problem shows the limitation of our approach (indeed the limits of pure general first-order proof procedures), but it will allow us to note the potential of selective use of automatically generated lemmas.

In this paper we report on METEOR, a parallel implementation of ME (or-parallelism, to be precise) and compare it to the sequential version of the same inference engine as well as comparing the implementations with others reported in the literature. (METEOR is derived from Model Elimination Theorem proving with Efficient OR-parallelism.) The significance of parallel ME in future theorem provers stems from the availability (and even low cost) of parallel machines and the decomposability of theorem-proving tasks, including search itself. The rapid progress of chip technology assures us of low-cost access to powerful parallel computers in the near future. The possibility of many specialized (and competing) components in a theorem prover suggests that parallel computers may be the preferred vehicles in the future. (Clearly, distributed processing is also appropriate here.) The immediate focus is to demonstrate that the inference rate is appropriately enhanced by parallelization and that added gain is possible by utilizing the flexibility of computation that parallel computation suggests.

For many years interest in ME was kept alive by Mark Stickel, who has demonstrated that the WAM architecture can be adopted to ME and yield a very high inference rate in this setting as well. This relationship to Prolog architecture is even recognized by the title in the recent theorem prover(s) developed by Stickel: the Prolog Technology Theorem Prover ([29, 30, 32]). When we began our project, Stickel's work was the sole guide for comparison. However, as we write this summary of an implementation of parallel ME, several years after initiation of the project, we see that we were not alone in seeing new opportunity in exploiting Model Elimination. Two other projects (that we know of) simultaneously have developed parallel ME implementations using the concepts employed in Prolog implementations, indeed using ideas from implementations of parallel versions of Prolog. PARTHENON, developed at CMU by Bose, Clarke, Long and Michaylov ([7], [8]) employs a modification of the SRI architecture for or-parallel Prolog designed by D. H. D. Warren [34]. PARTHENON has been run on a Encore MULTIMAX computer, using up to 16 processors, and on the IBM RP3, using up to 60 processors. Both systems are MIMD shared-memory systems. PARTHEO, developed at TUM (Munich) by Schumann and Letz [28], consists of a uniform network of sequential processors for ME communicating by message passing. The individual ME processors use an extension of the WAM architecture.

Our implementation has much in common with both of the implementations mentioned above for reasons given earlier; all have utilized the WAM architecture and the very close design relationship between Prolog (SLD-resolution) and ME. Our implementation is closer to PARTHENON in that it was originally based on Warren's SRI model and also runs on an MIMD shared-memory machine. (The current implementation is closer to the or-parallel system developed at SICS [1] than to the SRI model.) We use a BBN Butterfly GP1000 with up to 30 processors. We have also adapted the prover to run on a network of Sun SPARC stations each of which is a significantly more powerful inference engine than one node of the Butterfly.

Perhaps the most significant design difference with both other systems is our task allocator or scheduler. Whereas the other systems use a local task "stealing" mechanism, we use a top-level public sector/private sector task dispensing system (as in [21]) we call a SASS-pool, for Simultaneous Access of Selectively Shared objects pool, a type of concurrent pool. The notion of concurrent pool has been extensively studied by Ellis and Kotz (see [15]). This paper will focus on the differences between this system and the other implementations.

The most significant discovery from our METEOR project experimentation is the effect of different depth measures on the proof search size. The proof search uses iterative deepening, the now standard method of realizing completeness in a depth-first search procedure [14]. By iterative deepening we mean expanding the search tree fully to a specified depth, incrementing the depth bound and reexpanding the search tree, and continuing this until a proof is found or the search is terminated. Within this basic search strategy are significant variants determined by the measure of depth used. Using the number of inferences in a deduction as a depth measure, as done by PTTP [32] and PARTHENON [8], one is assured of a shortest proof if any proof is found. However, we have discovered that the original depth measure suggested for ME [18] very often finds a proof in a dramatically shorter time, although not necessarily a shortest proof. The SETHEO/PARTHEO [28] provers seem to use this measure uniformly, but with no reported data on comparison of depth measures. We have also discovered two different combination depth measures of interest, one that is a compromise of the two depth measures but is uniformly better than using the number of inferences as a depth measure, and one that is often strikingly good for reasons we do not fully understand.

After a presentation of the architecture of METEOR, we present performance figures on a collection of benchmark problems run on several other theorem provers, comparing previous performances with data from our sequential prover and from the METEOR parallel prover. We also provide performance figures on three of five Bledsoe challenge problems regarding the continuity of the sum of two continuous functions [5]. We begin with a brief review of the ME procedure, the WAM architecture and the METEOR architecture, a parallel version of the WAM architecture.

2.2 Background

We first outline the Model Elimination (ME) procedure. We assume the reader is familiar with basic resolution terminology. For such terminology, see Chang and Lee [10] or Loveland [20]. The latter book gives a thorough introduction to ME.

The ME procedure utilizes many of the same mechanisms as resolution, and can be roughly regarded as a linear input resolution procedure with the addition of an ancestor rule we call (ME) reduction. (However, the resultant resolution procedure corresponding to ME is not a linear input procedure. Linear input resolution procedures are complete only for the renamable Horn sets at the ground level.) The interest in the procedures is due to retention of the linear input property while providing a complete proof procedure. By a *linear* (resp., *input*) procedure we mean a procedure all of whose deductions are linear (resp., input) deductions. A *linear deduction* is a deduction where each line is derived from the preceding line (often using auxiliary information). An *input deduction* is a deduction where each line uses as auxiliary information only a given (i.e., input) formula. For example, an input resolution deduction restricts the resolution inference rule to one derived clause (at most) and one given clause as parents to the resolvent. It follows that input deductions are linear, since two lines of deduction can never join, but we often use the phrase *linear input* to emphasize the linearity (and because the phrase is in common use).

ME has two basic inference rules, the extension rule and the reduction rule. The rules manipulate formulas called chains, defined below. The extension rule (also defined below) is very similar to the resolution rule, using as parent formulas a derived chain and a given chain, and producing a chain. Reduction takes one chain as input and produces a modified chain. Hence the system has a (linear) input form.

ME is a refutation procedure, taking as pre-input a clause set prepared as is done for resolution. The actual input is a set of chains of a certain type. A *chain* is a sequence of literals; each literal is one of two types, A-literal (for ancestor) or B-literal. The ordering that determines acceptable sequencing is determined by the user subject to constraints given below. An input chain is a sequence of B-literals composed of the literals from a clause of the given clause set. The ordering of each input chain is determined entirely by the user except that for each literal in a given clause there must be a chain with that literal leftmost. Thus, for an n-literal given clause there are n input chains. Chains are accessed only by the leftmost literal so this restriction states that every literal of every given clause is accessible. (Whether or not n copies of an n-literal clause actually exist is an implementation matter.) (In [20], and in the earlier papers introducing ME [18, 19], chains had the rightmost literal as the accessible literal. That ordering is more convenient for presenting deductions. However, Prolog conventions utilize left access, and, as ME is now implemented using augmented Prolog architectures, we accommodate these conventions.)

Definition 2.2.1 *A chain is* admissible *if and only if (iff):*

1. *complementary B-literals are separated by an A-literal;*
2. *no B-literal is to the left of an identical A-literal;*
3. *no A-literal is identical or complementary to another A-literal; and*
4. *the leftmost literal is a B-literal.*

We need only deal with admissible chains. Any derived chain not admissible can terminate that search branch. These conditions, particularly condition 2, make useful pruning rules.

We give the two inference rules. Recall that two literals are complementary iff one is the negation of the other. By mgu we mean the "most-general unifier."

Definition 2.2.2 Extension*: Given admissible chain C_1 and input chain C_2 such that the leftmost B-literal l_1 of C_1 and the leftmost B-literal l_2 of C_2 can be made complementary by mgu σ, the resultant chain is $C_1\sigma$, with $l_1\sigma$ promoted to an A-literal augmented on the left by $C_2\sigma - \{l_2\sigma\}$, with $C_2\sigma - \{l_2\sigma\}$ reordered as the user wishes. Any leftmost A-literals are removed.*

Definition 2.2.3 Reduction*: Given admissible chain C with leftmost B-literal l_1 and A-literal l_2 (anywhere in C) such that l_1 and l_2 can be made complementary by mgu σ, then the resultant chain is $C\sigma - \{l_1\sigma\}$ with any A-literals leftmost in $C\sigma - \{l_1\sigma\}$ removed.*

Example: If $C_1 = Pxa[\neg Qy][Pf(x)b]Qx$ and $C_2 = \neg Paa$, then the extension of C_1 using C_2 is Qa, with intermediate step $[Paa][\neg Qy][Pf(a)b]Qa$ before the A-literals (in brackets) are removed as being leftmost (i.e., to the left of all B-literals).

Example: If $C = Pax[\neg Qb][\neg Pyz]$ then reduction can apply, yielding the empty chain, from intermediate step $[\neg Qb][\neg Pax]$ whereupon the A-literals are removed as being leftmost in the intermediate chain.

An ME refutation of a given clause set is a sequence of steps using the inference rules of extension and reduction from the set of associated input chains and with the last entry the empty chain, denoted $\Box$.

We note that there is no equivalent to factoring, although that can be added. SL-resolution (Kowalski and Kuehner [16]) is essentially ME with factoring. Incidentally, the selected literal capability of SL-resolution is present in the ME procedure under a different guise. In ME the user orders the input clause (freely) in the extension derived chain according to the order of selection desired. Literals between A-literals can be reordered at any point, with the justification that the resulting order could have been selected initially, if desired.

Before giving a short ME refutation we mention the lemma capability of ME, which has never been extensively explored. (Lemmas were used in [12] in an unconstrained manner, and found generally not helpful without further control.) There is a device for creating derived clauses during the search process that can be used to shorten proofs. Since these lemmas are not needed for completeness (unlike derived resolution resolvents) very strong restrictions can be applied regarding retention. The lemmas may by used to form new input chains. (We feel that most often only one-literal lemmas would be retained.) The use of lemmas and related devices ("caching") have the potential to greatly reduce the search needed to achieve refutations if the right mechanisms for use can be found. We are exploring this now.

A unit lemma may be created any time an A-literal is removed from the left of an intermediate chain during extension or reduction, provided that no reduction previously has used an A-literal to the right of the discarded A-literal. The unit lemma is simply the complement of the removed A-literal, as currently instantiated. (This is a simplification of the full rule for lemmas, given in [19, 20].)

In Figure 2.1 we present an ME refutation given in [20]. No ordering rule is needed here since the longest chain has two literals. The input chains are never explicitly listed, only the given clauses are listed; the input chains are irrelevant except conceptually because it is only the ordering after extension that really matters.

This ME refutation illustrates lemmas, and also that a chain with complementary B-literals separated by A-literals cannot be discarded. (Cf. the definition of admissible chain given earlier.)

(The label 1b means the second literal of clause 1.)

1.	$Px\neg Qy$	given clause
2.	$\neg PxRy$	given clause
3.	$\neg Pa\neg Ry$	given clause
4.	$QxPa$	given clause
5.	$Py[Qx]Pa$	extension with 1b
6.	$Rz[Py][Qx]Pa$	extension with 2a
7.	$\neg Pa[Rz][Py][Qx]Pa$	extension with 3b
8.	Pa	reduction

Unit lemmas formed:
$\neg Pa$
$\neg Qx$

(Note: If lemma $\neg Pa$ is used for extension then $\Box$ is immediately derived.)

9.	$\neg Ry[Pa]$	extension with 3a
10.	$\neg Px[\neg Ry][Pa]$	extension with 2b
11.	$\Box$	reduction

Figure 2.1: An ME Refutation

2.3 The Warren Abstract Machine

The success of Prolog as a logic programming language is due both to the declarative/procedural nature of logic as a programming language and to the efficiency of current implementations of Prolog. This efficiency is due to the input nature of Prolog deductions and to current realizations of such input procedures. This latter efficiency is due in large part to David H. D. Warren's design of an abstract machine for executing Prolog programs [33]. Most commercial and research Prolog implementations are based to a large degree on the WAM (Warren Abstract Machine) as are the Prolog technology class of theorem provers ([32, 8, 2]).

The WAM consists of both an architecture and an instruction set. Many Prolog implementations compile Prolog programs into WAM instructions (or some variant thereof) which can be run on different machines (e.g., Sun, VAX) by using different back ends for the Prolog compiler/interpreter. The architecture of Warren's virtual machine is designed to take advantage of traditional von Neumann architectures while optimizing execution of Prolog.

The architecture of our prover is based on that of the WAM although we do not compile a set of clauses into some variant of the WAM instruction set. Instead, we use the concepts underlying the design of the instruction set to compile the clauses into a data structure that is tuned to our implementation of parallel Model Elimination.

2.3.1 The WAM Architecture

We give a brief outline of the WAM architecture. For a more thorough explanation see [33, 22]. The WAM is a stack based machine. In addition to the code area (in which compiled Prolog procedures are stored) there are three stacks: the term stack, the trail stack and the runtime stack.

The term stack is often referred to as the heap. Although the first Prolog implementations employed structure sharing [9] for storing bindings of terms, current implementations use copy on use. When a term such as `f(X,g(a,Y))` is substituted for a variable during unification, a copy of the term is made on the term stack. If subsequent deductions cause this binding to be incorrect (and backtracking occurs), the space in the term stack will eventually be reclaimed. Note that the skeleton of a term as it appears in an input clause may be stored several times (with, perhaps, different instantiations of the variables in the term) on the term stack.

The trail stack is used to trail bindings of variables that may need to be undone when backtracking occurs. In a sequential setting, only the addresses of variables that are bound are stored in the trail stack. If backtracking occurs, these addresses are used to restore the state of the bound variable to unbound. In the SRI or-parallel extension to the WAM, both the address of a variable and the value bound to the variable are trailed. This is necessary since different processors may bind different values to the the same variable. Unlike the sequential WAM, in which variables are associated with an address, in the SRI extension to the WAM variables are associated with an index in a binding array (and an address). The use of the binding array enables different processors to bind values to the "same" variable by storing the value in the processor's binding array. In our original prover we also used such a binding array. Our current prover, however, uses the model of the sequential WAM in a parallel setting and no binding array is used.

The runtime stack stores both environments and choice points. An environment consists of the variables associated with a given clause and the bindings of those variables. When clauses are compiled, a choice point stores the values of the WAM registers that need to be restored when backtracking occurs. In our setting, a choicepoint stores these values as well as storing the information necessary to make the next "choice," i.e., which chain to try next. We thus fold the code area into the runtime stack, specifically into the choice points that comprise the METEOR runtime stack.

2.3.2 METEOR Architecture

The data structure diagrammed in Figure 2.2 shows how a clause set is compiled into a structure used by METEOR. Each chain corresponding to a clause in the clause set is represented in the section labeled *Chain Table.* Recall that in ME each literal of a clause is a candidate for the extension operation. Consequently, a given n-literal clause is represented n times in the chain table: once with each literal as the head (left-most literal) of a chain which corresponds to the clause. Each such instance of a clause is preprocessed to distinguish the first occurrence of each variable (for efficient operation of the occurs check) and indexed according to the first argument of the head literal.

Pointers representing each chain with a given literal as head reference another pointer (the *chain list pointer* section) according to the index assigned to the head literal's first argument. This level of indirection is essential as it avoids duplicate storage of a chain (each chain must "appear" in the *all* section of the chain table to match a goal whose first argument is a variable as well as in the hash bucket corresponding to the chain's first literal) while permitting alternative chains with the same head literal to be accessed sequentially.

The section labeled *chain list* stores information about a chain that is used during an extension operation. The number of variables in the chain is used in allocating an environment for the chain. The number of "body literals" (the term is derived from Prolog terminology) represents the number of new subgoals that must be solved if a chain is used for extension. This can be used to prune the search tree early using certain depth measures (see below). The literals that comprise a chain are accessed sequentially by following the head pointer as shown in the diagram.

2.3.3 METEOR Choice Points

At any time during a run of METEOR, the runtime stack stores information that allows the current deduction to be reconstructed. The stack is comprised of choice points, one choice point for each node of the current search tree. If the active branch of the search tree fails for some reason (e.g., depth bound is reached or no matching clauses are found), METEOR backtracks and tries an alternative extension or reduction. When run in parallel, the stack of each processor is divided into a public and private section in the same manner as [21]. This division is realized by putting one choice point from each processor's runtime stack into a publicly accessible data structure we call a SASS-pool (Simultaneous Access of Selectively Shared objects).

Access to this pool is supported by *CREW* (concurrent read exclusive write) locks [3], which permit simultaneous access to choice points while ensuring atomic conversion of a choice point from private to public status. When a "reading" or "stealing" process gains access to a public choice point, work is transferred to the reading process without interrupting the process from which the work is taken. By utilizing the atomic operators available on the

Butterfly we have been able to implement these locks so that, on the average, less than one one-thousandth of the total running time is spent acquiring locks for problems with sufficient parallelism. For full details on the locking mechanisms see [2].

2.3.4 Transferring Work

How work is transferred, from what part of the search tree the work is transferred, and the granularity of the work transferred are important details in an or-parallel system. In METEOR, we transfer work from nodes near the root of the search tree. In the original version of METEOR based on the SRI model, transferral of work included copying part of the global stack from the processor from which work is taken. In addition, a processor's logical stack physically resided on several processors. Although this model supports parallelism to some degree (it is used in PARTHENON and some models of or-parallel Prolog), we feel that current trends in parallel architectures will emphasize NUMA (non-uniform memory access) architectures and that such architectures preclude this approach.

The Butterfly GP1000 machine (on which METEOR has been developed) is a NUMA machine on which local memory accesses are roughly ten to fifteen times faster than remote accesses. Current trends towards cache based machines will make this penalty even more severe on other architectures. Developments in distributed computing also preclude an approach based on physically shared stacks. With these trends in mind, we have abandoned the SRI model and implemented an architecture in which stacks are not physically shared. Instead, when processor P steals work from processor Q, the stack of processor Q is reconstructed locally on processor P. In METEOR, the stack is not copied, but is recreated by copying the sequence of extensions and reductions that led to the state from which work was taken from processor Q. We have adopted this approach, rather than an approach of physically copying stack information, because recreating a stack is an extremely fast operation given our architecture. In addition, this approach has enabled us to implement a distributed version of METEOR on a network of Sun SPARC stations. The distributed version runs in two modes: a shared memory approach using the common file system as a shared pool, and a message passing approach using Unix sockets. The MUSE or-parallel Prolog system [1] also uses a non-shared stack approach, but physically copies the stack and attempts to optimize transferral of work so that stack copying is minimized. PARTHEO [28] uses a system like ours in which stacks are recreated locally. Kumar [17] discusses the parameters of a system that affect whether copying should be preferred to re-creation.

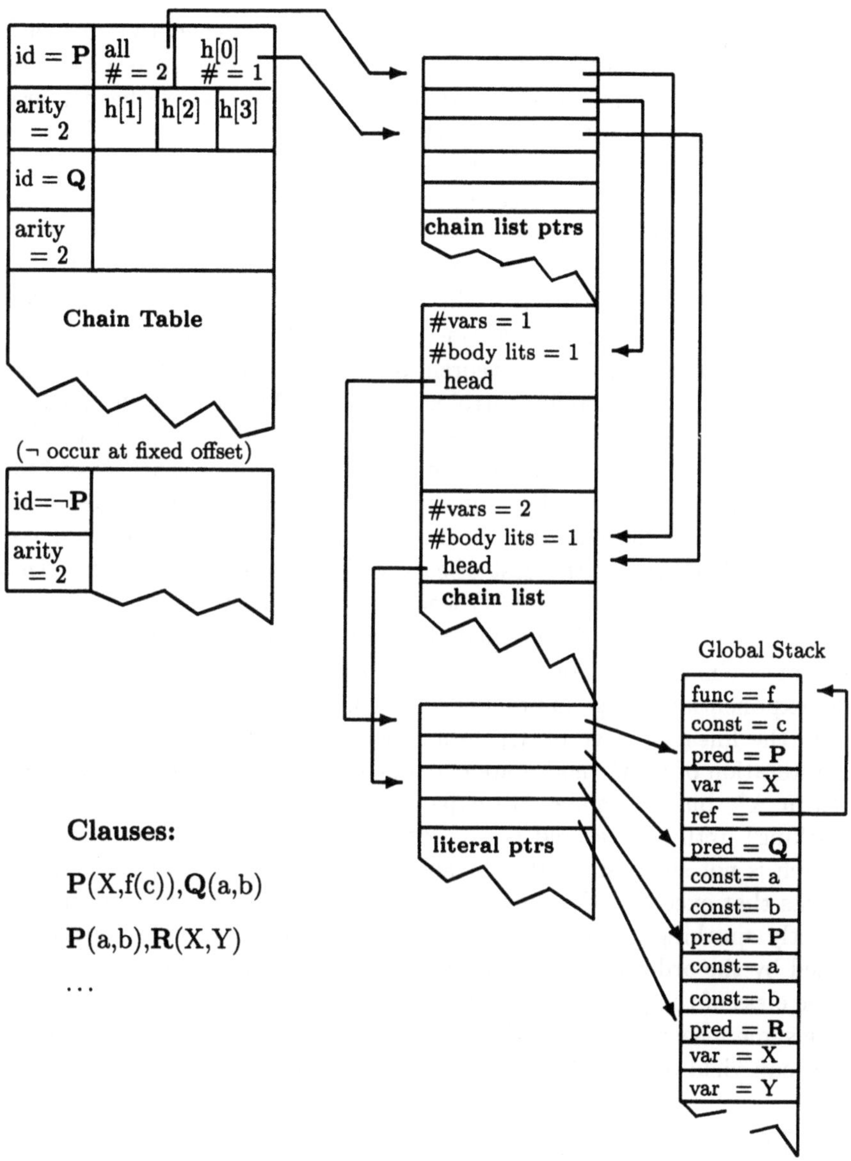

Figure 2.2: Principal Data Structures in METEOR

2.4 Discovering a Proof

In proof procedures based on Model Elimination, the search tree can be viewed as an AND-OR tree [24]. Each extension operation with an n-literal input chain yields an AND node in which there are $n-1$ branches (one for each literal l other than the head literal). Each of these branches links to an OR node from which there is a branch for each input chain whose head literal matches l (that may lead to an extension operation) and a branch for each matching ancestor literal (that may lead to a reduction operation). Thus alternating levels of the search tree consist solely of either AND or OR nodes with the OR nodes corresponding (loosely) to a WAM choice point. METEOR and other or-parallel provers attempt to explore the OR nodes of this tree in parallel. A successful search results in a proof tree which consists of AND nodes; several such trees are diagrammed below (2.3, 2.4, 2.5).

To ensure completeness, most theorem provers using depth-first search use iterative deepening as a search control strategy. METEOR (as well as PTTP and PARTHENON) employs an iterative form of the A* algorithm called IDA*.

In each iteration of the A* algorithm, nodes of lowest cost are chosen for expansion. Expansion of a node results in the children of the expanded node becoming candidates for expansion in the next iteration of the algorithm and requires the calculation of the cost of each child. The cost of a node, $f(n)$, is determined by $f(n) = g(n) + h(n)$ where g represents the cost of reaching node n and h represents the estimated cost of reaching a goal node from n (the empty chain in an ME deduction). Like other best-first searches, A* requires exponential storage in practice [25], which makes it prohibitive. IDA* combines the linear storage of depth-first search with the optimality of A*. At each iteration of IDA*, nodes are expanded according to the cost function $f(n)$ as in A*, but the nodes are expanded in a depth-first manner (with backtracking) until a threshold is exceeded. If no goal nodes are expanded (the search is not successful), the threshold is incremented (to the minimum of all costs that exceeded the previous threshold) and the search continues.

(Strictly speaking an IDA* algorithm is not used since node selection is controlled by the order in which clauses appear in the input file.)

In the class of theorem provers examined here, both g and h are computed from the same measure: the depth of a deduction. Different depth measures, however, can yield drastically different search times for a given problem.

2.4.1 Depth Measures

The depth measure employed in PTTP and PARTHENON is based on the number of inferences (extensions and reductions) used to reach a node of the search tree. For any node n we have

$$\begin{aligned} g(n) &= \text{total number of extensions and reductions to reach } n \\ h(n) &= \text{number of B literals in the chain at node } n \end{aligned}$$

To see that $h(n)$ is an admissible measure (i.e., never overestimates the cost to a goal node) note that the conversion of each body literal introduced by an extension operation to an A-literal requires at least one inference (either a reduction or extension with a unit chain). This measure will ensure that if a proof is found, the number of inference steps in the proof is minimal. We call this depth measure *cumulative inference* depth, or *cumulative* depth. Note that both g and h (indirectly) are defined in terms of the total number of inferences in a proof.

In many contexts we are concerned with finding proofs of minimal length. At times, however, we may be concerned with minimizing the time to find a proof and may be willing to sacrifice minimization of proof length for a "quicker" proof search. This leads us to also consider a different depth measure, one originally proposed in [18], which attempts to minimize the depth of the search tree. In this case for any node n we have

$$\begin{aligned} g(n) &= \text{number of A-literals in the chain at } n \\ h(n) &= 0 \end{aligned}$$

Note that the number of A-literals in the chain at n is the length of the path from goal to n in the (potential) AND proof tree. We assign h the value 0 since it is possible for each literal introduced at node n to be "solved" by the reduction operation, which does not introduce a new A-literal. This depth measure ensures that the maximum number of A-literals in any chain of the deduction is minimized, which corresponds to minimizing the depth of the AND-OR search tree. We call this depth measure *ME-depth* since it is the measure outlined in the original presentation of Model Elimination [18].

To see how these different measures can affect the search for a proof, consider the proof trees of two different group theory problems (wos10 and wos1) shown in Figures 2.3 and 2.4. The clause sets for these problems are given in Section 2.6 (in the search tree each number refers to the clause number in the input file). The ME-depth proof tree for wos10 indicates a proof is found at ME-depth four that consists of 16 inferences. There is, however, a ten inference proof found as shown in the cumulative inference depth proof tree. Comparison of the times taken to find these respective proofs (see Table 2.9) shows that the ME-depth proof takes 12 times as long to find.

In contrast, the ME-depth proof tree for wos1 indicates a proof is found at ME-depth three that consists of 10 inferences. The cumulative inference depth proof tree also consists of a 10 inference proof, but this proof takes roughly 8.5 times as long to find as the ME-proof.

In terms of time taken to find a proof, these problems are relatively simple (there is a large class of problems whose proofs are found in under 1 or 2 seconds by all provers of the PTTP class; we do not consider these here). Even so, we gain or lose almost a factor of 10 depending on our choice of depth measure for these problems. For "harder" problems (problems for which it takes more time to find a proof) this difference can be even more pronounced—so much so that the choice of depth measure can determine whether a proof is found.

Given the differences in time to find a proof using the different depth measures, the question naturally arises as to whether a choice of depth measure can be made *a priori*. As we shall see, there are *a priori* reasons for choosing one depth measure in lieu of another based on a static evaluation of the input clause set. We have also developed a heuristic depth measure that combines both ME-depth and cumulative inference depth. We consider the combined measure first.

The obvious methods of combining the two methods in an iterative deepening search involve searching to a given level of the search tree (ME-depth) while constraining the number of inferences permitted in the potential proof tree. As a heuristic, we have chosen to constrain the number of A-literals to be half the number of inferences allowed in exploring the search tree. We call this depth measure *heuristic depth*. As seen in Tables 2.9 and 2.12, this heuristic depth yields uniformly faster search times than cumulative inference depth on the examples run. For several problems, however, using ME-depth yields drastically shorter search times. We are currently investigating using a dynamic combination in which the number of A-literals is a function of both the branching factor in the search tree and the number of inferences.

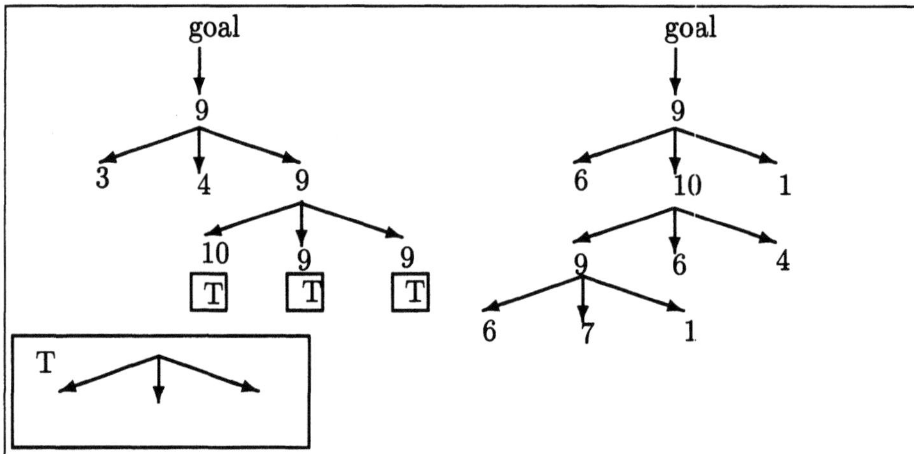

Figure 2.3: Proof Trees for wos10 (ME-depth on left, cumulative depth on right). The box labeled T represents three unit chain extensions.

2.4.2 An a Priori Choice of Depth Bound

There is a class of problems for which a static evaluation of the input clause set can suggest that ME-depth is the appropriate depth measure. Consider a

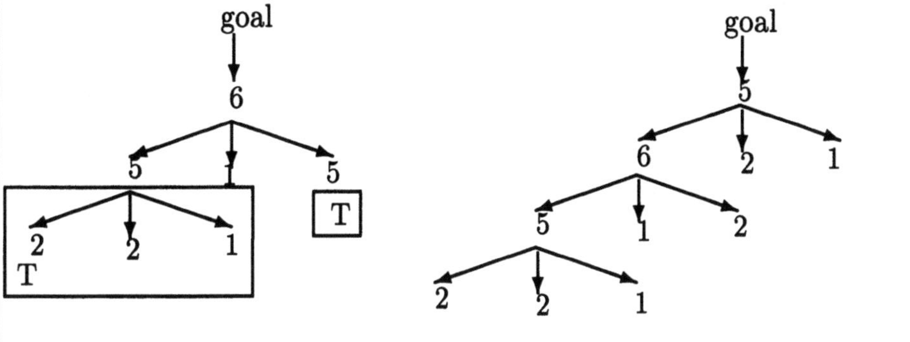

Figure 2.4: Proof Trees for wos1 (ME-depth on left, cumulative depth on right). The box labeled T represents the same extensions with three unit chains.

clause set which contains an n-literal clause for some "large" n (e.g., $n \geq 4$) or several such clauses. If such clauses are used several times in the proof tree, the tree will tend to be "bushy," i.e., several nodes will have a large branching factor. This is often an indication that the use of ME-depth will yield faster search times. Consider, for example, Schubert's Steamroller problem as formulated in [31]. The version of the problem suggested as the standard has a five literal goal and an eight literal clause among the input clauses. Although several forward chaining provers succeed quickly with this problem, the PTTP class of provers does not succeed when using cumulative search depth. Using ME-depth, however, (or rollback reduction—see below) METEOR finds a proof very quickly as indicated in Table 2.13. Examination of the proof tree shown in Figure 2.5 indicates that a proof tree of depth four exists, but that this tree consists of 69 inferences (all extensions). (Each □ in the figure represents an input chain that matches with either 1 or 2 unit clauses as indicated.)

As another class of problems for which ME-depth is well suited we consider the pigeonhole family of problems P_n, where P_n is a clausal form of the statement that $n+1$ pigeons cannot fit in n holes. This class of propositional problems is "hard" in the sense that the number of clauses in P_n is $O(n^3)$, but the shortest resolution refutation of P_n consists of $\Omega(c^n)$ clauses for some constant c ([13]), i.e., the size of the proof tree is exponential in n. The proofs of these theorems depend heavily on the reduction operation. For P_5 we note that the proof tree has depth 13 and that there are 4,178 inferences of which 1,568 are reductions (the proof is found in 25.67 seconds, see Table 2.13).

In comparing METEOR runtimes on the pigeonhole problems with results given for other provers (see Table 2.8) we have found a discrepancy in the use of the parameter n in the pigeonhole problem P_n. Some references ([23, 27]) use P_4 to refer to the same problem referred to by others as P_3([4]). We have adopted the meaning given in [13] and [4], which is the meaning given in the original presentation of the pigeonhole problem ([11]).

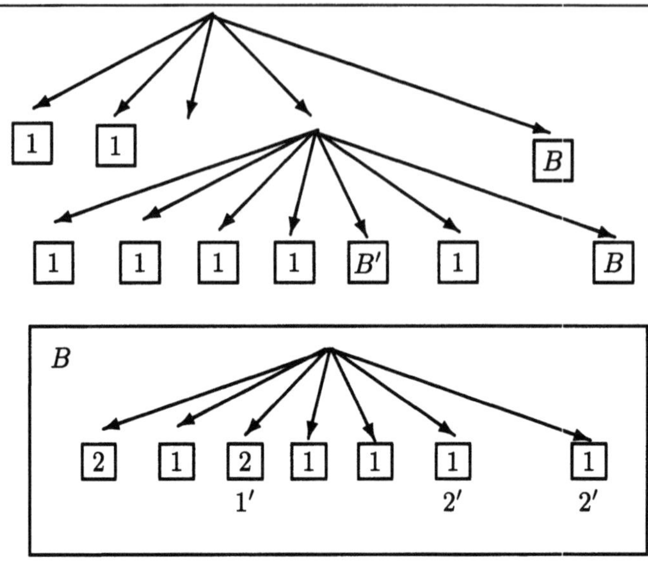

Figure 2.5: Proof Tree for Schubert's Steamroller

2.4.3 Rollback Reduction

As part of our examination of different depth measures we have discovered (somewhat serendipitously) a new depth measure which applies only to non-horn clause sets. In a sense this measure combines cumulative inference depth and ME-depth. With this measure, a depth is associated with each literal in a chain as well as with the chain itself. The depth of the initial chain (and the literals that comprise this chain) is zero. The depth of a successor chain depends on whether the chain is formed by extension or reduction. If the depth of a chain $\mathcal{C}$ is d, and the successor chain $\mathcal{C}'$ is formed by extension, then the depth of $\mathcal{C}'$ is $d+1$ and the depth of each literal (if any) introduced by the extension is also $d+1$. Thus for extensions, depth is calculated as it is using cumulative depth. If $\mathcal{C}'$ is formed from $\mathcal{C}$ by reduction, however, the depth of $\mathcal{C}'$ is the depth of the accessible (left most) literal of $\mathcal{C}'$.

Since the depth of a successor chain can decrease after a reduction operator (and never increases) we call this depth measure *rollback* reduction.

We have tried this measure on several non-horn problems and it is uniformly better than using cumulative inference depth on these problems. Using rollback reduction does not always result in a shortest proof. However, it appears that by allowing more reductions than would be allowed using cumulative inference depth that deeper proofs are found more quickly. For example, wos4 (another group theory problem) has a minimal length proof of 13 steps found in 1,083 seconds using cumulative depth (see Table 2.13) and an 18 step proof found in 0.78 seconds using rollback reduction.

The results for several non-horn problems are given in Table 2.13. For those problems for which no cumulative depth statistics are given, METEOR failed to find a proof using cumulative depth.

2.4.4 Weighted Depth

The depth measures we have mentioned may be set by runtime parameters when METEOR is invoked. We have also implemented a depth measure that may be used in conjunction with either cumulative depth or ME-depth, but which requires that input clauses be annotated with a weight. This weight should reflect the user's concept of how often a chain corresponding to the clause is used on any AND branch of the search tree ("heavy" clauses will be used less often). For example, in a group theory problem (e.g., wos1 and wos10, see pages 55 and 56) the axioms of associativity have many literals, none with ground terms. Unconstrained use of such clauses can result in chains that are not part of any proof, i.e., in searching dead ends in the search tree. One would thus weight these clauses more heavily than ground unit clauses.

When clauses are weighted and the appropriate runtime parameters are set to invoke weighting, the weight of a node N is the sum of the weights of the clauses used in the proof on the path from the root of the search tree to N. Typically, a weight bound is set by the user and this bound is used in conjunction with clause weighting to prune branches of the search tree that exceed this bound. If no proof is found with a given weight bound, the bound is incremented (automatically) and the search continued. We call this depth measure *weighted ME-depth* or *weighted depth.*

We use weighted depth in conjunction with either cumulative depth or ME-depth, not as a substitute for either. We have assigned weights to the clauses for many of the problems for which statistics are given in the tables below. For a given class of problems (e.g., Wos problems, Bledsoe problems), the weights assigned to common clauses are the same across problems in the same class. Statistics for the use of weighted depth with several problems are given in Table 2.9. The choice of weight bound and weight increment (the amount by which the bound is incremented on each search iteration) can have a profound effect on the search time. We choose the bound based on the weights of the heaviest clauses and an increment that permits progressively more instances of these heavy clauses to be used every three or four search iterations.

2.4.5 Other Pruning Techniques

There are several pruning techniques particular to Model Elimination that we incorporate in METEOR and which are used in other ME based provers ([28], [32], [8]). The most useful of these prunes (with failure) a branch of the search tree if the goal (literal eligible for extension and reduction) is identical to an ancestor literal. The other techniques prune (with success) a branch of the search tree if either reduction or extension with a unit chain does not specialize any variables in the goal. (These pruning rules are derived from Definition 2.2.1, the definition of an admissible chain.) The first of these techniques has a dramatic effect on the time needed to find a proof for many of the problems. In METEOR, the literals of a chain with the same sign and predicate are linked together so that this pruning method is relatively inexpensive to implement. The other techniques are also implemented efficiently due to the use of the trail stack. If a reduction or unit extension does not specialize any variables, no bindings will be trailed and a simple check of whether the top of the trail stack has changed is sufficient to determine whether pruning is permissible.

These pruning measures are active by default in METEOR although they can be deactivated at runtime. Note that in a parallel setting it is possible for the pruning techniques based on variable instantiation to be voided if a processor steals work prior to pruning.

Another pruning technique involves limiting the number of times a clause can participate in a deduction. This technique proved quite useful in the earliest implementation of Model Elimination [12] and appears to have been investigated by the PARTHEO group [28]. In METEOR, it is possible to limit the number of times a clause is used in a deduction by either annotating each clause in the input file (and specifying the appropriate runtime parameters) or by uniformly limiting clause use at runtime. The limit is incremented during each iteration of the search as are the other depth measures.

We also have implemented a mechanism that can control the complexity of the terms generated during a deduction. This is often useful in problems involving complex terms (e.g., the Bledsoe problems below) even though our mechanism is inefficient and reduces the inference rate by a significant factor. Our implementation is much cruder, for example, than that employed by OTTER [23]. We do not employ this pruning method nor the clause limiting method in the results reported here.

2.5 Results

In this section we provide the results of running METEOR on several problems from the literature using several different depth measures. We also compare these figures to those given for several other comparable theorem provers.

In these tables *sMETEOR* refers to our sequential implementation of METEOR. Statistics for this implementation are calculated from an average of at least five runs on a Sun SPARC Station 1+. Statistics for the parallel implementation of METEOR are also averaged over five runs on a Butterfly GP1000 using up to 30 nodes. (Each node consists of a Motorola 68020 processor, four megabytes of memory, and some custom hardware. There is no user level intranode multiprogramming.) We note that many of the problems can yield substantially different run times over several runs using the same number of processors due to the non-deterministic nature of parallel computation. Our distributed prover, *dMETEOR*, runs on a network of Sun SPARC and Sun-3 workstations. Since each "processor" in this network has a potentially different workload (unlike the Butterfly), and since our prover runs in a low priority background mode, the number of CPU seconds per processor varies greatly (across processors) during a distributed run.

Unless otherwise indicated, the problems we have used in testing our prover come from [32] (which pulled together problems from many other works, principally from Wilson and Minker [35]) and have been used as reported benchmarks in several other provers [8, 27].

It is difficult to compare the results reported for different automated reasoning systems because of differences in hardware. Ideally we would run the provers referenced in the tables below on the same machine and compare the runtimes, but we do not have access to all the provers nor are we expert at the use of many of the different provers. Nevertheless, we feel it is useful to report the comparisons we have made.

2.5.1 Bledsoe Challenge Problems

We have begun an investigation of the proofs of the problems Woody Bledsoe recently re-introduced [5]. The proofs of successive problems in this series are progressively deeper and harder. We have been able to prove the first two problems relatively easily with the appropriate depth measures, but have had trouble with the third problem in the series. The proof tree for the first problem is given in Figure 2.6.

The boxed regions of Figure 2.6 enclose the same proof tree. This proof tree can be developed only once and retained as a lemma. (The numbers of the nodes refer to clause numbers used in [5].) Similar multiple occurrences of subtrees, hence opportunity for lemma use, occur in proofs of the second and third problem. The proof search for the first and second Bledsoe challenge problems did not utilize lemmas. The third challenge problems has proven inaccessible to straight runs under any of the discussed depth measures, but has been solved when a lemma was used as an axiom. Likewise, we have been able to prove the lemma when stated as a goal (see Table 2.14). This is in no manner a demonstration that we can "solve" the third problem because lemma generation under several filters we have tried still returns too many lemmas to automate the passage from lemma generation to lemma use. (For the first two challenge problems, lemma generation is not so bountiful, but we have not yet explored coupling lemma generation and use systematically in any setting.) The third challenge problem illustrates the potential for added power if lemma selection can be done successfully for some deeper problems.

However, this third problem also demonstrates the limitation of purely high inference rates as this problem is combinatorially beyond the reach of linear input procedures without some augmentation similar to the use of lemmas.

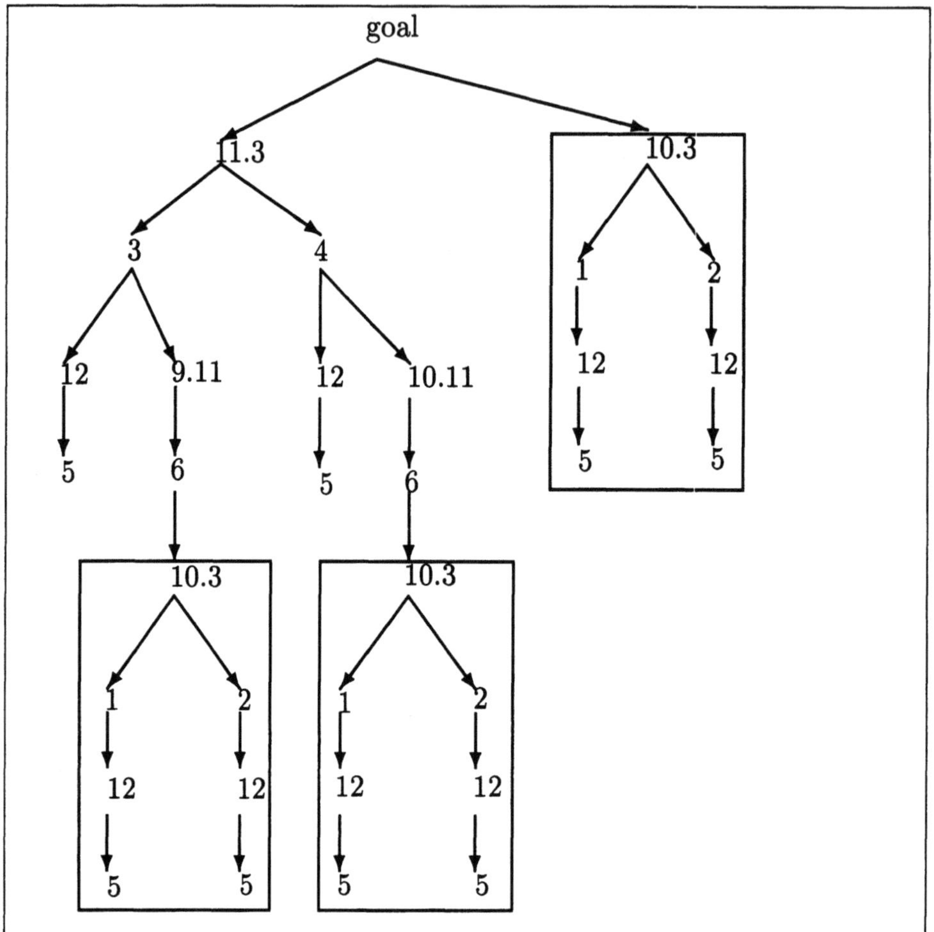

Figure 2.6: Proof Tree for First Bledsoe Challenge Problem

In Figures 2.11 and 2.12 we provide figures for several problems run on the Butterfly GP1000. The "superlinear" speedup (speedup of more than n on n processors) arises from the search at the last iteration of iterative deepening. If one of n processors chooses an alternative near the root of the search tree that leads directly to a proof, the time of this last iteration will be significantly less than the time of the last iteration using one processor. If completely searching all levels of the search tree prior to the last iteration is done efficiently, then a superlinear speedup results.

problem	Time (seconds) for several provers prover				
	PARTHEO†	PTTP	PARTHENON‡	sMETEOR	METEOR*
apabhp	na	873	2886/292	146.4	1,986/22
ls36	1666/352	548	2273/151	137.9	1,820/41
wos1	4.6/1.1	57.6	151/12	26.9	318/10.3
wos4	404/1.3	4,152	14,000/501	0.78	10.3/1.1
wos10	263/136	29.4	159/20	17.2	227/6.6
wos21	na	966	3315/373	503.8	6,373/274

†ME-depth, no iterative deepening, 1/16 processors
‡1/15 processors
* 1/30 processors

Figure 2.7: Comparison of Different Provers

Comparison with non-linear provers (runtime in seconds)			
problem	**OTTER†**	**Lee [27]‡**	**METEOR§**
apabhp	24	748	3.6
ls36	< 1	56.5	138
wos10	< .5	35.9	13.5
wos15	3	3,549	14,546
wos21	16	2,005	504
wos4	< 1	20.4	< 1
wos22	¶	73	381
pigeon 5	¶	7.28	25.67
ex 5	¶	97.9	15.62

†on Sun SPARC
‡on Sun 3
§optimal non-weighted depth measure
¶memory limit exceeded

Figure 2.8: Comparison with Non-linear Provers

problem	sMETEOR Timings (seconds) Depth Measure Used			
	cumulative inferences	ME-depth (# A-literals)	Heuristic (A-lits = inf/2)	Weighted
apabhp	146.4	3.6	16.6	1.1
ls36	137.9	31,023	109.4	11.1
wos1	26.9	3.1	3.9	0.7
wos4	0.78†	0.18	0.87†	0.87
wos10	17.2	206	13.48	9.2
wos21	503.8	> 190,600	344	249
bledsoe1		2,333		2.1
bledsoe2		7,984		194.09
steamroller	398†	3.41	400†	1.63

†non-horn, rollback reduction used

Figure 2.9: sMETEOR Timings (seconds) with Different Depth Measures

problem	Successful Inferences/second prover				
	PTTP	PARTHENON †	sMETEOR	wMETEOR	METEOR‡
apabhp	1,957	605/7,478	3,634	5,118	549/16,813
ls36	2,555	644/8,197	5,028	4,235	465/14,000
wos1	2,413	914/9,465	5,170	4,409	449/12,516
wos4	2,182	705/9,532	4,735	4,648	404/2,996
wos10	2,675	619/8,419	4,575	3,603	392/12,318
wos21	2,481	778/10,023	4,760	4,305	407/12,050
bledsoe1			1,224	1,682	440
bledsoe2			1,550	2,691	326
steamroller			10,943	12,403	960

†1/15 processors
‡1/30 processors

Figure 2.10: Successful Inferences per Second

problem	Time in seconds (speedup) Cumulative Inference Depth Measure Number of processors				
	1	2	10	20	30
apabhp	1,986	341 (5.82)	61 (32.55)	32 (62.06)	22 (90.27)
ls36	1,820	888 (2.04)	99 (18.38)	55 (33.09)	41 (44.39)
wos1	318	153 (2.07)	29 (10.96)	15 (21.2)	10.5 (30.28)
wos4	10.3	6 (1.71)	2.28 (4.51)	1.33 (7.74)	1.1 (9.36)
wos10	227	112 (2.02)	21 (10.8)	13 (17.46)	6.6 (34.39)
wos21	6,373	3290 (1.93)	848 (7.51)	335 (19.02)	274 (23.25)

Figure 2.11: Butterfly Execution Results (cumulative depth measure)

problem	Time in seconds (speedup) Heuristic Depth Measure Number of processors				
	1	2	10	20	30
apabhp	234.8	40.17 (5.84)	8.1 (28.98)	5.75 (40.83)	4.32 (54.35)
ls36	1,483	724.7 (2.04)	78.7 (18.84)	45.6 (32.52)	35.2 (42.13)
wos1	53.3	21.85 (2.43)	4.36 (12.22)	3.6 (14.8)	3.0 (17.76)
wos10	188	92.6 (2.03)	20.8 (9.03)	6.8 (27.64)	6.06 (31.02)
wos21	4,678	2,396 (1.95)	680.5 (6.87)	248 (18.86)	204 (22.93)

Figure 2.12: Butterfly Execution Runtime (heuristic depth measure)

problem	Some harder problems time (secs)	depth measure	prover
wos22	381.14	ME-depth	sMeteor
	11,703	heuristic	sMeteor
	17,797	inference	sMeteor
	26,766	inference	Butterfly†
wos15	14,546	inference	sMeteor
	10,604	heuristic	sMeteor
	5,629	heuristic	Butterfly †
non-horn problems			
ex5	16,889	heuristic	sMeteor
	1,662	ME-depth	sMeteor
	15.62	(no reductions)	sMeteor
wos4	1,083	inference	sMeteor
	0.78	rollback	sMeteor
steamroller	3.41	ME-depth	sMeteor
	398	rollback	sMeteor
	38.85/2.97	ME-depth	Butterfly ‡
pigeon 3	0.04	ME-depth	sMeteor
	223.9	inference	sMeteor
pigeon 4	0.51	ME-depth	sMeteor
	1.21	rollback	sMeteor
pigeon 5	25.67	ME-depth	sMeteor
nonobvious	1.51	ME-depth	sMeteor
(see [26])	7.41	rollback	sMeteor

†20 processors
‡1/20 processors

Figure 2.13: Some Harder Problems

Bledsoe Third Challenge Problem (using lemma as axiom)		
time (secs)	prover	depth measure
31,649	sMETEOR	ME-depth
≈ 70	dMETEOR (10 procs)	ME-depth
165	METEOR (20 procs)	ME-depth

Figure 2.14: Bledsoe Third Challenge Problem

2.6 Clause Sets

2.6.1 wos1

This is an example from group theory. Given left identity and inverses, there is a right identity. In the clauses below, `e` represents the left identity, `g` is the left inverse function, `h` is the right inverse function, and `f` represents the result of operating on two group elements. The predicates used are `p` for the result of operating on two objects ("product") and `r` for equality of two objects. Taken from [35].

```
1.  p(e,X,X)
2.  p(g(X),X,e)
3.  p(X,Y,f(X,Y))

4.  r(X,X)

5.  -p(X,Y,U), -p(Y,Z,V), -p(U,Z,W), p(X,V,W)
6.  -p(X,Y,U), -p(Y,Z,V), -p(X,V,W), p(U,Z,W)

7.  -r(X,Y), r(Y,X)
8.  -r(X,Y), -r(Y,Z), r(X,Z)

9.  -p(X,Y,U), -p(X,Y,V), r(U,V)
10. -r(U,V),   -p(X,Y,U), p(X,Y,V)
11. -r(U,V),   -p(X,U,Y), p(X,V,Y)
12. -r(U,V),   -p(U,X,Y), p(V,X,Y)

13. -r(U,V), r(f(X,U),f(X,V))
14. -r(U,V), r(f(U,Y),f(V,Y))
15. -r(U,V), r(g(U),g(V))
16. -r(U,V), r(h(U),h(V))

goal
17. -p(h(X),X,h(X))
```

2.6.2 wos10

Another group theory problem. The constants, functors, and predicates are as in wos1. This is a proof of: if every group element is idempotent (e.g., $X^*X = e$), then the group is commutative.

```
1.  p(e,X,X)
2.  p(g(X),X,e)
3.  p(X,Y,f(X,Y))
4.  p(X,e,X)
5.  p(X,g(X),e)
6.  p(X,X,e)
7.  p(a,b,c)

8.  r(X,X)

9.  -p(X,Y,U), -p(Y,Z,V), -p(U,Z,W), p(X,V,W)
10. -p(X,Y,U), -p(Y,Z,V), -p(X,V,W), p(U,Z,W)

11. -r(X,Y), r(Y,X)
12. -r(X,Y), -r(Y,Z), r(X,Z)

13. -p(X,Y,U), -p(X,Y,V), r(U,V)
14. -r(U,V),   -p(X,Y,U), p(X,Y,V)
15. -r(U,V),   -p(X,U,Y), p(X,V,Y)
16. -r(U,V),   -p(U,X,Y), p(V,X,Y)

17. -r(U,V), r(f(X,U),f(X,V))
18. -r(U,V), r(f(U,Y),f(V,Y))
19. -r(U,V), r(g(U),g(V))

goal
20. -p(b,a,c)
```

References

[1] Khayri A. M. Ali and Roland Karlsson (1990): The Muse Or-Parallel Prolog Model and Its Performance. North American Conference on Logic Programming, 757–776.

[2] Owen Astrachan (1989): METEOR: Model Elimination Theorem-proving for Efficient OR-parallelism. Master's thesis. Duke University.

[3] R. Bayer and M. Schkolnick (1977): Concurrency of Operations in B-trees. Acta Informatica 9, 1–21.

[4] W. Bibel (1990): Short Proofs of the Pigeonhole Formulas Based on the Connection Method. Journal of Automated Reasoning 6, 287–297.

[5] W. W. Bledsoe (1990): Challenge Problems in Elementary Calculus. Journal of Automated Reasoning 6(3), 341–359.

[6] W. W. Bledsoe and L. Hines (1980): Variable Elimination and Chaining in a Resolution-based Prover for Inequalities. Fifth Conference on Automated Deduction, Springer-Verlag, 281–292.

[7] Soumitra Bose, Edmund Clarke, David E. Long, and Spiro Michaylov (1989): Parthenon: A Parallel Theorem Prover for Non-Horn Clauses. Symposium on Logic in Computer Science.

[8] Soumitra Bose, Edmund Clarke, David E. Long, and Spiro Michaylov (1990): Parthenon: A Parallel Theorem Prover for Non-Horn Clauses. Journal of Automated Reasoning, (to appear).

[9] R. S. Boyer and J. Moore (1972): The Sharing of Structure in Theorem Proving Programs. In: B. Meltzer and D. Michie, eds.: Machine Intelligence 7. Edinburgh: Edinburgh University Press. 101–116.

[10] C. Chang and R. Lee (1973): Symbolic Logic and Mechanical Theorem Proving. New York: Academic Press.

[11] S. A. Cook and R. A. Reckhow (1979): On the Lengths of Proofs in the Propositional Calculus. Journal of Symbolic Logic 44(1), 15–22.

[12] S. Fleisig, D. Loveland, A. Smiley, and D. Yarmash (1974): An Implementation of the Model Elimination Proof Procedure. JACM 21, 124–139.

[13] A. Haken (1985): The Intractability of Resolution. Theoretical Computer Science 39, 297–308.

[14] Richard E. Korf (1985): Depth-first Iterative Deepening: An Optimal Admissible Tree Search. Artificial Intelligence 27, 97–109.

[15] David Kotz and Carla Ellis (1989): Evaluation of Concurrent Pools. Ninth International Conference on Distributed Computing Systems, 378–385.

[16] R. A. Kowalski and D. Kuehner (1971): Linear Resolution with Selection Function. Artificial Intelligence 2, 227–260.

[17] Vipin Kumar and V. Nageshwara Rao (1990): Scalable Parallel Formulations of Depth-first Search. In: Vipin Kumar, P. S. Gopalakrishnan, and Laveen N. Kanal, eds.: Parallel Algorithms for Machine Intelligence and Vision. Berlin: Springer-Verlag. 1–41.

[18] Donald W. Loveland (1968): Mechanical Theorem Proving by Model Elimination. JACM 15(2), 236–251.

[19] Donald W. Loveland (1969): A Simplified Format for the Model Elimination Procedure. JACM 16(3), 349–363.

[20] Donald W. Loveland (1978): Automated Theorem Proving: A Logical Basis. Amsterdam: North-Holland.

[21] Ewing Lusk, David H. D. Warren, Seif Haridi, et al. (1990): The Aurora Or-parallel Prolog System. New Generation Computing 7, 243–271.

[22] David Maier and David S. Warren (1988): Computing with Logic. Benjamin/Cummings.

[23] William W. McCune (1990): OTTER 2.0 Users Guide. Argonne National Laboratory, Argonne, IL, USA.

[24] Nils Nilsson (1980): Principles of Artificial Intelligence. Tioga Press.

[25] Judea Pearl (1984): Heuristics: Intelligent Search Strategies for Computer Problem Solving. Addison-Wesley.

[26] F. J. Pelletier and P. Rudnicki (1986): Non-obviousness. AAR Newsletter 6, 4–5.

[27] David Plaisted and Shie-Jue Lee (1990): Inference by Clause Linking. Technical Report TR90-022, University of North Carolina, Department of Computer Science, Chapel Hill, NC, USA.

[28] J. Schumann and R. Letz (1990): PARTHEO: A High Performance Parallel Theorem Prover. Tenth International Conference on Automated Deduction, 40–56.

[29] Mark E. Stickel (1984): A Prolog Technology Theorem Prover. New Generation Computing 2(4), 371–383.

[30] Mark E. Stickel (1986): A Prolog Technology Theorem Prover: Implementation by an Extended Prolog Compiler. Eight International Conference on Automated Deduction, 573–587. Berlin: Springer.

[31] Mark E. Stickel (1986): Schubert's Steamroller Problem: Formulations and Solutions. Journal of Automated Reasoning 2, 89–100.

[32] Mark E. Stickel (1988): A Prolog Technology Theorem Prover: Implementation by an Extended Prolog Compiler. Journal of Automated Reasoning 4, 343–380.

[33] David H. D. Warren (1983): An Abstract Prolog Instruction Set. Technical Report 309, SRI International, Menlo Park, CA, USA.

[34] David H. D. Warren (1987): The SRI Model for OR-Parallel Execution of Prolog—Abstract Design and Implementation Issues. IEEE Symposium on Logic Programming, 92–102.

[35] Gerald A. Wilson and Jack Minker (1976): Resolution, Refinements, and Search Strategies: A Comparative Study. IEEE Transactions on Computers, C-25(8), 782–801.

Chapter 3

The Metatheorist: Automatic Proofs of Theorems in Analysis Using Non-Standard Techniques, Part II[1]

A. M. Ballantyne

Austin Lab for Software Engineering and Computer Science
EDS Research and Development
1601 Rio Grande, Suite 451
Austin, Texas

Abstract

We present a version of the theorem prover IMPLY [2] for proving theorems in the theory of non-standard analysis. The theorems proved are statements of the equivalence between the standard and non-standard definitions of concepts from analysis. Crucial to the success of the program are two heuristics, the *Meta-heuristic* and the *F-heuristic*, which are based on metatheorems relating the standard and non-standard theories of the reals. Equally important are heuristics which relate inequalities to types.

[1]The work described here was conducted at the University of Texas and supported by NSF Grant DCR74-12866.

3.1 Introduction

The original formulation of the calculus was based on the intuitive notion of an "infinitesimal" or an infinitely small number. The concepts of analysis were defined in terms of infinitesimals. For example, a continuous function was defined to be a function which mapped numbers infinitely close together onto numbers infinitely close together.

Infinitesimals provided a powerful conceptual tool for early investigations into analysis, but they could not, until recently, be placed on firm, formal ground. In 1960 Abraham Robinson devised the theory of non-standard analysis which finally provided a formal basis for the vague and controversial notion of an infinitely small quantity. The use of non-standard analysis appears to greatly simplify many proofs of theorems in analysis. In fact Keisler [5] based an elementary calculus text on these methods.

In the early 70's, the UT theorem-proving group was actively pursuing mechanical proofs of theorems in analysis. Ballantyne and Bledsoe [1] built a version of the UT Prover which incorporated a large number of heuristics specific to non-standard analysis. This prover mechanically generated proofs of a large number of theorems from elementary analysis.

A good reference for non-standard analysis is [7]. The paper [1] provides a concise but reasonably comprehensive introduction to the subject. The introductory material in that paper should be considered a prerequisite to the material presented here.

3.2 Non-standard Analysis

In this section we will briefly outline the non-standard theory of the reals. If the reader is at all familiar with non-standard analysis, we suggest he skip this section.

In the following we will denote the standard theory of analysis by $\mathcal{T}$. The non-standard reals $\mathcal{R}^*$ are a proper extension of the reals which behave *formally* like the reals. By formally we mean that $\mathcal{R}^*$ is a model of analysis in the following sense. Let $\mathcal{L}$ be the language in which the theory of analysis $\mathcal{T}$ is expressed. Then for any sentence S in $\mathcal{L}$, S is "true" about $\mathcal{R}$ if and only if S is true about $\mathcal{R}^*$. We assume that for every element in $\mathcal{R}$ there is a corresponding constant symbol in $\mathcal{L}$. Since one can show the structure $\mathcal{R}^*$ is non-Archimedean, there exist elements which are smaller than $1/n$ for every n. These elements are said to be infinitely small or infinitesimals. Every element in $\mathcal{R}^*$ which is also in $\mathcal{R}$ will be called a standard real number. Since $\mathcal{R}^*$ is a field, each infinitesimal has an inverse and each such inverse is greater than every natural number. Such large real numbers are said to be infinitely large. If a number is not infinitely large, it is said to be finite.

Within the language $\mathcal{L}$, one cannot state that a number is infinitely large or infinitely small or standard. Let $\mathcal{L}^*$ be an extension of $\mathcal{L}$ which includes the predicates

1. IRL(x) — x is infinitely large,
2. STANDARD(x) — x is a standard number,
3. FINITE(x) — x is finite,
4. ISR(x) — x is infinitely small
5. $x \simeq y$ — x is infinitely close to y.

The theory of non-standard analysis defines standard concepts in analysis in terms of $\mathcal{L}^*$. We will denote the non-standard theory of analysis by $\mathcal{T}^*$. The theory $\mathcal{T}^*$ is a conservative extension of $\mathcal{T}$ in that if S is any sentence in $\mathcal{L}$, then S is in $\mathcal{T}$ if and only if S is in $\mathcal{T}^*$.

Some non-standard definitions of familiar concepts are:

1. A function f is said to be continuous at the standard point x if for all y such that $y \simeq x$, $f(y) \simeq f(x)$.
2. The sequence x_n converges to the standard point p if for every infinitely large integer ω, $x_\omega \simeq p$.
3. The standard point p is a limit point of the set S if there is some y in S such that $y \simeq p$.
4. A set S is open if for every standard p in S and for all y, if $y \simeq p$ then $y \in S$.

While these definitions seem to capture the intuitive meaning of standard concepts from analysis, there seems to be some magic involved. How, for example, are we to reason about terms like x_ω when the original definition of a sequence only refers to finite entities. You must consult the references for the answer to this question—for the time being we assert that the construction of the non-standard model was done in such a way that standard functions and sets are interpreted in the correct way. Thus, for example, the sine function is defined for every element in $\mathcal{R}^*$ whether it be standard or not.

3.3 A Sample Proof of Equivalence

To motivate our presentation of the heuristics added to the prover, we give an informal proof of the equivalence between the standard and non-standard definitions of continuous function. In a previous section we presented the non-standard definition of continuous function. For the sake of completeness we now provide the standard one.

A function f is *continuous* at the point x_0 if for all $\epsilon > 0$ there exists a $\delta > 0$ such that for all y, if $|x_0 - y| < \delta$ then $|f(x_0) - f(y)| < \epsilon$.

We will first prove the standard implies non-standard half of the equivalence. As a notational convention, we will often emphasize that a term is a constant by writing it with a subscript. Pick a b_0 such that $b_0 \cong x_0$. In order to show that $f(x_0) \cong f(b_0)$ we must show that for any standard ϵ, $|f(x_0) - f(b_0)| < \epsilon$. Let ϵ_0 be any fixed standard number. By the standard definition of continuity, there is therefore some standard δ_0 such that for any x where $x > 0$ and $|x_0 - x| < \delta_0$, $|f(x_0) - f(x)| < \epsilon_0$. The fact that $b_0 \cong x_0$ implies that $|x_0 - b_0| < \delta$ for any standard number δ, in particular δ_0. Thus the conclusion follows.

We now prove the converse, that the non-standard definition implies the standard one. Given any standard ϵ_0 and any positive infinitesimal b_0, the statement

$$\epsilon_0 > 0 \rightarrow \forall x(|x_0 - x| < b_0 \rightarrow |f(x_0) - f(x)| < \epsilon_0) \tag{3.1}$$

holds in $\mathcal{R}^*$. Consequently the statement

$$\epsilon_0 > 0 \rightarrow \exists \delta(\delta > 0 \wedge \forall x(|x_0 - x| < \delta \rightarrow |f(x_0) - f(x)| < \epsilon_0)) \tag{3.2}$$

also holds in $\mathcal{R}^*$. Since the constants ϵ_0 and x_0 denote arbitrary standard real numbers, 3.2 also holds in $\mathcal{R}$. Since ϵ_0 is an arbitrary (standard) constant, the statement

$$\forall \epsilon(\epsilon > 0 \rightarrow \exists \delta(\delta > 0 \wedge \forall x(|x_0 - x| < \delta \rightarrow |f(x_0) - f(x)| < \epsilon))) \tag{3.3}$$

holds in $\mathcal{R}$.

3.4 The Meta-heuristic

We now make a few observations about the relationship between $\mathcal{T}$ and $\mathcal{T}^*$. As we mentioned earlier, any sentence S of $\mathcal{L}$ is in $\mathcal{T}$ if and only if S is in $\mathcal{T}^*$.

Let S be the sentence $\forall x P(x)$ where P is some wff in $\mathcal{L}$. Let S^* be the sentence $\forall x[\text{STANDARD}(x) \rightarrow P(x)]$ in $\mathcal{L}^*$. We can easily show that the sentence $S^* \rightarrow S$ is in $\mathcal{T}^*$. Assuming otherwise, then S^* is true and S is false. Then $\neg S$ is in $\mathcal{T}^*$ and since $\neg S$ is in $\mathcal{L}$, then $\neg S$ is in $\mathcal{T}$. Hence for some constant symbol a in $\mathcal{L}$, we have $\neg P(a)$ in $\mathcal{T}$. This implies that $\neg P(a)$ is in $\mathcal{T}^*$. Since a denotes a standard real number under the standard interpretation, it denotes a standard number under the non-standard interpretation. Hence the sentence STANDARD(a) is in $\mathcal{T}^*$. Since STANDARD(a) and S^* are both in $\mathcal{T}^*$, so is $P(a)$. This gives a contradiction. We will see in a moment how to use the fact that S^* implies S.

We now say for some standard predicate C and for its standard and non-standard definitions $D(C)$ and $D^*(C)$, that $D(C)$ and $D^*(C)$ are equivalent if $D(C) \leftrightarrow D^*(C)$ is in $\mathcal{T}^*$. We can now illustrate the importance of the observation made in the previous paragraph. Suppose that we are trying to establish $D^*(C) \rightarrow D(C)$ where $D(C)$ is of the form

$$\forall x_1 \ldots x_n S(x_1, \ldots, x_n).$$

Since $D(C)$ is a sentence in $\mathcal{L}$, we can substitute for it the modified sentence

$$\forall x_1 \ldots x_n (\text{STANDARD}(x_1) \wedge \ldots \wedge \text{STANDARD}(x_n) \rightarrow S(x_1, \ldots, x_n)).$$

The modified form of $D(C)$ will be denoted by $D(C)^*$. Whenever PROVER is asked to prove $D^*(C) \rightarrow D(C)$ it attempts to prove instead $D^*(C) \rightarrow D(C)^*$. Clearly a proof of the latter supplies a proof of the former. As we shall see, this heuristic is of crucial importance in what follows. We shall call it the "Meta-heuristic". It is by means of an implicit use of the Meta-heuristic that we are able to infer (3.3) from (3.2) in the informal proof of equivalence between the standard and non-standard definitions of continuous function.

3.5 The F-heuristic

Let t be any ground term in $\mathcal{L}$. Then the sentence STANDARD(t) is true in $\mathcal{T}^*$. This fact can often be very useful when attempting a proof, hence PROVER keeps track of which constants and function symbols are in $\mathcal{L}$.

As an illustration, suppose we were trying to prove that the standard definition of bounded set implies the non-standard definition; in other words that

$$D(\text{BOUNDED}) \rightarrow D^*(\text{BOUNDED}) \tag{3.4}$$

is in $\mathcal{T}^*$. The standard definition of bounded is

$$\exists r \forall x (x \in S \rightarrow x < r) \tag{3.5}$$

where S is the set we are asserting is bounded. The Skolemized form of (3.5) is

$$x \in S \rightarrow x < r_0 \tag{3.6}$$

where r_0 is a Skolem constant. If S is bounded then under the standard interpretation, (3.5) is true in $\mathcal{R}$. Hence there is an interpretation of r_0 which makes (3.6) true in $\mathcal{R}^*$. Since the non-standard interpretation is an extension of the standard interpretation, and since r_0 defines a standard number in the standard interpretation, r_0 denotes a standard number in the non-standard interpretation. Thus we are allowed to assert

$$\text{STANDARD}(r_0)$$

and use that fact in the proof of 3.4.

More generally, suppose PROVER is trying to prove $D(C) \rightarrow D^*(C)$ for some concept C. Suppose also that the Skolemized form of $D(C)$ contains the term $f(x_1, \ldots, x_n)$, where f is either a Skolem function or is a symbol in $\mathcal{L}$. Reasoning as above, we are justified in treating f as a standard function in that whenever a term $f(t_1, \ldots, t_n)$ appears in the course of a proof, and the terms t_i have type STANDARD, then we can assert that $f(t_1, \ldots, t_n)$ also has type STANDARD.

When PROVER attempts to prove $D(C) \rightarrow D^*(C)$, it first Skolemizes $D(C) \rightarrow D^*(C)$ and then for each term $f(x_1, \ldots, x_n)$ in the hypothesis, it adds the additional hypothesis

$$\text{STANDARD}(x_1) \wedge \ldots \wedge \text{STANDARD}(x_n) \rightarrow \text{STANDARD}(f(x_1, \ldots, x_n)).$$

For each constant c occuring in the hypothesis, PROVER adds the hypothesis

$$\text{STANDARD}(c).$$

This process of adding these new hypotheses we will call the F-heuristic. An implicit use of the F-heuristic in our example proof enabled us to assert that δ_0 denoted a standard number.

3.6 Other Modifications to PROVER

In addition to the Meta-heuristic and F-heuristic, PROVER was modified to include procedures for handling the inequalities that arise in the metatheory of non-standard analysis. More specifically, these procedures coordinate typing information with inequality information.[2] As in the program described in [1], this version of PROVER processes the statement of the theorem to extract all possible typing information about every constant, and every variable that occurs in a goal. Thus if the hypothesis contained the statements ISR(a) and $0 < a$ for some constant a, PROVER would note that a was of type ISR and positive.

We will now give a few examples to illustrate the interplay between types and inequalities. Suppose, for example, the terms $t_1 < t_2$ and $0 < t_1$ occur in the hypothesis of some theorem. Suppose in addition that the term t_2 is of type ISR (an infinitesimal). We can then infer that t_1 is also of type ISR. As another example, suppose the atomic formula $a < x$ occurs as a goal in the conclusion of some theorem. Suppose also that there is no "obvious" formula in the hypothesis against which we could match the goal to provide a solution for x. We could use the axioms of the reals and propose the substitution $a+1$ for x.

Assume on the other hand that the goal was $t(x) < x$ and that there was no matching hypothesis. The substitution $t(x) + 1$ for x is illegal since

[2]The version of PROVER cited in [1] incorporated procedures for coordinating typing information with the arithmetic operators. Inequalities did not occur in the theorems which that version of PROVER attempted—the non-standard characterization had replaced them by the more tractable notion of $\simeq$.

it fails the occurs check. If $t(x)$ is declared to be of a certain (small) type, then sometimes we can use the properties of the non-standard reals to solve the goal. If $t(x)$ is declared to be of type FINITE for all values of x, then a solution for x would be any number of type ILR. In this situation, PROVER would create a new constant c, proclaim it to be of type ILR and return the substitution c for x. This is a valid operation since there exists an infinite number of infinitely large reals and each infinitely large real is greater than any finite real.

Inequalities can be used to solve typing goals. When trying to establish the goal ISR(t) where $t > 0$ and t is some non-variable term, it is often useful to search for another term r which is infinitely small and which bounds t from above. In other words, when proving

$$\mathrm{ISR}(t)$$

where t is known to be positive, PROVER will eventually try to establish

$$\exists r(t < r \wedge \mathrm{ISR}(r)).$$

It is usually the case that there is not enough information explicit in the statement of the theorem to solve the second conjunct ISR(r). PROVER will then make implicit use of the following fact about the non-standard reals:

$$\exists r(0 < r \wedge \mathrm{ISR}(r))$$

Thus, PROVER solves ISR(r) by creating a new constant c_0 of type ISR. PROVER also asserts $0 < c_0$ and ISR(c_0). In fact, PROVER makes these assertions prior to solving the $t < r$ subgoal, as they are sometimes useful in the proof of that goal. For lack of better terminology, we call these actions the ISR-heuristic.

We briefly outline the steps that PROVER takes to solve a goal of the form $a < b$.

1. If both a and b are typed terms, PROVER tries to solve $a < b$ by means of the following table. (We assume here that $0 < a$.)

		b			
		ISR	FINITE	STANDARD	ILR
	ISR		T	T	T
a	FINITE				T
	STANDARD	F			T
	ILR	F	F	F	

   If only one of the terms is typed, then PROVER will try to establish the type of the other term in accordance with the table. For example, if the type of b is known to be STANDARD and the type of a is unknown, then PROVER will (eventually) try to establish that the type of a is ISR.

2. If b is a variable x and a is some term $t(x)$ which contains x, and the type of $t(x)$ is either ISR or FINITE, then return c_0 for x where c_0 is a new constant with type ILR.

3. If a is the variable x and b is some term $t(x)$ which contains x and the type of $t(x)$ is either FINITE or ILR, then return c_0 for x where c_0 is a new constant of type ISR.

We will refer to these heuristics in the proof traces as $<(1)$, $<(2)$, and $<(3)$ respectively.

3.7 A Brief Introduction to PROVER

In the proofs that follow, we assume familiarity with the UT PROVER [3]. For those unfamiliar with PROVER we now provide an extremely terse description. The following few sentences will provide little more than some of the background terminology. Basically, PROVER is goal oriented; it works backwards from the conclusion. A goal is established when it unifies with some hypothesis. We denote this by "Match" in the following proofs. Conjunctive subgoals are established by solving in turn the individual conjuncts with the satisfying substitution from each subgoal being applied to the next. In the following traces, we represent this splitting of conjuncts by a new level of indentation.

If PROVER does not find a matching hypothesis, it will try to "Backchain". If the current goal is C and there is a hypothesis of the form $H \to C'$ and C and C' unify with substitution θ, then PROVER will try to solve the goal $H\theta$.

Sometimes PROVER will use the hypotheses in a forward direction. If there is a hypothesis of the form

$$H_1 \wedge H_2 \ldots \wedge H_n \to C$$

and ground literals H_1', H_2', ..., H_n' which match the H_i with substitution θ_i then PROVER will assert

$$C\theta_1\theta_2\cdots\theta_n.$$

This assertion of new facts is called forward chaining ("F-chain" in the proof traces).

Often PROVER will have to replace a defined concept with its definition. This will be denoted by the label "Define(*concept*)" in the proof step.

Expressions of the form

$$P \to (Q \to R)$$

are rewritten as

$$P \wedge Q \rightarrow R$$

so as to expose the conclusion. This action is labelled "Promote" in the proof traces. Note that it adds a new hypothesis. Whenever a new sentence is added to the hypothesis, either by forward chaining or by promotion, we annotate this addition with the label **adding** rather than display the whole new hypothesis.

3.8 Machine Proofs

In the following proof traces, only those goals that lead to a successful proof are displayed. The label to the right of selected proof steps gives the reason the step was taken.

3.8.1 Continuous Function

STANDARD Implies NON-STANDARD

$$\begin{array}{l} \forall\epsilon(\epsilon > 0 \rightarrow \exists\delta(\delta > 0 \wedge \forall y(|x_0 - y| < \delta \rightarrow |f(x_0) - f(y)| < \epsilon))) \\ \quad \rightarrow \forall z(x_0 \simeq z \rightarrow f(x_0) \simeq f(z)) \end{array}$$

$$\begin{array}{l} (\epsilon > 0 \rightarrow \delta(\epsilon) > 0 \wedge (|x_0 - y| < \delta(\epsilon) \\ \qquad \rightarrow |f(x_0) - f(y)| < \epsilon) \wedge x_0 \simeq z_0) \\ \quad \rightarrow f(x_0) \simeq f(z_0) \end{array}$$

		Skolemize
		Promote
adding	$\text{STANDARD}(x) \rightarrow \text{STANDARD}(\|x\|)$	F-heuristic
adding	$\text{STANDARD}(x) \rightarrow \text{STANDARD}(\delta(x))$	F-heuristic
adding	$\text{STANDARD}(x) \rightarrow \text{STANDARD}(f(x))$	F-heuristic
adding	$\text{STANDARD}(x_0)$	F-chain
adding	$\text{STANDARD}(\|x_0\|)$	F-chain
adding	$\text{STANDARD}(f(x_0))$	F-chain

$$\begin{array}{l} (\epsilon > 0 \rightarrow (\delta(\epsilon) > 0 \wedge (|x_0 - y| < \delta(\epsilon) \rightarrow |f(x_0) - f(y)| < \epsilon)) \\ \quad \wedge (q > 0 \wedge \text{STANDARD}(q) \rightarrow |y_0 - x_0| < q) \\ \quad \wedge p_0 > 0 \\ \quad \wedge \text{STANDARD}(p_0)) \\ \quad \rightarrow |f(x_0) - f(y_0)| < p_0 \end{array}$$

		Define ($\simeq$)
		Promote
adding	$\text{STANDARD}(\delta(p_0))$	F-chain
adding	$\text{STANDARD}(\|p_0\|)$	F-chain
adding	$\text{STANDARD}(f(p_0))$	F-chain
adding	$\|y_0 - x_0\| < p_0$	F-chain
adding	$\delta(p_0) > 0$	F-chain

goal $|f(y_0) - f(x_0)| < p_0$
goal $|y_0 - x_0| < p_0$ Backchain
goal $0 < \delta(p_0) \wedge \text{STANDARD}(\delta(p_0))$ Backchain
goal $0 < \delta(p_0)$
TRUE T Match
goal $\text{STANDARD}(\delta(p_0))$ Match
TRUE T
TRUE T

NON-STANDARD Implies STANDARD

$$\forall y(y \simeq x_0 \rightarrow f(y) \simeq f(x_0)) \rightarrow \forall \epsilon(\epsilon > 0 \rightarrow \exists \delta(\delta > 0 \wedge \forall z(|z - x_0| < \delta \rightarrow |f(z) - f(x_0)| < \epsilon)))$$

$$y \simeq x_0 \rightarrow f(y) \simeq f(x_0) \wedge \text{STANDARD}(\epsilon_0) \wedge \epsilon_0 > 0 \rightarrow \delta > 0 \wedge (|z(\delta) - x_0| < \delta \rightarrow |f(z(\delta)) - f(x_0)| < \epsilon_0)$$ Skolemize
Promote
Meta-heurist

COMMENT: The hypothesis $\text{STANDARD}(\epsilon_0)$ is added by the Meta-heuristic

goal $(|z(\delta) - x_0| < \delta \rightarrow |f(z(\delta)) - f(x_0)| < \epsilon_0) \wedge \delta > 0$

COMMENT: PROVER has a rudimentary ability to sort its goals, hence the rearrangement of the conclusion.

goal $|z(\delta) - x_0| < \delta \rightarrow |f(z(\delta)) - f(x_0)| < \epsilon_0$
adding **adding** $|z(\delta) - x_0| < \delta$ Promote

COMMENT: This new hypothesis pertains only to the present goal.

goal $|f(z(\delta)) - f(x_0)| < \epsilon_0$
goal $\text{ISR}(|f(z(\delta)) - f(x_0)|)$ <(1)
goal $f(z(\delta)) \simeq f(x_0)$
goal $z(\delta) \simeq x_0$ Backchain
goal $\text{ISR}(|z(\delta) - x_0|)$
goal $|z(\delta) - x_0)| < w \wedge \text{ISR}(w)$ ISR-heuristic
goal $|z(\delta) - x_0| < w$

TRUE δ for w Match
goal ISR(δ)

adding ISR(d_0) <(3)
adding $\delta_0 > 0$ <(3)

TRUE d_0 for δ <(3)
TRUE d_0 for w
goal $d_0 > 0$

COMMENT: This goal derives from the goal $\delta > 0$ by substituting d_0 for δ.

TRUE T
TRUE d_0 for δ

3.8.2 Open Set

STANDARD Implies NON-STANDARD

$$\forall p(p \in S_0 \rightarrow \exists\epsilon(0 < \epsilon \wedge \ \forall y(|p-y| < \epsilon \rightarrow y \in S_0))) \\ \rightarrow \forall p(\text{STANDARD}(p) \wedge p \in S_0 \rightarrow \forall y(y \simeq p \rightarrow y \in S_0))$$

$$p \in S_0 \rightarrow (0 < \epsilon(p) \wedge (|p-y| < \epsilon(p) \rightarrow y \in S_0)) \\ \wedge \text{STANDARD}(p_0) \\ \wedge p_0 \in S_0 \\ \wedge (0 < \delta \wedge \text{STANDARD}(\delta) \rightarrow |p_0 - y_0| < \delta) \\ \rightarrow y_0 \in S_0$$

Skolemize
Define ($\simeq$)
Promote

adding $\text{STANDARD}(x) \rightarrow \text{STANDARD}(|x|)$ F-heuristic
adding $\text{STANDARD}(x) \rightarrow \text{STANDARD}(\epsilon(x))$ F-heuristic
adding $(\text{STANDARD}(x) \wedge \text{STANDARD}(y))$
$\rightarrow \text{STANDARD}(x+y)$ F-heuristic
adding $\text{STANDARD}(|p_0|)$ F-chain
adding $\text{STANDARD}(\epsilon(p_0))$ F-chain
adding $0 < \epsilon(p_0)$ F-chain
adding $|p_0 - y| < \epsilon(p_0) \rightarrow y \in S_0$ F-chain

goal $y_0 \in S_0$
goal $|p_0 - y_0| < \epsilon(p_0)$ Backchain
goal $0 < \epsilon(p_0) \wedge \text{STANDARD}(\epsilon(p_0))$ Backchain
goal $0 < \epsilon(p_0)$

TRUE T Match
goal STANDARD($\epsilon(p_0)$)
TRUE T
TRUE T

NON-STANDARD Implies STANDARD

$$\forall p(\text{STANDARD}(p) \wedge p \in S_0 \rightarrow \forall q(q \simeq p \rightarrow q \in S_0))$$
$$\rightarrow \forall p(p \in S_0 \rightarrow \exists \epsilon(0 < \epsilon \wedge \forall y(|p - y| < \epsilon \rightarrow y \in S_0)))$$

STANDARD(p) $\wedge\, p \in S_0 \wedge q \simeq p \rightarrow q \in S_0$
$\wedge\, p_0 \in S_0$ Skolemize
$\wedge$ STANDARD(p_0) Promote
$\rightarrow (|p_0 - y(\epsilon)| < \epsilon \rightarrow y(\epsilon) \in S_0) \wedge 0 < \epsilon$ Meta-heuristic

COMMENT: The hypothesis "STANDARD(p_0)" was added by the Meta-heuristic. Note that again the conclusion is sorted

adding $q \simeq p_0 \rightarrow q \in S_0$ F-chain

goal $(|p_0 - y(\epsilon)| < \epsilon \rightarrow y(\epsilon) \in S_0) \wedge 0 < \epsilon$
goal $|p_0 - y(\epsilon)| \rightarrow y(\epsilon) \in S_0$

adding $|p_0 - y(\epsilon)| < \epsilon$ Promote

COMMENT: This new hypothesis is local to the current goal

goal $y(\epsilon) \in S_0$
goal $y(\epsilon) \simeq p_0$ Backchain
goal ISR($|p_0 - y(\epsilon)|$)
goal $|p_0 - y(\epsilon)| < \delta \wedge$ ISR(δ) ISR-heuristic
$|p_0 - y(\epsilon)| < \delta$
TRUE ϵ for δ Match
goal ISR(ϵ)

adding ISR(d_0) <(3)
adding $0 < d_0$ <(3)

TRUE d_0 for ϵ Match
TRUE d_0 for ϵ
TRUE d_0 for ϵ

3.8.3 Limit Point

STANDARD Implies NON-STANDARD

$\forall\epsilon(0 < \epsilon \rightarrow \exists q(q \in S_0 \wedge q \neq p_0 \wedge |p_0 - q| < \epsilon))$
$\quad \rightarrow \exists r(r \in S_0 \wedge r \neq p_0 \wedge r \simeq p_0)$

$(0 < \epsilon \rightarrow q(\epsilon) \in S_0 \wedge q(\epsilon) \neq p_0 \wedge |p_0 - q(\epsilon)| < \epsilon)$
$\quad \rightarrow r \in S_0 \wedge r \neq p_0 \wedge r \simeq p_0$ — Skolemize

$(0 < \epsilon \rightarrow q(\epsilon) \in S_0 \wedge q(\epsilon) \neq p_0 \wedge |p_0 - q(\epsilon)| < \epsilon)$
$\quad \rightarrow r \in S_0 \wedge r \neq p_0$
$\qquad \wedge (\text{STANDARD}(e(r)) \wedge 0 < e(r) \rightarrow |p_0 - r| < e(r))$ — Define ($\simeq$)

adding STANDARD(p_0) — F-heuristic

goal $(\text{STANDARD}(e(r)) \wedge 0 < e(r) \rightarrow |p_0 - r| < e(r))$
$\quad \wedge r \in S_0 \wedge r \neq p_0$

COMMENT: PROVER does not order the goal into its present form at the beginning. It does so only after failing on the goal as ordered in the original statement of the theorem

goal $\text{STANDARD}(e(r)) \wedge 0 < e(r) \rightarrow |p_0 - r| < e(r)$

adding STANDARD($e(r)$) — Promote
adding $0 < e(r)$ — Promote

goal $|p_0 - r| < e(r)$
goal ISR($|p_0 - r|$) — <(1)
goal $|p_0 - r| < \delta \wedge \text{ISR}(\delta)$ — ISR-heuristic
goal $|p_0 - r| < \delta$

adding ISR(d_0) — <(3)
adding $0 < d_0$ — <(3)
adding $q(d_0) \in S_0$ — F-chain
adding $q(d_0) \neq p_0$ — F-chain
adding $|p_0 - q(d_0)| < d_0$ — F-chain

TRUE $q(d_0)$ for r, d_0 for δ — Match
goal ISR(d_0)
TRUE T — Match
goal $q(d_0) \in S_0$
TRUE T — Match
goal $q(d_0) \neq p_0$

TRUE T Match

TRUE $q(d_0)$ for r, d_0 for δ

3.9 Conclusion

In addition to the proofs shown, PROVER obtained proofs of the equivalence of many other concepts in analysis. Among these were proofs of uniform continuity, limit of a sequence, and derivative.

We do not claim that this paper advances the art of mechanical theorem proving in any general way. In fact, it appears that the proofs obtained are so simple that a good resolution prover such as OTTER [6], armed with the Meta-heuristic, the F-heuristic, and the typing heuristics, could probably make quick work of the theorems presented. Probably the most important lesson from this exercise is that sometimes a few well-chosen heuristics based on general principles can enable one to prove a large class of theorems in a particular domain. This in itself is no revelation; witness the reliance of the Boyer-Moore prover [4] on its induction heuristic. What is interesting in this particular domain is that all the relevant knowledge about the relationship between the standard and nonstandard theories (at least for the purpose of proving equivalence of concepts) is captured by the Meta and F-heuristics.

References

[1] A. M. Ballantyne and W. W. Bledsoe (1977): Automatic Proofs of Theorems in Analysis Using Non-Standard Techniques. JACM 24(3), 353–371.

[2] W. W. Bledsoe and P. Bruell (1974): A Man Machine Theorem-proving System. Artificial Intelligence 5(1), 51–72.

[3] W. W. Bledsoe and M. Tyson (1975): The UT Interactive Prover. Memo ATP 17, Dept. of Mathematics, Univ. of Texas, Austin.

[4] Robert S. Boyer and J Moore (1979): A Computational Logic. New York: Academic Press.

[5] H. J. Keisler (1976): Elementary Calculus. Boston: Prindle, Weber and Schmidt.

[6] William McCune (1990): OTTER 2.0 Users Guide. Technical Report ANL–90/9, Argonne National Labs, Argonne Illinois.

[7] A. Robinson (1966): Non-Standard Analysis. Amsterdam: North-Holland.

Chapter 4

Perspectives on Automated Deduction[1]

W. Bibel

Technische Hochschule
Darmstadt, Germany

Abstract

A two-dimensional space of approaches to automated deduction is presented here. It is the result of an analysis of basic characteristics and representational techniques of proof systems. As a consequence, the incorporation into deductive systems of connection structures, of various instantiations of the principle of compression (such as cycle unifications), and of the use of lemmas are identified as major challenges for future research. The background of the presentation consists of an attempted summary of the field's basic assumptions and goals.

4.1 Introduction

These days, anything related to Germany naturally reminds us of the revolutionary events that happened a short while ago in the middle of our country and in the middle of Europe. We tend to interpret these events not least as a proof for the strength of the western market-oriented economy, Adam Smith style.

[1]This is the revised text of an invited talk presented to the Tenth Conference on Automated Deduction held July 1990 in Kaiserslautern, Germany.

Scientists typically feign an attitude of indifference towards economic considerations (except that we all like and expect to get large amounts of research money). Unfortunately, competition for such money is tough. Hence, while at present the flow of cash into the field of ATP (Automated Theorem Proving) is not that bad, things may change rather quickly if the results are not up to the expectations raised by our promises. In more concrete terms, the chance for major projects such as those at Argonne, Austin, Kaiserslautern, and so forth, which cost a lot of money, will not arise that often. We thus better think of using these scarce resources in the best possible ways. I think we should therefore be concerned *as a community* about which way we direct our research.

Now you may say, "You just reminded us of the strength of market forces, so why should we worry as a collective body? In the market of ideas the best ones will eventually win anyway." The problem is that markets work successfully only with prices measuring values appropriately. Is there any way in which contributions to our field could be measured as to their relevance for progress? If there is such a way, the community would still have to agree on it.

You also might counter-argue that our field enjoys a relatively high reputation. But let's be frank and admit that our successes are not really startling. Sam's lemma still remains the best-known (not the best, fortunately) of our successes, an achievement that after twenty-five years looks a bit dated, doesn't it? In a sense one of the biggest successes of automating deduction is Prolog, which, however, is not generally seen as an achievement of the ATP community (though it originated from ATP). Also we have not been able so far to offer any more generally useful tools (beyond Prolog) to the AI community. Perhaps some reflection on why we lack more generally accepted successes and on what it is that we are doing might be worthwhile, and I personally believe, indeed it is.

I therefore decided to make here an attempt towards evaluating where we are, where we should go from here, and how one could measure contributions as to whether they bring us closer to the goals set. Obviously, I cannot claim to speak for the community as a whole. Rather, I mean to stimulate thought and discussion. The hope is that in cases where we can agree we make it explicit and known to others as well. Logic with its many different forms of representation is particularly prone to wasting talent in reinventing the wheel, which is bad for remaining competitive. Such waste could be prevented by an agreement of the sort just mentioned. All this cannot be achieved unless we understand the *basics* of our field. Most of what follows will therefore be concerned with what I think these basics are.

The drawback of this kind of essay is that its breadth prevents entering into details for any particular subject. While this will frustrate some of you, others might take it as an incentive for looking into the more detailed writings of authors listed in the references.

4.2 Goals and Assumptions

In this section we briefly reflect on the goals and underlying basic assumptions that our community of automated deduction seems to share.

4.2.1 Basic Assumptions of AD

Any goal depends on beliefs and assumptions. Before discussing the goals in ATP and evaluating the work we are doing, we should first make clear what our basic assumptions actually are. Let us therefore state and briefly discuss the following five basic assumptions.

Assumption 1

Anything a machine could possibly achieve in simulating human thought processes can be formally described.

Assumption 2

First-order logic along with its extensions (such as higher-order, modal, etc., logic) provides at least the core of a formalism rich enough to serve for realizing descriptions of the kind mentioned in assumption 1.

Assumption 3

Any solution to a formalized problem can be generated by a proof of some logical statement.

Assumption 4

While other formalisms may offer similar representational power, logic (in its various disguises) has advantages with respect to its closeness to non-formal representations (i.e., its naturalness), its unmatched canonical nature and its unprecedented wealth of knowledge.

Assumption 5

There is an *adequate* general proof method that can automatically discover any proof done by humans provided the problem (including all required knowledge) is stated in appropriately formalized terms.

Assumption 1 underlies all of Intellectics[2], not just ATP. We may thus refer to the literature (such as [26]) for a more detailed discussion. Some researchers in ATP might criticize it for its generality, arguing that ATP has much narrower goals. I would disagree with them because the history of our field in my view demonstrates that with automating mathematics the more general goal of automating human rational thought has actually been envisioned. While it could be that eventually connectionist techniques might allow us to dispense with formal descriptions, it is unlikely that these will be available in the near future.

[2]The field of Artificial Intelligence and Cognitive Science

Assumption 2 reflects some uncertainty as to which logic is actually to be used. This uncertainty is a consequence of the unsatisfactory state of the art in this respect. We feel only certain that the basic features of first-order logic are so fundamental that any future logic will include them in some way.

Assumption 3 seems pretty obvious since we may always put a problem in the form of claiming the existence of a solution under the stated assumptions.

That it should be an advantage from many points of view to use logic rather than any of the other known formalisms, as stated in Assumption 4, is clearly hotly debated in the AI community.

Assumption 5 is probably the most controversial. First note that it presumes a formalized version of the problem by which much of the burden is shifted into this part. At the disposal of a mathematician there is a lot of knowledge that we do not know at present how to get into a computer. The assumption then states, in other words, that a mathematician's work can be simulated by an algorithmic mechanism, once this knowledge is at its disposal in some formalized way.

We have emphasized the adequateness of the method. By this property we understand roughly that, for any given knowledge base, the method solves simpler problems faster than more difficult ones. Thereby simplicity is measured under consideration of all (general) formalisms available to capture the problem. In this sense probably all known proof methods are not adequate. For instance resolution proofs for the pigeon-hole formulas have exponential length although these formulas are relatively simple to prove in other formalisms. In other words, I believe in a formalism that has all the strengths of resolution as well as of any other (general) formalism. This notion of adequateness obviously must be viewed in relation to the available knowledge since there is always a one-step proof when the solution is added to the knowledge base.

Intrinsic in this assumption is a belief in the existence of an algorithm that is feasible (from a complexity point of view) for the set of problems humans can solve. In regard to that the assumption is not in contradiction with complexity results connected with NP-completeness. First of all, the P–NP–problem is an *open* problem. But even if indeed there is no polynomial algorithm for solving, say, all tautology problems, this would still allow feasibility with respect to all problems done by humans (and hence many more).

4.2.2 General Goals of AD Research

Let us now make an attempt to state, on the basis of the assumptions just given, the goals by which our work is governed. Once again let me emphasize that this is the view of an individual and thus in no way a representative statement. A more official version could only be the result of discussions hopefully triggered by this provocation. I expect such discussions to be quite controversial.

In brief one could say, our goal is to provide evidence for the above assumptions to hold true. In more detail this breaks down into the following subgoals.

Goal 1

For the logics under consideration general and adequate proof methods and techniques are sought.

Goal 2

On the basis of experiences made in pursuing goal 1 an appropriate logic is sought.

Goal 3

The viability of representing problems from various areas within the framework of such a logic as mentioned in goal 2 and of solving them with the methods mentioned in goal 1 is to be demonstrated.

Goal 4

As a by-product of the previous goal, tools for other areas should come out from our work.

Goal 5

While retaining generality (implicit in goals 1 and 2), the resulting systems should eventually outperform humans to a certain extent.

These goals will become clearer as we proceed; the emphasis of our discussions will be on goal 1. At this point we restrict ourselves to a few preliminary explanations.

Adequateness mentioned in goal 1 has the computational aspect discussed in the previous subsection (and further below in Section 4.4.4). As a goal we want to combine it with two further aspects. Proof methods differ greatly as to whether they support our intuition towards inventing strategies which we consider an equally important aspect. Further, since we are aiming at general methods it is important that the techniques lend themselves easily towards integration within a larger system, an issue discussed in Section 4.4.4.

It is part of the search strategy involved in achieving goal 2 to take first-order logic as a point of departure and restrict or extend it as seems appropriate according to the results achieved in pursuit of goal 1. For instance it could well happen that an adequate proof procedure will be found that turns out to be feasible for a logic not covering all of first-order logic but at the same time exceeding it in certain aspects.

There has been some success in pursuit of goals 3 (e.g., feasible logics) and 4 (e.g., Prolog), but much of it was achieved by researchers who would not consider themselves as belonging to the ATP community. While there is no problem in this observation as such, it has the consequence of little interaction of the gained experiences with the work in ATP. I strongly believe that this lack of interaction weakens ATP research. At the very least such interaction would generate a more vital interest by other communities in our work.

Outperforming humans, by (initially) specialized and (eventually) general systems at least under some aspects is required since otherwise the usefulness of our systems will not easily be demonstrated to the rest of the world. A proof of $x^3 = x$ *in a ring implies commutativity* does not impress anyone at all except those involved in the experiment. The same is true for theorems in logics that seem strange to most mathematicians, let alone to other scientists.

4.3 Evaluation of Approaches to ATP

Note that starting out with some formal system $\mathcal{X}_{n+1}$ and reinventing everything in its terms that is already known in terms of system $\mathcal{X}_n$ as a goal in itself is not necessarily subsumed by any of the goals set out in the previous section. On the other hand a particular proof system such as resolution may not reveal easily everything about the nature of proof. Further perspectives might be extremely helpful in finding an answer to the question "What *makes* a logical formula valid?" [2]. How little interest there is in our community in such different perspectives may be demonstrated with the following experiment.

Consider the following figure which shows a derivation of the matrix

$$\{\{P,Q\},\{\neg P,Q,R\},\{\neg Q,R\},\{\neg R\}\}$$

and determine which formal system is used for this derivation.

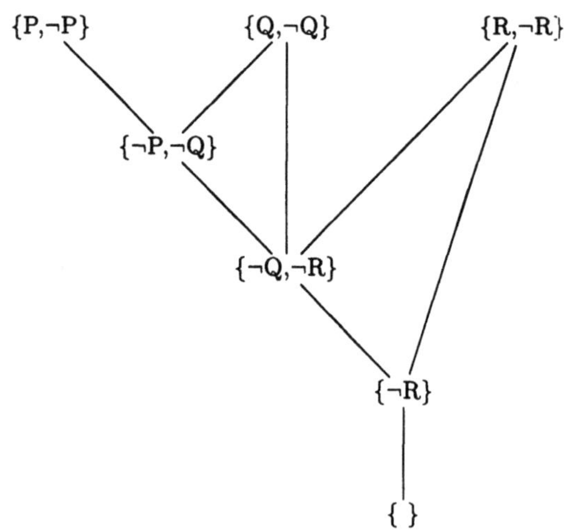

Casual surveys indicate that less than 1% of the researchers in the ATP community know the answer. The formal system is called the *inverse method* and was developed by Maslov [20] more than two decades ago (see [3] for a short introduction). Whether or not Maslov's system (or any other for that matter) allows shorter proofs than resolution or directs a prover to *find* such proofs in a shorter time is not only interesting in itself, but a satisfactory answer necessarily involves a deeper understanding of the nature of proofs. To most researchers it appears much more rewarding to attach one's name to a "new" method rather than to wade painstakingly through the details of such a comparison.

My own work is characterized to a large extent by attempts to get a common understanding of different proof methods (see, e.g., [6]). In a sense the development of the connection method may be seen as a side-product of this kind of study. In fact one of the strengths of this particular method is the fact that it provides a unique approach to understanding virtually any other method. Let us therefore take a glimpse at its features.

4.3.1 The Connection Method

It is well known that proofs can be done in an affirmative way or by contradiction. The difference is negligibly small. For historical reasons in ATP everyone prefers proof by contradiction, which has led to the wide-spread, though false, belief that this is an inherent feature of automated proofs. Originating from natural deduction proof systems, the connection method has taken the affirmative route. But, as its sibling, Andrews's matings method [1], demonstrates, the difference is indeed negligible. In fact, the same is true for resolution, the affirmative version of which is Quine's consensus rule [33]. Nevertheless the reader has to be aware of the fact that we use this affirmative (or positive) representation.

One of the basic notions of the connection method is that of a *path* through a formula F. This notion is best understood by considering the disjunctive normal form of F. Then the set of literals in any of its disjuncts is such a path, and this way we obtain all such paths. Note however that the transformation need not be performed explicitly in order to consider paths. Another way of illustrating paths is by representing formulas as matrices, i.e., sets of clauses displayed in the following two-dimensional way.

$$\begin{matrix} & \neg P & \neg Q & \neg R \\ P & Q & R & \end{matrix}$$

This matrix, which only for simplicity is a propositional one, has the paths $\{P, \neg P, \neg Q, \neg R\}$, $\{P, \neg P, R, \neg R\}$, $\{P, Q, \neg Q, \neg R\}$, and $\{P, Q, R, \neg R\}$ easily recognized as "paths" traversing the matrix from left to right. A *connection* is a subset of a path of the form $\{P, \neg P\}$. A *mating* is a set of connections. It is said to *span* the matrix if each of its paths contains (at least) one of its connections. The three connections shown in the following display of our example indeed span it in this sense.

The fundamental theorem on which the connection method is based says that a formula is valid iff there is a spanning mating. On the first-order level are needed indices for the distinction of different instances of clauses with an appropriate generalization of the notion of a path through a matrix with more than one instance of its clauses as well as a substitution making all connections complementary. This is illustrated in the following matrix.

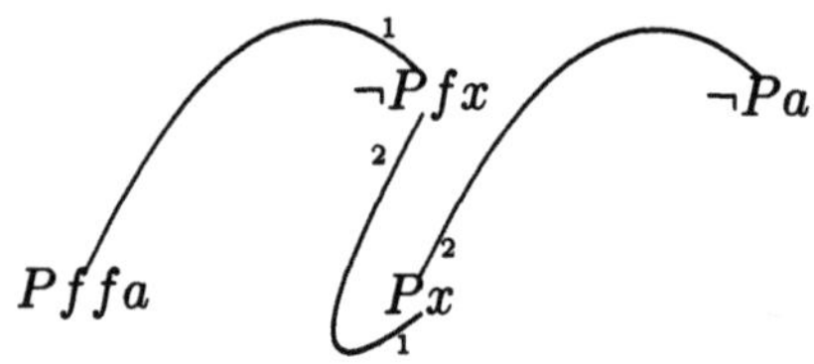

It may be regarded as a shorthand for the following amplified matrix in which the two instances of the second clause are shown explicitly.

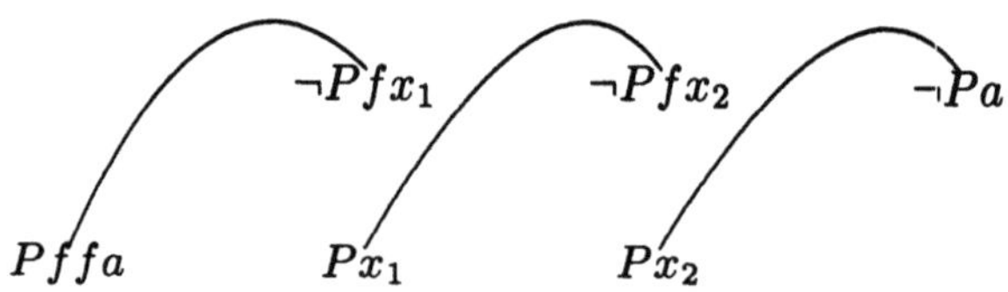

There are many possible ways of determining spanning matings. This is why we speak of the connection *method*. Each of these ways may lead to some particular connection *calculus*. It may be a calculus for formulas in normal form, but also for general formulas as the connection method naturally applies to the general case.

One of the characteristics of such connection calculi in comparison with resolution is the clear separation of the formula to be proved on the one side and the control information accumulated during the proof process on the other. Such control information in resolution is encoded into the resolvent which becomes part of the formula, thus making such a distinction impossible. It is for this reason that resolution still is badly understood. For instance, what exactly is achieved by performing a single resolution step? In consequence of this lack of understanding, there are still open completeness problems, e.g., for connection graph resolution (see [5, 16]). One may expect that they are solvable more easily now with the results reported further below.

No such difficulties are known for the connection method. Rather it enables the understanding not only of itself, but also of other methods (such as resolution). Research into these two directions has provided an insight into basic characteristics of connection proofs, which turn out to be characteristics of any proofs, in one dimension, and a comparative understanding of representational differences of various calculi in another dimension. These results bring us closer to a satisfactory answer to Andrews's question posed at the outset of the present section. The remainder of this section is meant to give a quick overview of these results, beginning with those along the first dimension.

4.3.2 Linear Chains

The first example of the last subsection easily suggests a simple way of proving theorems by way of the connection method. We pick some connection such as $\{P, \neg P\}$ there, proving all paths complementary that contain it, and this way reduce the problem to that of checking all remaining paths. The various calculi differ as to which connection will be selected next. There are good arguments for choosing one that has one of its literals in common with the previously connected clauses. In general there is no advantage in preferring either of the two clauses. Most calculi have opted for the clause containing the negative literal of the previously selected connection (a closer look at Maslov's inverse method reveals that it is the only exception in this regard).

Repeated selection of connections this way results in a linear chain of connected clauses as illustrated by the example of the previous subsection. In more abstract terms we may look at this as imposing a direction to the connections containing the literals of a given clause as illustrated in the following picture.

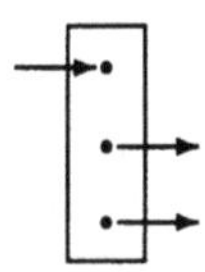

The chaining of two such clauses in this abstraction looks as follows.

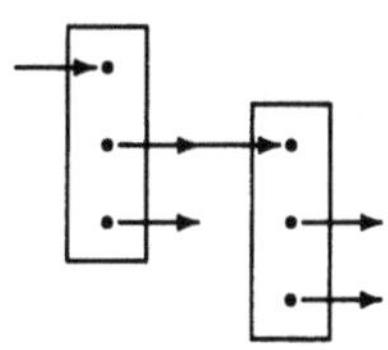

The operation is also known as *extension*. The open ends are stacked for treatment after the current linear chain has been completed.

Virtually all proof procedures in existence in some way or another use this technique (linear resolution, model elimination, etc.). In fact for Horn logic this is all we need do. For full first-order logic this is true as well if each clause with n literals is represented in n copies with the incoming arrow (in our illustration) pointing at a different literal in each of these, referred to as the *contrapositives* of the given clause. PTTP [31] has taken advantage of this generalization. The technique of linear chaining lends itself to compilation which has made Prolog such a successful programming language (despite its various weaknesses). SETHEO [19] extended the compilation technique to full first-order logic and, like PTTP, uses contrapositives. It is amazing how much can be achieved even with this simple idea as we will see further below.

4.3.3 Reductions

A similarly simple idea in proving theorems suggests to start out with well-known reduction (or simplification) operations, such as *purity*, *factoring* or *merging*, *tautology*, *unit* resolution, and *subsumption*, which are to a large extent independent of the particular deductive method under consideration. Their usefulness and importance lie in the fact that they *simplify* the stated problem so that the resulting problem requires strictly less deductive effort.

One may uniformly characterize such *simplifications* as operations that

- can be performed fast (typically in linear time);
- reduce the size of the formula to be tested for validity; and thereby
- retain the formula's original truth value, or at least its validity.

On the first-order level simplifications have to be handled with some care in so far as they require non-trivial substitutions in order to become applicable. This is because the calculus resulting from integration of reductions usually is no more confluent, so that sophisticated backtracking is needed in case of failure.

The reductions mentioned above do not, by far, exhaust the possibilities of simplifying a given problem. In [12] isolated connections (and literals) in combination with so-called database reductions have been discussed as a further powerful source for simplification. And with all this we have not even touched upon the topic of *dynamic* rather than static simplifications that, at certain points in the proof process (not only during preprocessing), check the applicability of reduction operations. For a detailed analysis of reductions to be applied in a resolution framework see [27].

While the importance of reductions is obvious, I think their role within the general proof mechanisms is not yet well understood. In fact a more desirable solution would be a uniform mechanism that takes no regress to any reductions at all, simply because these would automatically be part of the mechanism itself.

4.3.4 SETHEO

SETHEO is a sound and complete theorem prover for first-order logic developed at the Technical University Munich [19]. It basically uses the two simple

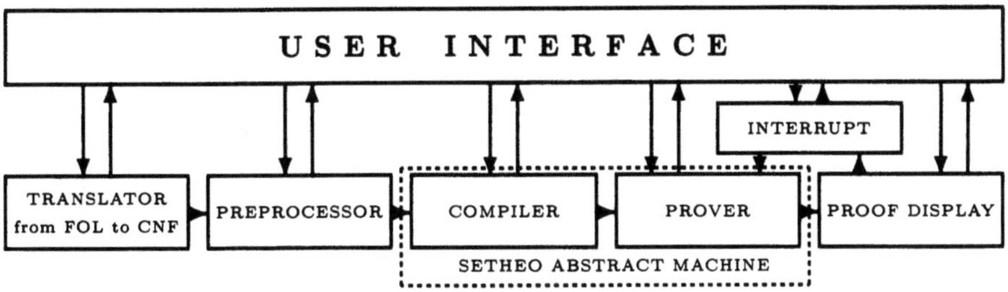

Figure 4.1: SETHEO's Architecture

techniques described in the previous two subsections, viz., linear chaining and (static) reductions (for which reason we briefly interrupt our line of thought to present this as an example). While this characterization is not exactly fair, since there are a number of additional features in SETHEO described in the reference above, it nevertheless gives an approximately correct view of this prover.

The system accepts first-order formulas in various forms and transforms them into clause form. This transformation may optionally be performed using the linear transformation presented in [13]. The reductions are applied in a preprocessing phase. Then the resulting formula is extended to include all contrapositives, and subgoals are reordered automatically for search optimization. The result is compiled into extended Warren abstract machine (WAM) code. Its instructions are finally interpreted by a program written in C. A depth-first iterative-deepening search is employed for depth-bounding with three different possible modes. An output module allows for the display of the proof tree found as well as for other useful information. The architecture is shown in figure 4.1. The remarkable experience with SETHEO is that such a simple-minded proof machine can already perform so well on a wide range of problems. A similar experience has been had with Stickel's PTTP [31] with which SETHEO favorably compares. This comparison is demonstrated in Table 4.1 which compares the performances on those examples from a set of about one hundred that differ by a factor ten or more. The running times are measured in seconds.

None of the improvements deemed necessary for truly powerful deductive reasoning discussed below are incorporated into SETHEO in its present version. In fact, most of them are not present in any prover. It seems not surprising then that breakthroughs have not been made recently in this area. The main obstacle seems to be the enormous investment necessary to develop complex tools such as SETHEO. Once developed it is even more difficult for such a system to be modified in order to incorporate more advanced features. This is true even for SETHEO, which has been developed in a strictly modular

Problem	SETH.	PTTP	Problem	SETH.	PTTP	Problem	SETH.	PTTP
Prim	**<.03**	.646	Wos19	**.048**	3.70	Ls75	**<.03**	1.28
Apabhp	**1.53**	872	Wos22	**1143**	36694	Ls87	**.549**	23.3
Ex6	**.648**	41.7	Wos25	4.53	**.077**	Ls103	**<.03**	.443
Wos1	**2.05**	57.6	Wos27	.900	**.045**	Ls108	201	**27.4**
Wos2	**1.01**	6.54	Wos32	**<.03**	.670	Ls112	—	**31.2**
Wos4	**29.0**	4152	Ls17	.648	**.030**	Ls118	—	**48.6**
Wos6	**.450**	4.23	Ls28	**.048**	.843	Ls121	52.5	**2.47**
Wos11	37.5	**3.57**	Ls29	**.048**	.622	Ch.-Lee8	**.048**	.652
Wos15	**2445**	39560	Ls65	**.048**	6.16			

Table 4.1: Run-time Differences. A comparison of SETHEO (on a SUN SPARCstation1) and PTTP (on a Symbolics 3600 with Instruction Fetch Unit).

way following standard software development techniques.

4.3.5 Hinged Loops

While the use of contrapositives allows the application of standard Prolog compilation techniques in a prover for full first-order logic (rather than just Horn clause logic), this approach has the enormous drawback of introducing a lot of redundancy into the proof process. Recall that each clause with n literals is kept in n different copies, thus blowing up the matrix and with it *a fortiori* the search space.

The idea was born from want of a compilation technique for handling cycles, which are crucial in the general case, as the following example, resulting from the previous one by a few added literals, shows.

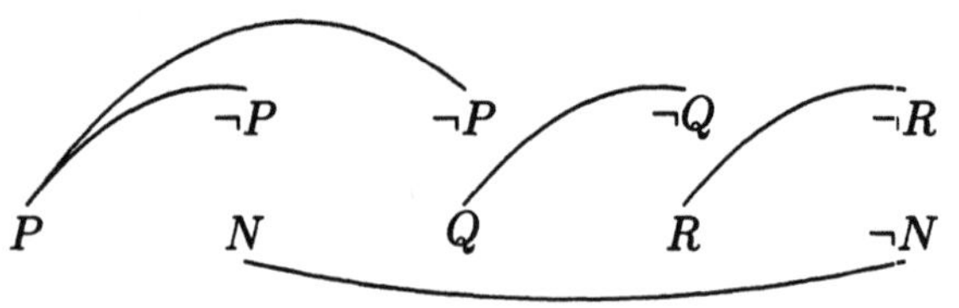

Here we still have the chaining structure from the previous proof. In addition, however, there are two further connections that close a loop, called a *hinged loop*. Such a loop contains a unique literal, here the goal P, that (statically) occurs in two different connections, which we call the *hinge*. Starting with the hinge in linear chaining mode may eventually lead us back to it. Because of this latter property we see that there must at least be one clause involved where the chain does not switch signs. In our example this is the case with the rightmost clause where both literals are negative ones (thus making the clause non-Horn). From there a so-called *body connection* leads us back towards the hinge.

Hinged loops of this sort have been studied extensively in a number of papers, including [18, 1, 28, 29, 7], under different names (such as *merge loops* in [29]). It turns out that any proof (i.e., spanning set of connections) consists of linear chains and possibly hinged loops, so that the general case has been attained with this additional feature (without use of contrapositives). These two, then, are among the basic characteristics of proofs.

It seems obvious that a convincing treatment of hinged loops would amount to a less redundant process than that with contrapositives. But the many ideas around have yet to be carved into a uniform compilation technique. Since such a technique suggests itself in a rather obvious way, it is a mystery why no one so far has done it. As in linear chaining one would impose a direction to the connections. In a non-Horn clause with more than one negative literal, one could be chosen to play the role of the head while the others would be prepared for establishing body connections. In our previous way of illustration, for a clause with two negative and one positive literals, this looks as follows.

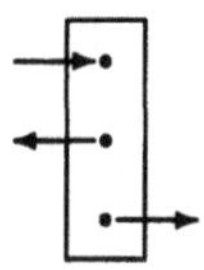

More than one linear chain has to be handled and pointers established to all possible mates whenever a literal in a potential body connection is reached, a remark which is meant to encourage someone to work out the details.

4.3.6 Factoring

Apart from reductions, we have so far isolated two basic characteristics of proofs, viz., linear chains and hinged loops. As a third one, present in all resolution systems, there is the use of at least two forms of factoring. The first form is illustrated by the clauses $\{K, L\}, \{\neg K, L\}, \{\neg L\}$. The resolvent of the first two clauses by factoring becomes a unit clause, so that the two subgoals may be solved by a single proof.

In the connection method this can be achieved by static or dynamic techniques. The static one consists of actually factoring the matrix into a non-normal-form—one resulting in the two clauses $\{\{K, \neg K\}, L\}, \{\neg L\}$. This case has been treated as a static form of reduction in this paper. For the dynamic ones see [7].

There is a dynamic form of factoring parts of proofs (rather than literals or formulas) in resolution that is not covered by these techniques which will be illustrated by the following matrix. Altogether these forms of factoring are variations of the same idea and amount to our third basic characteristic.

$$\begin{array}{ccc} & Px & \\ \neg Pa & & Pf^8a \\ & \neg Pfx & \end{array}$$

In the following connection proof the internal factoring structure is illustrated by boxing the factors.

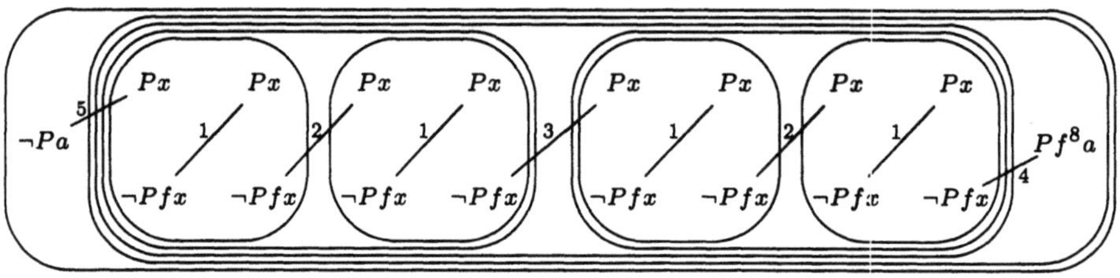

The smallest item (with the connection labeled by 1) occurs four times, two of these making up a larger item (labeled by 2) that in turn occurs twice. Taking advantage of this structure only five proof steps are needed as opposed to the 9 connections. Note that in general the difference may be an exponential one (as is the case for this example). If the right strategy is applied resolution will take advantage of this possibility, resolving the second clause with itself, then twice the resolvent with itself, and finally with each of the unit clauses, five steps as well. This identity in the number of steps is no coincidence as Eder has shown [14, 15].

He has developed a calculus, the *connection structure* calculus, in which structures like those indicated with the boxes above may formally be described without expanding the entire connection proof, so that full use of the factoring structure may be made and no exponential representation is needed as in simple connection calculi. Otherwise the calculus is like a connection calculus. Although the idea is simple, the technical details are not, for which reason we cannot show them here. Eder has then proved that the connection structure calculus can linearly simulate any resolution proof. In other words, it is at least as powerful as the calculus dominating the field. Our third basic characteristic has thus brought us to the level of present proof systems represented by resolution.

While this in a sense is the end of the story of existing uniform systems, it is far from the end of the story of our field, as we will discuss in Section 4.4. However, before we turn our attention to those further, so far unexploited, characteristics, we will first have a look at the variations possible on the basis of the characteristics discussed so far.

4.3.7 A Taxonomy of Calculi

At the end of Section 4.3.1 we spoke of the dimension of basic characteristics. This is what has been discussed since. Now we turn our attention to the other dimension mentioned there, viz., the one along representational differences of the various calculi with respect to proof search. Although we will establish a linear order among the categories of calculi discussed in the following, the arguments for this ordering admittedly are less convincing than those presented for the previously discussed dimension. Still they seem to underlie the decisions made in practice, and perhaps they might one day be based on firmer grounds. Together with the results discussed so far and in anticipation of the discussion in the next section, this altogether leads to the two-dimensional taxonomy of calculi shown in Figure 4.2.

First there is the class of calculi that have modus ponens as one of their rules. In terms of searching for a proof of a given formula G this means that F and $F \rightarrow G$ have to be considered as premises for arbitrary F, i.e., we have a search tree with a potentially infinite branching factor. Therefore automation seems hardly possible in these cases, for which reason such calculi rank lowest within this dimension (but they have virtues in aspects other than automation of proof search). Frege-Hilbert calculi, Gentzen's calculus of natural deduction and sequent calculi, both with cut, fall into this category. It is shown in [14] that these calculi may polynomially simulate each other, and that extended resolution may polynomially simulate each of them. Hence their respective places in the diagram of Figure 4.2 whereby proofs in these calculi are considered in dag form.

Next there is the class of sequent calculi without cut and thus more suitable. However, they allow "detours" in proofs by introducing and eliminating logical symbols. While this may shorten proofs (in comparison with the calculi of the next category, say), it seems again rather impractical for proof search.

Third we have those calculi in which the inference rules relate formulas with subformulas whereby the cut is not allowed. Derivations of these are structured either as trees or as dags. The Schütte calculus without cut and tableau calculi are of this category. While, in contrast to those of the previous two categories, they lend themselves to a systematic search for proofs, the redundancy inherent in the form of the proof rules, which rewrite essentially the entire formula, makes them less suitable than the calculi of the remaining two categories.

Then come the calculi with some sort of resolvent. Resolution, Maslov's inverse method, and extended resolution belong to this category. They do not suffer from the redundancy mentioned in the previous category.

Finally there are the connection calculi (and, further, the topic of compression discussed in the next section). In contrast to any of the previous categories, no rewriting of the given formula itself is needed at all. Rather the

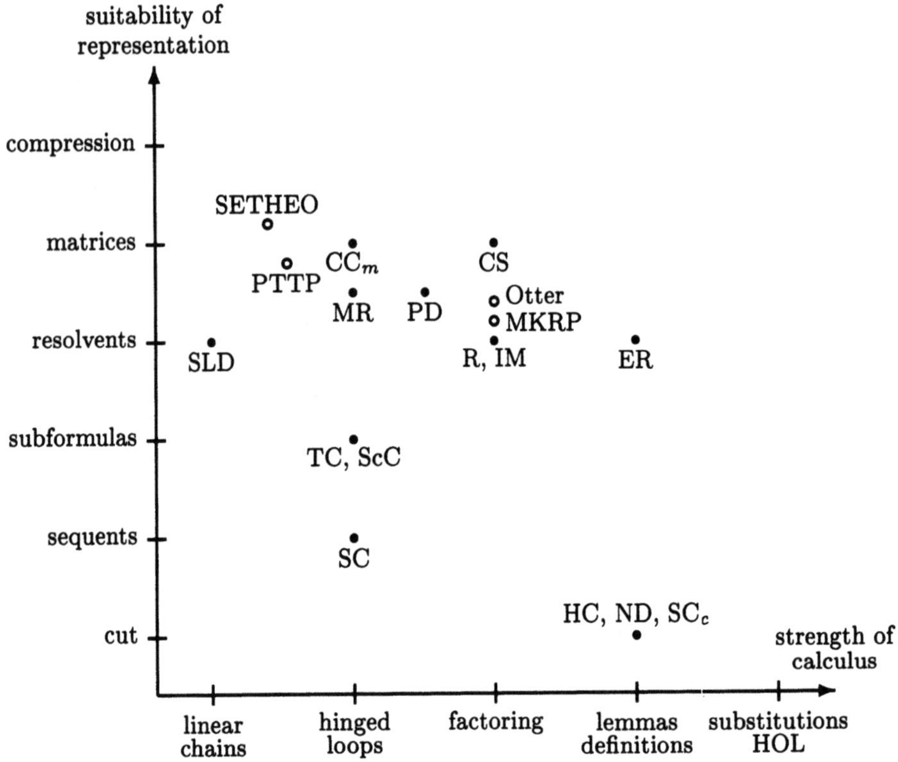

SLD	– SLD resolution	**HC**	– Hilbert calculus
R	– Resolution w linear NF trafo	**ND**	– Gentzen's natural deduction
IM	– Maslov's inverse method	**SC$_c$**	– Sequent calculus w cut
MR	– Prawitz' matrix reduction	**SC**	– Sequent calculus (trees) w/o cut
PD	– M &R path dissolution	**ScC**	– Schütte calculus w/o cut
CC$_m$	– ME-like connection calculus	**TC**	– Tableau calculus
CS	– Connection structure calculus	**ER**	– Resolution w extensions

Figure 4.2: A Taxonomy of Calculi

proof process proceeds by rewriting structures imposed on the matrix that seem to amount to a minimum of information needed for keeping track of the proof process. For this reason they are ranked best among all these categories. There is another good reason for this preference. Recall the adequateness of a method deemed important in goal 1, one of its aspects being to support our intuition towards inventing good strategies. In this respect the connection method is a clear winner among all existing methods due to its transparency discussed further above.

Note that the categories are not fixed in their relative position since implementational details also play an important role. For instance, implementing resolution with full use of structure sharing lifts it closer to the structural category (but it still lacks transparency).

It is shown in [14] that the cut-free sequent calculus, the tableau calculus and the version of a connection calculus that is a close relative of model elimination, are all just representational variants of each other, i.e., they may simulate each other in linear time. It is also shown there that resolution has a clear advantage in comparison to them. In as far as these calculi are restricted to normal-form formulas, it is taken for granted that prior to their application a *linear* transformation to normal form like the one described in [13] is employed. Prawitz' matrix reduction [25] may be regarded as a version of the connection calculus that affords a more redundant form of representation (explosion of storage). Path dissolution [21] generalizes matrix reduction to non-normal form matrices, in this way coping with all kinds of formula factoring, but not with factoring proof structures discussed in Section 4.3.6 in connection with connection structures. All this explains the positions of the respective calculi in the diagram of Figure 4.2.

The diagram admittedly suffers from a lot weaknesses. Each of the dimensions would require more refinement and justification as already pointed out above. Also it neglects important features of proof systems such as strategies in the first place. The rationale behind this particular neglect is as follows. If any proof in some calculus may be transformed in linear time into a proof in some other, then also any search strategy applicable in the first may be transformed into one in the second. In this sense search strategies are to some extent independent of the calculi within a given class. On the other hand, the best search strategy is of no use at all for some calculus if proofs in it are exponentially longer than in some other, since the proofs could not even be printed for lack of the time this would take, let alone be discovered.

Tentatively we have allocated places in the diagram also to four selected proof systems, some well known. Because of its highest place in the diagram, SETHEO (as a theorem prover) has a good potential, but only if it is developed towards the right. At present it clearly cannot cope with the performance of systems like OTTER in problems that either are non-Horn or require a significant amount of factoring. For the latter reason SETHEO fails to discover the proof of the Łukasiewicz-single-axiom theorem [24] (see Section 4.4.2). A second potential of SETHEO is providing a tool (as requested in goal 4), viz., the computational basis for a programming language LOP covering full first-order logic.

4.4 Tasks Ahead

The evaluation of proof calculi has prepared ground for better judgment. If some new proof technique is proposed—as happens about every month—the author should be required to carry the burden and embed the proposal within this taxonomy with detailed and precise arguments. Since authors would become frustrated by realizing they are reinventing wheels, this would seem to me to free plenty of talent for attacking the more challenging tasks ahead. These are typically located in the right upper corner of the diagram. Some of them will be discussed in the present section.

4.4.1 Compression

Assume a matrix (in positive representation, as always) that, among other clauses, contains three facts, $\neg L\ell m\,,\neg L\ell t\,,\neg L\ell w$, supposed to express that some lion lies Mondays through Wednesdays, and $\neg Dxyz\,,Lxy\,,\neg Lxz$, a clause supposed to express something about lying-habits. Then this simple example illustrates the luxury that logic affords in representing this information. First, the three facts could as well be represented in the more compressed form $\neg L\ell\{m,t,w\}$ without loss of any information. Further the rule pretends to be applicable to any kind of object, while in fact the context restricts it to $\neg D\ell\{m,t,w\}z, \neg L\ell z$ which is a similar shorthand for three different clauses obtained by resolution in an obvious way.

The example is taken from [23]. Altogether it contains nine more facts, a second rule obtained by switching the negation to the other L-literal, and a query (if the problem is represented in logic in the most straightforward way).

In [12] reduction rules have been introduced that, if applied to our example, result in exactly the kind of compression just illustrated. The proof search carried out on the reduced problem is fairly trivial. For a similar problem—Schubert's steamroller—it takes SETHEO about 3ms (preprocessing about 0.3s). Without the compression by way of these reduction rules the proof search requires orders of magnitude more time. The reason for the speed-up lies in our bringing the relevant information right at the places where the search process needs it for taking the right options.

There are three important lessons to be learned from this example. First, the luxury of logic just mentioned is a luxury indeed. It allows us to add information to the knowledge-base at any given time without caring about its state of contents (except for consistent naming). This is a crucial feature if we think of knowledge-bases with tens of thousands of entries.

Second, while this luxury should be offered to the user, it can by no means be afforded in terms of the proof process. So prior to any proof search the problem must be transformed in an as much compressed form as possible.

Third, such compressions are possible *without* new concepts such as sorted logic, special constraint satisfaction representations (cf. [9]), or what have you, but simply with *abbreviations* of the kind just illustrated. Working with such abbreviations suggests, as a consequence, different deductive computations (e.g., unification algorithms[3]).

The last lesson seems to come as a surprise to many of those working in sorted logic, generalized unification, constraint satisfaction, and so forth. An example like the one above usually is represented in a sorted logic. People even argue that no other way would be able to solve it in reasonable time. While there is nothing wrong with using sorted logic, it is definitely wrong to argue for its exclusiveness in handling such problems efficiently—with the following caveat.

What we have illustrated above in a sense *is* nothing else than using sorted logic except that we access it by way of an abbreviation from uniform logic. The difference in itself might be regarded as negligible, but it matters with regard to our goal 1, which included the request for an easy way of integration within larger systems. Much of the work in ATP presently is characterized by designing specialized forms for handling a certain restricted class of problems. We are talking on the order of hundreds of different forms. No perspective emerges as to how these specialist systems could be (truly) integrated into an adequate uniform deductive system. How much more effective would it be for the progress of our work were general techniques for treating uniform logic distilled out of comparisons of such special techniques rather than were new special techniques added each month. Here lies one of the major tasks I see.

As we just indicated the *principle of compression* illustrated in the example above has many applications, nearly as many as there are theories (i.e., satisfiable sets of formulas). In view of this wealth of variety, once again, it seems mandatory to come up with general principles rather than treating each of them separately. Equality is a particularly important example whereby the principle of compression here leads to equality connections [7], an abbreviation of a connection structure involving the equality axioms. The algorithmic treatment has recently been perfected in [17].

The same idea has later been generalized to theory resolution [30], which in our terminology amounts to theory connections [7], of which the theory of equality is one out of many examples. Again it could and, as we just argued, should be regarded as an abbreviation rather than a new concept. All work on rewriting, Knuth-Bendix, generalized unification, and so forth may be subsumed under this uniform principle (see [7]) with the benefit of a perspective towards our goal 1. Such a perspective is not clear to me in observing the field producing hundreds of specialized techniques.

---

[3]Part of what is needed in the case of our example is included in as old a work as [32].

4.4.2 Cycles

By [24] the attention of the community was brought to a class of problems originally studied by Łukasiewicz. The hardest of these is presented in the following matrix.

$$\begin{matrix} & \neg Pw & \neg Pi(i(ixy,z),i(izx,iux)) \\ Pi(iab,i(ibc,iac)) & Pivw & \\ & Pv & \end{matrix}$$

The terms represent implicational formulas and P asserts derivability of its argument. Thus the second clause represents modus ponens. It causes exponential growth of the search space since each of its subgoals may instantiate it again.

Few provers are able to find a proof for this problem. The parallel version of OTTER at Argonne is among the exceptions.[4] In about half an hour it generates 6.5 million clauses to obtain a hyperresolution proof with about 150 steps.

While OTTER's performance in terms of sheer power is indeed impressive, the example clearly exposes the stupidity of all our provers (including SETHEO of course). Łukasiewicz has found a 29 step proof in a search space that most likely never exceeded a total of a few hundred steps (perhaps even a few tens only). When I myself played with the problem I soon conjectured that a proof would never require a linear chain of more than one consecutive instance of the connection depicted in the matrix. SETHEO has verified this conjecture for the easier problems out of the same class.

There is no reason why considerations like those I made could not be automated to a certain degree. Here is where I see the second major task ahead of us. What it amounts to is a study of the cyclic structure (including effects on terms) of the given problem prior to any search. In [8] I have discussed this task in greater detail and sketched a solution to a restricted case. Just to give an idea here, let us consider the cycle defined by the depicted connection and, for simplicity, ignore the third literal in the second clause entirely for the moment.

Say the second clause is entered through its top literal (or head) by some connection, instantiating the first instance of the variable w as $w_1 = i(iab, i(ibc, iac))$. If the second literal (as a subgoal) is solved by calling the head of this clause again, which is to say that we chain along the depicted connection, we get $w_2 = iv_1w_1$. Continuing to proceed this way for $k-1$ steps would result in $w_k = i(v_k, i(v_{k-1}, \ldots, i(v_1, w_1)\cdots))$. The only other way to eventually stop this cycle is to solve the subgoal with the third clause

[4]Private communication with E. Lusk.

which yields $w_k = i(i(ixy, z), i(izx, iux))$. From these three equations the only remaining unknown k may be determined. Let us refer to this technique as *cycle unification*. It attempts to unify entire matings including parameterized cycles.

If the third literal excluded so far is now taken into consideration, a second such cycle (and with it a second parameter like k) enters the scene. Moreover both cycles may interplay with each other (resulting in indices for these parameters). But with the help of a machine the resulting equations of the nature just illustrated may still be established. Then just solve them, or, alternatively (in cases where solving is too hard), derive restrictions from them that constrain the remaining proof search. This way not 6.5 million clauses would need be generated in a blind way, rather we would just unify a set of terms. True, cycle unification is much harder than first-order unification (even undecidable), but still it allows us to proceed in the goal-oriented fashion inherent in unification.

As an additional complication, the idea behind the connection structure calculus (discussed in Section 4.3.6) would possibly have to be integrated into the form of these equations. As we discussed there, for reasons of efficiency we cannot dispense with this feature. Whether it is subsumed by the technique discussed here is, however, an open question.

Note that the task described in this subsection is a form of compression as well. I would in fact conjecture that all forms of compression discussed in the preceding subsection could be attacked in the unificational way just illustrated.

4.4.3 Lemmas

In Section 4.3.6 we have discussed forms of reusing parts of a proof. This is a form of lemma application that is proved once to be used wherever it applies. In the special case of Section 4.3.6 the lemma is applied in each instance in identical form.

A more general case is to allow the symbols occurring in the lemma to vary from one application to another. We could look at this as generalizing the lemma by taking its closure, i.e., by all-quantifying over all its freely occurring symbols (which results in a second-order formula).

There are important mathematical formulas that are infeasible for automation unless this general form of lemma application is employed. None of the methods discussed so far work for them. One class of problems of this nature is the "pigeon hole" class $\{P_i \mid i = 1, 2, \ldots\}$.

Figure 4.3 at its top node shows a coded form of P_3. The uncoded matrix is obtained from it by (i) taking as the first set of clauses of P_3 all the given clauses but with all variables occurring negated, and (ii) adding for each pair of variables in any row a clause consisting of these two variables thus obtaining the remaining set. The punched matrices at all other nodes are to be uncoded

$$\begin{array}{cccc} x_{11} & x_{12} & x_{13} & x_{14} \\ x_{21} & x_{22} & x_{23} & x_{24} \\ x_{31} & x_{32} & x_{33} & x_{34} \end{array}$$

$$\begin{array}{cccc} x_{11} & x_{12} & x_{13} & \bullet \\ x_{21} & x_{22} & x_{23} & \bullet \\ \bullet & \bullet & \bullet & \bullet \end{array} \qquad \begin{array}{cccc} x_{11} & x_{12} & x_{13} & x_{14} \\ x_{21} & x_{22} & x_{23} & x_{24} \\ x_{31} & x_{32} & x_{33} & \bullet \end{array}$$

$$\begin{array}{cccc} x_{11} & x_{12} & \bullet & x_{14} \\ x_{21} & x_{22} & \bullet & x_{24} \\ \bullet & \bullet & \bullet & \bullet \end{array} \qquad \begin{array}{cccc} x_{11} & x_{12} & x_{13} & x_{14} \\ x_{21} & x_{22} & x_{23} & x_{24} \\ x_{31} & x_{32} & \bullet & \bullet \end{array}$$

$$\begin{array}{cccc} x_{11} & \bullet & x_{13} & x_{14} \\ x_{21} & \bullet & x_{23} & x_{24} \\ \bullet & \bullet & \bullet & \bullet \end{array} \qquad \begin{array}{cccc} \bullet & x_{12} & x_{13} & x_{14} \\ \bullet & x_{22} & x_{23} & x_{24} \\ \bullet & \bullet & \bullet & \bullet \end{array}$$

Figure 4.3: The Reduction of P_3 to P_2

in exactly the same way. In [10] it is shown that the figure is a binary tree of valid reduction steps. The point now is that all leaves of this tree are P_2 but with different symbols. Proving it once suffices to establish the entire proof which then becomes linear in length. However if the prover fails to recognize this feature then the proof necessarily has exponential length and thus can never be found for larger i.

Testing whether two matrices may become identical by renaming, necessary for this feature to be included in a prover, may be regarded as a version of the graph isomorphism problem[5], for which no polynomial algorithm is known. But this is true for many parts of theorem proving. As we have just seen there is no choice other than including the handling of lemmas in this general form in a proof system unless we content ourselves with excluding all such problems from treatment.

Another powerful technique used in mathematics all the time is the use of definitions for complex terms in proofs. Extended resolution includes this feature (see [15] for more details). It could be that this feature is of the same deductive power as the use of (generalized) lemmas described in the present section.

Even more powerful is the use of substitutions of symbols by arbitrary terms or formulas, not just renaming. This could be regarded as using for the closure of the lemma existential (rather than all) quantifiers (of higher order).

4.4.4 The Composition Problem

Had we solved all the tasks discussed so far then first of all there would yet be several more which could not be mentioned in this paper. Treating induction [7], taking advantage of semantic information [22], getting serious with higher-order logic [2] and modal logics [34] are just some of them. Beyond that a major remaining task would be to integrate all solutions for these particular tasks into one single uniform system. This is a task of tremendous (extensive) complexity which may stretch human capabilities beyond their limits, at least with present system building techniques.

It is for this reason that we have emphasized in goal 1 ease of integration. Would the community put more emphasis on this particular aspect, we would not see so many scattered techniques being developed which ultimately might turn out short-lived in view of this grander goal of a uniform system.

An important issue deserves special attention in this compositional task. Not only is the entire problem of proof search hard in terms of complexity, but also a number of subproblems share this property. Determining spanning matings and connection structures, constraint satisfaction, subsumption, complementarity of eq-connections, cycle unification, recognizing common lemmas, all are among these.

---

[5]This observation was communicated to the author by N. Pippenger.

For instance think of the generalized lemma feature of the previous subsection. At the one extreme one could consider the problem too hard and leave it out of consideration for the global system. At the other extreme one could give it algorithmically so much emphasis that typically the given resources would be used up by this subproblem alone. Both extremes have the same bad effect, which is failure in proving theorems that otherwise would have relatively easy proofs. What is first needed then in the global system is a fair share of the given resources for each of these "wolverines." One might even think of determining such shares with something like market mechanisms from economics. Alternatively, different processors might become specialized for such subproblems within a multi-processor system. While this seems like a manageable problem, the trouble with the intricacy of theorem proving is that even separating apart those various subproblems is a challenge.

Thus restricted in its share of resources, each subproblem would be forced to spend them wisely. Some sort of an iterative deepening technique seems most appropriate to secure a computationally adequate behavior as requested in goal 1. This means that, if there is a relatively simple solution for a problem with *some* existing technique, the prover will find it relatively quickly. Only truly harder problems will take longer to prove. It is an idea that this author has pursued since 1970, initially under the label "systematic" and restricted to a narrow range of subproblems [4]. All existing provers are far from computational adequacy in this sense.

4.5 Conclusions

The goal of a uniform proof system has experienced ups and downs over the decades. En vogue are specialized techniques for particular theories nowadays. I have argued in this paper for generalizing such techniques to upgrade existing uniform provers to computationally adequate ones which compete with any of these specialists.

The principle of compression underlying such generalizations has been identified as one of several deficits in existing provers. Cycle unifications have been discussed as one of the instances of this principle. As another such deficit the general use of lemmas has been presented.

A two-dimensional taxonomy of calculi has been designed which is meant to focus the direction of future research and help avoid frequent reinventions of the wheel, a problem that logic is particularly prone to suffer from.

On the basis of these considerations a uniform prover based on some version of the connection structure method, enhanced by the principle of compression and by extensive use of lemmas, and engineered with the implementational know-how behind OTTER, say, may be expected to perform orders of magnitude better than what we see today.

There is some tradition in this conference to put forth challenge problems.

In the spirit of this essay the challenge I offer is for one and the same system to prove a wide variety of well-known challenge problems that today can be done by just one or two systems. These should include the Łukasiewicz theorem as well as the pigeon-hole formula with ten pigeons, both discussed in this paper, with a proof obtained in each case in time of the order of a second. But the point is that the system should be similarly strong in many other areas and thus easily prove "$x^3 = x$ *implies commutativity in a ring*" as well. My prediction is that the system with the characteristics just outlined would range in that category.

Only if this level is reached will things get truly exciting. The mathematical knowledge accumulated over thousands of years could become involved in a way that Bledsoe has successfully demonstrated [11]. Concepts like analogy, use of examples and counterexamples, special cases, and much more of this kind would account for an additional dimension to be added to our space of methods. It is only then that mathematicians would begin to get truly interested in using such systems in their daily work and discover proofs perhaps even for long-standing and well-known conjectures in a cooperative way. As a consequence of such a success the attitude of Intellecticians would be changed towards integrating uniform tools resulting from such research within their problem solving systems, from where the penetration of all science would proceed.

Acknowledgments. Many discussions with Elmar Eder have deepened my insights into the relations of various proof systems. Also I appreciate the strong commitment of my former group at the Technische Universität München which completed the projects underway at the point of my departure, with SETHEO being one of the fine results. Especially, I would like to thank Professor Eike Jessen, who made this possible in a difficult situation, as well as Reinhold Letz and Johannes Schumann who did most of the final design, encoding, and testing of SETHEO. Bernd Jünger provided technical assistance in producing this typescript, and Robert Boyer helped in cleaning up my faulty English grammar.

References

[1] Peter B. Andrews (1981): Theorem Proving via General Matings. Journal of the ACM 28, 193–214.

[2] Peter B. Andrews (1989): On Connections and Higher–order Logic. Journal of Automated Reasoning 5, 257–291.

[3] W. Bibel and E. Eder (1990). Methods and Calculi for Deduction. In: D. M. Gabbay, C. J. Hogger, and J. A. Robinson, eds.: Handbook of Logic in Artificial Intelligence and Logic Programming. Oxford: Oxford University Press, to appear.

[4] W. Bibel (1974): An Approach to a Systematic Theorem Proving Procedure in First-order Logic. Computing 12, 43–55. First presented to the GI Annual Conference in 1971.

[5] W. Bibel (1980): A Strong Completeness Result for the Connection Graph Proof Procedure. Technical Report ATP–3–IV–80, FG Künstliche Intelligenz, Technische Universität München, Munich, Germany.

[6] W. Bibel (1982): A Comparative Study of Several Proof Procedures. Artificial Intelligence 12, 269–293.

[7] W. Bibel (1987): Automated Theorem Proving. Vieweg Verlag, Braunschweig, second edition.

[8] W. Bibel (1988): Advanced Topics in Automated Deduction. In: R. Nossum, ed.: Advanced Topic in Artificial Intelligence, LNCS 345, 41–59. Berlin: Springer.

[9] W. Bibel (1988): Constraint Satisfaction from a Deductive Viewpoint. Artificial Intelligence Journal 35, 401–413.

[10] W. Bibel (1990): Short Proofs of the Pigeonhole Formulas Based on the Connection Method. Journal of Automated Reasoning, 1990.

[11] Woodrow W. Bledsoe (1984): Some Automated Proofs in Analysis. In: W. W. Bledsoe and D. W. Loveland, eds.: Automated Theorem Proving: After 25 Years, 89–118. Providence, RI: American Mathematical Society.

[12] W. Bibel, R. Letz, and J. Schumann (1987): Bottom–up Enhancements of Deductive Systems. In: I. Plander, ed.: Proceedings of 4th International Conference on Artificial Intelligence and Information–Control Systems of Robots, 1–10, Smolenice, CSSR North–Holland.

[13] Elmar Eder (1985): An Implementation of a Theorem Prover Based on the Connection Method. In: W. Bibel and B. Petkoff, eds.: Artificial Intelligence, 121–128. Amsterdam: North-Holland.

[14] Elmar Eder (1989): A Comparison of the Resolution Calculus and the Connection Method. In: E. Börger, H. Kleine Büning, and M. M. Richter, eds.: CSL'88—Second Workshop on Computer Science Logic, 80–98. Berlin: Springer.

[15] Elmar Eder (1990): Relative Complexities of First Order Calculi. Habilitation Thesis, Universität Dortmund, Germany. To appear as a book in Vieweg Verlag, Braunschweig.

[16] Norbert Eisinger (1989): Completeness, Confluence, and Related Properties of Clause Graph Resolution. Ph. D. thesis, Universität Kaiserslautern, Germany.

[17] Jean Gallier, Paliath Narendran, Stan Raatz, and Wayne Snyder (1990): Theorem Proving Using Equational Matings and Rigid E-unification. To appear in the Journal of the ACM.

[18] D. W. Loveland (1969): A Simplified Format for the Model Elimination Theorem-proving Procedure. Journal of the ACM 16(3), 349–363.

[19] Reinhold Letz, Johannes Schumann, Stephan Bayerl, and Wolfgang Bibel (1990): SETHEO—A High-performance Theorem Prover for First-order Logic. Journal of Automated Reasoning 8(2), to appear.

[20] S. J. Maslov (1968): The Inverse Method for Establishing Deducibility for Logical Calculi. Trudy Mathemat. Inst. AN USSR 98, 26–87.

[21] Neil V. Murray and Erik Rosenthal (1987): Inference with Path Resolution and Semantic Graphs. Journal of the ACM 34, 225–254.

[22] Xumin Nie and David A. Plaisted (1990): A Complete Semantic Back Chaining Proof System. In: Mark Stickel, ed.: Tenth International Conference on Automated Deduction, 16–27. Berlin: Springer.

[23] H. J. Ohlbach and M. Schmidt-Schauß (1985): The Lion and the Unicorn. Journal of Automated Reasoning 1(3), 327–332.

[24] Frank Pfenning (1988): Single Axioms in the Implicational Propositional Calculus. In: Ewing Lusk and Ross Overbeek, eds.: Ninth International Conference on Automated Deduction, 710–713. Berlin: LNCS 310, Springer.

[25] Dag Prawitz (1970): A Proof Procedure with Matrix Reduction. In: M. Laudet et al., eds.: Symposium on Automatic Demonstration, Lecture Notes in Mathem. 125, 207–214. Berlin: Springer.

[26] Zenon W. Pylyshyn (1986): Computation and Cognition—Toward a Foundation for Cognitive Science. Cambridge, MA: The MIT Press.

[27] Rolf Socher-Ambrosius (1990): Simplification and Reduction for Automated Theorem Proving. Ph. D. thesis, Universität Kaiserslautern, Germany.

[28] Robert E. Shostak (1976): Refutation Graphs. Artificial Intelligence 7, 51–64.

[29] Sharon Sickel (1976): A Search Technique for Clause Interconnectivity Graphs. IEEE Transactions on Computers 35, 823–835.

[30] Mark E. Stickel (1985): Automated Deduction by Theory Resolution. Journal of Automated Reasoning 1, 333–355.

[31] M. E. Stickel (1988): A Prolog Technology Theorem Prover: Implementation by an Extended Prolog Compiler. Journal of Automated Reasoning 4, 353–380.

[32] Mabry Tyson and Woodrow W. Bledsoe (1979): Conflicting Bindings and Generalized Substitutions. In: W. S. Joyner, ed.: Proceedings of the Fourth Workshop on Automated Deduction, 14–18, Austin, TX.

[33] Willard van Orman Quine (1955): A Way to Simplify Truth Functions. American Mathematical Monthly 62.

[34] Lincoln A. Wallen (1990): Automated Deduction in Non-Classical Logics. Cambridge, MA: MIT Press.

Chapter 5

MJRTY—A Fast Majority Vote Algorithm[1]

Robert S. Boyer and J Strother Moore

Computer Sciences Department
University of Texas at Austin
and
Computational Logic, Inc.
1717 West Sixth Street, Suite 290
Austin, Texas

Abstract

A new algorithm is presented for determining which, if any, of an arbitrary number of candidates has received a majority of the votes cast in an election. The number of comparisons required is at most twice the number of votes. Furthermore, the algorithm uses storage in a way that permits an efficient use of magnetic tape. A Fortran version of the algorithm is exhibited. The Fortran code has been proved correct by a mechanical verification system for Fortran. The system and the proof are discussed.

[1]The work described here was conducted in the Computer Science Laboratory of SRI International and suported in part by NASA Contract NAS1-15528, NSF Grant MCS-7904081, and ONR Contract N00014-75-C-0816 1981. A brief history of this work is given in the concluding section.

5.1 Introduction

Reliability may be obtained by redundant computation and voting in critical hardware systems. What is the best way to determine the majority, if any, of a multiset of n votes? An obvious algorithm scans the votes in one pass, keeping a running tally of the votes for each candidate encountered. If the number of candidates is fixed, then this obvious algorithm can execute in order n. However, if the number of candidates is not fixed, then the storage and retrieval of the running tallies may lead to execution time that is worse than linear in the number of votes—such an algorithm could run in order n^2.

If the votes can be simply ordered, an algorithm with order n execution time can be coded first to find the median using the Rivest-Tarjan algorithm [7] and then to check whether the median received more than half the votes. The Rivest-Tarjan algorithm is bounded above by $5.43n - 163$ comparisons, when $n > 32$.

In this paper we describe an algorithm that requires at most $2n$ comparisons. The algorithm does not require that the votes can be ordered; only comparisons of equality are performed.

5.2 The Algorithm

Imagine a convention center filled with delegates (i.e., voters) each carrying a placard proclaiming the name of his candidate. Suppose a floor fight ensues and delegates of different persuasions begin to knock one another down with their placards. Suppose that each delegate who knocks down a member of the opposition is simultaneously knocked down by his opponent. Clearly, should any candidate field more delegates than all the others combined, that candidate would win the floor fight and, when the chaos subsided, the only delegates left standing would be from the majority block. Should no candidate field a clear majority, the outcome is less clear; at the conclusion of the fight, delegates in favor of at most one candidate, say, the nominee, would remain standing—but the nominee might not represent a majority of all the delegates. Thus, in general, if someone remains standing at the end of such a fight, the convention chairman is obliged to count the nominee's placards (including those held by downed delegates) to determine whether a majority exists.

Thus our algorithm has two parts. The first part pairs off disagreeing delegates until all remaining delegates agree. We call this the "pairing" phase. Perhaps nonobviously, pairing can be done with n comparisons. If pairing leaves any delegates standing then those delegates unanimously favor a single candidate—the nominee—who must be in the majority if a majority exists. The second part of the algorithm, called the "counting" phase, determines whether the nominee received more than half the votes. The counting phase obviously requires at most n comparisons. The focus of this paper is on the pairing phase.

Here is a bloodless way the chairman can simulate the pairing phase. He visits each delegate in turn, keeping in mind a current candidate *cand* and a count k, which is initialized to 0. Upon visiting each delegate, the chairman first determines whether k is 0; if it is, the chairman selects the delegate's candidate as the new value of *cand* and sets k to 1. Otherwise, the chairman asks the delegate whether his candidate is *cand*. If so, then k is incremented by 1. If not, then k is decremented by 1. The chairman then proceeds to the next delegate. When all the delegates have been processed, *cand* is in the majority if a majority exists.

Proof: Suppose there are n delegates. After the chairman visits the ith delegate, $1 \leq i \leq n$, the delegates he has processed can be divided into two groups: a group of k delegates in favor of *cand*, and a group of delegates that can be paired in such a way that paired delegates disagree. From this invariant we may conclude, after processing all of the delegates, that *cand* has a majority, *if* there is a majority. For suppose there exists an x different from *cand* with more than $n/2$ votes. Since the second group can be paired, x receives at most $(n-k)/2$ votes from that group. Thus, x must have received a vote from the first group, contradicting the fact that all votes in the first group are for *cand*.

Here is a proof by simple induction on i that the delegates polled may always be divided into two such groups after the chairman has processed the first i delegates. After the chairman has processed the first delegate, k and i are both 1: the group of delegates passed has 1 vote for *cand*. So suppose the invariant holds after the ith candidate, and suppose the i delegates processed so far may be divided into two groups, U and P, with the aforementioned properties. If after processing the ith delegate k is 0, then *cand* is reset to the candidate preferred by the $i+1^{\text{st}}$ delegate and k is set to 1. But when k is 0 the invariant tells us that P contains all the first i delegates. Thus the first $i+1$ delegates may be divided into two groups: one containing only the $i+1^{\text{st}}$ delegate and one that is P. If after processing the ith delegate k is not 0, there are two cases: the $i+1^{\text{st}}$ delegate votes for or against *cand*. If the $i+1^{\text{st}}$ delegate votes for *cand*, k is incremented; the first $i+1$ delegates may be divided into two groups: U plus the $i+1^{\text{st}}$ delegate and P. If the $i+1^{\text{st}}$ delegate votes against *cand*, k is decremented; the first $i+1$ delegates may be divided into two groups as follows. Let j be any one of the delegates in U. Let the first group be U minus j, and let the second group be P together with both j and the $i+1^{\text{st}}$ delegate.

5.3 Examples

Suppose there are three candidates, `A`, `B`, and `C`, and suppose that the delegates are polled by the chairman in the following order:

```
A A A C C B B C C C B C C.
```

After the chairman has visited the third delegate, candidate A is leading with 3 votes. In the diagram below we put a vertical bar after the third delegate to indicate the position of the chairman.

```
                                          CAND      K
    A A A|C C B B C C C B C C              A        3
```

In processing the next three delegates, the chairman pairs off the three A votes against three other votes (two for C and one for B). After the sixth delegate has been visited, k is 0 and the vote of the seventh delegate makes B the leading candidate.

```
                                          CAND      K
    A A A C C B B|C C C B C C              B        1
```

The next delegate, however, cancels out B's short-lived ascendancy and the next two delegates give C the lead by two votes.

```
                                          CAND      K
    A A A C C B B C C C|B C C              C        2
```

The next delegate diminishes C's lead by one, but the last two raise it to 3 by the time the pairing phase terminates. The claim is that if any candidate has a majority, it is C.

Here is a simple example of the final state of the pairing phase on a ballot in which no candidate has a majority:

```
                                          CAND      K
    A A A B B B C|                         C        1
```

The votes for A and B cancel one another out and C wins the pairing phase by default. Had the delegates been polled in a different order, A or B might have won.

5.4 The Fortran Implementation

Suppose the delegates are in an array A of length N. The subroutine MJRTY shown in Figure 5.1 takes A and N as input and sets BOOLE and CAND to communicate the results. BOOLE will be set either to .TRUE. or to .FALSE.. If BOOLE is set to .TRUE., there is one (and only one) majority element in A and CAND is set to that element. If BOOLE is set to .FALSE., there is no majority element in A.

Note that the algorithm fetches the elements of A in linear order. Thus, the algorithm can be used efficiently when the number of votes is so large that they must be read from magnetic tape. One tape rewind may be necessary after the first phase.

```
      SUBROUTINE MJRTY(A, N, BOOLE, CAND)
      INTEGER N
      INTEGER A
      LOGICAL BOOLE
      INTEGER CAND
      INTEGER I
      INTEGER K
      DIMENSION A(N)
      K = 0
C     THE FOLLOWING DO IMPLEMENTS THE PAIRING PHASE. CAND IS
C     THE CURRENTLY LEADING CANDIDATE AND K IS THE NUMBER OF
C     UNPAIRED VOTES FOR CAND.
      DO 100 I = 1, N
      IF ((K .EQ. 0)) GOTO 50
      IF ((CAND .EQ. A(I))) GOTO 75
      K = (K - 1)
      GOTO 100
50    CAND = A(I)
      K = 1
      GOTO 100
75    K = (K + 1)
100   CONTINUE
      IF ((K .EQ. 0)) GOTO 300
      BOOLE = .TRUE.
      IF ((K .GT. (N / 2))) RETURN
C     WE NOW ENTER THE COUNTING PHASE. BOOLE IS SET TO TRUE
C     IN ANTICIPATION OF FINDING CAND IN THE MAJORITY. K IS
C     USED AS THE RUNNING TALLY FOR CAND. WE EXIT AS SOON
C     AS K EXCEEDS N/2.
      K = 0
      DO 200 I = 1, N
      IF ((CAND .NE. A(I))) GOTO 200
      K = (K + 1)
      IF ((K .GT. (N / 2))) RETURN
200   CONTINUE
300   BOOLE = .FALSE.
      RETURN
      END
```

Figure 5.1: Fortran Implementation of the Algorithm

In some applications it may be assumed that a majority candidate exists. For example, in the SIFT aircraft control system [11], where reliability is achieved with redundant processors and software voting, the failure rate of the individual processors is sufficiently low to permit the assumption that on a given flight a majority of the working processors agree on each vote. If it can be assumed that a majority exists, the counting phase may be eliminated. More importantly, the algorithm can then be implemented to poll the delegates in real time (rather than store the votes for batch processing).

The Fortran code above contains one minor improvement not mentioned in the convention floor-fight analogy. After the pairing phase has terminated, we test k against $n/2$. If k is greater than $n/2$, we announce that *cand* is the majority candidate without bothering with the counting phase, because we know there are at least k votes for *cand*. Indeed, one could make such a test every time k is incremented in the first loop. This would sometimes allow the algorithm to avoid making the second pass through the votes. Whether the running time of the algorithm is improved when such a test is inside the loop depends upon the distribution of the votes.

We have failed to find a variation of the first phase of the algorithm that obviates the second phase.

5.5 The Fortran Verification System

The informal proof sketched above may be convincing evidence that the algorithm computes the majority element if one exists. Of more practical importance, however, is whether the Fortran code implements the algorithm correctly and executes without error on all Fortran processors. There are many potential sources of error in the code that are completely ignored by the "proof" above. Is the program really a legal ANSI Fortran program? Does it violate any of the rules about aliasing and second level definition? Have we correctly analyzed the flow of control? Have we considered all the possibilities at run time? For example, ANSI Fortran permits individual elements of an array to be "undefined" (e.g., uninitialized). In such cases, even the meaning of an equality test is left unspecified by ANSI. A more obvious run time worry is that n might be so large that one of the arithmetic operations causes an overflow. Furthermore, the proofs are very informal. Are they correct? Have cases been ignored? Have false or unwarranted properties about "unanimity" and "majority" been assumed?

To permit the reliable verification of many Fortran programs we have implemented a mechanical verification system for Fortran. That system has been used to verify `MJRTY` and other subprograms. Before presenting the formal specifications that were verified, we briefly sketch our verification system.

The system handles a subset of both ANSI Fortran 66 [10] and ANSI Fortran 77 [1]. The subset is described precisely in [4]. Informally stated, the subset includes all the statements of Fortran 66 except the i/o, `DATA`, `BLOCK DATA`, and `EQUIVALENCE` statements. Certain restrictions, however, are placed on some of the remaining statements. For example, we allow only named `COMMON` blocks, we prohibit `REAL` arithmetic because we do not have a machine independent semantics for floating point operations, and we require that all arithmetic statements be fully parenthesized to permit straightforward overflow analysis. (ANSI permits the compiler to associate `A+B+C` to either the left or right. The overflow analysis is different for the two cases. We therefore require the programmer to write `(A+(B+C))` or `((A+B)+C)`, which, according the ANSI standard, determines the run time association. We have implemented this requirement in a simple but conservative way: all arithmetic expressions must be fully parenthesized. Thus the code for `MJRTY` contains unnecessary parentheses, e.g., in `K=(K+1)`. A more elaborate expression grammar could eliminate the unnecessary parentheses.)

Here, expressed informally, is what we mean when we say that our system has established the "correctness" of a subprogram:

> If a Fortran subprogram is accepted and proved by our system and the program can be loaded onto a Fortran processor that meets the ANSI specification of Fortran and certain parameterized constraints on the accuracy of arithmetic, then any invocation of the program in an environment satisfying the input condition of the program will terminate without run time errors and will produce an environment satisfying the output condition of the program.

This statement is made more precise in [4].

Our Fortran verifier is a standard Floyd-King-style system [5, 6, 2, 8] consisting of two parts: a Fortran analyzer (syntax checker and verification condition generator) and a mechanical theorem-prover. For those readers unfamiliar with Floyd-King-style verification, we briefly describe our system below.

Input to the analyzer consists of the Fortran subprogram (function or subroutine) to be verified, the mathematical specification of the subprogram, and all the subprograms somehow referenced by the candidate program. Each referenced subprogram must have been previously verified by the system. A specification consists of two mathematical formulas, called the "input assertion" and the "output assertion." The first describes those states in which the program may be properly invoked. The second describes the states produced by the program. In addition to the input/output assertions, each loop in the subprogram must be cut by an inductive assertion—a mathematical formula describing the machine state each time execution arrives at the indicated point in the program. All the formulas are written in the formal logical language described in [3].

The analyzer checks that the program satisfies all our syntactic requirements and then generates mathematical formulas called "verification conditions." If these can be proved—i.e., derived symbolically from a certain set of axioms using certain rules of inference—then, whenever the program is invoked in an input state satisfying the input assertion it produces a state satisfying the output assertion.

In general, there is one such formula for each assertion-free path between any two assertions. The formula for such a path requires proving that, if the assertion at the beginning of the path is true and one is led down the path by the tests, then the assertion at the end of the path is true. In addition, formulas are generated to establish that no array bound errors, overflows, or other run time errors occur, and that the program terminates. (See [4].)

To permit consideration of arithmetic overflow, our verification system permits formal talk about the "least inexpressible positive integer" and the "greatest inexpressible negative integer" on the host Fortran processor. Typical input assertions for programs must specify the relations between the input variables and these otherwise unspecified constants. We assume that ANSI Fortran processors compute the correct results and cause no arithmetic overflow on primitive `INTEGER` arithmetic operations (i.e., `+`, `-`, `*`, `/`, and `**`) in which the inputs and the mathematically defined result are all strictly between the least and greatest inexpressible integers.[2]

The second part of the verification system is a mechanical theorem-prover that attempts to prove the formulas generated by the analyzer. The theorem-prover, which is described in [3], is entirely responsible for the correctness of each proof.

5.6 Formal Specification

Below we denote the initial values of the Fortran variables `A` and `N` by A and n respectively. The precise input assertion for `MJRTY` is that n is a positive integer, that $n+1$ is strictly less than the least inexpressible positive integer, and that every element of A is defined. The reason $n+1$, rather than merely n, must be expressible is that the ANSI standard permits `I` to obtain the value $n+1$ immediately before the termination of `DO 100 I = 1, N`.

The output assertion for `MJRTY` is

- The final value of `BOOLE` is `.TRUE.` or `.FALSE.` (that is, `BOOLE` may not be returned "undefined").

- The elements of `A` are not changed.

- If `BOOLE` is set to `.TRUE.`, then `CAND` is defined and the number of occurrences in A of its final value, *cand*, is more than $n/2$. (By "/" we denote the integer "floor" of the real quotient.)

[2] In addition, for division we require that the denominator be nonzero.

- If BOOLE is set to .FALSE., then for all x, the number of times x occurs in A is less than or equal to $n/2$.

We phrase these requirements in terms of the mathematical function cnt of four arguments. One may read $\text{cnt}(x, A, i, j)$ as "the number of times x occurs in A from i through j inclusive." The function cnt is a typical example of a concept that must be introduced into one's underlying logical theory to specify a program. We may define cnt recursively for all $i \geq 0$ and $j \geq 0$ as follows:

$$\text{cnt}(x, A, i, j) = \begin{cases} 0 & \text{if } j = 0 \lor j < i \\ 1 + \text{cnt}(x, A, i, j-1) & \text{if } 0 < j \not< i \land x = A(j) \\ \text{cnt}(x, A, i, j-1) & \text{otherwise} \end{cases}$$

Our mechanical theorem-prover verifies that there exists a function satisfying the above equation before the equation is added as a new axiom. Without such a check, the user of a verification system might inadvertently "overspecify" a concept and permit correctness proofs based on contradictions in the underlying specification.

We cut the first DO-loop in MJRTY with an invariant at the bottom of the loop, just before I is incremented and tested against N. In our informal proof the invariant required that the i delegates processed thus far could be divided into a unanimous group for *cand* of size k and a group that could be paired into disagreeing delegates. Since the algorithm does not explicitly keep track of any such division of the delegates, we reformulated the invariant in a slightly weaker fashion. The reformulation is based on the observation that, if a collection of delegates can be paired in such a way that paired delegates disagree, then the collection has no majority.[3]

Here is the actual invariant used:

(1) $0 < i$ and $0 \leq k \leq i \leq n$.

(2) CAND is always defined and has some value *cand*.

(3) The number of times *cand* occurs in A from 1 through i is at least k.

(4) The number of times *cand* occurs in A from 1 through i, minus k, is no greater than $(i - k)/2$.

(5) For all x other than *cand*, the number of times x occurs in A from 1 through i is no greater than $(i - k)/2$.

Although conjuncts (1) and (2) were ignored in our informal proof, they are essential in a careful proof. Conjunct (3) establishes that we have at least k votes for *cand*. Let those k delegates constitute the "unanimous group." The $i - k$ remaining delegates are the "majority-free group." Conjunct (4)

[3]The converse also holds for collections with an even number of members.

says that *cand* does not have a majority in the majority-free group; ignoring the k votes in the unanimous group, the number of votes for *cand* thus far encountered is less than $(i-k)/2$. Conjunct (5) says that no other candidate has a majority in the majority-free group. We count the votes for candidates other than *cand* over the entire interval processed, rather than just over the majority-free group, since we do not really know where the majority-free group is. But we know that the unanimous group contributes nothing to the tally of a candidate other than *cand*. (It is easy to see by the construction of a counterexample that (4) and (5) do not imply (3). Nevertheless, if one modifies the code so that k is not tested against $n/2$ before entering the counting phase, one can omit conjunct (3) of this invariant. That is, unless the program exits early when k exceeds $n/2$, a demon within the first loop is permitted to raise k above the count of *cand* (within the constraint imposed by (5)) without causing the algorithm to perform incorrectly. We do not know how to interpret this lack of constraint.)

As the counting phase is trivial, we shall not discuss it.

5.7 The Formal Proofs

The Fortran analyzer produced 61 verification conditions for `MJRTY`. Most of the conjectures established that array bounds are not violated, that arithmetic operations cause no overflows, and that variables and array elements are defined when required.

The mechanical theorem-prover proved all 61 conjectures. Most of the proofs were immediate either from the axioms and definitions in the "basic Fortran theory" [4] (e.g., the definition of the negative integers in terms of the Peano numbers), from the definition of cnt (e.g., if x is $A(i+1)$ and $i \geq 0$ then $\mathrm{cnt}(x, A, 1, i+1)$ is $1+\mathrm{cnt}(x, A, 1, i)$), or from elementary arithmetic lemmas (e.g., the theorem that for all naturals m and n, $n/2 < m$ iff $n < 2m$). Several of the paths to the invariants and to the output assertion required that the user help the system.

The user of our system can help the system prove a "hard" theorem by suggesting that it first prove some key lemmas. When the system proves a theorem for the user, it stores the theorem for use in future proofs. Thus, by bringing to the theorem-prover's attention previously unrecognized truths, the well-trained user of our system can get the theorem-prover to prove formulas that would otherwise be beyond the system's competence. The user of our system, however, does not have to be trusted. The machine—not the human—is responsible for the validity of the final proof; the user cannot maliciously or inadvertently cause the system to accept falsehoods, because the system proves for itself every fact used.

To get all 61 theorems proved, we had to instruct the theorem-prover to prove five lemmas about cnt. The two most interesting ones were as follows:

- cnt is monotonic: the number of times x occurs from 1 through i is less than or equal to the number of times it occurs from 1 through j if $0 \leq i \leq j$. Without knowing this, the theorem-prover could not approve our exiting from the counting phase as soon as k exceeds $n/2$ lest subsequent processing of the remaining delegates decrease k.

- The number of times x occurs from 1 through i ($i \geq 0$) is no greater than i. This ensures that k in the second loop will never exceed i (and thus incrementing k will never cause an overflow).

These two lemmas are proved by the system with mathematical induction on the length of the interval scanned.

The other three lemmas we proved were required because of inadequacies in the theorem-prover itself. For example, when `MJRTY` exits because k is 0 at the end of the counting phase, the theorem-prover knows that *cand* has no majority and that no x other than *cand* has a majority. It must prove that no x has a majority. The proof is obvious if one merely asks, "Is x equal to *cand* or not?" and considers the two cases. Without an explicit theorem stated by the user, the theorem-prover failed to consider such a case split. The other two lemmas were necessary for similar reasons and indicate inadequacies in our system that we hope to repair in the future.

The entire effort of specifying `MJRTY` and getting the 61 verification conditions proved required about 20 man hours. Most of the time was spent identifying problems caused by incorrectly written invariants, overcoming inadequacies in the theorem-prover by identifying appropriate lemmas, and struggling with the still awkward interface to our Fortran verification condition generator. It requires about 55 minutes of computer time to prove the final list of 66 theorems. The time was measured on a Foonly F2 Computer (about 30 percent as fast as a DEC 2060) running small INTERLISP-10. A total of 42 minutes was required for theorem-proving, 8 minutes for garbage collection, and 5 minutes for printing out the proofs.

Readers interested in obtaining the system's complete English description of its proofs may contact the authors.

5.8 Postscript 1991

In this paper we have described a linear time majority vote algorithm and discussed the mechanically checked correctness proof of a Fortran implementation of it. This work has a rather convoluted history which we would here like to clarify.

The algorithm described here was invented in 1980 while we worked at SRI International. A colleague at SRI, working on fault tolerance, was trying to specify some algorithms using the logic supported by "Boyer-Moore Theorem Prover." He asked us for an elegant definition within that logic of the notion of the majority element of a list. Our answer to this challenge was the recursive expression of the algorithm described here.

In late 1980, we wrote a Fortran version of the algorithm and proved it correct mechanically. In February, 1981, we wrote this paper, describing that work. In our minds the paper was noteworthy because it simultaneously announced an interesting new algorithm and offered a mechanically checked correctness proof. We submitted the paper for publication.

In 1981 we moved to the University of Texas. Jay Misra, a colleague at UT, heard our presentation of the algorithm to an NSF site-visit team. According to Misra (private communication, 1990): "I wondered how to generalize [the algorithm] to detect elements that occur more than n/k times, for all k, $k \geq 2$. I developed algorithm 2 [given in Section 3 of [9]] which is directly inspired by your algorithm. Also, I showed that this algorithm is optimal [Section 5, *op. cit.*]. On a visit to Cornell, I showed all this to David Gries; he was inspired enough to contribute algorithm 1 [Section 2, *op. cit.*]." In 1982, Misra and Gries published their work [9], citing our technical report appropriately as "submitted for publication."

However, our paper was repeatedly rejected for publication, largely because of its emphasis on Fortran and mechanical verification. A rewritten version emphasizing the algorithm itself was rejected on the grounds that the work was superceded by the paper of Misra and Gries!

When we were invited to contribute to the Bledsoe festschrift we decided to use the opportunity to put our original paper into the literature. We still think of this as a minor landmark in the development of formal verification and automated reasoning: here for the first time a new algorithm is presented along with its mechanically checked correctness proof—eleven years after the work.

References

[1] American National Standards Institute, Inc. (1978): American National Standard Programming Language Fortran, ANSI X3.9–1978, 1430 Broadway, New York, N. Y.

[2] R. B. Anderson (1979): Proving Programs Correct. New York: John Wiley and Sons.

[3] R. S. Boyer and J S. Moore (1979): A Computational Logic. New York: Academic Press.

[4] R. S. Boyer and J S. Moore (1981): A Verification Condition Generator for Fortran. In R. S. Boyer and J S. Moore, eds.: The Correctness Problem in Computer Science. London: Academic Press.

[5] R. W. Floyd (1967): Assigning Meanings to Programs. In: Mathematical Aspects of Computer Science, Proc. Symp. Appl. Math., 19. Providence, RI: American Mathematical Society.

[6] J. C. King (1969): A Program Verifier. Ph. D. thesis, Department of Computer Science, Carnegie-Mellon University, Pittsburgh, PA.

[7] D. E. Knuth (1973): The Art of Computer Programming, Volume 3: Sorting and Searching. Reading, MA: Addison-Wesley.

[8] Z. Manna (1974): Mathematical Theory of Computation. New York: McGraw-Hill.

[9] J. Misra and D. Gries (1982): Finding Repeated Elements. Science of Computer Programming 2, 143–152.

[10] United States of America Standards Institute (1966): USA Standard Fortran, USAS X3.9–1966, 10 East 40th Street, New York, N. Y.

[11] J. Wensley, et al. (1978): SIFT: Design and Analysis of a Fault Tolerant Computer for Aircraft Control. Proc. IEEE 66(10), 1240–1255.

Chapter 6

How the Brain Adjusts Synapses—Maybe

Hans J. Bremermann

Division of Biophysics, Department of Molecular and Cell Biology and
Department of Mathematics
University of California at Berkeley

Russell W. Anderson

Graduate Group in Bioengineering
University of California at Berkeley and San Francisco

6.1 Introduction

The notion that the synapse is the site of lasting change in memory and learning has had wide acceptance for decades. Hebb [46] postulated that when one neuron repeatedly excites another, the synaptic knobs are strengthened. Verification has taken time, but there is now ample evidence that Hebbian type long term potentiation (with some modifications of the original hypothesis) does indeed occur [61].

For anything but special cases Hebb's rule is insufficient as a learning rule. This was noted by Rosenblatt [80]. Consider a circuit diagram of several neurons with weighted inputs (the weights representing synaptic strengths). Given a discrimination or motor control task, how must the values of the weights be changed in order for the neural circuit to perform the given task?

There is no general method for *finding* learning rules, but once a particular one has been proposed, one can analyse it by proving mathematically that it converges (and how fast), or one can explore it through computer simulation or hardware implementation.

In 1956 Frank Rosenblatt formulated his "Perceptron" model and set out to explore all three approaches. Analytically, Frank described an "error correction procedure" for certain neural net configurations, that can be proven mathematically to converge if there is a solution to a classification problem within the constraints of the given neural net. Today this algorithm for changing synaptic weights is also known as the "Widrow-Hoff rule." It can be proven to converge, it is reasonably efficient, but unfortunately it works only for *"simple perceptrons,"* which in today's terminology are equivalent to neural nets without "hidden units."

Frank Rosenblatt himself was aware of the limited power of *simple perceptrons*. Buried in his "Principles of Neurodynamics" (1962) is a calculation of the number of units (neurons) that would be required to do some rather straightforward discriminations (involving size and shift invariance), with a simple perceptron: it is astronomical.

From the beginning, Frank was interested in exploring "multilayer perceptrons," that is neural nets with "hidden units." The input-response function of such nets is inherently nonlinear and mathematically intractable. The "error correction procedure" does not apply, and Hebb's rule is not a viable training procedure either. The problem *can* be solved, however, by numerical minimization of the "squared error performance function." Most numerical methods of optimization are methods of steepest descent: compute the gradient and take a step along the gradient. Alas, most numerical problems that yield to gradient methods have very few variables, while the number of weights in modest model nets runs into the hundreds, and real brains with trillions of neurons and tens of thousands of synapses per neuron are in a complexity category by themselves. Moreover, the linear threshold elements (McCulloch and Pitts model neurons) that Rosenblatt employed are discontinuous and don't even have a gradient. Against these odds Frank and others carried out computer simulations, but computer time in the early sixties was much too expensive for extensive experimentation and no encouraging results were obtained.

Frank's critics, notably Marvin Minsky [69], pointed out that "hill climbing" is inherently inefficient, slow, and perhaps futile (because of local maxima), to solve problems of intelligence. They saw Turing machines and mathematical logic as the proper approach towards artificial intelligence. Once the problem solving algorithms would be better understood one could begin to search for their implementation in real, biological neural circuits. Ultimately, this was an act of faith in the primacy of logic and mathematical principles which has proven so fruitful from Newton to Einstein.

Woody Bledsoe has made major contributions to this approach in his work on automation of proofs through logical deduction. Around 1960-61, however, we both experimented with the idea of applying "genetic algorithms" to optimizing the performance of perceptrons: Treat the synaptic weights like nucleotides of DNA; mutate, recombine, cross over and select, as in Darwinian evolution. HJB spent several weeks during the summer of 1961 with Frank in Ithaca. Subsequently Woody and HJB tried it in earnest, but on simpler and well-defined objective functions [20, 16, 17, 21, 22, 23]. The method works *in principle* (and is currently a popular method in Artificial Intelligence [49, 34, 71, 8]) but for neural nets it turned out to be an optimization problem of extraordinary computational requirements, due to the very large numbers of synaptic weights that are involved.

Can ever more powerful computers be built with the capability of solving very computation intensive problems not only *in principle* but *in practice*? In 1961 it occurred to us that the answer is *no*. HJB (1962) formulated a quantum limit that generally applies to information transmission in computers, and Woody transformed it into a more stringent limit that applies to all sequential machines [18]. Problems that fall beyond these definite bounds are "transcomputational." (For a recent review, including criticisms and proofs in the context of contemporary physics, see [13]). For perceptrons and other neural nets with N input nodes which are such that they require training that increases *exponentially* with N, (as does the N bit parity problem discussed in the following) the quantum limits imply that computing requirements for these nets, with increasing N quickly outgrow all conceivable computing resources in the universe.

In the early sixties many groups experimented with perceptrons, which they tried to apply to anything from recognition of handprinted characters, speech, geological data indicative of oil bearing strata, and more. Results with simple perceptrons proved disappointing, and practical or analytical results for multilayer perceptrons never emerged. Then came Minsky and Papert's analysis in their book "Perceptrons" [70], that could have been entitled "What Simple Perceptrons Can't Do." This book helped lay to rest research on simple perceptrons, but it had little to say one way or the other about multilayer perceptrons.

More than a decade later Rumelhart, Hinton, McClelland, Sejnowski, and others [81] had surprising successes with neural nets with "hidden units" (which correspond to multi-layer perceptrons). They replaced the discontinuous step function of the linear threshold elements (McCulloch and Pitts neurons) that Rosenblatt had employed by a continuous activation function [38], and were able to compute gradients. With steepest descent along gradients they were able to obtain some startling results: Pronunciation of English words [84], elucidation of 3-Dimensional shape from 2-D shaded images [59], predicting secondary structure in protein conformation [77], computing stereo disparity [60], selecting moves in "backgammon" [92]. The method behind all these achievements is back-propagation. Surprisingly, moderate size nets performed some remarkable tasks. The nature of neural nets seems to make local minima less of a problem than they are in some other optimization problems of comparable size.

More recently, multilayer neural networks have yielded to some theoretical analysis. Cybenko [30] showed that linear combinations of sigmoidal functions (often used as the transfer functions in neural networks) can uniformly approximate any continuous function. Poggio and Girosi [76] have related neural networks to regularization theory. These studies have shed light on the capabilities of such networks as powerful interpolation devices.

Unfortunately, however, computation of the gradient and adjustment of synaptic weights, individually through "back propagation of error," would require a neural network as complex as the "feed forward net," and the method becomes inefficient and incredibly complex in network architectures that allow lateral and feedback connections. Hinton, one of the pioneers of neural networks, acknowledges that "as a biological model, back-propagation is implausible" [48]. Mel [64, p. 34] asks, "[i]s it, then, a fundamental law that neural associative learning algorithms must be either representationally impoverished or mechanistically overcomplex?"

In his article "The Recent Excitement about Neural Networks," Francis Crick [28] writes:

> It is hardly surprising that such achievements have produced a heady sense of euphoria. But is this what the brain actually does? Alas, the back-prop nets are unrealistic in almost every respect... *Obviously what is really required is a brain-like algorithm which produces results of the same general character as back propagation* [emphasis added].

In this paper we describe such an algorithm: We have experimented with a method that is so simple and straight forward that it seems ideal for implementation in a biological structure. DNA can generate neural circuits through morphogenesis (though evolution would have taken many millions of years). Thus, in principle, DNA could have organized back-propagation networks in the brain. But, there is a trade-off between simplicity and whatever increased performance might be obtained from complex neural circuitry. The algorithm that we propose is of utmost simplicity. This is its merit. It does not require much "genetic information" (as would back-propagation).

The algorithm that we are proposing is a random walk through the space of synaptic weights. It relies on Gaussian trial fluctuations, which can be generated locally at the synapses, and which are then fixed by a global signal "To all synapses: Freeze the last trial fluctuation and make it the starting point for new trials." How can such a simple rule possibly be substitutable for back-propagation? We were able to show that this algorithm, augmented by a few additional features, is indeed a candidate for the synapse adjustment.

We have successfully used Gaussian random walks and "chemotaxis" for the training of nets for a variety of tasks: discrimination of earthquakes from underground explosions ([31] and RWA, Ph. D. thesis in preparation), generation of time-varying signals from "recurrent networks," which allow lateral and feedback connections (RWA, Ph. D. thesis in preparation), and the generation of control signals for open-loop, dynamical systems ([5] and RWA, Ph. D. thesis in preparation). For these tasks no comparative performance data for the back-propagation algorithm are available in the literature. To obtain a comparison we have applied our method to two benchmark problems: "n-bit parity" and the "encoder problem." The former is of maximum difficulty and statistics of the performance of back-propagation (with fine tuned "momentum term" and "learning rate" [93]) are available in the literature. For "n-bit parity" performance of our algorithm is slower by a factor of 5 to 10 (independently of increasing n); for the encoder problem it is ahead, especially if there are several intermediate layers. This is very encouraging since the essential point of Gaussian random walk descent is its utter simplicity in terms of biological implementation.

The encoder can also be seen as a model of the problem of how neural connections are established that project from the retina, through the optic nerve, the lateral geniculate and to several cortical projection areas. We propose that a Gaussian random walk may help explain how these maps can be "self-adjusting" so as to refine and maintain topographic projections.

6.2 Training for Performance

6.2.1 Error Function and Back-Propagation

Neural nets consist of elements called "neurons" or "neurodes" which respond, according to an "activation function" to weighted sums of inputs. We call the collection of all the weights the "weight vector" $\vec{w}$. The outputs of the neurodes become inputs of other neurodes, etc., till eventually a response emerges on a set of designated output neurodes. Given a collection P of inputs p (input "patterns") and associated "target" outputs t_k on $k = 1, \dots, K$ output elements, then an error function

$$E_{\text{error}} = \sum_{p}^{P} \sum_{k}^{K} (t_k - o_k)^2$$

can be computed. The task is to minimize E_{error} by optimizing over the weight vector $\vec{w}$.

When there are no constraints on the architecture of the neural net, then this optimization task can be enormously time consuming. We therefore consider in the following "feed-forward nets" with N input units, one or several layers of hidden units, and one or several output units. A "N-M-L-K" net, for example, is a net with N inputs, M hidden units in the first layer, L

units in the second, and K output units. Each unit has a sigmoid activation function, given by the "squashing function"

$$f = \frac{1}{(1+\exp(-z))},$$

where z is the sum of the weighted inputs to the unit. Let w_{ijk} denote the synaptic weight of the kth unit in the jth layer with input $y_{(j-1)i}$ from the ith unit in the previous $(j-1)^{\text{st}}$ layer. Then

$$z_{jk} = \sum w_{ijk} y_{(j-1)i}$$

is the weighted sum of inputs to the kth unit in the jth layer. The thresholds are represented by w_{ojk} and the convention $y_{o(j-1)} = -1$.

The dimension of the weight space is the sum of all the weights and thresholds in the net. Training consists of a series of weight changes $\Delta\vec{w}$ until the squared error function E_{error} is reduced to an acceptably small value. The algorithm has to find a path in weight space, descending from and tunneling through, graphically speaking, a very ragged mountain landscape (defined by the objective function $E_{\text{error}}(\vec{w})$). It has to overcome not only local minima but find its way out of saddle points in which escape directions are extremely rare. (This problem can become exponentially worse with the dimension N and thus is poorly represented by 3-dimensional intuition.)

The back-propagation algorithm computes

$$\Delta\vec{w}(n) = \lambda\nabla E + \mu\Delta\vec{w}(n-1)$$

where ∇E is the gradient of the error function, λ is the learning rate, and $\mu\Delta\vec{w}(n-1)$ is a "momentum term" that adds a portion of the previous increment $\Delta\vec{w}(n-1)$ to the gradient. The momentum term helps to overcome local minima. For some problems, conjugate gradient methods have proven faster than back-propagation [53].

6.2.2 Search Algorithms

Every optimization algorithm combines prediction with search. When the objective function is easy to predict (through interpolation or extrapolation), then search is less important. Back-propagation is based upon the first order Taylor approximation of the objective function, which is only locally predictive. We tried to predict the objective function more globally, by polynomial and spline approximation along randomly chosen lines. This method can be an effective method of optimization [24, 29], but in our experiments with training nets to classify n-bit binary strings for "parity" the "landscape" defined by the objective function is so rugged that the approximations in question were useless.

When prediction is impossible, then search is all that remains. In a fully connected neural net we are talking about very large dimensions of the weight space, even for nets with few units. Our algorithms thus must search spaces of high dimension and exhaustive search is out of the question. This leaves only one option: Search through random sampling.

In our case, we have for any given point $\vec{w}^0$ a set of "target" points. These are the points to which the algorithm can jump. They are the points $\vec{w}^0 + \Delta\vec{w}$ such that $E(\vec{w}^0 + \Delta\vec{w})$ is less than $E(\vec{w}^0)$. If the objective function is effectively unpredictable, then the sets of target points, averaged over the weight space, have an average density which is Gaussian. The proof is analogous to the proof of the central limit theorem.

Given an average density of targets, how should the search effort be allocated? There exists an elaborate theory for the allocation of search efforts for targets whose location is given by a probability distribution [90]. In our case it implies that the search effort should be proportional to the target density. In other words, the trial steps $\Delta\vec{w}$ should follow the same Gaussian that represents the average target density. The variances of coordinate components of the Gaussian *a priori* need not be the same, but they can be normalized by scaling.

This leaves a single parameter to be determined: the common standard deviation. We generated random steps from a Gaussian distribution with standard deviation equal to one, then multiplied them with a stepsize parameter H. This parameter can then be adjusted interactively or automatically by means of feedback from successful steps (see Section 6.2.5).

6.2.3 Mutational Versus Gaussian Random Walks

Recently a number of authors have studied "mutational random walks in rugged landscapes," among them Kauffman and Levin [55], and in the context of macromolecular evolution and the origin of life: [35, 36, 83, 39, 63, 74, 63] have explored random walks on "rugged landscapes" that are generated by single point mutations in the context of affinity maturation of antibodies. The mutations affect the coding region for the variable part of antibody molecules, the quantity selected is the binding affinity of antibodies for the antigen to which they respond. During the primary immune response those clones of the B-cell repertoire are expanded whose immunoglobulin receptors have the highest affinity for the invading antibody. Subsequently the affinity (equilibrium binding constant) of antibodies for the immunizing antigens increases, typically ten- to fifty-fold. This is due to *somatic point hypermutations* which occur at the high rate of 10^{-3} per basepair per generation. After an initial improvement of affinity and typically 6-to-8 muations in the V- (for "variable") region of the immunoglobulins further mutations tend to lead to no further improvement in affinity.

Macken and Perelson have simulated this process by representing genomes by *binary strings* and assigning affinities at random. This gives rise to a "worst case scenario," a totally *rugged landscape.* They consider only single point mutations and describe the random walk as follows:

> The germline sequence defines a starting point on the landscape. At any stage on a walk, single-mutant variants of the current antibody are tested in random order until the first neighbor having a higher fitness is attained. The walk then moves to this new point in sequence space (i.e., new antibody sequence), and the testing process starts anew ... If no fitter variant is found among these different one-mutant neighbors, the process stops, as the walk has reached a local maximum.

They then analyzed the probability that a random walk ends (gets trapped) at the kth step. This depends, of course, upon the number of mutable bases involved. Macken and Perelson estimate that for the V-region about 700 basepairs are involved, and that about 75% of the mutations (due to the degeneracy of the genetic code) lead to changes in affinity, which amounts to 1575 one-mutant neighbors. Rounding this figure to 1500 they found that the "trapping probability" reaches a maximum at seven steps, and that very few of these mutational walks extend beyond fifteen steps.

Some 30 years ago Woody Bledsoe [18] and HJB [21] came to similar conclusions, except that the above mentioned facts about immunology were not yet known and we applied mutational random walks to binary representations of the integer solutions of ill-conditioned systems of linear equations. Formulated this way a fairly "rugged fitness landscape" is generated and we observed the same kind of trapping phenomenon reported by Macken and Perelson. "Simulated sex" (recombination through "crossing over") did not solve the trapping problem [23]. HJB realized then that trapping was a byproduct of the lack of symmetry of the random-step distribution that was generated by the mutational process.

When the underlying space consists of digital strings (as it necessarily does in protein evolution), then there is little that can be done about trapping. However, when the underlying space consists of vectors of real variables, then it is possible to draw random steps to form a *spherically symmetric* probability distribution. The n-variate Gaussian (with equal variances) is spherically symmetric, and it has the added advantage to vary *step length* in a way that is an optimal search strategy when the "points of opportunity" themselves have a Gaussian distribution (previous section). HJB went on to develop this insight into a stochastic optimizing algorithm [24] (which recently has been compared with other stochastic optimizers by Cutler [29]).

Originally HJB [20] had envisioned to apply *mutational random walks* or *genetic algorithms* to optimizing the weights of multilayer perceptrons. By 1970, however, perceptron research had withered away and the complexity of optimizing over hundreds of weights looked forbidding. It took an enthusiastic new generation (Hinton, Rumelhart, McClelland, Sejnowski, et al.) and the miracle of affordable computer work stations to revive the subject.

Inspired by this revival both of us attempted to apply the interpolating random ray optimization algorithm of HJB [24] to the "fitness landscape" of the n-bit parity problem. We found that the "fitness landscape" of this problem is so unpredictable that interpolation is useless. In other words it seems to be the continuous equivalent of a "rugged landscape" in string space. We found interactively that nevertheless Gaussian random walks rarely get trapped completely, though sometimes escape from a narrow valley or saddle point may take time. Subsequently we found that the Gaussian random walk is an optimal strategy of search (previous section). If nature optimizes through random walks, then, we conjecture, for the vertebrate brain, it should be Gaussian random walks.

6.2.4 Learning the Stepsize

Besides biased random walks as described above we have also experimented with an algorithm that resembles bacterial chemotaxis. It is a shade less pessimistic about the predictability of the objective function. As before, random trial steps are taken with a Gaussian distribution, but when the algorithm takes a successful step, it attempts to keep going in the same direction until the objective function no longer decreases. In other words this algorithm will attempt to follow a declining slope if it finds one.

The steps of our algorithm are analogous to the motions that bacteria perform during chemotaxis [58, 15]. Bacteria are too small to be able to measure spatial concentration gradients of chemoattractants. When swimming in a medium with varying concentrations they generate random directions instead and keep going as long as concentration increases. If attractant concentration does not or no longer increases, then they stop, tumble, then emerge in a new direction at random angles to the old direction. In this way they move towards larger and larger concentration values of the attractant. In other words, they optimize the function that describes the concentration of the chemoattractant in the medium.

How much distance do bacteria have to swim under chemotaxis versus gradient ascent? That depends strongly upon the shape of the equiconcentration surfaces in the medium. When these are spherical (a condition that is ideal for gradient descent), they have to swim an average of only 39% longer as compared with following the gradient [25]. In our case the error function takes the place of the chemoattractant concentration, and the weight space (of high dimension) corresponds to the 3-dimensional physical space in which bacteria swim.

Algorithms, involving random walks, have long been the numerical method of choice for many particle problems of mathematical physics. "Monte Carlo" algorithms were introduced by Metropolis and Ulam in 1949 [67, 85]. Recently

Ceperley and Alder [26] reported that the many particle Schrödinger equation can be solved by a "quantum Monte Carlo method" that solves the equation by random walks in the many-dimensional state space. The method is faster and more accurate than conventional variational methods, which for high dimension are too slow and cannot be refined to give sufficiently accurate results. In the limit, for short stepsize and many trajectories, random walks are described by the *diffusion equation* (cf. [58, 14, 15, 73]). Chemotaxis, too, in the limit is described by a reaction-diffusion equation [56, 4, 72]. Equations for selection-diffusion have been analyzed by Ebeling et al. [33]. Ceperley and Alder note that the many-particle Schrödinger equation, written in Euclidean time, is equivalent to a bacterial chemotaxis equation.

6.2.5 The Chemotaxis Algorithm

In the Gaussian random walk algorithm, the network is initialized with an arbitrary set of weights, $\vec{w}^0$, and performance $E(\vec{w}^0)$ is evaluated. A random vector $\Delta\vec{w}$ is chosen from a multivariate Gaussian distribution with a zero mean and a unit standard deviation. This random vector is added to the current weights to create a 'tentative' set of weights $\vec{w}^t$:

$$\vec{w}^t = \vec{w}^0 + h\Delta\vec{w}$$

where h is a stepsize parameter. Performance $E(\vec{w}^t)$ is then calculated for the tentative weights. If the error of the new configuration is lower than the original configuration, the tentative changes in the weight vector are retained; otherwise, the system reverts to its original configuration. A new random vector is chosen, and the process is repeated.

In the chemotaxis algorithm, if a successful direction in weight space is found, an additional step is taken along *the same random vector* and the stepsize parameter, h, is increased automatically. Weight modifications continue along this same random vector until progress ceases. In addition, if after several trials, a successful direction is *not* found, h is reduced by 50%. This allows for automatic adjustment of the single learning parameter, which otherwise has to be found empirically. Because of these automatic features and the ease of implementation, the chemotaxis algorithm was used in all computer simulations reported here.

6.3 Random Walk Learning in Actual Neural Systems

6.3.1 Gaussian Fluctuations of Neuronal Connections and Synaptic Strength

A Gaussian distribution can easily be generated in biological nerve nets for two reasons:

1) The product of individual Gaussians (with standard deviations σ_1 to σ_n) is the multivariate Gaussian distribution with multivariate standard deviations σ_1 to σ_n. In other words if individual synaptic weights fluctuate *independently,* each according to a Gaussian distribution, then the entire weight vector fluctuates according to a multivariate Gaussian. Not only is the multivariate Gaussian distribution required by the the optimal search theorem, but it can be generated *without interaction between the weight changes.*

Glanzman et al. [43] have observed structural change *in vitro* of a coculture of *Aplysia* sensory neurons and their target (L7 motor) neurons over a 24 hour period. They note that morphological changes (varicosities and new synaptic processes) at the junctions between the sensory and motor cells appeared to be normally (Gaussian) distributed in the control group. In the cocultures repeatedly treated with serotonin, however, structural change was shown to be highly biased toward increases in connectivity. Furthermore, they showed that these observed structural changes corresponded to measurable changes in monosynaptic EPSP produced in L7 motor cells by firing the sensory neuron. Perhaps the random variation observed in the control group serves a vital role in learning by generating tentative, exploratory variation in neural structure.

2) When synaptic strength is the result of many components (channels in the presynaptic membrane, transmitter release mechanisms, postsynaptic receptors, etc. [61, 57]) then fluctuations in the total mechanism, because of the central limit theorem are distributed approximately Gaussian, even if individual fluctuations are not.

Learning is likely to occur not only at the level of individual neurons but could involve populations of neurons (Edelman [34] calls them "groups"). Plasticity in contact between axons and spines (for which there is evidence, cf. [7]) would lead to fluctuations in the strength of contact between groups. Such fluctuation, because of the central limit theorem, would also follow a Gaussian distribution. *Therefore the same basic learning algorithm would apply to learning by individual units as well as learning by large and small groups of neurons.*

6.3.2 Fixation of Successful Modifications

Our algorithm requires that all fluctuations in the strengths of weights (or synaptic contact between groups) remain tentative and reversible unless made permanent by a special "hold" signal (which is analogous to Grossberg's postulated "print" signal [44]). It is important to note that this signal does not require any local computations whatsoever, it is a global signal addressed "to all synapses" within the net. It could be an electrical signal propagating through the net (such as tetanization), or chemical, or an interaction of both. This signal would fit Crick's [28] description of a signal "to tell the system when something is worth remembering" and he points out that "there are many return pathways in the brain" [95].

Our proposed modification rule seems to be consistent with current knowledge about molecular mechanism of synaptic modification [54, 89, 96]. Both back-propagation and our algorithms require that synaptic strength can be increased as well as decreased. Associative long term depression (LTD) has been reported by Stanton and Sejnowski [88]. They show that this is consistent with a "Hebbian covariance learning rule," but such phenomena are consistent with several reinforcement learning rules—including random walks.

6.3.3 Origin of the Global Signal

The algorithm requires that the performance of a net be evaluated. This evaluation could be accomplished by other brain circuits. We do not consider this requirement problematic, since evaluation of performance tends to be computationally easier than improvement. For example, throwing a ball requires precise coordination and timing of numerous muscles. It is hard to achieve and may require extensive training. How close a ball comes to hitting the target is evaluated by the visual cortex and is a separate process. One portion of the brain thus could act for another system as "supervisor."

Other learning schemes also require a global signal. In the Boltzmann machine weights are adjusted according to the activities of the neurons that are connected [1]. The local adjustment mechanism, however, must receive information when the neural net is free running and when it is "clamped." Barto et al. employ a rather complex 'adaptive critic element' in their reinforcement learning schemes [11, 12]. In Edelman's "Neural Darwinism" [34] a global "value" of the net activity is required.

6.4 Training Networks on Benchmark Problems

6.4.1 N-bit Parity

Here we report results of computer simulations of the chemotaxis algorithm applied to the "n-bit parity" problem. n-bit parity can be considered a generalization of the "exclusive or" (XOR) which is known to be problematic for single layer perceptrons [70]. Parity is considered a "difficult" test case for

Dimension (N)	Chemotaxis Average (Low/High)	Back-Prop (cycles through training set)
2	113 (35/188)	24
3	251 (96/421)	33
4	962 (649/1660)	75
5	1259 (820/1951)	130
6	4169 (2494/7107)	310
7	5789 (3159/10169)	800

Table 6.1: Comparison of the 'Chemotaxis' Algorithm and Back-propagation. Average, best and worst training times for the chemotaxis algorithm. Net architecture: N-($2N + 1$)-1. Weights were updated after performance was evaluated over the entire input set. Back-propagation averages from Tesauro and Janssens [93]. For comparison of *computational effort,* the figures of Tesauro and Janssens should be doubled (see text).

neural networks since its solution requires complete knowledge of the input vector in order to determine a correct response. In addition, neural network training time is known to scale exponentially with N [93]. However, local minima do not seem to present a significant obstacle when more than the minimal number of hidden units are used. Michael Conrad has noted that evolutionary optimization is easier on a system with redundant variables [27].

The network used for this study had an 'N-($2N+1$)-1' architecture where N represents the number of bits in the input layer. In all the explorations reported in the following binary inputs were represented as 'one' and 'minus one'. In this way we avoid multiplying weights by zero.

The output unit's response to a given input is defined as 'perfect' if it is within .2 of the the target output. A training cycle consisted of testing all 2^N input vectors on a given trial weight vector. Training was continued until 100% of the input vectors had 'perfect' responses. Best, worst and average training times as a function of input vector dimension are shown in Table 6.1. These training times are roughly an order of magnitude longer than back-propagation. Note that the adjustment of weights according to back-propagation is computation intensive, while ours is trivial. A back-propagation cycle requires about twice the amount of computation as our algorithm. Also, Tesauro and Janssens fine-tuned their version of back-propagation by choosing optimal values for the learning rate and momentum parameters for each N. This required considerable trial and error. Altogether they expended several weeks of SUN workstation CPU-time compiling their statistics (Table 6.1). There are indications that training times for our method could be further improved through fine-tuning of the algorithm. For example, one could add a "momentum term" $\mu\Delta\vec{w}(n-1)$, where $\Delta\vec{w}(n-1)$ is the previous weight adjustment. Chemotaxis adjusts the stepsize automatically. Automatic expansion and contraction of net architecture might also help improve convergence (see below).

We also have experimented with back-propagation by substituting it for our own optimization algorithm, minimizing the *sum* of squared errors of *all* test patterns rather than, as Tesauro and Janssens, adjusting weights after each *individual* test pattern presentation. Tesauro and Janssens' performance figures are much better than the results that we obtained in this way with constant learning rate λ and momentum parameter μ. Tesauro and Janssens' method depends critically upon λ and μ. We did not attempt to fine-tune these parameters, which would have been very computation intensive. Our "chemotaxis algorithm," in contrast, is automatic and does not require any tuning.

Back-propagation likewise can be modified to automatically adjust parameters. When we incorporated a rule for adjusting the learning rate, (omitting the momentum term), back-propagation performed slightly better than our chemotaxis algorithm on 4-bit parity.

6.4.2 The 4-bit Encoder Problem

This problem was introduced by Ackley, Hinton, and Sejnowski to demonstrate a learning algorithm for Boltzmann machines [1]. It has since become a standard problem in neural networks (cf. [81, 10, 48, 28]). The net architecture chosen for this study takes the form 4-N-4: four input nodes are connected to N hidden units, which then feed into four output units (in this study N=2, 3, 4). Four input patterns: (0,0,0,1), (0,0,1,0), (0,1,0,0), (1,0,0,0) are then trained to elicit a corresponding set of patterns at the output units. For N less than 4, the network is forced to find a more compact representation of the input vector in order to communicate all of the input patterns to the output layer, thus the name *encoder*.

Twenty training runs were performed for each architecture (4-4-4, 4-3-4, 4-2-4). Weights were updated after each cycle through all four input patterns. Training was continued until the sum of the squared errors had been reduced to 0.1. All runs converged except one run of a 4-2-4 net, which learned only 3 of the four patterns. Average training times are given in Table 6.2. Restriction of the number of hidden units did not appear to present a significant obstacle to the chemotaxis algorithm in this case.

Architecture	Average Training Times	
	(to 1.0)	(to 0.1)
'4-4-4'	259	496
'4-3-4'	352	648
'4-2-4'	328*	963*

Table 6.2: Performance of the Chemotaxis Algorithm on the Encoder Problem with Communication "Bottlenecks." Average performance on the encoder problem as described by Ackley et al. [1]. In this problem, the hidden units are fewer in number than the number of input and output units (forcing coding in the hidden layer). Averages are taken from 20 simulation runs. One run of the 4-2-4 encoder failed to learn all 4 patterns and was omitted from that average.

6.4.3 The Multilayer Encoder

The problem was extended to *several* layers of hidden units [10]. Following Ballard's convention, the architecture chosen consisted of four input units, feeding into four hidden units, etc. (see Figure 6.1). Note that no 'bottleneck' was imposed for simplicity. Performance was averaged over 20 training runs for networks with 1 to 5 layers of hidden units (3 to 7 layers of *nodes*).

The algorithm converged for all runs. Average training times to a squared error of 1.00 are given in Table 6.3 (for 1 and 3 layers of hidden units, errors were further reduced for comparison). Comparison with Ballard's study shows that the chemotaxis algorithm is 2 to 5 times slower than back-propagation for *one* hidden layer. But, *for nets with three or more layers of hidden units, our method matches or outperforms back-propagation.* A comparison with an evolutionary optimization algorithm on this problem is given in [87].

6.4.4 Boltzmann Factor

In the spirit of the Metropolis [68] algorithm and the Boltzmann machine [50, 51, 52, 1, 42, 19, 2, 45] we modified the algorithm so as to allow the occasional acceptance of values of the objective function that were *higher* than the current value in the hope that this would help overcome local minima. Acceptance of higher values was governed by a probability equal to a negative exponential of a "temperature" factor and the difference in values of the objective function (Boltzmann factor). Unfortunately this did not speed up convergence but had detrimental effects, allowing a slow deterioration of the function value, unless the "temperature" factor was virtually zero, which takes us back to the original algorithm.

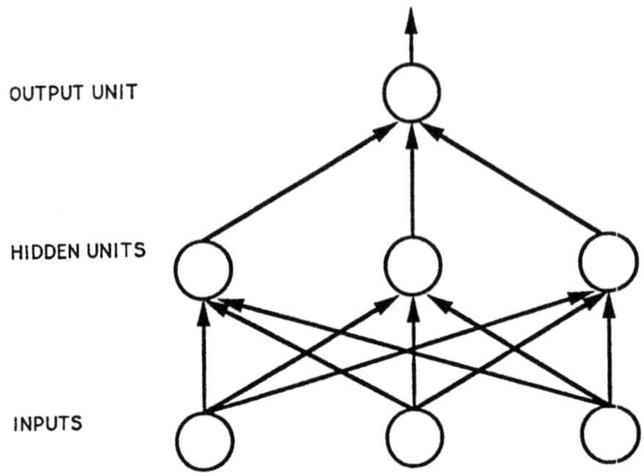

6.1.A

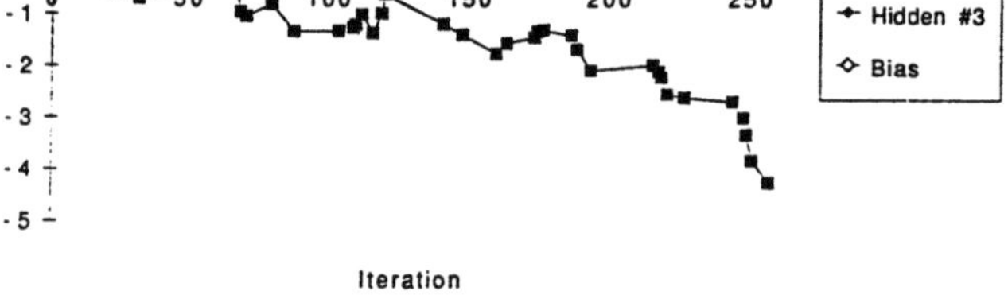

6.1.B

Figure 6.1: Graphical Presentation of the Random Walk in Weight Space. A 3-3-1 network was trained using the 'chemotaxis' algorithm for 3-bit parity. Training continued until the network responses to all 8 patterns were within 10% of their 'target' values. Figure B shows the evolution of the individual weights during training from the three hidden units and bias to the output unit. Each marker identifies a point in training where a successful random step was found.

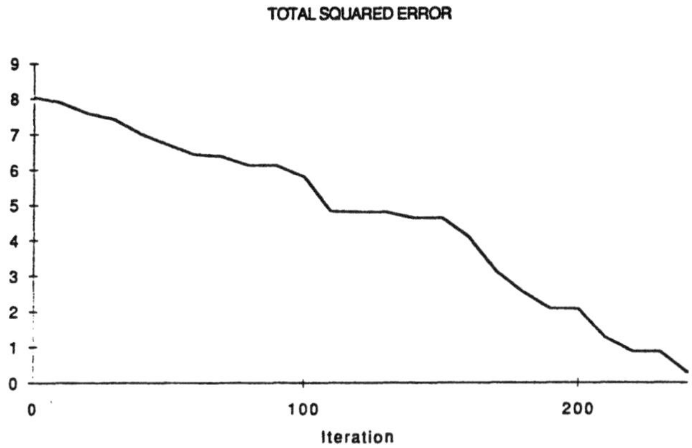

6.1.C

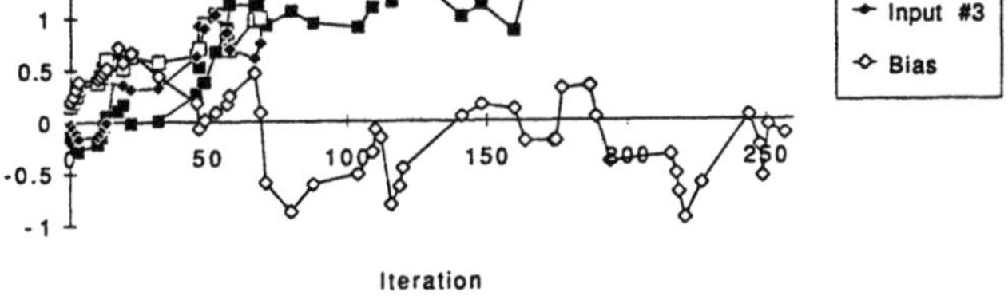

6.1.D

Figure 6.1 Continued. Total system error versus iterations of the algorithm is shown in C. Evolution of the weights from the input nodes to the three hidden units are shown in Figures D-F. Random walks in the weight space of larger nets look similar, but would require too much space for graphical presentation.

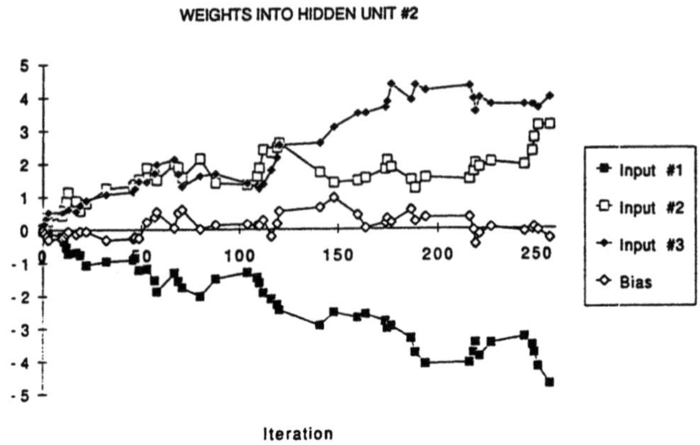

6.1.E

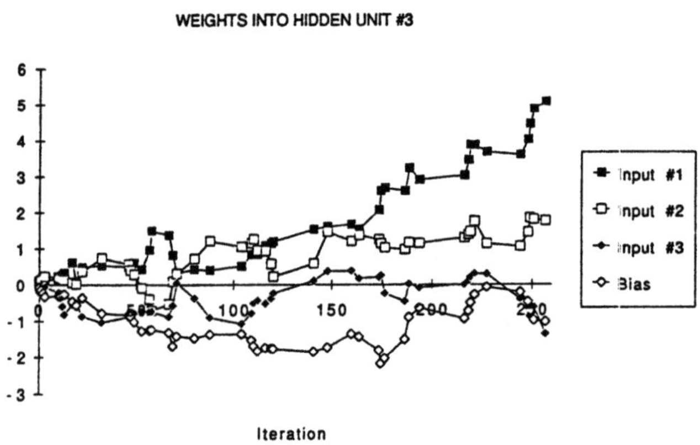

6.1.F

Figure 6.1 Concluded.

Number of Layers of Hidden Units	Chemotaxis Algorithm Training		Back-propagation Training Times to Squared Errors:	
	(to 1.0)	(to 0.1)	(1.0)	(0.1)
1	259	497	50	200
2	584	—	—	—
3	829	1837	2000	3000
4	2404	—	—	—
5	3560	—	—	—

Table 6.3: Comparison of Training Times for the Multilayer Encoder Problem. Averages taken over 20 simulation runs. Back-propagation numbers are estimated from [10].

6.5 Random Walks and Chemotaxis Performance in Other Tasks

Since the tests of the algorithm on the n-bit parity benchmark problem one of us (RWA) has applied the chemotaxis algorithm to other problems: classification of seismic signals [31]; "recurrent" networks (RWA, Ph. D. thesis in preparation); and to the generation of control signals for movements [5]. The results will be reported elsewhere (RWA, Ph. D. thesis in preparation). Of these the seismic classification problem (with no hidden units) is the easiest, and chemotaxis is about an order of magnitude slower than back-propagation.

6.5.1 Recurrent Networks

Recurrent networks allow unconstrained connections between units such as lateral connections and feedback. Such networks can show dynamic behavior and are capable of preserving internal states, self-activation, lateral inhibition/excitation, refractory periods and delays. Pattern recognition in real cortical tissues typically involves transitory states of chaos and coherent oscillations [9, 40, 86]. Clearly, to understand intrinsically sequential neural processes, it is necessary to employ dynamical neural network models. Recent efforts to train recurrent networks to generate or classify temporal patterns (for example adding sequential inputs [94] or memorizing oscillatory patterns [32]) have relied on extensions of the back-propagation learning algorithm [3, 75, 97, 37]. For a review, see Williams and Zipser [97]. Some of these methods require restrictions on network architectures while others place artificial limits on the network dynamics. The most general algorithm, developed by Williams and Zipser, places no constraints on how the network is connected or the dynamics of the neural signals. It does, however, require that gradients be calculated and summed over the duration of the simulation.

One advantage of our algorithm is that it does not require modification for recurrent networks other than in the definition of the objective function. In this respect, *it is a general learning rule.* We have successfully trained networks to oscillate at various frequencies and to mimic a digital "flip-flop" (RWA, Ph. D. thesis).

6.6 Cortical Mappings

6.6.1 Refinement of Coarse Maps

The principles of cortical self-organization have recently been reviewed by Stryker et al. [91]. In this context, topographic maps, especially from the retina to the visual cortex (and other areas of the cortex, as many as 24 in the rhesus monkey), are of special importance. Von der Malsburg and Singer in [78] have suggested that the establishment of the topological map from the retina to target areas in the cortex is "an important paradigm case."

> There are many topological fiber projections between thalamic areas and also between and within cortical areas. Retinotopy is one of the few biological model problems experimentally addressed with enough intensity that theoretical issues can be decided.

The authors propose the following three mechanisms:

- A) There is a mechanism to guide fibers to tectum.
- B) There is a mechanism to position fiber terminals within tectum. This mechanism is responsible for rigid constraints on the mapping.
- C) There is a fiber-sorting mechanism which improves the precision of the mapping over that attained by mechanism B alone and which is activity-dependent.

They further state that "the fiber-sorting mechanism C, since it deals with activity-dependent network organization ... can be generalized to other interesting cases, in particular to neocortex."

A bundle of fibers ascending from the retina, via the optical nerve, through the lateral geniculate, to the visual cortex resembles the "encoder problem."

There are many reverse pathways in the brain [95]. The corticofugal fibers carry information from the cortex to the dorsal lateral geniculate nucleus. Could it be that these fibers return information to the lateral geniculate where it is compared for quality of matching? A topological map is entirely determined by the maps of points and their neighborhoods such that the mappings in neighborhood intersections are consistent. According to postulates A and B by Von der Malsburg and Singer, morphogenesis locates fibres coarsely in correct neighborhoods. The problem thus is to account for refinement of this

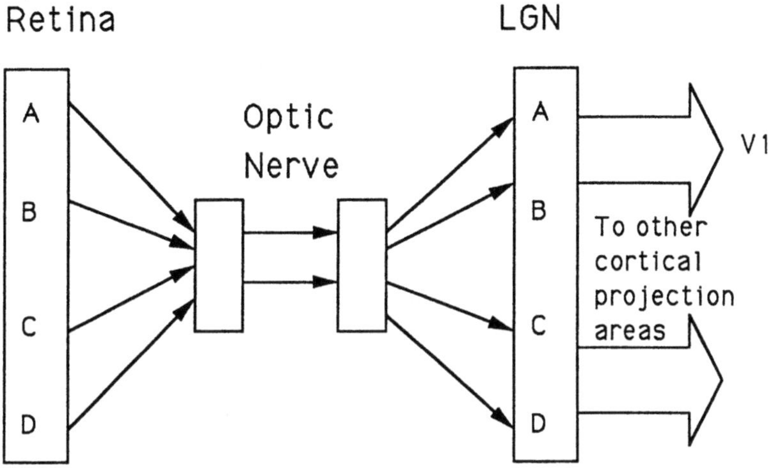

Figure 6.2: Schematic Representation of the Map from the Retina to the Lateral Geniculate Nucleus. Pathways originating from the retina project via the optic nerve through the Lateral Geniculate Nucleus (LGN) on to the primary visual cortex (V1) as well as other cortical projection areas. The mapping on V1 is a point-to-point map of 2-D visual space. This characteristic organization is typical of most mappings in the brain and resembles a multilayer encoder. (Figure adapted from [79].)

map. Suppose a cell in the originating layer creates a ring of inhibition when it fires. Then its projection should be surrounded by a "ring of silence." If the point to point map is not consistent in the projection area, a distorted ring will result. Suppose that such distortion would generate an error signal that is returned to the originating layer. The random walk algorithm could then adjust weights and or connections until the error signal is minimized. Repeated over and over again for all projecting points, the initially coarse map could be improved to the limit of resolution.

Another problem is: how do the reverse fibers "know" exactly where to go? Here is a suggestion for a possible solution of this dilemma: *The projections from the lateral geniculate nucleus to the visual cortex and back to the geniculate are trained as a single encoder.* The input is sent on ascending pathways to the cortex and returned on reverse channels to its origin where the original pattern and its "echo" are compared. This configuration would constitute an encoder with the cortex functioning like a layer of "hidden" units. Initially the two patterns might match only very coarsely. The encoder would then be trained with varying inputs till the identity map becomes better and better.

To ensure that the map is a one-to-one map we suggest that signals originate alternately in the geniculate and are trained to return to the point of origin, and conversely: signals that originate in the projection areas are trained to return to the point of origin in the cortex.

Our model would also be consistent with multiple projections onto cortical areas, each projection being fine-tuned separately. The lateral geniculate would thus be a distribution area for multiple retinotopical maps.

6.6.2 Net Expansion and Contraction

We conducted experiments for n-bit parity (interactively, with simple random walks as well as chemotaxis) where during the training of a net additional units are added or removed. This process does not unduly disturb training progress. In fact, when a task proved too difficult for a net (not enough hidden units), adding units would help achieve ultimate convergence. Also, when training reached a prolonged plateau, the algorithm could be made to converge by training a smaller net (on a simpler version of the problem), then expanding the net and training it on the full problem again.

The algorithm thus is consistent with observations by Merzenich et al. [66, 65], who showed that loss of sensory input leads to reallocation of the vacated space in the cortex.

Acknowledgement. We thank Terrence Sejnowski for helpful comments about an earlier version of this paper. RWA was supported in part by NIH Training Grant Number GM 08155.

References

[1] David H. Ackley, Geoffrey E. Hinton, and Terrence J. Sejnowski (1985): A Learning Algorithm for Boltzmann Machines. Cognitive Science 9, 147–169.

[2] David H. Ackley (1987): A Connectionist Machine for Hillclimbing. Boston: Kluwer Academic Publishers.

[3] L. B. Almeida (1987): A Learning Rule for Asynchronous Perceptrons with Feedback in a Combinatorial Environment. Proc. IEEE First Int'l. Conf. Neural Networks.

[4] Wolfgang Alt (1980): Biased Random Walk Models for Chemotaxis and Related Diffusion Approximations. J. Mathem. Biology 9, 147–177.

[5] Russell W. Anderson and V. Vemuri (1990): Neural Networks Can Be Used For Open-Loop, Dynamic Control. To appear in: International Journal of Neural Networks: Research and Applications.

[6] Russell W. Anderson (1991): Ph. D. dissertation in progress, U. C. Berkeley.

[7] Chiye Aoki and Philip Siekevitz (1988): Plasticity in Brain Development. Scientific American 259(6), December, 56–64.

[8] Scott Austin (1990): Genetic Solutions To XOR Problems. AI Expert, December, 52–57.

[9] Bill Baird (1990): Ph.D. Thesis, U. C. Berkeley.

[10] Dana H. Ballard (1987): Modular Learning in Neural Networks. AAAI National Conference on Artificial Intelligence, 279–284.

[11] Andrew G. Barto, Richard S. Sutton, and Peter S. Brouwer (1981): Associative Search Network: A Reinforcement Learning Associative Memory. J. Biological Cybernetics 40, 201–211.

[12] Andrew G. Barto and Richard S. Sutton (1983): Neuronlike Adaptive Elements That Can Solve Difficult Learning Control Problems. IEEE Transactions on Systems, Man, and Cybernetics SMC–13(5), 835–846.

[13] Jacob D. Bekenstein and Marcello Schiffer (1990): Quantum Limitations on the Storage and Transmission of Information. International Journal of Modern Physics C. (in press).

[14] Howard Berg (1975): How Bacteria Swim. Scientific American 233(2), 36–44.

[15] Howard Berg (1983): Random Walks in Biology. Princeton: Princeton University Press.

[16] W. W. Bledsoe (1961): The Use of Biological Concepts in the Analytical Study of Systems. Technical Report, Panoramic Research Inc., Palo Alto, CA.

[17] W. W. Bledsoe (1961): Lethally Dependent Genes Using Instant Selection. Technical Report, Panoramic Research Inc., Palo Alto, CA.

[18] W. W. Bledsoe (1961): A Quantum-Theoretical Limitation of the Speed of Digital Computers. IRE Trans. Elec. Comp. EC–10 (3).

[19] T. Boseniuk, W. Ebeling, and A. Engel (1987): Boltzmann and Darwin Strategies in Complex Optimization. Physics Letters A 125, 307–310.

[20] Hans J. Bremermann (1958): The Evolution of Intelligence. ONR Technical Report No. 1, Contract Nonr 477(17), University of Washington, Seattle.

[21] Hans J. Bremermann (1962): Optimization Through Evolution and Recombination. In: Yovits, Jacobi, Goldstein, eds.: Self-Organizing Systems. Washington, D. C.: Spartan Books.

[22] Hans J. Bremermann and M. Rogson (1964): An Evolution-Type Search Method for Convex Sets. Technical Report, Contracts Nonr 222(85) and 3656(08), Berkeley, CA.

[23] Hans J. Bremermann, M. Rogson, and S. Salaff (1966): Global Properties of Evolution Processes. In: H. H. Pattee, E. A. Edelsack, Louis Fein, and A. B. Callahan, eds.: Natural Automata and Useful Simulations. Washington, D. C.: Spartan Books. 3–41.

[24] Hans J. Bremermann (1970): A Method of Unconstrained Global Optimization. Mathematical Biosciences 9, 1–15.

[25] Hans J. Bremermann (1974): Chemotaxis and Optimization. J. of the Franklin Institute 297, 397–404. (Special Issue: Mathematical Models of Biological Systems).

[26] David Ceperley and Berni Alder (1986): Quantum Monte Carlo. Science 231, 555–560, 7 Feb.

[27] Michael Conrad (1983): Adaptability. Chapter 10, Plenum Press, N.Y.

[28] Francis Crick (1989): The Recent Excitement about Neural Networks. Nature 337, 129–132, 12 January.

[29] Adele Cutler (1988): Optimization Methods in Statistics. Ph. D. Thesis, Department of Statistics, University of California, Berkeley, CA.

[30] G. Cybenko (1989): Approximation by Superpositions of a Sigmoidal Function. Math. Contr., Signal and Sys. 2, 303–14.

[31] Farid U. Dowla, Steven R. Taylor, and Russell W. Anderson (1990): Seismic Discrimination with Artificial Neural Networks: Preliminary Results with Regional Spectral Data. Bulletin of the Seismological Society of America 80(5), 1346–1373. October.

[32] Kenji Doya and Shuji Yoshizawa (1989): Memorizing Oscillatory Patterns in the Analog Neuron Network. Internat. Joint Conf. Neural Networks. I–27–32, Washington, D. C.

[33] W. Ebeling, A. Engel, B. Esser, and R. Feistel (1984): Diffusion and Reaction in Random Media and Models of Evolution Processes. J. Statistical Physics 37(3/4), 369–384.

[34] G. M. Edelman (1987): Neural Darwinism. New York: Basic Books.

[35] M. Eigen (1988): Macromolecular Evolution: Dynamical Ordering in Sequence Space. In: D. Pines, ed.: Emerging Synthesis in Science. Redwood City, CA: Addison-Wesley. 21–42.

[36] M. Eigen, J. McCasgill, and P. Schuster (1991): Dynamics of Darwinian Molecular Systems. J. Phys. Chem. (in press).

[37] J. L. Elman (1988): Finding Structure in Time. Technical Report 8801, La Jolla: University of California, San Diego, Center for Research in Language.

[38] J. A. Feldman (1981): A Connectionist Model of Visual Memory. In: G. E. Hinton and J. A. Anderson, eds.: Parallel Models of Associative Memory. Hillsdale, N. J.: Erlbaum. 49–81.

[39] Walter Fontana, W. Schnabl, and Peter Schuster (1989): Physical Aspects of Evolutionary Optimization and Adaptation. Phys. Rev. A 40, 3301–21.

[40] Walter J. Freeman (1991): The Physiology of Perception. Scientific American 264(2), 78–85, February.

[41] S. Geman and D. Geman (1984): Stochastic Relaxation, Gibbs Distribution, and Bayesian Restoration of Images. IEEE Transactions on Pattern Analysis and Machine Intelligence 6, 721–741.

[42] S. Geman and D. Geman (1988): In: James A. Anderson and Edward Rosenfeld, eds.: Neurocomputing: Foundations of Research. Cambridge, MA: MIT Press. Reprint of [41].

[43] David L. Glanzman, Eric R. Kandel, and Samuel Schacher (1990): Target-Dependent Structural Changes Accompanying Long-Term Synaptic Facilitation in Aplysia Neurons. Science 249, 799–802, 17 August.

[44] Stephen Grossberg (1988): Neurocomputing: Foundations of Research. In: James A. Anderson and Rosenfeld, eds.: Neurocomputing: Foundations of Research. Cambridge: MIT Press. paper number 24.

[45] H. Haken (1988): Neural and Synergetic Computers. Berlin and Heidelberg: Springer-Verlag.

[46] D. O. Hebb (1949): The Organization of Behavior. New York: Wiley.

[47] G. E. Hinton and J. L. McClelland (1988): Learning Representations by Recirculation. In: D. Z. Anderson, ed.: Neural Information Processing Systems. New York: American Institute of Physics.

[48] G. E. Hinton (1989): Connectionist Learning Procedures. Artificial Intelligence 40(1), 143–150.

[49] John H. Holland (1975): Adaptation in Natural and Artificial Systems. Ann Arbor: University of Michigan Press.

[50] J. J. Hopfield (1982): Neural Networks and Physical Systems with Emergent Collective Computational Abilities. PNAS (USA) 79, 2554–2558, April.

[51] John J. Hopfield and David W. Tank (1985): "Neural" Computation of Decisions in Optimization Problems. Biological Cybernetics 52, 141.

[52] John J. Hopfield and David W. Tank (1986): Computing with Neural Circuits: A Model. Science, 625–633, 8 Aug.

[53] Eric M. Johansson, Farid U. Dowla, and D. M. Goodman (1990): Backpropagation Learning for Multi-Layer Feed-Forward Neural Networks Using the Conjugate Gradient Method, submitted to IEEE Transactions on Neural Networks, Technical Report UCRL–JC–1850, Lawrence Livermore National Laboratory, September 26.

[54] J. A. Kauer, R. C. Malenka, and R. A. Nicoll (1988): NDMA Application Potentiates Synaptic Transmission in the Hippocampus. Nature 334, 250–252. 21 July.

[55] Stuart A. Kauffman and S. Levin (1987): Towards a General Theory of Adaptive Walks on Rugged Landscapes. J. Theoret. Biol. 128, 11–45.

[56] Evelyn Fox Keller and Lee Segel (1970): J. of Theoretical Biology 26, 399.

[57] Mary B. Kennedy (1988): Synaptic Memory Molecules. Nature 335, 770–772, 27 Oct.

[58] Daniel Koshland (1980): Bacterial Chemotaxis as a Model Behavioral System. New York: Raven Press.

[59] S. R. Lehky and Terrence J. Sejnowski (1988): Computing 3-D Curvatures from Images of Surfaces Using a Neural Model. Nature 333, 452.

[60] S. R. Lehky and Terrence J. Sejnowski (1990): Neuronal Model of Stereoacuity and Depth Interpolation Based on a Distributed Representation of Stereo Disparity. Journal of Neuroscience 10(7), 2281–2299, July.

[61] Gary Lynch (1986): Synapses, Circuits, and the Beginnings of Memory. Cambridge, MA: Bradford/MIT Press.

[62] Catherine A. Macken and Alan S. Perelson (1989): Protein Evolution on Rugged Landscapes. PNAS(USA) 86, 6191–5, August.

[63] Catherine A. Macken, Patrick S. Hagan, and Alan S. Perelson (1991): Evolutionary Walks on Rugged Landscapes, SIAM J. Appl. Math., in press.

[64] Bartlett W. Mel (1990): Connectionist Robot Motion Planning. Boston, San Diego: Academic Press.

[65] M. M. Merzenich, G. Recanzone, W. M. Jenkins, T. T. Allard, and R. J. Nudo (1988): Cortical Representational Plasticity. In: [78]. 41–67.

[66] M. M. Merzenich, R. J. Nelson, J. H. Kaas, M. P. Stryker, W. M. Jenkins, J. M. Zook, M. S. Cynader, and A. Schoppman (1987): Variability in Hand Surface Representations in Areas 3b and 1 in Adult Owl and Squirrel Monkeys. J. of Comparative Neurology 258(2), 281–96, April 8.

[67] N. Metropolis and S. Ulam (1949): The Monte Carlo Method. J. Amer. Statistical Association 44(247), 335–341.

[68] N. Metropolis, A. Rosenbluth, M. Rosenbluth, A. Teller, and E. Teller (1953): Equation of State Calculations for Fast Computing Machines. J. of Chemical Physics 21, 1087–1092

[69] M. Minsky (1961): Steps Toward Artificial Intelligence. Proc. IRE. 49, 8–30, Jan.

[70] M. Minsky and S. Papert (1969): Perceptrons: An Introduction to Computational Geometry. Cambridge, Mass.: M. I. T. Press.

[71] D. Montana and L. Davis (1989): Training Feedforward Neural Networks Using Genetic Algorithms. Proc. 11th IJCAI.

[72] Ralph Nossal (1980): Mathematical Theories of Topotaxis. Lecture Notes in Biomathematics 38, Springer-Verlag, 410–439.

[73] A. Okubo (1980): Diffusion and Ecological Problems: Mathematical Models Biomathematics 10.

[74] Alan S. Perelson and Stuart A. Kauffman (1991): Molecular Evolution and Rugged Landscapes: Proteins, RNA and the Immune System, volume IX. Redwood City, CA: Addison-Wesley.

[75] Fernando J. Pineda (1988): Generalization of Backpropagation to Recurrent and Higher Order Neural Networks. Physics Review Letters, 602–611.

[76] T. Poggio and F. Girosi (1990): Regularization Algorithms for Learning That are Equivalent to Multilayer Neural Networks. Science 247, 978–82, 23 February.

[77] Ning Qian and Terrence J. Sejnowski (1988): Predicting the Secondary Structure of Globular Proteins Using Neural Network Models. J. Molec. Biol. 202, 865–884.

[78] P. Rakic and W. Singer (1988): Neurobiology of Neocortex. Dahlem Conferences Report No. 42. New York: Wiley-Interscience.

[79] Anna W. Roe, Sarah L. Pallas, Jong-On Hahm, and Mriganka Sur (1990): A Map of Visual Space Induced in Primary Auditory Cortex. Science 250, 818–20. 9 November.

[80] Frank Rosenblatt (1962): Principles of Neurodynamics. Washington, D. C.: Spartan Books.

[81] David E. Rumelhart, Geoffrey E. Hinton, and R. J. Williams (1986): Learning Internal Representations by Error Propagation In: D. E. Rumelhart and J. L. McClelland, eds.: Parallel Distributed Processing Vol.1, 318–362. Cambridge, MA: MIT Press.

[82] Peter Schuster and K. Sigmund (1985): Dynamics of Evolutionary Optimization. Ber. Bunsenges. Phys. Chem. 89, 668–682.

[83] Peter Schuster and Jorg Swetina (1988): Stationary Mutant Distributions and Evolutionary Optimization. Bulletin of Mathematical Biology 50(6), 635–660.

[84] Terrence J. Sejnowski and Charles R. Rosenberg (1987): Parallel Networks that Learn to Pronounce English Text. Complex Systems 1, 145–168.

[85] Y. A. Shreider (1966): The Monte Carlo Method. Pure and Applied Mathematics 87. Translation from Russian. Oxford: Pergamon Press.

[86] Christine A. Skarda and Walter J. Freeman (1987): How Brains Make Chaos In Order To Make Sense of the World. Behavioral and Brain Sciences 10(2), 161–195.

[87] Robert Smalz and Michael Conrad (1990): A Credit Apportionment Algorithm for Evolutionary Learning with Neural Networks. Dept. of Computer Science, Wayne State University, Detroit (preprint).

[88] Patric K. Stanton and Terrence J. Sejnowski (1989): Associative Long-term Depression in the Hippocampus Induced by Hebbian Covariance. Nature 339, 215–218, 18 May.

[89] Charles F. Stevens (1989): Strengthening the Synapses. Nature 338, 460–461, 6 April.

[90] Lawrence D. Stone (1975): Theory of Optimal Search. New York: Academic Press.

[91] M. P. Stryker, J. Allman, C. Blakemore, J. M. Greuel, J. H. Kaas, M. M. Merzenich, P. Rakic, W. Singer, G. S. Stent, T. N. Wiesel, and H. van der Loos (1988): Group Report: Principles of Cortical Self-Organization. In: [78], 115–136.

[92] G. Tesauro and Terrence J. Sejnowski (1989): A Parallel Network that Learns to Play Backgammon. Artificial Intelligence 39(3), 357–390. July.

[93] Gerald Tesauro and Bob Janssens (1988): Scaling Relationships in Back-propagation Learning. Complex Systems 2, 39–44.

[94] Fu-Sheng Tsung and Garrison W. Cottrell (1989): A Sequential Adder Using Recurrent Networks. Internat. Joint Conf. Neural Networks. II–133–39, Washington, D. C.

[95] D. C. Van Essen (1985): In: A. Peters and F. G. Jones, eds.: Cerebral Cortex Vol. 3, 259–324. New York: Plenum Press.

[96] J. H. Williams, M. L. Errington, M. A. Lynch, and T. V. P. Bliss (1989): Arachidonic Acid Induces a Long-Term Activity-Dependent Enhancement of Synaptic Transmission in the Hippocampus. Nature 341, 739–42, 26 October.

[97] R. J. Williams and D. A. Zipser (1988): A Learning Algorithm for Continually Running Fully Recurrent Neural Networks. Technical Report ICS–8805, University of California, San Diego.

Chapter 7

The Use of Proof Plans for Normalization

Alan Bundy

Department of Artificial Intelligence
University of Edinburgh
Edinburgh, Scotland

Abstract

We propose using proof plans to implement expression normalizers in automatic theorem proving. We outline some general-purpose proof plans and show how these can be combined in various ways to yield some standard normalizers. We claim that using proof plans facilitates the flexible application of these normalizers so that they can interact with the theorem prover in which they are embedded. We intend to extend this technique to decision procedures.

7.1 Introduction

In [1], Boyer and Moore investigate a case study in the use of decision procedures in automated deduction, namely a decision procedure for linear arithmetic. Their conclusions were as follows.

- Such decision procedures have a vital role to play in reducing the combinatorial explosion, but
- they cannot be treated as black boxes. In practice, few sub-goals exactly fit the requirements of a decision procedure, but many almost do.

In these cases it is necessary to augment the decision procedure with additional lemmas. So the operations of the decision procedure must be interleaved with those of the theorem prover, i.e., as well as the theorem prover calling the decision procedure, the decision procedure must be able to call the theorem prover.

- Organising this interaction between decision procedure and theorem prover dominates both the implementation time and the run time of the decision procedure. So the efficiency of the decision procedure itself is of secondary importance.

Boyer and Moore's original implementation took a decision procedure of Hodes, [2], and allowed their theorem prover to call it when a sub-goal fell within its scope. Unfortunately, in a run of their standard corpus of several hundred theorems only a handful of sub-goals were of suitable form and the decision procedure had no significant impact on the performance of the theorem prover.

They then re-implemented Hodes' procedure so that it could interact with the theorem prover. For instance, their decision procedure applies a series of linear rules to its input expression. It can call on an expression rewriter to establish the hypotheses of these linear rules. This more flexible implementation was successfully used many times during a run on their standard corpus. It reduced the overall run-time of the theorem prover by 40%. Of this overall run-time, less than 3% was spent in the decision procedure itself, but 22% was spent in interaction between the decision procedure and the theorem prover. Thus the speed of the decision procedure was of relatively low significance. Far more important was having an implementation that simplified the process of interaction. The changes required to both Hodes' procedure and the theorem prover, in order to integrate them, were extensive and complex, and time-consuming to implement. It would not be a simple matter to modify this decision procedure or to swap it for another one.

In this paper we initiate an investigation into the use of *proof plans*, [3], for implementing decision procedures that might help solve these problems. The hypothesis we are trying to test is that proof plans will provide a modular representation of decision procedures that will facilitate their synthesis and modification and their interaction with the theorem provers in which they are embedded. The cost may be some inefficiency in the implementation of the decision procedures, but the Boyer-Moore experience seems to suggest that this cost is not significant within the context of the complete system.

We focus on the use of proof plans for the implementation of *normalizers*, i.e., procedures for putting expressions into normal form. A normalizer takes an expression belonging to some syntactic class and finds an equivalent (in some sense) expression belonging to another class (usually a strict sub-class of the original). This equivalent expression is the *normal form* of the original expression.

Normalization plays an important role in theorem proving; many proofs include steps in which expressions are put into normal form. In particular, normalization plays an important role in many decision procedures. For instance, some decision procedures consist of the application of a sequence of normalization steps which reduce the original expression to something which is evaluable. So implementing normalizations would solve an important part of the problem of implementing decision procedures. At the end of the paper we will discuss what more is required to extend the ideas to a complete decision procedure, such as the Hodes one. We will restrict the discussion to the implementation of normalizers by the exhaustive or selective application of sets of rewrite rules. A large subset of normalizers can be implemented in this way, and it will simplify the application of proof plans to make this restriction. The cost will be in the efficiency of the implementation, but as discussed above, this cost is less important than is normally assumed in this context. It may be possible to lift the restriction at some future time.

7.2 Proof Plans

A proof plan is a representation of the structure of all or part of a mathematical proof. In [4] we describe the use of proof plans to control the search for proofs of mathematical conjectures. The key idea is that many proofs share common structure, either in whole or in part. We analyse the structure of a family of proofs and identify any common structure. We represent these common structures in general-purpose proof plans, and then use them as blueprints to guide the search for a proof in the same family. A special-purpose proof plan is developed for this new proof by putting together general-purpose proof plans, using techniques of plan formation.

A proof plan consists of two parts: a *tactic* and a *method*. A tactic is a computer program which constructs a proof by applying rules of inference to a conjecture. A method is a partial specification of a tactic. It consists of preconditions for the application of its tactic and a description of the effects of the successful application of the tactic. Both preconditions and effects are written in a meta-logic, whose domain of discourse includes expressions of the object-logic, whose functions manipulate these expressions and whose predicates describe syntactic properties of, or relations between, the expressions. AI plan formation techniques can be used to combine smaller proof plans into larger ones. Proof plans are linked together by inferring the preconditions of later ones from the effects of earlier ones. Plan formation can be used in two modes:

on-line To prove a particular conjecture, the planner can put together some general-purpose proof plans into a special-purpose proof plan, customised for this conjecture. Our planner, CLAM, does this, [4].

off-line To synthesise a new general-purpose proof plan without regard to any particular conjecture. Desimone has built a planner of this form, [5].

In this paper we will outline some general-purpose proof plans which we claim are of general utility in putting expressions into normal form. By instantiating these proof plans with different sets of rewrite rules and by combining them in different ways we can represent a wide variety of different normalizers. To represent the preconditions and effects of these proof plans we will make extensive use of Backus-Naur Form (henceforth, BNF). We will write BNFs using the following notation:

$$Cls \quad := \quad Base_1 | \ldots | Base_k | F_1(Cls_1^1, \ldots, Cls_{m_1}^1) | \ldots | F_n(Cls_1^n, \ldots, Cls_{m_n}^n)$$

which means that expressions of class Cls are either of class $Base_i$ for some $1 \leq i \leq k$ or of the form $F_i(E_1^i, \ldots, E_{m_i}^i)$, for some $1 \leq i \leq n$, where E_j^i is of class Cls_j^i for all $1 \leq j \leq m_i$. Note that F_i need not be a function symbol, but might be a compound expression. In practice, the classes Cls_j^i are usually Cls, but in principle can be different classes.

The effect of most of the proof plans will be to take an expression belonging to one BNF and transform it to an equivalent (in some sense) expression belonging to another one. For instance, the procedure for putting propositional expressions into conjunctive normal form takes expressions of class, fm, defined by:

$$fm \quad := \quad prop | \neg fm | fm \vee fm | fm \wedge fm | fm \rightarrow fm | fm \leftrightarrow fm$$

and puts them into the class, fm', defined by:

$$\begin{aligned} fm' \quad &:= \quad clause | fm' \wedge fm' \\ clause \quad &:= \quad literal | clause \vee clause \\ literal \quad &:= \quad prop | \neg prop \end{aligned}$$

Thus our meta-logic will be mostly concerned with defining such syntactic classes, manipulating them and describing their relationships with each other and with syntactic expressions.

We will discuss both the on-line and off-line synthesis of such proof plans. In off-line synthesis a new general-purpose normalizer will be built for a class of expressions. In on-line synthesis a new normalizer will be custom-built for a particular expression. This normalizer will be able to use any available definitions or lemmas about the symbols in the expression to be normalized. This will constitute our proposed solution to the problem of integrating decision procedures and theorem provers.

7.3 Removing 'Defined' Functions

We will start by considering one of the simplest proof plans: the removal of particular functions[1] with the aid of rewrite rules of the form:

$$F(X_1,\ldots,X_m) \quad \Rightarrow \quad T(X_1,\ldots,X_m) \tag{7.1}$$

where F is a function, the X_i are variables and $T(X_1,\ldots,X_m)$ is a term not containing F. Such rewrite rules might be based on either explicit definitions or lemmas of similar form, and applied left to right or right to left, as appropriate. Examples of such rewrite rules are:

$$\begin{aligned}
p \leftrightarrow q &\quad\Rightarrow\quad p \rightarrow q \wedge q \rightarrow p \\
p \rightarrow q &\quad\Rightarrow\quad \neg p \vee q \\
x \geq y &\quad\Rightarrow\quad y \leq x \\
x \leq y &\quad\Rightarrow\quad \neg y < x \\
\forall x.\, p &\quad\Rightarrow\quad \neg \exists x.\, \neg p
\end{aligned}$$

We will call this proof plan, *remove*. It will take *one* such rule and apply it exhaustively to the current expression until no occurrences of the function F remain. By repeatedly applying *removes* a set of functions can be removed. Care must be taken not to reintroduce an earlier function during the removal of a later one. This can be done by partially ordering the functions in such a way that the right hand side, T, of the remove rewrite rule, (7.1), for function F, contains only functions which are lower in the order than F.

Preventing such loops is the job of the planner using the method of the proof plan. This means that the preconditions and effects of *remove* should note both the function removed and those in terms of which it is removed. This will be done implicitly by the BNFs describing the input and output of the tactic. Recall that we define such BNFs using the notation:

$$Cls \quad := \quad Base_1|\ldots|Base_k|F_1(Cls_1^1,\ldots,Cls_{m_1}^1)|\ldots|F_n(Cls_1^n,\ldots,Cls_{m_n}^n)$$

So to remove a function, F, from an expression, Exp, the *remove* tactic will take two inputs: Exp and the rewrite rule used to remove F, and output an expression, Exp', free of F. The input expression, Exp, will belong to some syntactic class, Cls, containing F, i.e.:

$$Cls \quad := \quad Base_1|\ldots|Base_k|F(Cls_1^F,\ldots,Cls_m^F)|\ldots G(Cls_1^G,\ldots Cls_n^G)|\ldots$$

where the F disjunct has been placed first for notational convenience, but without loss of generality, and the G disjunct represents one of the other

[1] We use the word 'function' here in the lambda calculus sense to include predicates, connectives, quantifiers, etc.

non-base disjuncts. The output expression, Exp', will belong to the syntactic class, Cls', which is Cls with F deleted, i.e.,

$$Cls' \quad := \quad Base_1|\dots|Base_k|\dots G(Cls_1^G\phi,\dots Cls_n^G\phi)|\dots$$

where $\phi = \{Cls'/Cls\}$.

The remove rewrite rule for F must have the form:

$$F(X_1,\dots,X_m) \quad \Rightarrow \quad T(X_1,\dots,X_m)$$

where if X_i is of syntactic class $Cls_i\phi$ for all $1 \le j \le n$ then $T(X_1,\dots,X_m)$ is of syntactic class Cls'.

Note that the above description does not restrict F to be a function symbol. It could be a compound expression.

Hodes' decision procedure for linear arithmetic contains an example of a more complex form of removal. The rewrite rules:

$$\neg x = y \quad \Rightarrow \quad x < y \lor y < x \tag{7.2}$$
$$\neg x < y \quad \Rightarrow \quad x = y \lor y < x \tag{7.3}$$

are used to transform expressions of the form fm, defined by the BNF:

$$\begin{aligned} fm &:= literal|fm \lor fm|fm \land fm \qquad (7.4)\\ literal &:= prop|\neg prop \\ prop &:= tm = tm|tm < tm \end{aligned}$$

into the expressions of the form fm', defined by the BNF:

$$fm' \quad := \quad prop|fm' \lor fm'|fm' \land fm' \tag{7.5}$$
$$prop \quad := \quad tm = tm|tm < tm \tag{7.6}$$

This transformation can be viewed in various ways. The following account establishes a clear connection to *remove*. The input BNF, (7.4), is first rewritten to the equivalent BNF:

$$fm \quad := \quad tm = tm|tm < tm|\neg tm = tm|\neg tm < tm|fm \lor fm|fm \land fm$$

Rewrite rule (7.2) is then used to *remove* $\neg tm = tm$ and rewrite rule (7.3) is used to *remove* $\neg tm < tm$. The new BNF is:

$$fm' \quad := \quad tm = tm|tm < tm|fm' \lor fm'|fm' \land fm'$$

which is equivalent to (7.5). Here we take advantage of the possibility of F being a compound expression.

What might motivate this transformation of the BNF for fm which enables *remove* to work? There seem to be at least three possibilities.

1. We could incorporate transformations of the BNFs into the search space at the meta-level. Unfortunately, this would significantly increase the amount of search.

2. We could put all BNFs into normal form. Unfortunately, it is not clear that we could devise a normal form that would satisfy the preconditions of all the various versions of *remove* that we want.

3. We could devise more powerful preconditions for *remove* that would be transparent to the various equivalent forms in which a BNF might appear. Unfortunately, these preconditions would be hard to write and might only internalise the additional search involved in possibility 1 above.

7.4 Stratification of Classes

Another simple and common proof plan is to stratify a class into a number of layers, with each layer containing only one function. For instance, when converting to conjunctive normal form the class of formulae, fm, defined by:

$$fm \quad := \quad prop|\neg fm|fm \vee fm|fm \wedge fm$$

is stratified into the three layers:

$$\begin{aligned} fm' &:= clause|fm' \wedge fm' \\ clause &:= literal|clause \vee clause \\ literal &:= prop|\neg literal \end{aligned}$$

by moving all $\neg$s beneath all $\vee$s and $\wedge$s and all $\vee$s beneath all $\wedge$s. The fm' layer contains only $\wedge$, the *clause* layer contains only $\vee$, and the *literal* layer contains only $\neg$. This requires two different stratification operations.

Stratification requires a set of rewrite rules. Its object is to move one function, F, beneath each of the other functions in the class. So if the 'before' BNF class is of the form:

$$Cls \quad := \quad Base_1|\ldots|Base_k|F(Cls_1^F,\ldots,Cls_m^F)|\ldots G(Cls_1^G,\ldots Cls_n^G)|\ldots$$

then it will be broken into two 'after' classes of the form:

$$\begin{aligned} Cls' &:= NewBase|\ldots G(Cls_1^G\phi,\ldots,Cls_n^G\phi)|\ldots \\ NewBase &:= Base_1|\ldots|Base_k|F(Cls_1^F\psi,\ldots,Cls_m^F\psi) \end{aligned}$$

where $\phi = \{Cls'/Cls\}$ and $\psi = \{NewBase/Cls\}$.

One distributive law is required for each argument of F, for each of the other functions, G, in the definition of Cls, i.e., for each function, G, and for the ith argument position of F, a distributive law is required of the form:

$$F(X_1,\ldots,G(Y_1,\ldots,Y_n),\ldots,X_m) \quad \Rightarrow \quad T(X_1,\ldots,Y_1,\ldots,Y_n,\ldots,X_m)$$

where $T(X_1, \ldots, Y_1, \ldots, Y_n, \ldots, X_m)$ is in the syntactic class Cls'. Examples of such distributive law sets are the following ones for moving $\neg$ beneath $\vee$ and $\wedge$,

$$\begin{aligned} \neg(p \vee q) &\Rightarrow \neg p \wedge \neg q \\ \neg(p \wedge q) &\Rightarrow \neg p \vee \neg q \end{aligned}$$

$\vee$ beneath $\wedge$,

$$\begin{aligned} p \vee (q \wedge r) &\Rightarrow (p \vee q) \wedge (p \vee r) \\ (q \wedge r) \vee p &\Rightarrow (q \vee p) \wedge (r \vee p) \end{aligned}$$

and $\exists$ beneath $\vee$.

$$\exists x.\,(p \vee q) \Rightarrow \exists x.\,p \vee \exists x.\,q$$

Using these ideas we can design a tactic, *stratify*, which takes an expression and a complete set of distributive laws, applies the rules exhaustively to the expression and returns a stratified expression.

Note that it is usually convenient to precede *stratify* with *remove* since this reduces the number of functions in the syntactic class and, hence, the number of distributive laws required in the *stratify* rule set.

7.5 Reorganising Within a Class

Repeated stratification will often convert the BNF into a number of layers with only one function in each layer. We might then want to reorganise the arguments and applications of this function within the layer. The standard way to do this is by selective applications of laws like commutativity and associativity. There are a variety of ways of doing this according to the arity of the function and which laws are true of it. Below we consider some of the most common cases.

7.5.1 Left(Right) Association

Suppose a layer contains only one function, F, which is both binary and associative. This layer can be put into left or right associative form. Without loss of generality we consider a proof plan for left associative form, *left_assoc*. This proof plan will use the associative law as a rewrite rule in the form:

$$F(X, F(Y, Z)) \Rightarrow F(F(X, Y), Z)$$

Exhaustive application of which will have the effect of converting a BNF class:

$$Cls := Base_1 | \ldots | Base_k | F(Cls, Cls)$$

into the class:

$$Cls' := Base_1 | \ldots | Base_k | F(Cls', Base_1) | \ldots | F(Cls', Base_k)$$

7.5.2 Reordering

If F is also commutative then we can reorder its arguments. Suppose $\prec$ is some order on the members of $Base_1 \cup \ldots \cup Base_k$. We can use the rewrite rules:

$$\begin{aligned} F(X,Y) &\Rightarrow F(Y,X) \\ F(X,F(Y,Z)) &\Rightarrow F(F(X,Y),Z) \\ F(F(X,Y),Z) &\Rightarrow F(X,F(Y,Z)) \end{aligned}$$

selectively, to ensure that Exp is both left associated and for all base elements S and T that $S \prec T$ whenever S occurs before T in Exp.

Note that we cannot apply these three rules exhaustively because the left/right and right/left associativity rules will cause a loop. Finding the right selective ordering is tricky. Such reordering is not usually implemented by the application of rewrite rules. It is usually done by some meta-function that, for instance, puts the arguments into a list and calls a sort algorithm on them. However, in this paper we have restricted ourselves to implementing tactics by the application of rewrite rules. In this case, reordering can be done by a set of 11 commutated associative laws, of which some typical examples are:

$$\begin{aligned} F(T,F(S,A)) &\Rightarrow F(F(S,T),A) \\ F(T,F(A,S)) &\Rightarrow F(F(S,T),A) \\ F(F(T,S),A) &\Rightarrow F(F(S,T),A) \end{aligned}$$

The right hand side of each of these 11 rules is $F(F(S,T),A)$. The 11 different left hand sides are the 6 permutations of $F(T,F(S,A))$ and the 6 permutations of $F(F(T,S),A)$ except $F(F(S,T),A)$. They are to be applied whenever S and T are in $Base_1 \cup \ldots \cup Base_k$ with $S \prec T$ and A is in Cls. Applied in this way they implement a kind of bubble sort. The 11 rules could be automatically generated from the standard associativity and commutativity laws.

We will call this proof plan *reorder*. It will have the effect of converting a BNF class:

$$Cls \ := \ Base_1|\ldots|Base_k|F(Cls,Cls)$$

into the class:

$$Cls' \ := \ Base_1|\ldots|Base_k|F(Cls',Base_1)|\ldots|F(Cls',Base_k)$$

just as with *left_assoc*, but with the additional condition that for all base elements S and T, whenever S occurs before T in any member of Cls', then $S \prec T$. So in this case it is not possible to describe the effect of the proof plan using standard BNF notation alone. To describe the ordering effect we need either an expanded BNF notation or additional meta-formulae.

7.5.3 Thinning

If F is unary and obeys a thinning rewrite rule, i.e., one of the form:

$$F(F(X)) \Rightarrow X$$

then we can remove multiple applications of F from the layer. Examples of such functions are $\neg$ and $-$, which obey the rules:

$$\neg\neg p \Rightarrow p$$
$$--x \Rightarrow x$$

The *thin* proof plan will apply such rules exhaustively. This will convert the BNF form:

$$Cls := Base_1|\ldots|Base_k|F(Cls)$$

into the form:

$$Cls' := Base_1|\ldots|Base_k|F(Base_1)|\ldots|F(Base_k)$$

Thinning could be generalised to the use of any rewrite rule of the form:

$$F^n(X) \Rightarrow F^m(X)$$

where $F^n(X)$ means $F(F(\ldots F(X)\ldots))$, n times, and $m < n$, but the case $n = 2$ and $m = 0$ seems to the only common one.

Thinning can also be seen as removing $\lambda x.F(F(x))$[2], but this requires a hard to motivate transformation of Cls to:

$$Cls := Base_1|\ldots|Base_k|F(Base_1)|\ldots|F(Base_k)|F(F(Cls))$$

so it is unclear whether this view of thinning is helpful.

7.6 Normalization

The proof plans outlined above are rather small scale. To obtain useful normalizers we must combine them together. This can be done in a variety of ways.

Special-Purpose: We can construct special-purpose proof plans, e.g., for conjunctive normal form or polynomial normal form, by instantiating the rewrite rule arguments of our tactics with particular rules and combining the resulting tactics in a fixed combination. For instance, a disjunctive normal form procedure could be constructed by combining *remove* $\leftrightarrow$, *remove* $\rightarrow$, *stratify* $\neg$ over $\vee$ and $\wedge$, *stratify* $\vee$ over $\wedge$, and *thin* $\neg$.

[2] I am indebted to Toby Walsh for this observation.

General-Purpose: We can construct a general-purpose normalizer by combining our tactics together into a super-tactic, but without instantiating their rewrite rule arguments. For instance, we might form a general-purpose super-tactic that applied repeated *removes*, followed by repeated *stratifys*, followed by an appropriate reorganising tactic: *left_assoc*, *reorder* or *thin* to each layer. The preconditions of this super-tactic would have to determine how many repeats of each sub-tactic were required and with what rewrite rules.

Dynamic: We can leave it to the planner to determine how to combine the sub-tactics according to the situation it confronts at run-time. It would use the preconditions and effects of these sub-tactics to determine how best to combine them. See Section 7.10.1 for an example.

Each of these combination techniques is valid and has its role to play. Our CLAM system already employs aspects of each of them in the domain of inductive proofs.

7.7 The Meta-Logic

The preconditions and effects of the proof plans described above will be formulae of a meta-logic in which most of the functions and predicates will manipulate and describe expressions and BNFs. Among the functions and predicates required will be the following:

- $Exp{:}Cls$: means that expression Exp is in syntactic class Cls.
- $in(Disj, Cls)$: means that $Disj$ is one of the disjuncts of class Cls.
- $delete(Disj, Cls)$: is the result of deleting disjunct $Disj$ from Cls.
- $subst(New, Old, Cls)$: is the result of substituting New for Old in Cls.

Using this meta-language, the preconditions of:

$$remove(Exp, F(X_1, \dots, X_m) \Rightarrow T(X_1, \dots, X_m))$$

can be written:

$$\begin{array}{l} \exists Cls, Cls_1, \dots, Cls_m. \\ \qquad\qquad Exp{:}Cls \wedge in(F(Cls_1, \dots, Cls_m), Cls) \wedge \\ \forall X_1{:}Cls_1, \dots, X_m{:}Cls_m.\ T(X_1, \dots, X_m){:}delete(F(Cls_1, \dots, Cls_m), Cls) \end{array}$$

This says that Cls, the syntactic class of Exp, contains a disjunct of the form $F(Cls_1, \dots, Cls_m)$, and that the right hand side of the removal rule belongs to the class formed from Cls by removing this disjunct. We follow the standard naming convention that object-level variables are meta-level constants and that object-level expressions serve as their own meta-level names.

7.8 Linear Algebra

The Hodes decision procedure for linear arithmetic starts by applying a sequence of normalizers, which transform the input expression into a variable free form which can be evaluated. These normalizers can be implemented using the proof plans described above.

By linear arithmetic, we mean the expressions generated by the following BNF.

$$\begin{aligned}
fm &:= prop|\neg fm|fm \vee fm|fm \wedge fm|fm \rightarrow fm|fm \leftrightarrow fm|\exists var.\ fm|\forall var. \\
prop &:= tm = tm|tm < tm|tm > tm|tm \leq tm|tm \geq tm \\
tm &:= var|rat| - tm|tm + tm|tm - tm|rat.tm \\
var &:= x|y|z|\ldots \\
rat &:= nat/nat|(-nat)/nat \\
nat &:= 0|succ(nat)
\end{aligned}$$

where fm stands for formulae, $prop$ for propositions, tm for terms, var for variables, rat for rational numbers and nat for natural numbers.

Hodes' procedure works by eliminating all quantifiers from formulae of class fm, without changing their free variables. If the input formula contains no free variables then the output formula will contain neither free nor bound variables, and so can be evaluated to determine its truth value. At each stage of the procedure an innermost existentially quantified sub-formula is chosen, i.e., a sub-formulae of the form $\exists x.\ p(x)$, where $p(x)$ contains no quantifiers. This existential quantifier is then eliminated by transforming the sub-formulae into an equivalent one not containing x. If necessary, an innermost universally quantified formula is transformed into an innermost existentially quantified formula by removing $\forall$ in favour of $\exists$ and $\neg$.

The transformation of the sub-formula can be described using our proof plans as follows.

1. *remove* is used to remove $\leftrightarrow$, $\rightarrow$, $>$, $\geq$ and $\leq$, in that order.
2. *stratify* is used to move $\neg$ inside $\vee$ and $\wedge$.
3. *thin* is used to eliminate multiple $\neg$s.
4. The extended version of *remove*, described at the end of Section 7.3, is used to remove $\neg$.
5. *stratify* is used to put formulae in disjunctive normal form.
6. *stratify* is used to move $\exists$ inside $\vee$.
7. Various equation and inequality solving procedures are used to eliminate x from the sub-formula, so that the quantifier becomes redundant and can be dropped.

7.9 Equation and Inequality Solving

As can be seen from step 7 of the description above, in order to implement Hodes' linear arithmetic decision procedure using proof plans we will need to devise proof plans for solving equations and inequalities. Fortunately, much of this work has already been done. Our PRESS equation solving program, [6], was implemented using a precursor of proof plans, which we called PRESS *methods*. The PRESS methods of *polysolve*, *collection* and *isolation* are more than adequate for the simple linear equation solving required in Hodes' procedure. They could easily be realised as proof plans. This would require the development preconditions and effects in the meta-language used for the normalization proof plans above. In particular, a normalizer for polynomial normal form, used by *polysolve*, could be implemented using *remove*, *stratify* and *reorder*. PRESS also contains some simple methods for manipulating inequalities, but these will need extending.

7.10 Facilitating Flexible Interaction

In devising proof plans for normalization we have taken trouble to make these as general-purpose as possible. For instance, rather than design separate proof plans for disjunctive normal form and polynomial normal form, we have recognised the common structure underlying these normalizers and represented these as the general-purpose proof plans *remove*, *stratify*, *thin* and *reorder*, from which these special-purpose normalizers can be constructed. The benefit of this organisation is that these general-purpose proof plans can be reused to build normalizers for new domains and can be applied flexibly to take account of additional information. To see this consider the following examples. For explanatory purposes they have been kept as simple as possible and are, as a consequence, rather artificial.

7.10.1 Application to New Domains

The following BNF describes a subset of list processing expressions.

$$tm \quad := \quad nil|var|palindrome(tm)|app(tm, tm)|rev(tm)$$

The existence of the rewrite rule:

$$rev(app(l, k)) \quad \Rightarrow \quad app(rev(k), rev(l))$$

suggests stratification of *app* above *rev*, provided *palindrome* can be removed first. This can be done with the rewrite rule:

$$palindrome(l) \quad \Rightarrow \quad app(l, rev(l))$$

So removing *palindrome* transforms the BNF to:

$$tm \;:=\; nil|var|app(tm,tm)|rev(tm)$$

and stratification transforms it to:

$$\begin{aligned} tm &:= newbase|app(tm,tm) \\ newbase &:= nil|var|rev(newbase) \end{aligned}$$

Finally, multiple applications of *rev* can be thinned with the rewrite rule:

$$rev(rev(l)) \;\Rightarrow\; l$$

to get the BNF:

$$\begin{aligned} tm &:= newbase|app(tm,tm) \\ newbase &:= nil|var|rev(nil)|rev(var) \end{aligned}$$

This example shows that our general-purpose proof plans can be combined to make a normalizer for a new area of mathematics. This could be done on-line, i.e., built up dynamically, in response to a particular normalization problem, using previously proved lemmas or even the hypotheses of the current conjecture, as the required rewrite rules.

7.10.2 Application to Extended Domains

Suppose we have constructed a special-purpose normalizer, *poly_form*, for polynomial normal form, which is designed to transform expressions of the BNF class:

$$tm \;:=\; var|rat|-tm|tm+tm|tm-tm|rat.tm \tag{7.7}$$

into the BNF class:

$$\begin{aligned} tm &:= summand|tm+summand \\ summand &:= var|rat|rat.var \end{aligned}$$

The proof plan, *poly_form*, is not restricted to expressions in the class (7.7). Consideration of the preconditions for *remove* given in Section 7.7 will show that the class of the input expression must have certain properties, but is not, in general, constrained to be a particular class. We design the preconditions of proof plans to be the minimum required for the tactic application to make sense. In particular, *poly_form* will apply to the expression[3] $max(a)+k$, where a is an array of numbers and $max(a)$ is the maximum element of this

[3]Taken from [1].

array. Suppose a is an expression of class $array\_tm$. The BNF of the input expression might have the form:

$$tm \quad := \quad var|rat| - tm|tm + tm|tm - tm|rat.tm|max(array\_tm)$$

The proof plan $poly\_form$, because it does not make a recursive call to tm, will treat $max(array\_tm)$ as if it were an additional base class, which will be carried through the sequence of BNFs, so that $summand$ is defined as:

$$summand \quad := \quad var|rat|rat.var|max(array\_tm)$$

Thus the use of weakest preconditions in proof plans will enable $poly\_form$ to treat terms that lie outside its domain of knowledge. In this case such terms are treated as if they were variables.

This example shows that special-purpose proof plans can be applied outside their intended syntactic class, and will still behave as required.

7.10.3 Bridging Between Proof Plans

If an additional disjunct in the BNF of a class *does* make a recursive call to the class, then it cannot be treated as a base case. For instance, if *tm* were defined as:

$$tm \quad := \quad var|rat| - tm|tm + tm|tm - tm|rat.tm|double(tm)$$

where $double(x) = x + x$, then $double(tm)$ cannot be treated as a base class. Thus the preconditions of $poly\_form$ will fail, so that it cannot be applied. In particular, the parts of its precondition which are inherited from one of its *stratify* sub-proof-plans will fail because the set of distributive laws will not be complete. At this stage the planner will back up and try to plan its way around the blockage. One way that it might find to do this will be to *remove double*, before the calls to *stratify*, using the rewrite rule:

$$double(x) \quad \Rightarrow \quad x + x$$

This example shows that special-purpose proof plans can fail if they are applied to expressions too far outside their intended syntactic class. In this case plan formation can be used, dynamically, to patch holes in the old proof plans. These patches can be composed of any combination of other proof plans. This permits a flexible interaction between normalizers and other parts of the theorem prover. This is similar to the interaction allowed in the Boyer-Moore implementation of the Hodes' procedure, but in a more modular form. In particular, we could add or delete proof plans from the store of those available to the proof planner and it will use the methods of those available to combine them in the most suitable form it can find. No major reprogramming will be required.

7.11 Conclusion

In this paper we have proposed the use of proof plans for representing normalizers, as a first step in using proof plans to represent decision procedures. We have outlined some general-purpose proof plans: *remove*, *stratify*, *left_assoc*, *reorder* and *thin*, from which a wide variety of normalizers can be constructed. The advantages of this method of representing normalizers are as follows:

- We can represent a wide variety of special-purpose normalizers in a uniform way.
- These special-purpose normalizers can be successfully applied to expressions just outside their intended domain of application.
- When such applications fail, these special-purpose normalizers can be dynamically repaired, by inserting new proof plans as patches. This allows a flexible interaction between those proof plans concerned with normalization and those concerned with other aspects of theorem proving.
- New normalizers can be constructed for new domains, either 'on-line', i.e., for a particular expression during a proof, or 'off-line' for a whole new class of expressions.

These advantages go some way to address the issue, raised in [1], of how to integrate decision procedures into theorem provers in a way that facilitates their flexible interaction. In particular, it suggests a modular way to do this that facilitates modification of either decision procedure or theorem prover without major reprogramming.

However, these proposals are still some way from realisation, and the advantages listed above are only predictions. Much remains to be done before they can become a reality.

- The proof plans outlined above must be implemented in a system such as CLAM. Both the tactics and their associated methods must be built. The CLAM meta-language must be extended to cover the description and manipulation of BNFs required in these methods.
- In order to implement decision procedures like that of Hodes, our set of proof plans must be extended. Some of the required proof plans can be adapted from the PRESS equation solver, but others will also be needed.
- Further normalizers and decision procedures must be analysed to discover additional general-purpose proof plans.
- Empirical testing must be conducted to see whether the flexibility gained by implementing normalizers and decision procedures declaratively justifies the increased cost over the usual procedural implementation. It

might be possible to avoid this cost by compilation of the declarative representation into a procedural one.

Nevertheless, even at this early stage this appears to be a very fruitful application of proof plans.

Acknowledgements

The research reported in this paper was supported by SERC grant GR/E/44598 and an SERC Senior Fellowship to the author. I am grateful for feedback on this paper from David Basin and Toby Walsh and for many conversations with other members of the mathematical reasoning group at Edinburgh.

References

[1] R. S. Boyer and J. S. Moore (1988): Integrating Decision Procedures into Heuristic Theorem Provers: A Case Study of Linear Arithmetic. In: J. E. Hayes, J. Richards, and D. Michie, eds.: Machine Intelligence 11, 83–124.

[2] L. Hodes (1971): Solving Problems by Formula Manipulation. Proceedings of the Second IJCAI, The British Computer Society, 553–559.

[3] A. Bundy (1988): The Use of Explicit Plans to Guide Inductive Proofs. Proceedings of CADE-9, R. Lusk and R. Overbeek, eds. Lecture Notes in Computer Science 310, Springer-Verlag, 111–120. Longer version available from Edinburgh as Research Paper No. 349.

[4] A. Bundy, F. van Harmelen, J. Hesketh, and A. Smaill (in press): Experiments with Proof Plans for Induction. Journal of Automated Reasoning. Earlier version available from Edinburgh as Research Paper No. 413.

[5] R. V. Desimone (1989): Learning Control Knowledge within an Explanation-Based Learning Framework. Ph. D. thesis, Dept. of Artificial Intelligence, University of Edinburgh.

[6] L. Sterling, A. Bundy, L. Byrd, R. O'Keefe, and B. Silver (1989): Solving Symbolic Equations with PRESS. J. Symbolic Computation 7, 71–84. Also available from Edinburgh as Research Paper No. 171.

Chapter 8

What Are the Limitations of the Situation Calculus?

M. Gelfond

Department of Computer Science
University of Texas at El Paso
El Paso, Texas

V. Lifschitz

Department of Computer Sciences
and Department of Philosophy
University of Texas at Austin
Austin, Texas

A. Rabinov

Apple Computer, Inc.
Cupertino, California

8.1 Introduction

The situation calculus [8] is a methodology for expressing facts about action and change in formal languages of mathematical logic. It involves expressions for situations and actions (events), and the function *Result* that relates them to each other. The possibilities and limitations of this methodology have never been systematically investigated, and some of the commonly accepted views on this subject seem to be inaccurate.

In this paper we argue that the attractive syntax of the situation calculus is not tied to the primitive ontology that is usually associated with its use,[1] and that many of the usual criticisms of this formalism are unfounded. It is true that most of the recent work on formalizing temporal reasoning in the situation calculus ignores the issues related to the duration of actions, continuous time and concurrency. We believe, however, that this is merely the price that researchers had to pay for their determination to come up with complete solutions, with all details worked out, including those related to the use of nonmonotonic reasoning, and not to be satisfied with sketching a few intuitively appealing axioms that may be applicable to more difficult examples. The relative simplicity of the examples formalized so far in the situation calculus is not a reflection of its intrinsic limitations.

In the first half of the paper (Section 8.2) we review some common criticisms of the situation calculus and present our objections to many of them. In the second half (Section 8.3), we discuss the problem of formalizing commonsense knowledge about the concurrent execution of actions, and show how some possible axioms can be expressed in the situation calculus.

8.2 Some Common Criticisms of the Situation Calculus

Here are examples of what has been said about limitations of the situation calculus. (Some of these issues are discussed below, and the boldface numbers in parentheses point to the relevant subsections of this section.) In [6], the situation calculus is described as the "sequential case," appropriate only for problems that do not involve concurrent actions (**6**). According to [1], it is not clear, for instance, how the situation calculus can be used to formalize the sentence: *I walked to the store while juggling three balls*. "Furthermore, since the action in the situation calculus is equated with change, actions that involve no activity, or restore the world to its original state (e.g., running around a track), cannot be modeled" (**3**). According to [3], the situation calculus cannot deal with "partially ordered events (**1**). In the usual formalizations, it is difficult also to deal with incomplete information about a situation (**1**), and therefore to assimilate new information about the past." The survey paper [12] gives a long list of limitations of the "change-based approach" to temporal reasoning, the situation calculus being the main example of this approach. It asserts that actions in the situation calculus "...do not have any duration (**2**). Sentences such as *The robot should move the vase to the table slowly so that it does not break, but move the book to the table fast to save as much time as possible* cannot be expressed because we cannot

[1]Similar ideas are developed by Schubert[11].

define the notion of performing an action slow or fast" (**2**, **4**). Another alleged limitation is the inability to handle actions with delayed effects: "...there is no way to express in the system the sentence: *30 seconds after you press the button at the crosswalk, the pedestrian light turns to green*" (**5**). The phenomenon of "natural death" is also mentioned in this connection, illustrated by the example: *When you press the button of the hand-drying machine at the airport, it emits hot air for 30 seconds* (**5**). "Simultaneous actions cannot be expressed in the change-based formalism... Similarly, overlapping actions are not expressible" (**6**). Finally, continuous processes are said to cause difficulties, because in the situation calculus "the only notion of time possible is the discrete view, with states corresponding to different time points" (**2**, **4**).

8.2.1 "Unknown" Situations and Actions

Many examples of the use of the situation calculus for formalizing temporal reasoning involve (i) an intitial situation S_0 that is "completely known," in the sense that the axioms specify the initial values of all relevant fluents,[2] and (ii) the result s of executing a completely specified sequence of actions in the initial situation, for instance:

$$s = Result(Move(Block1, Top(Block2)), Result(Move(Block2, Table), S_0)).$$

Using axioms that describe the effects of actions, we may be able then to determine the values of fluents in the situation s. This kind of reasoning is termed "temporal projection" [2].

But there is no reason, of course, why more complex forms of temporal reasoning should not be expressible in the same language. If the initial values of some of the fluents are not given, and the axioms provide the values of some fluents in the situation s instead, then we may be able to use the effect axioms to find the missing initial values. Moreover, in order to refer to a future situation in this language, one does not have to give a complete description of the actions that lead to it. For instance, the formula

$$\exists a[s = Result(a, S_0)]$$

says that s is the result of performing an unknown action a in the situation S_0. We can express that the s is the result of moving *some* block onto the table by writing

$$\exists b[s = Result(Move(b, Table), S_0)].$$

It may be possible to determine the unknown action using given values of fluents in the situations S_0 and s.

[2] A *fluent* is a function defined on the space of situations [8]. For instance, the location of a movable object depends on the situation and hence can be viewed as a fluent.

"Partially ordered events" can be handled in a similar way. The fact that s is the result of moving *Block*1 and *Block*2 onto the table in some unknown order can be expressed by the formula

$$\exists a_1 a_2 [F(a_1, a_2) \wedge s = Result(a_1, Result(a_2, S_0))],$$

where $F(a_1, a_2)$ stands for

$$\begin{array}{l}(a_1 = Move(Block1, Table) \wedge a_2 = Move(Block2, Table)) \\ \vee (a_1 = Move(Block2, Table) \wedge a_2 = Move(Block1, Table)).\end{array}$$

8.2.2 The Duration of Actions

To talk about the duration of an action, we need the concept of time. We can treat time as a fluent; actually, this is one of the first examples of fluents given in [8]. We can think of the values of *Time* as integers or reals, or, in a more abstract way, as elements of a totally ordered group. Here are possible axioms for durations:

$$Duration(a) > 0,$$

$$Time(Result(a, s)) = Time(s) + Duration(a).$$

Then we can have two different actions, $SlowMove(x, l)$ and $FastMove(x, l)$, such that

$$Duration(SlowMove(x, l)) > Duration(FastMove(x, l)).$$

We can even introduce $Move(x, l, t)$, the action of moving x to l within t seconds:

$$Duration(Move(x, l, t)) = t.$$

It can be argued that axioms like these do not adequately express the difference between moving slowly and moving fast, because they say nothing about how uniform the motion is. The idea of constant motion is discussed in Section 8.2.4 below.

8.2.3 The Actions that Involve No Activity

Using the situation calculus for formalizing actions that involve no activity is quite common these days, because such an action, *Wait*, is a part of the celebrated Yale shooting scenario [2]. Notice that the result of waiting, or of running around a track, is not exactly identical to the initial situation, because executing these actions takes time. The action of waiting for t seconds can be defined by the axiom

$$Duration(Wait(t)) = t.$$

It follows then that

$$Time(Result(Wait(t), s)) = Time(s) + t.$$

Appropriate frame axioms (or their nonmonotonic equivalents)[3] will guarantee that some fluents other than *Time*, such as the locations and colors of objects, do not change their values while you wait.

8.2.4 Divisible Actions

Let us assume that the values of *Time* are axiomatized as real numbers. Then infinitely divisible actions can be described using the following three operations:

$a_1; a_2$	— the action which is performed by doing a_1 and then a_2;
$Head(a, t)$	— doing a for t seconds ($t < Duration(a)$);
$Tail(a, t)$	— doing the rest of a (which takes $Duration(a) - t$ seconds).

Here are some of their properties:

$$Result((a_1; a_2), s) = Result(a_2, Result(a_1, s)),$$

$$Duration(a_1; a_2) = Duration(a_1) + Duration(a_2),$$

$$0 < t < Duration(a) \supset Duration(Head(a, t)) = t,$$

$$0 < t < Duration(a) \supset Duration(Tail(a, t)) = Duration(a) - t,$$

$$0 < t < Duration(a) \supset a = Head(a, t); Tail(a, t).$$

$$Head((a_1; a_2), Duration(a_1)) = a_1,$$

$$Tail((a_1; a_2), Duration(a_1)) = a_2.$$

Using this notation, we can define the concept of constant motion. If locations are points in $\mathbf{R}^3$, then the action $CM(x, v, t)$ which moves x with constant velocity v for t seconds can be described as follows:

$$t > 0 \supset Duration(CM(x, v, t)) = t,$$

$$t > 0$$
$$\supset Value(Location(x), Result(CM(x, v, t), s)) = Value(Location(x), s) + v \cdot t,$$

$$0 < t' < t \supset Head(CM(x, v, t), t') = CM(x, v, t').$$

[3] A *frame axiom* is an axiom expressing that some fluent is not affected by some action. For instance, the axiom saying that moving an object does not change its color is a frame axiom. The frame problem is the problem of replacing the frame axioms by a formal expression of the "commonsense law of inertia," according to which a fluent can be assumed by default to remain unchanged when an action is performed. The frame problem has been partially solved using formal nonmonotonic reasoning.

8.2.5 Delayed Effects and "Natural Death"

Typical situation calculus axioms describing the effects of some action a specify the values of fluents in the situation $Result(a, s)$. For instance, the effect of the action *Load* (load the gun) on the fluent *Loaded* is to make it true:

$$Holds(Loaded, Result(Load, s)). \tag{8.1}$$

To describe an action with delayed effects, we need to specify the values of fluents in situations *in the future* of the result of the action. This can be easily done using the notation introduced above. For instance, the sentence "30 seconds after you press the button at the crosswalk, the pedestrian light turns to green" can be represented by the formula

$$Duration(a) = 30 \supset Holds(Green, Result((Press; a), s)).$$

In case of "natural death," the effects are observed during a certain period of time after the resulting situation. The example "When you press the button of the hand-drying machine at the airport, it emits hot air for 30 seconds" can be formalized as follows:

$$Duration(a) \leq 30 \supset Holds(EmitsAir, Result((Press; a), s)).$$

8.2.6 Concurrency

The concurrent execution of actions can be described using the expression $a_1 + a_2$ for the action which is performed by doing a_1 and a_2 concurrently. Here are some properties of this operation:

$$Duration(a_1 + a_2) = \max(Duration(a_1), Duration(a_2)),$$

$$\begin{array}{l} 0 < t < \min(Duration(a_1), Duration(a_2)) \supset \\ \qquad Head(a_1 + a_2, t) = Head(a_1, t) + Head(a_2, t). \end{array}$$

The last formula tells us that when $a_1 + a_2$ is performed, the execution of a_1 and a_2 begins simultaneously.

Partially overlapping actions can be represented by expressions like

$$a_1 + (Wait(t); a_2).$$

To be able to formalize reasoning about concurrent actions, we need axioms describing how the effect of two actions performed simultaneously is related to the effects of these actions performed separately. This is an important unsolved problem in the theory of commonsense reasoning. Some situation calculus axioms that may be useful are discussed in the next section. In this semi-formal discussion of concurrency, we ignore time and the possibility of overlapping actions.

8.3 Axioms for Concurrent Actions

We will use atomic formulas of the form $Causes(a, f, v)$ to express that the action a causes the fluent f to take on the value v. In this notation, introduced in [4], the fact that *Load* makes *Loaded* true can be expressed by the axiom

$$Causes(Load, Loaded, True).$$

The formula (8.1) follows from this axiom and the first of two general "laws of motion":

$$Causes(a, f, True) \supset Holds(f, Result(a, s)), \tag{8.2}$$

$$Causes(a, f, False) \supset \neg Holds(f, Result(a, s)). \tag{8.3}$$

The "commonsense law of inertia" expresses that any change in the value of a fluent is caused by an action:

$$\neg \exists v\, Causes(a, f, v) \supset (Holds(f, Result(a, s)) \equiv Holds(f, s)). \tag{8.4}$$

In this simplified form, the laws of motion and inertia do not take into account that an attempt to perform an action can be unsuccessful. We will return to this question in Section 8.3.4.

8.3.1 Independent Actions

Consider first an example of two concurrent actions whose effects are completely independent.

Example 1. Mary is lifting a bowl of soup from the kitchen table, while John is opening the door to the dining room.

The effects of the actions *Lift* (lift the bowl) and *Open* (open the door) can be expressed by the formulas

$$Causes(Lift, Lifted, True), \tag{8.5}$$

$$Causes(Open, Opened, True). \tag{8.6}$$

The sum of the two actions will "inherit" these two effects, so that we want to have also:

$$Causes(Lift + Open, Lifted, True), \tag{8.7}$$

$$Causes(Lift + Open, Opened, True). \tag{8.8}$$

In more complex examples, it is convenient to distinguish between the case when the effect is caused by the action itself, as in (8.5) and (8.6), and the more general case, when it is possibly inherited from a proper subaction, as in (8.7) and (8.8). We will use the symbol $Causes_0$ in the first case, and *Causes* in the second; the former will be a primitive predicate, and the latter will be an abbreviation, defined in terms of $Causes_0$. The definition will imply that

$$Causes_0(a,f,v) \supset Causes(a,f,v). \tag{8.9}$$

Then our axioms for *Lift* and *Open* will be not (8.5) and (8.6), but

$$Causes_0(Lift, Lifted, True) \tag{8.10}$$

and

$$Causes_0(Open, Opened, True). \tag{8.11}$$

Formulas (8.5)–(8.8) will be theorems. Using the laws of motion (8.2), (8.3), we will be able to conclude, furthermore, that both *Lifted* and *Opened* are true in the situation $Result(Lift + Open, S_0)$.

How will we define *Causes* in terms of $Causes_0$? The simplest possibility is to take $Causes(a,f,v)$ to mean

$$\exists a'[a' \leq a \wedge Causes_0(a',f,v)]. \tag{8.12}$$

Here $\leq$ is the "subaction" relation, which can be defined as usual in lattice theory: $a' \leq a$ stands for $a' + a = a$. This definition says that an action inherits all causal effects of its subactions. In combination with the usual lattice axioms, it would allow us to prove (8.9) and to derive (8.7) and (8.8) from (8.10) and (8.11).

However, this definition eliminates the exceptional cases, when the effect of a subaction of a is "cancelled" by other subactions and is not inherited by a. This phenomenon is discussed in the next section.

8.3.2 Cancellation

Example 2. Whenever Mary tries to lift the bowl with one hand, she spills the soup. When she uses both hands, she does not spill the soup.[4]

Lifting the bowl with two hands can be viewed as lifting it with the right hand and with the left hand simultaneously:

$$Lift = Lift\text{-}left + Lift\text{-}right. \tag{8.13}$$

As before, the effect of *Lift* on *Lifted* is described by (8.10); in addition, we assume:

$$Causes_0(Lift\text{-}left, Spilled, True) \tag{8.14}$$

and

$$Causes_0(Lift\text{-}right, Spilled, True). \tag{8.15}$$

[4]This is essentially Pednault's example of two agents lifting the opposite ends of a table [9].

But we do not want the sum of the two actions to inherit this effect from any of them; we do not want to get the conclusion *Causes*(*Lift*, *Spilled*, *True*).

The action *Lift* can be said to "cancel" the effects of its proper subactions on *Spilled*:

$$Cancels(Lift, Spilled). \tag{8.16}$$

The definition (8.12) of *Causes* needs to be modified to take into account the possibility of the cancellation of the effect of a' on f. The improved definition says that a causes a fluent f to take on a value v, if this is an immediate causal effect of a subaction a' of a which is not cancelled by any intermediate subaction a'':

$$\begin{aligned}&Causes(a, f, v)\\ &\quad \equiv \exists a'[a' \leq a \wedge Causes_0(a', f, v) \wedge \neg\exists a''(a' < a'' \leq a \wedge Cancels(a'', f))].\end{aligned} \tag{8.17}$$

With the new definition of *Causes*, (8.9) will still follow (take a' to be a).

If a is postulated to affect f, then we may wish to disregard the possible effects of any subactions of a on f. This can be achieved by the axiom

$$Causes_0(a, f, v) \supset Cancels(a, f). \tag{8.18}$$

Axioms like (8.14)–(8.16) and (8.18) give positive information about the predicates *Causes* and *Cancels*. Instead of including negative facts about these predicates in the axiom set, we can postulate the closed world assumption for them, using circumscription [7] or some other nonmonotonic formalism.

In Example 2 formalized along these lines, we will not be able to prove that *Spilled* is false in the situation *Result*(*Lift*, S_0); perhaps the soup has been spilled earlier. But the law of inertia (8.4) will imply that the value of *Spilled* in this situation is the same as its value in the initial situation S_0.

8.3.3 Conflicting Subactions

Example 3. John wants to go for a walk, and is trying to open the door; Mary wants him to stay home, and is trying to close the door.

The effects of the actions *Open* and *Close* are described by axioms (8.11) and

$$Causes_0(Close, Opened, False). \tag{8.19}$$

Assuming that the effects of the subactions of *Open* + *Close* on *Opened* are not cancelled, we can conclude:

$$Causes(Open + Close, Opened, True), \tag{8.20}$$

$$Causes(Open + Close, Opened, False). \tag{8.21}$$

These formulas are inconsistent with our formulation of the laws of motion (8.2), (8.3).

In order to accomodate examples of this kind, we need to formalize the idea of *conflicting effects* and make the laws of motion inapplicable to such cases.

It is clear that Example 3 has something in common with Example 2, where an action cancels the effects of its subactions. In both cases, the effect of the sum of two subactions is not the same as the sum of its effects. But there is an important difference. In the cancellation example, we wanted the effect of the action to be determined by the law of inertia; the value of *Spilled* remained *unchanged*. In the conflict example, we would like the outcome to be *undecided*; we do not know whether Mary or John is going to prevail, and the formalization of the problem should reflect this incompleteness.

We will represent the fact that an action a has conflicting effects on the fluent f by writing $Conflict(a, f)$. In case of truth-valued fluents, the existence of conflicts can be proved using the axiom

$$Causes(a, f, False) \wedge Causes(a, f, True) \supset Conflict(a, f). \tag{8.22}$$

For instance, (8.20), (8.21) and (8.22) imply

$$Conflict(Open + Close, Opened).$$

More generally, we can postulate

$$Causes(a, f, v_1) \wedge Causes(a, f, v_2) \wedge v_1 \neq v_2 \supset Conflict(a, f).$$

In the new formulation of the laws of motion, the formula $\neg Conflict(a, f)$ is included in the antecedents of both axioms as an additional assumption:

$$Causes(a, f, True) \wedge \neg Conflict(a, f) \supset Holds(f, Result(a, s)), \tag{8.23}$$

$$Causes(a, f, False) \wedge \neg Conflict(a, f) \supset \neg Holds(f, Result(a, s)). \tag{8.24}$$

Axioms like (8.22) give sufficient conditions for *Conflict*. In order to be able to prove the absence of conflicts, the closed world assumption needs to be applied to this predicate.

In Example 3 formalized in this way, neither the laws of motion nor the law of inertia will be applicable, and the value of *Opened* in the situation $Result(Open + Close, S_0)$ will remain undecided.

8.3.4 Qualifications

So far, we have ignored the fact that the success of an action may be qualified. What happens when two actions are performed concurrently, and one of them is not successful?

To answer this question, we need to distinguish between two ways in which an action can be unsuccessful in a given situation. First, it can be physically impossible. Second, it can be unsuccessful in the weaker sense of failing to produce a particular effect.[5]

We can illustrate this difference by the familiar example of the action *Shoot* [2]. If we take the word "shoot" to mean "send forth bullets from a gun," then *Shoot* is unsuccessful in the strong sense unless the gun is loaded. Alternatively, we can understand *Shoot* as "pull the trigger." Then performing this action when the gun is not loaded will be unsuccessful only in the weak sense, relative to the fluent *shot*. It will be still unsuccessful in the strong sense, however, if the gun is locked in a safe.

If a is unsuccessful in a situation s in the strong sense, then it would be natural to treat the expression $Result(a, s)$ as undefined, like $1/0$ in arithmetic. However, the standard semantics of classical logic presumes that every function constant represents an everywhere defined function. One way out is to single out "fictional" elements in the domain of situation variables, that do not correspond to actual physical situations. We can introduce, for instance, the unary predicate *Nonexistent*, and indicate the impossibility of doing a in a situation s by writing $Nonexistent(Result(a, s))$.

If, on the other hand, a is physically possible in s, but unsuccessful in the weak sense relative to some fluent f, then the term $Result(a, s)$ represents a physically possible situation, and we want to be able to conclude, by the law of inertia, that the value of f in this situation is the same as its value in the situation s.

The difference between the two meanings of the word "unsuccessful" is crucial for the question posed at the beginning of this section. If an action is physically impossible, then, in general, performing it concurrently with another action will not be possible either. But if an action a_1 merely fails to affect a certain fluent, then the effect of $a_1 + a_2$ on that fluent will generally coincide with the effect of a_2 on it.

8.3.5 Unknown Actions and Concurrency

In conclusion, we would like to discuss a form of reasoning in which known and unknown actions are performed concurrently.

Example 4. Mary picked up a bowl of soup from the table and noticed that the door to the dining room, closed a moment before, was open. She concludes that John opened the door while she was not looking.

In order to formalize this example in a theory based on the situation calculus, we propose to extend the language by an action constant ϵ, representing the unobserved component of the action that has led to the new situation. Then the available facts can be expressed by the formulas

$$\neg Holds(Opened, S_0)$$

[5]This is similar to the distinction between "primary" and "secondary" preconditions in [10].

and

$$Holds(Opened, Result(Lift + \epsilon, S_0)).$$

In the presence of the law of inertia (8.4), these axioms imply

$$\exists v\ Causes(Lift + \epsilon, Opened, v).$$

If *Open* is assumed to be the only action satisfying the condition

$$Causes_0(a, Opened, True),$$

then we will be able to conclude that $\epsilon \geq Open$. Thus opening the door is at least one of the actions performed simultaneously with *Lift*.

How can we express the assumption that ϵ is the "most economical" explanation of the event described in Example 4? Intuitively, this assumption would lead to the conclusion that $\epsilon = Open$. One possible approach is to "minimize" ϵ by circumscribing the predicate $\lambda a(a \leq \epsilon)$.

It is interesting to compare this discussion of Example 4 with the treatment of similar examples based on the theory of miracles [5]. In that theory, the symbol *Result* is given a somewhat different meaning; *Result*(a, s) stands there for a situation that results from s when *a set of concurrent actions* is performed that includes a and *by default* nothing else. The theory of miracles can be viewed as a weak theory of concurrent actions, in which unobserved actions cannot be named explicitly. In our theory, "miracles" corresponds to the cases when $\epsilon \neq \emptyset$.

Acknowledgements

We are grateful to Robert Boyer, Michael Georgeff, Fangzhen Lin, Norm McCain, John McCarthy, Leora Morgenstern, Ray Reiter, Johanna Seibt, Murray Shanahan and Yoav Shoham for useful discussions on the subject of this paper and for comments on preliminary drafts. The first author acknowledges the support of NSF grant IRI-89-06516. The second and third authors started their work on this paper when they were affiliated with Stanford University and were supported in part by NSF grant IRI-89-04611 and by DARPA under Contract N00039-84-C-0211.

References

[1] James Allen (1984): Towards a General Theory of Action and Time. Artificial Intelligence 23, 123–154.

[2] Steve Hanks and Drew McDermott (1987): Nonmonotonic Logic and Temporal Projection. Artificial Intelligence 33(3), 379–412.

[3] Robert Kowalski and Marek Sergot (1986): A Logic-based Calculus of Events. New Generation Computing 4, 67–95.

[4] Vladimir Lifschitz (1987): Formal Theories of Action (preliminary report). Proc. IJCAI-87, 966–972.

[5] Vladimir Lifschitz and Arkady Rabinov (1989): Miracles in Formal Theories of Actions. Artificial Intelligence 38(2), 225–237.

[6] John McCarthy (1977): Epistemological Problems of Artificial Intelligence. Proc. IJCAI-77, 1038–1044.

[7] John McCarthy (1986): Applications of Circumscription to Formalizing Common Sense Knowledge. Artificial Intelligence 26(3), 89–116.

[8] John McCarthy and Patrick Hayes (1969): Some Philosophical Problems from the Standpoint of Artificial Intelligence. In: B. Meltzer and D. Michie, eds.: Machine Intelligence 4. 463–502. Edinburgh: Edinburgh University Press.

[9] Edwin Pednault (1987): Formulating Multi-agent, Dynamic World Problems in the Classical Planning Framework. In: Michael Georgeff and Amy Lansky, eds.: Reasoning about Actions and Plans. 47–82. San Mateo, CA: Morgan Kaufmann.

[10] Edwin Pednault (1988): Extending Conventional Planning Techniques to Handle Actions with Context-dependent Effects. Proc. AAAI-88, 55–59.

[11] Len Schubert (1990): Monotonic Solution of the Frame Problem in the Situation Calculus: an Efficient Method for Worlds with Fully Specified Actions. In: H.E. Kyberg, R.P. Loui, and G.N. Carlson, eds.: Knowledge Representation and Defeasible Reasoning. Boston: Kluwer Academic Press. 23–67.

[12] Yoav Shoham and Nita Goyal (1988): Temporal Reasoning in AI. In: Howard Shrobe, ed.: Exporing Artificial Intelligence. San Mateo, CA: Morgan Kaufmann.

Chapter 9

Reasoning In Paraconsistent Logics

James J. Lu
Lawrence J. Henschen

Northwestern University
Evanston, Illinois

V. S. Subrahmanian

University of Maryland
College Park, Maryland

Newton C. A. da Costa

University of Sao Paulo, Brazil

9.1 Introduction

Databases and knowledge bases could be inconsistent in many ways. For example, if a programmer is constructing an expert system ES relating to a domain D, he does so by consulting several experts (say n in number) in the field D. From each expert d_i, $1 \leq i \leq n$, he obtains some information, and this may be represented in logic as a collection of sentences, ES_i, for

$1 \leq i \leq n$. The simplest way of combining the resulting knowledge is to set

$$ES = \bigcup_{i=1}^{n} ES_i.$$

However, experts ES_i and ES_j may *disagree* on some proposition p. In that case, p may be a logical consequence of ES_i, while $\neg p$ is a logical consequence of ES_j. As a consequence, in classical logic, our expert system ES would be inconsistent, and hence meaningless by virtue of its lack of models.

Nevertheless, our point is that ES is *not* meaningless. Some arguments in favour of this point of view are:

1. A large subset of ES may well be consistent and convey meaningful information. This meaningful information must not be lost.

2. The very existence of a disagreement between experts ES_i and ES_j may be meaningful. For instance, if doctor D_1 diagnoses patient X as being fatally afflicted with cancer, while doctor D_2 also diagnoses X as being afflicted by cancer, but by a benign variety of cancer, then X, the patient, would like to know about this disagreement. The *very existence* of this disagreement is meaningful because it may be cause for X to take appropriate action (seek a third medical opinion, etc.).

The point being made in (2) above is that finding ad-hoc means of discarding formulas identified as the cause(s) of the inconsistency in ES is not always appropriate because it may often obliterate important information. In many cases, the existence of the inconsistency is itself important.

In the late 1950's, a thorough systematic characterization of the logics of inconsistency was conducted by Newton da Costa [7, 8]. da Costa introduced a series of logics, known by the generic name *paraconsistent logics*, which formalize schemes for mechanical reasoning in inconsistent formal systems.

Annotated logics are a family of paraconsistent logics that were proposed initially by Subrahmanian [23]. Subsequently, they were studied, from the point of view of logic programming by Blair and Subrahmanian [3] and from a foundational viewpoint by da Costa, Subrahmanian, and Vago [10]. Simultaneously, Kifer and Lozinskii [17] extended the annotated logic framework by the addition of new logical symbols. They have also been applied towards the development of a declarative semantics for inheritance networks [16] and object-oriented databases [18]. In view of the applicability of annotated logics to these differing formalisms in computer science, it seems necessary to study these logics more carefully. In this paper, we develop a resolution style proof procedure (cf. [19, 20]) that is proven to be sound and complete for theorem proving in finitely axiomatized paraconsistent theories.

The logics we study are multivalued logics based on a complete lattice of truth values. In other words, given a complete lattice $\mathcal{T}$, and any unary function $\neg : \mathcal{T} \to \mathcal{T}$, we show how to define a paraconsistent logic, denoted $\mathbf{Q}\mathcal{T}$. As such, the theoretical framework is robust in that it applies to a large class of multivalued logics.

The organization of this paper is as follows. In Section 9.2, we first describe the syntax and semantics of annotated logics. (Subsection 9.2.1 basically recapitulates the basic syntax and model theory of annotated logics developed in [10, 4] and does not constitute new work.) We also present the inference rule p-resolution, which is shown to be a refutation complete resolution proof procedure for annotated logics in clause form. We discuss an implementation of p-resolution in Section 9.3. A modified version of p-resolution has also been implemented. Reports on the experimental results of the two versions of p-resolution are given. Finally in Section 9.4, we introduce an extension to clausal annotated logics which facilitates parallel search.

9.2 Paraconsistent Resolution in Annotated Logics

9.2.1 Syntax and Semantics

Throughout this paper, we assume that $\mathcal{T}$ is some arbitrary, but fixed, complete lattice of objects called *truth values* with the associated ordering $\leq$. The least and greatest element of $\mathcal{T}$ are denoted, respectively, by the symbols $\bot$ and $\top$. We also assume that we have a language L consisting of finite sets of constant symbols, functions symbols (with associated arities), and predicate symbols (with associated arities). L also contains infinitely many variable symbols, together with the binary connectives $\&, \vee, \Leftarrow$ and the unary connectives $\neg, \sim$, and two quantifiers, $\forall$ and $\exists$. Atoms are defined in the usual way (cf. Chang and Lee [5]).

Definition 1 An *interpretation* I for the language L consists of a non-empty domain, $dom(I)$, and:

1. a function η_I that maps constants in L to $dom(I)$ and
2. a function ζ_I that assigns to each n-ary function symbol ρ in L a function from $(dom(I))^n$ to $dom(I)$ and
3. a function κ_I that assigns to each n-ary predicate symbol in L a function from $(dom(I))^n$ to $\mathcal{T}$.

Suppose I is an interpretation for L. A *variable assignment* v for L w.r.t. I is a map from the set of variable symbols of L to $dom(I)$. The *denotation*, $d_{I,v}(t)$, of a term t (w.r.t. interpretation I and variable assignment v) is defined as follows:

1. If t is a constant symbol, then $d_{I,v}(t) = \eta_I(t)$.
2. If t is a variable symbol, then $d_{I,v}(t) = v(t)$.
3. If $t = f(t_1, \ldots, t_n)$, then $d_{I,v}(t) = (\zeta_I(f))(d_{I,v}(t_1), \ldots, d_{I,v}(t_n))$

For a tuple $(t_1, \ldots, t_n) \in \mathcal{T}^n$, we extend the denotation naturally as:

$$d_{I,v}(t_1, \ldots, t_n) = (d_{I,v}(t_1), \ldots, d_{I,v}(t_n))$$

Definition 2 If A is an atom, and $\mu \in \mathcal{T}$, then

1. $A : \mu$ is an *annotated atom.*
2. $\underbrace{\neg \cdots \neg}_{k \text{ times}}(A : \mu)$ is called a *hyper-literal* of order k, for $k \geq 0$ and
3. $\underbrace{\sim \cdots \sim}_{k \text{ times}}(A : \mu)$ is called a *mega-literal* of order k, $k \geq 0$.

These notations are the same used in previous works on annotated logic.

Definition 3 Every hyper-literal is a formula. Suppose F, G are formulas and x is a variable symbol. Then: $\sim (F)$, $(F \,\&\, G)$, $(F \vee G)$, $(F \Leftarrow G)$, $(\forall x)F$, and $(\exists x)F$ are all formulas. We will often abuse notation and omit parentheses whenever possible.

Note that our logic contains two species of negation: the square negation $\neg$ which is interpreted as a lattice function, and a squiggly negation $\sim$ which is interpreted in a more classical way (cf. Definition 4 below). Thus even if $A : \mu$ and $B : \mu'$ are annotated atoms, $\neg(A : \mu \,\&\, B : \mu')$ is not a formula, but $\sim (A : \mu \,\&\, B : \mu')$ is.

Definition 4 *(Satisfaction)* Suppose I is an interpretation for L, and v is a variable assignment w.r.t. I. In the sequel, A is any atom, $\mu \in \mathcal{T}$, and F, G, H are formulas. In addition, we assume that $\neg$, when written in front of an annotation, represents some arbitrary, but fixed, unary function from $\mathcal{T}$ to $\mathcal{T}$.

1. $I, v \models (A : \mu)$ iff $A = p(t_1, \ldots, t_n)$ and $\kappa_I(p)(d_{I,v}(t_1, \ldots, t_n)) \geq \mu$.
2. $I, v \models \neg^k(A : \mu)$, $k \geq 1$ iff $I, v \models \neg^{k-1}(A : \neg\mu)$.
3. $I, v \models (F \,\&\, G)$ iff $I, v \models F$ and $I, v \models G$.
4. $I, v \models (F \vee G)$ iff $I, v \models F$ or $I, v \models G$.
5. $I, v \models (F \Leftarrow G)$ iff $I, v \models F$ or $I, v \not\models G$.
6. $I, v \models \sim F$ iff $I, v \not\models F$.

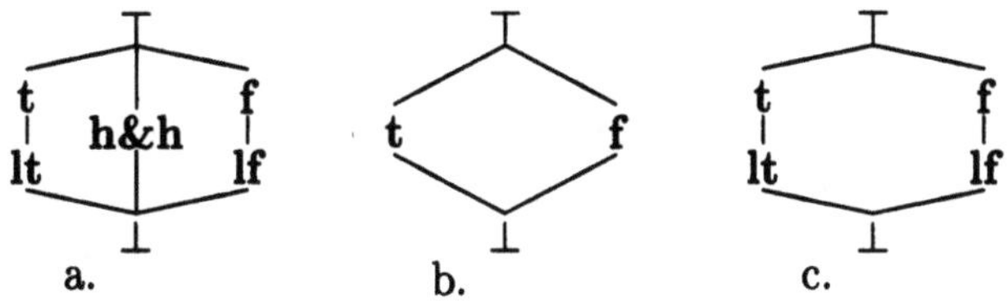

Figure 9.1: The Complete Lattices *SEVEN*, *FOUR*, and *SIX*

7. $I, v \models (\exists x)H$ iff for some variable assignment v' such that for all variable symbols y different from x, $v(y) = v'(y)$, it is the case that $I, v' \models H$.

8. $I, v \models (\forall x)F$ iff for all variable assignments v' such that for all variable symbols y different from x, $v(y) = v'(y)$, it is the case that $I, v' \models H$.

We say $I \models H$ iff for all variable assignments v, $I, v \models H$.

The philosophical motivation behind the negation symbol $\neg$ can be found in [10]. However, as Theorem 1 shows (Normal Form Theorem), $\neg$ is not necessary for the development of our theory.

A recursively enumerable axiomatization of the *propositional logic* fragment of $\mathbf{Q}\mathcal{T}$ was proven sound and complete in [10]. However, this axiomatization is not amenable to a computer implementation because it contains sixteen rules of inference. Many theorem-proving researchers would agree that the more rules of inference an automated reasoning program is allowed to apply, the more "irrelevant" inferences it will generate, thus leading to the notorious combinatorial explosion problem. Thus, our aim in this paper is to develop a resolution-like proof procedure for paraconsistent logics which is amenable to a computer implementation. Moreover, this procedure will be applicable to first order $\mathbf{Q}\mathcal{T}$ rather than just its propositional fragment. We end this section with an example that shows some rules written in annotated logic.

Example 1 The lattice *SEVEN* shown in Figure 9.1a can be described informally as follows. $\top$ and $\bot$ indicate respectively, "inconsistent" (w.r.t. the intuition of classical two valued logic) and "unknown". **t**, **lt**, **f**, and **lf** denote true, likely to be true, false, and likely to be false, respectively. The meaning of the truth value **h & h** is a bit more subtle. Intuitively, a proposition p : **h & h** says the likelihood of p being true is the same as the likelihood of p being false. Note that this is not the same as saying p is unknown, written $p : \bot$, which is less informative than p : **h & h**. The justification for **h & h** being a separate truth value from $\bot$ can best be explained by a brief example. If an expert in meteorology states that the likelihood of rain tomorrow is about 50 percent, or half and half (hence the notation **h & h**), he is presumably basing his forecast on meteorological evidence, say for example, there is a weak cold front coming through. Although one could argue that the expert's prediction is not very useful in that it does not tell us much about how the weather might turn out tomorrow, few would disagree that his statement conveys meaningful information and that to be cautious, one might pack an umbrella before leaving the next day. On the other hand, if a person who has absolutely no knowledge about meteorology says that he does not know what the likelihood for rain is tomorrow, in some sense, he is telling us as much about tomorrow's weather as the expert, but certainly one would not expect his statement to be a sufficient cause for packing rain gear.

The intuition behind the ordering of the truth values in *SEVEN* is that first of all, true and false are incompatible with respect to our normal intuition. Secondly, true is more informative than likely to be true, hence true is "greater" than likely to be true in the ordering. Similarly, false is "greater" than likely to be false. Thirdly, saying that a proposition p is true or likely to be true is quite different from saying p is just as likely to be false as it is true. Hence the truth value **h & h** is incompatible with either **t** or **lt** and is also incompatible with **f** or **lf**. It follows, based on this intuition, that the meaning of negation $\neg$ in *SEVEN* is the function that assigns **t** to **f**, **f** to **t**, **lt** to **lf**, and **lf** to **lt**. All other truth values are assigned themselves.

Our example consists of a small medical knowledge base. Suppose the construction of this knowledge base involved consulting two doctors, DOC_1 and DOC_2. The knowledge base is aimed at diagnosing three diseases d_1, d_2 and d_3. Suppose DOC_1 provided us with the following general rules:

1. If an individual tests positive for symptom s_1 and negative for s_2, then he or she is likely to have disease d_2.

2. No individual is likely to have both diseases d_1 and d_3.

These rules translate into the following formulae.

R1 :	$(\forall X)(d_2(X) : \mathbf{lt} \Leftarrow s_1(X) : \mathbf{t} \,\&\, s_2(X) : \mathbf{f})$
R2 :	$(\forall X)(d_1(X) : \mathbf{lf} \Leftarrow d_3(X) : \mathbf{lt}) \,\&\, (d_3(X) : \mathbf{lf} \Leftarrow d_1(X) : \mathbf{lt}))$

On the other hand, suppose DOC_2 gave the following general rules:

1. If a patient tests negative for s_2 and positive for s_3, then the likelihood of his or her having disease d_2 is about 50 percent.

2. If a patient tests positive for s_4, then it is likely that the patient has disease d_1.

The translations of these rules are given below.

R3 :	$(\forall X)(d_2(X) : \mathbf{h} \,\&\, \mathbf{h} \Leftarrow s_2(X) : \mathbf{f} \,\&\, s_3(X) : \mathbf{t})$
R4 :	$(\forall X)(d_1(X) : \mathbf{lt} \Leftarrow s_4(X) : \mathbf{t})$

9.2.2 Clausal Subset of Annotated Logics

Similar to classical resolution [5], the resolution proof procedure we will describe in the next subsection works on formulae in *clause form*.

A *mega-clause* is a formula of the form:

$$(\forall x_1)\dots(\forall x_m)(L_1 \vee \cdots \vee L_n)$$

where $x_1, \dots, x_m$ are all the variable symbols that occur in $L_1 \vee \dots \vee L_n$ and each L_i is a mega-literal of order less than or equal to 1.

Theorem 1 (Normal Form Theorem) Suppose F is a closed formula. Then there is a finite set $MCL(F)$ of mega-clauses containing no occurrence of the annotation $\perp$ such that F has a model iff $MCL(F)$ has a model.

Proof. We demonstrate the existence of a set $MCL(F)$ of mega-clauses that preserves the satisfiability of F.
Step 1. For every occurrence of a negated literal of the form $\underbrace{\neg\dots\neg}_{k \text{ times}} A : \mu$ in F, replace the occurrence by $A : \neg^k \mu$. Let F_1 be the formula obtained from F after all such replacements. Clearly, F_1 contains no occurrences of $\neg$ because, by definition of formulas, $\neg$ may apply only to either annotated atoms or formulas of the form G where G is of the form $\neg \cdots \neg(A : \mu)$ where $A : \mu$ is an annotated atom. Moreover, F_1 and F evaluate the same way in all interpretations (this is easily proved by induction on the structure of F).
Step 2. All other steps are identical to the procedure in classical logic for transformation of formulas into clausal form (cf. Chang and Lee [5]). Let the resulting set of mega-clauses be denoted by $MCL(F)$. The proof that these preserve satisfiability is the same as the classical case.
Step 3. It is easy to see that $MCL(F)$ is a finite set of mega-clauses. So suppose $MCL(F) = \{C_1, \dots, C_n\}$ and let I be an interpretation that satisfies $MCL(F)$. For each $i = 1, \dots, n$ do the following.

(1) If the literal $A : \perp$ occurs in C_i, then remove C_i from $MCL(F)$. (Deletion of C_i does not affect the satisfiability of $MCL(F)$ in I because $A : \perp$ always evaluates to true, and thus so does C_i.)

(2) Otherwise delete all occurrences of annotated literals of the form $\sim (A : \perp)$ from C_i. ($\sim (A : \perp)$ always evaluates to false, so deleting it from C_i does not change the truth value of C_i in I.)

In each case, I is a model of the given formula iff it is a model of the transformed formula. Thus, the theorem follows. □

A consequence of this theorem is that given any (finite) set S of formulas, there is a set of mega-clauses, denoted $MCL(S)$ such that S is satisfiable iff $MCL(S)$ is satisfiable. Moreover, if S is finite, then $MCL(S)$ is finite. In addition, $MCL(S)$ doesn't contain any occurrences of the annotation $\perp$. We call sets of mega-clauses that contain neither hyper-literals of order greater than 0 nor occurrences of $\perp$ *normal*. In the rest of the paper, we work only with normal mega-clauses.

Herbrand interpretations are defined in the usual way—the domain of any such interpretation is the set of all ground terms that can be expressed using the non-logical symbols of our language L. The assignments to constant and function symbols are identical to the classical logic case. However, predicate symbols of arity n are assigned *functions* from HU^n to $\mathcal{T}$ where HU is the Herbrand Universe of our language.

Theorem 2 ([10]) Suppose T is any set of mega-clauses containing no occurrence of $\neg$ and such that the set of annotations, $\mu \in \mathcal{T}$, occurring in T is finite. Then T has a model iff every finite subset of the set of ground instances of T has a model. □

A similar theorem was established by Kifer and Lozinskii [17]. Note that the restriction that T be free of occurrences of $\neg$ is essential for the above theorem to hold even if T contains only finitely many annotations from $\mathcal{T}$. To see this, consider the following example:

Example 2 Let $\mathcal{T}$ be the set of ordinals up to, and including ω, that is, $\mathcal{T}$ contains all the natural numbers together with ω. $\mathcal{T}$ is ordered in the usual way, i.e., all natural numbers are less than ω, the less-than-or-equals relation between natural numbers is the usual one. Now define the negation function $\neg$ as follows: $\neg(x) = (x+1)$ if x is an integer, and $\neg(\omega) = \omega$. Now take T to be the set consisting of the elements $\{A_1, A_2, \ldots\} \cup \{\sim A\}$ where $A = p : \omega$ and $A_i = \underbrace{\neg\neg\cdots\neg}_{i \text{ times}}(p : 0)$. T contains exactly two members of $\mathcal{T}$, viz., 0 and ω, but T is unsatisfiable even though no finite subset of T is unsatisfiable. □

Note that the restriction in Theorem 2 that theories not contain occurrences of $\neg$ is not unreasonable. After all, Theorem 1 tells us that we can eliminate $\neg$ anyway.

There are two common situations under which T contains only finitely many annotations. The first is when $\mathcal{T}$ contains finitely many truth values. There are many useful logics when $\mathcal{T}$ is finite. For example, the logic based on the set *FOUR* (cf. Figure 9.1b, [2, 3, 12, 13, 14]) of truth values is one such logic, while the logic based on the set *SIX* (cf. Figure 9.1c, [24]) of truth values is another. The logic *SEVEN* in the example of the previous subsection is also finite. The second situation is when T is any finite set of clauses; then it automatically contains only finitely many annotations. In general, this is the case that arises ordinarily in logic programming and in artificial intelligence, where storage requirements dictate that only finitely many sentences can be stored at any given point in time.

Example 3 The mega-clauses shown below are the translations of $R1$ to $R4$ given previously. From $R1$ and $R2$ we obtain three mega-clauses.

$C1:$	$d_2(X) : \mathbf{lt} \vee \sim (s_1(X) : \mathbf{t}) \vee \sim (s_2(X) : \mathbf{f})$
$C2:$	$d_1(X) : \mathbf{lf} \vee \sim (d_3(X) : \mathbf{lt})$
$C3:$	$d_3(X) : \mathbf{lf} \vee \sim (d_1(X) : \mathbf{lt})$

$R3$ and $R4$ translate into $C4$ and $C5$ shown below.

$C4:$	$d_2(X) : \mathbf{h}\,\&\,\mathbf{h} \vee \sim (s_2(X) : \mathbf{f}) \vee \sim (s_3(X) : \mathbf{t})$
$C5:$	$d_1(X) : \mathbf{lt} \vee \sim (s_4(X) : \mathbf{t})$

9.2.3 Resolution Procedure for $Q\mathcal{T}$

In this section, we develop the basic theory of resolution in paraconsistent theories. We recall that throughout this paper $\mathcal{T}$ will denote some arbitrary, but fixed, complete lattice of truth values, and $\neg$ will be some arbitrary, but fixed, unary function from $\mathcal{T}$ to $\mathcal{T}$. As described in the previous section (cf. Definition 4), this will define a particular annotated logic. We use $\sqcup$ and $\sqcap$ to denote least upper bound and greatest lower bound, respectively.

Definition 5 *(Mega-Resolvent)* Suppose C_1 and C_2 are the mega-clauses shown below with no variables in common

$$C_1 \equiv \quad (L_1 \vee \ldots \vee L_n \vee C_1')$$

$$C_2 \equiv \quad (L_1' \vee \ldots \vee L_m' \vee C_2')$$

and suppose $L_i = (A_i : \mu)$ and $L_j' = \sim (A_j' : \rho)$ and $A_1, \ldots, A_n, A_1', \ldots, A_m'$ are unifiable (via mgu θ) and $\rho \leq \mu$. Then the *mega-resolvent* of C_1 and C_2 *on* mega-literals L_i and L_j', $1 \leq i \leq n$, $1 \leq j \leq m$ is defined to be $(C_1' \vee C_2')\theta$. The substitution θ is called the *unifier* of the mega-resolution step. L_i and L_j' are the mega-literals resolved on.

We say that either C_1 resolves with C_2 or C_2 resolves with C_1 with the understanding that the annotation of the positive mega-literal that is resolved on is greater than or equal to the annotation of the negative mega-literal resolved on.

Definition 6 *(Clones)* Suppose C_1 and C_2 are the following mega-clauses:

$$C_1 : \quad A_1 : \mu_1 \vee \ldots \vee A_n : \mu_1 \vee E_1$$

$$C_2 : \quad B_1 : \mu_2 \vee \ldots \vee B_m : \mu_2 \vee E_2$$

and $A_1, \ldots, A_n, B_1, \ldots, B_m$ are unifiable via mgu θ and μ_1 and μ_2 are incomparable. Then the mega-clause $(A_1 : \mu \vee E_1 \vee E_2)\theta$ where $\mu = \sqcup\{\mu_1, \mu_2\}$ is called a *clone* of C_1 and C_2. The set of all clones of a pair of mega-clauses C_1, C_2 together with C_1 and C_2 themselves is denoted by $\mathbf{clone}(C_1, C_2)$.

Definition 7 *(P-Resolvent)* Suppose C_1 and C_2 are mega-clauses. Then the mega-clause C is said to be a *paraconsistent resolvent* (p-resolvent, for short) of C_1 and C_2 iff

1. C is a mega-resolvent of C_1 and C_2, or
2. C is a clone of C_1 and C_2.

$E1$:	$\sim (d_3(ann) : \mathbf{lf})$	(Initial Query)
$E2$:	$\sim (d_1(ann) : \mathbf{lt})$	(By $E1$ and $C3$)
$E3$:	$\sim (s_4(ann) : \mathbf{t})$	(By $E2$ and $C5$)
$E4$:	$\Box$	(By $E3$ and $C9$)

Figure 9.2: p-Refutation of $\{C1, \ldots, C9\} \cup \{\sim (d_3(ann) : \mathbf{lf})\}$

Definition 8 *(P-Deduction)* A *paraconsistent deduction sequence* from a set S of mega-clauses is a sequence $C_1, \ldots, C_n$ such that:

1. $C_1 \in S$ (or is a renaming of a clause in S) and
2. for $i > 0$, C_{i+1} is a p-resolvent of C' and C'' where $C', C'' \in S \cup \{C_1, \ldots, C_i\}$.

The deduction sequence $C_1, \ldots, C_n$ is a *paraconsistent refutation* (p-refutation for short) of S iff $C_n = \Box$, the empty mega-clause.

The definition of p-deduction given here is somewhat different from the one given previously in [9] where clones do not appear explicitly in a deduction sequence. In Section 9.3, a version of p-resolution where cloning is performed implicitly is presented. A comparison of the performance of the two versions is also given.

We end this subsection by considering several p-deductions using our "doctor" example. Suppose in addition to the rules (C1) through (C5) given by the two doctors, we have some facts about Ann, a patient.

$C6$:	$s_1(ann) : \mathbf{t}$
$C7$:	$s_2(ann) : \mathbf{f}$
$C8$:	$s_3(ann) : \mathbf{t}$
$C9$:	$s_4(ann) : \mathbf{t}$

It is easy to see that based on rules $C3$ and $C5$ and the fact $C9$ which says Ann shows symptom s_4, we should be able to conclude that Ann is not likely to have disease d_3. Hence the query $\sim (d_3(Ann) : \mathbf{lf})$ should have a p-refutation. This is shown in Figure 9.2. In this example, no inconsistencies arise.

However, there is an apparent inconsistency in our example that involves the other clauses. From $C1$, $C6$, and $C7$, we are able to conclude that Ann is likely to have disease d_2. On the other hand, from $C4$, $C7$, and $C8$, we are able to conclude that the likelihood of Ann having disease d_2 is only about 50 percent. Hence according to our semantics, the only model of all the clauses together must assign $\top$, or inconsistent, to $d_2(Ann)$. Figure 9.3 shows a p-refutation of the query $\sim (d_2(Ann) : \top)$ that uses cloning.

$G1$:	$\sim (d_2(ann) : \top)$	(Initial Query)
$G2$:	$\sim (s_3(X) : \mathbf{t}) \vee \sim (s_2(X) : \mathbf{f}) \vee \sim (s_1(X) : \mathbf{t}) \vee$ $d_2(X) : \top$	(clone of $C4$ and $C1$)
$G3$:	$\sim (s_2(ann) : \mathbf{f}) \vee \sim (s_1(ann) : \mathbf{t}) \vee d_2(ann) : \top$	(By $G2$ and $C8$)
$G4$:	$\sim (s_1(ann) : \mathbf{t}) \vee d_2(ann) : \top$	(By $G3$ and $C7$)
$G5$:	$d_2(ann) : \top$	(By $G4$ and $C6$)
$G6$:	$\Box$	(By $G1$ and $G5$)

Figure 9.3: p-Refutation of $\{C1, \ldots, C9\} \cup \{\sim (d_2(ann) : \top\}$

It should also be pointed out that the existence of an inconsistency does not allow us to draw arbitrary conclusions. For example, we would not be able to deduce $d_1(ann) : \mathbf{f}$. On the other hand, as the inconsistency relates to whether or not Ann has disease d_2, it follows that we are able to deduce every possible truth value with respect to $d_2(ann)$. In other words, $d_2(ann) : \mathbf{f}$, $d_2(ann) : \mathbf{lf}$, and so on, are all logical consequences of our knowledge base, and one can easily verify that $\sim (d_2(ann) : \mu)$, for any truth value μ in *SEVEN*, has a p-refutation. It is this feature of our logic, and paraconsistent logics in general, that causes us to refer to them as locally consistent logics.

9.2.4 Soundness and Completeness of P-resolution

Lemma 1 *(Soundness of Mega-Resolution)* Mega-resolution is sound.

Proof. Let I be a model of mega-clauses C_1 and C_2, where C_1, C_2 have the forms

$$C_1 \equiv (A_1 : \mu \vee \ldots \vee A_n : \mu \vee D_1)$$

$$C_2 \equiv (\sim (B_1 : \rho) \vee \ldots \vee \sim (B_m : \rho) \vee D_2)$$

where A_i, B_j, for $1 \le i \le n$, $1 \le j \le m$ are simultaneously unifiable via mgu θ. I satisfies $C_1\theta$ and $C_2\theta$. Further, I satisfies every ground instance C_1' and C_2', respectively, of $C_1\theta$,$C_2\theta$. Suppose C_1' and C_2' have the forms

$$C_1' \equiv \quad (A : \mu \vee D_1')$$

$$C_2' \equiv \quad (\sim (A : \rho) \vee D_2')$$

1. If $I \not\models A : \mu$, then $I \models D_1'$. In this case I satisfies the mega-resolvent $D_1' \vee D_2'$.
2. $I \models A : \mu$. $\mu \le I(A)$. Since $\rho \le \mu$, it follows by the transitivity of $\le$ that $\rho \le I(A)$. Thus $I \not\models \sim (A : \rho)$. But then $I \models D_2'$ and hence I satisfies the mega-resolvent $D_1' \vee D_2'$. $\Box$

Lemma 2 *(Soundness of Cloning)* Suppose C is a clone of C_1 and C_2. Then $\{C_1, C_2\} \models C$.

Proof. Let I be a model of C_1 and C_2, where C_1 and C_2 have the forms

$$C_1 \equiv (A_1 : \mu_1 \vee \ldots \vee A_n : \mu_1 \vee D_1)$$

$$C_2 \equiv (B_1 : \mu_2 \vee \ldots \vee B_m : \mu_2 \vee D_2)$$

where A_i, B_j, for $1 \leq i \leq n$, $1 \leq j \leq m$ are simultaneously unifiable via mgu θ and μ_1, μ_2 are incomparable. I satisfies $C_1\theta$ and $C_2\theta$. Let C_1' and C_2' be ground instances of $C_1\theta$ and $C_2\theta$, respectively. C_1' and C_2' have the forms

$$C_1' \equiv A : \mu_1 \vee D_1'$$

$$C_2' \equiv A : \mu_2 \vee D_2'$$

If $I \not\models A : \mu_1$ or $I \not\models A : \mu_2$, then either I satisfies D_1' or I satisfies D_2'. In either case, I satisfies the clone $(D_1' \vee D_2' \vee A : \mu)$, where $\mu = \sqcup\{\mu_1, \mu_2\}$. Otherwise $I \models A : \mu_1$ and $I \models A : \mu_2$. Then $I(A) \geq \mu_1$ and $I(A) \geq \mu_2$. It follows that $I(A) \geq \mu$. Whence $I \models A : \mu$ and so I satisfies the clone. □

Theorem 3 *(Soundness of P-Resolution)* P-resolution is sound.

Proof. Follows from Lemma 1 and Lemma 2. □

Theorem 4 *(Ground Completeness of P-resolution)* Suppose S is an unsatisfiable set of ground mega-clauses. Then S has a p-refutation.

Proof. The proof follows the well-known Anderson and Bledsoe technique [1] for showing completeness. Let $k(S)$ be the number of excess literals for a set S of unsatisfiable ground mega-clauses ($k(S)$ is the total number of occurrences of mega-literals in S minus the number of mega-clauses in S). By induction on $k(S)$.

(Base Case) $k(S) = 0$. S is a set of unit ground mega-clauses. Since S is unsatisfiable iff there exists a ground atom A such that $S \models A : \mu$ and $S \models \sim (A : \rho)$ and $\mu \geq \rho$, we consider the conditions under which this occurs.

1. S contains $A : \mu$ and $\sim (A : \rho)$. In this case, mega-resolving $A : \mu$ and $\sim (A : \rho)$ yields a p-refutation of S.

2. S contains $A : \mu_1, \ldots, A : \mu_n$ and $\sim (A : \rho)$ and $\sqcup\{\mu_1, \ldots, \mu_n\} \geq \rho$. In this case, take the clone $A : \sqcup\{\mu_1, \mu_2\}$ of $A : \mu_1$ and $A : \mu_2$. Next clone $A : \sqcup\{\mu_1, \mu_2\}$ with $A : \mu_3$ to obtain $A : \sqcup\{\sqcup\{\mu_1, \mu_2\}, \mu_3\}$. Repeating this for $A : \mu_4$ through $A : \mu_n$ we obtain $A : \sqcup\{\mu_1, \ldots, \mu_n\}$. Mega-resolving it with $\sim (A : \rho)$, we obtain the empty clause.

(Inductive Step) $k(S) = n+1$. Let $C \equiv (A \vee L)$ be a clause containing more than one mega-literal. Consider the two sets of clauses $S_1 \equiv (S - \{C\}) \cup \{A\}$ and $S_2 \equiv (S - \{C\}) \cup \{L\}$. S_1 and S_2 are unsatisfiable. By the induction hypothesis, there exist p-refutations R_1, R_2 of S_1, S_2 respectively. By performing the same p-resolutions steps as in R_1, we derive either the empty clause or L from S. In the first case, we are done. Otherwise perform the p-resolution steps as in R_2 with $\{L\} \cup (S - \{C\})$, which is simply the set S_2; we obtain a p-refutation of S. □

Lemma 3 *(Lifting Lemmas)*

1. (Mega-Resolvent) Suppose C_1' and C_2' are instances of mega-clauses C_1 and C_2 respectively. If C' is a mega-resolvent of C_1' and C_2', then there is a mega-resolvent C of C_1 and C_2 such that C' is an instance of C.

2. (Clones) If C_1' and C_2' are instances of C_1 and C_2, respectively, then for each clone C' of C_1' and C_2', there is a clone C of C_1 and C_2 such that C' is an instance of a factor of C.

3. (P-Resolvent) Suppose C_1' and C_2' are instances of mega-clauses C_1 and C_2 respectively. If C' is a p-resolvent of C_1 and C_2, then there is a p-resolvent C of C_1 and C_2 such that C' is an instance of C.

Proof. Similar to lifting lemma for ordinary resolution. □

Theorem 5 *(General Completeness of P-resolution)* If a finite set S of mega-clauses has no model, then there is a p-refutation of S.

Proof Outline. By Herbrand's Theorem (Theorem 2), there exists a finite unsatisfiable set S' of ground instances of S. By ground completeness of p-resolution, there exists a p-refutation R' of S'. Applying the lifting lemma to R' we obtain a p-refutation R of S. □

9.3 Implementation and Experiments

Efficiency is the major concern in any implementation of a theorem prover [26]. Implementing p-resolution is no exception. Many strategies that have been developed for increasing the efficiency of theorem provers in classical logic are also useful in p-resolution. These include, for example, subsumption check, where subsumption in the context of annotated logics is defined as follows. Suppose C_1 and C_2 are mega-clauses of the form:

$$C_1 \equiv \quad A_1 : \mu_1 \vee \ldots \vee A_n : \mu_n \vee \sim (B_1 : \rho_1) \vee \ldots \vee \sim (B_m : \rho_m)$$

$$C_2 \equiv \quad G_1 : \psi_1 \vee \ldots \vee G_r : \psi_r \vee \sim (H_1 : \phi_1) \vee \ldots \vee \sim (H_s : \phi_s)$$

C_1 is said to *subsume* C_2 iff there is a substitution θ such that:

1. for all $1 \leq i \leq n$, there is a $1 \leq j \leq r$ such that $G_j = A_i\theta$ and $\mu_i \leq \psi_j$ and

2. for all $1 \leq \ell \leq m$, there is a $1 \leq w \leq s$ such that $H_w = B_\ell\theta$ and $\phi_w \leq \rho_\ell$.

Guidance strategies such as the unit preference [25] and restriction strategies like linear resolution and set-of-support [5] can also be incorporated into p-resolution. In this section, we discuss an implementation (TiPPLe) of our deduction system for annotated logics. We also present experimental results of p-resolution (as given in the previous section) and a slightly modified version of p-resolution, which we will refer to as p*-resolution. Informally, the difference between p-resolution and p*-resolution is that clones do not appear explicitly as a part of a deduction sequence in p*-resolution. The formal definition of p* is as follows.

Given a set of mega-clauses S and a mega-clause C, define $\mathbf{clone}_S(C)$ as the set

$$\bigcup\{\mathbf{clone}(C, C')|C' \in S\}.$$

The following is an alternative definition of p-deduction.

Definition 9 A *restricted paraconsistent deduction sequence* (abbreviated *p*-*deduction) from a set S of mega-clauses is a sequence $C_1, \ldots, C_n$ such that:

1. $C_1 \in S$ (or is a renaming of a clause in S) and

2. Whenever $D', D'' \in (S \cup \{C_1, \ldots, C_i\})$, it follows that for $i > 0$, C_{i+1} is a mega-resolvent of C' and C'' where $C' \in \mathbf{clone}_{S\cup\{C_1,\ldots,C_i\}}(D')$ and $C'' \in \mathbf{clone}_{S\cup\{C_1,\ldots,C_i\}}(D'')$.

The deduction sequence $C_1, \ldots, C_n$ is a *restricted paraconsistent refutation* of S iff $C_n = \Box$.

It is important to note that each C_i is obtained by an application of *mega-resolution* rather than p-resolution. The proof that p*-resolution is complete is straightforward.

Theorem 6 If S is a finite unsatisfiable set of mega-clauses, then there is a *p**-refutation of S. □

A first version of TiPPLe has been implemented using Quintus Prolog 2.4.2 running on Sun workstations. An input mega-clause is represented by a fact with two arguments. The first is a clause number and the second a list of mega-literals representing the clause. The clause number is used to take advantage of first argument indexing of Quintus Prolog. Each positive mega-literal is prefixed by a + and each negative mega-literal by −. The input to TiPPLe consists first of all of a file containing the clause set in the format just described. A second file which specifies the lattice over which the clause set is defined is also provided by the user. The lattice is defined by a set S of facts of the form $less(\alpha_1, \alpha_2)$ for each pair of truth values α_1 and α_2 such that α_1 is covered by α_2 (x is covered by y if $x < y$ and $x \leq z < y$ implies $z = x$ [11]). A preprocessor has been implemented to compute the transitive closure of S. This eliminates such computations during deductions.

Currently, the selection of which clauses are to be resolved is based on "Linear preference". Hence a newly generated clause is usually tried first as one of the parents for the next deduction step. But one could always backtrack to previous clauses if necessary, and the eventual proof need not be linear. The selection of the other parent in a deduction step is by preferring those clauses with fewer predicate symbols. To do this, a table to keep track of which clauses each predicate occurs in is maintained. The table also uses first argument indexing. For p-resolution, mega-resolution is always attempted before cloning. The rationale behind this is simply that mega-resolution is the only way an empty clause can be derived. Both backward and forward subsumption has been incorporated in TiPPLe. The user may also specify a depth bound on the search of TiPPLe.

The results of several experiments conducted with p- and p*-resolutions are presented in Figure 9.4. The test problems are generally small (up to 13 clauses). They are written over five lattices of varying lengths and widths including *FOUR* and *SEVEN*. (The length of the lattice is the length of the longest chain while the width is the size of the largest antichain in the lattice. [11]). The test problems presented are referred to by names. Complete details of the clause sets can be found in [21]. In general, the first letter of a test problem corresponds to the first letter of the lattice over which the problem is defined. So for example, *F1* is a set of clauses written over the lattice *FOUR*.

The cpu time reported is from running on a Sun SPARC station. Since the implementation of TiPPLe is still in its infant stage, the only comparison we make is in the cpu time and the number of resolvents generated that are not forward subsumed. As more sophistications are incorporated, other parameters for measurements such as the number of subsumptions will be included as well.

The most extensive experiments are carried out over the lattice *SEVEN*. In almost all cases for all lattices, p*-resolution generates fewer resolvents than p-resolution. However, the time required varies widely depending on both the clause set and the order of the clauses. As an example, the two sets *S6a* and *S6b* contain exactly the same clauses but are ordered differently. This results in considerable difference in the number of clauses generated as well as the runtime for both p and p* resolution.

Clause set	P*-resolution		P-resolution	
	cpu	# clauses	cpu	# clauses
$F2$	0.16 sec	10	0.28 sec	11
$F3$	3.12 min	31	2.20 min	49
$F4$	1.27 sec	21	0.67 sec	21
$F4a$	1.30 sec	21	28.83 sec	87
$S2$	0.18 sec	13	0.35 sec	15
$S4$	0.33 sec	14	0.83 sec	19
$S5a$	3.13 min	31	2.23 min	49
$S5a1$	3.31 min	31	2.32 min	52
$S5b$	16.88 sec	26	12.05 sec	26
$S5c$	0.73 sec	11	0.87 sec	14
$S6a$	3.616 sec	21	29.48 sec	87
$S6b$	62.50 sec	51	14.02 sec	56
$W5$	1.10 sec	12	0.57 sec	13
$W6$	1.23 sec	24	1.57 sec	26
$D2$	2.00 sec	16	0.62 sec	17
$B1$	0.42 sec	16	0.75 sec	18
$B4$	3.53 sec	11	0.72 sec	13
$B5$	3.02 sec	19	0.92 sec	20

Figure 9.4: Experimental Results of P Resolution and P^* Resolution

9.4 Extending Annotated Logics for Parallel Search

Wos [26] has emphasized the importance of utilizing parallelism to improve the performance of theorem-proving programs. The most common approach to parallel search is by regarding deductions as searches over AND/OR trees. This leads naturally to AND/OR parallelism. Other approaches include for example partitioning the clauses into subsets, each stored in the local memory of a processor that performs deductions independently of other processors [6]. Communications occur in the form of resolvents being passed between the processors. Roberge [22] has developed a different approach by decomposing the initial unsatisfiable set of clauses into smaller subsets each of which is also unsatisfiable. The decomposition process involves breaking up clauses into subclauses. The resulting refutation from each of the smaller subsets can be performed in parallel and then combined to form the refutation proof of the original set. In this section, we extend the syntax of clausal annotated logics to allow for set annotations. The extension provides a compact way of representing mega-clauses. This in turn facilitates a form of parallel search naturally by embedding it in the proof procedures, thus reducing the searching and backtracking process of p-resolution significantly. We present one such rule of inference for these extended clauses. As compared to related research on parallel deductions, our work is most similar to the work of Fishman and Minker [15]. There an extended form of first order clauses, called Π-clauses, is also used to efficiently represent sets of clauses.

9.4.1 Extended Mega-Clauses

The formal description of the extended syntax is as follows.

Definition 10 *(Extended Mega-Literals)* Given a finite set $S \subseteq \mathcal{T}$ and an atom A, $A : S$ and $\sim (A : S)$ are called *extended mega-literals* (em-literals, for short).

Obviously, a mega-literal may be considered to be an em-literal whose annotation set is a singleton. An *em-clause* is a set of em-literals.

Intuitively, an em-literal A with annotation set S means that A is a mega-literal whose truth value is known to be greater than or equal to at least one of the values in S. But the exact one is unknown.

Definition 11 Suppose A is a ground atom and I is an interpretation. If S is a finite subset of $\mathcal{T}$, then $I \models (A : S)$ iff $I \models (A : \mu)$ for some $\mu \in S$. The extension of the satisfaction relation for complex formulas generated by the connectives $\&, \vee, \sim$ (but not $\neg$) and the quantifiers $\forall$ and $\exists$ is defined in the obvious way. We do not allow $\neg$ to be applied to em-literals.

Note that it follows that for a negative em-literal $\sim (A : S)$, $I \models \sim (A : S)$ iff $I \not\models A : \mu$ for all $\mu \in S$. As a special case, $A : \emptyset$ is unsatisfiable while $\sim (A : \emptyset)$ is a tautology, for all atoms A.

From the above definition, one can see that extended mega-literals do not have an effect on the expressive power of positive literals in clauses. In other words, an em-clause $L_1 \vee A : \{s_1, \ldots, s_n\} \vee L_m$ is simply equivalent to $L_1 \vee A : s_1 \vee \ldots \vee A : s_n \vee L_m$. However, they do enhance the expressive power of negative literals that occur in clauses since a negative em-literal $\sim (A : \{s_1, \ldots, s_n\})$ is by definition equivalent to $\sim (A : s_1) \wedge \ldots \wedge \sim (A : s_n)$. So a negative em-literal with an annotation set of cardinality greater than one occurring in a clause C "embeds" a conjunction of negative literals in C. On the other hand, as we shall see in the next subsection, positive em-literals are still useful for reducing the search space of p-resolution.

Example 4 Consider the positive mega-clause C over the lattice *SEVEN*

$$p(X,Y) : \mathbf{t} \vee p(X,Y) : \mathbf{f} \vee p(X,Y) : \mathbf{h} \,\&\, \mathbf{h}.$$

C contains three positive mega-literals. In the extended notation, C could be represented very succinctly as the em-clause $p(X,Y) : \{\mathbf{t}, \mathbf{f}, \mathbf{h} \,\&\, \mathbf{h}\}$. □

Example 5 Suppose we have the set S of mega-clauses containing the following three negative unit mega-clauses over the lattice *SEVEN*.

$$\{\sim (p(X,Y) : \mathbf{lt}), \sim (p(X,Y) : \mathbf{lf}), \sim (p(X,Y) : \mathbf{h} \,\&\, \mathbf{h})\}.$$

S could be represented in em-clauses as the singleton negative em-unit

$$\{\sim (p(X,Y) : \{\mathbf{lt}, \mathbf{lf}, \mathbf{h} \,\&\, \mathbf{h}\})\}.$$

□

These simple examples illustrate that, similar to the Π-clauses of Fishman and Minker [15], em-clauses enable mega-clauses to be efficiently stored.

9.4.2 Gp-resolution

We now present an inference rule which is an extension of p-resolution for making deductions over em-clauses.

Definition 12 *(Finite Decomposability)* Let $\mathcal{T}$ be a complete lattice. We say that $\mathcal{T}$ is finitely decomposable iff whenever μ and δ are in $\mathcal{T}$ and $\mu \not\geq \delta$, then the set of $\rho \in \mathcal{T}$ where $\sqcup\{\rho, \mu\} \geq \delta$ is finite.

For the remainder of the paper, we assume that the lattice of truth values $\mathcal{T}$ is finitely decomposable. The reason for this restriction will be explained. We should first note, however, that many useful lattices, including all those that we have seen thus far, are finitely decomposable. Hence the extensions that we will describe are still widely applicable. Further, many infinite lattices may often be accommodated. For instance, consider the theory T written over the lattice $EV = [0,1] \times [0,1]$, i.e., each member of EV is a pair (x, y) where x, y are real numbers in the unit interval. EV is a complete lattice when ordered as follows: $(x, y) \leq (x', y')$ iff $x \leq x'$ and $y \leq y'$. Now, if T is finite, and free of $\neg$ (as is often the case), then the set of annotations in T, when closed under least upper bounds and greatest lower bounds, generates a *finite sub-lattice* of EV. For our purposes, we may consider T to be written over this finite sub-lattice, rather than over EV. Of course, this finite sub-lattice is obviously finitely decomposable.

Given a set D of truth values, we denote by D^* the subset of D such that for every $\mu \in D^*$, there does not exist a $\lambda \in D$ where $\mu > \lambda$. Thus, D^* is the set of minimal elements, if any, of D. Note that if D is finite, then D^* is non-empty iff D is non-empty. For infinite $\mathcal{T}$, D^* may be empty even though D is non-empty. This may occur, for example, if D is an infinitely descending chain that has no bottom element.

Definition 13 Let $S_1, \ldots, S_n$ be sets of truth values. Define $\coprod\{S_1, \ldots, S_n\}$ to be the set

$$\{\sqcup\{s_1, \ldots, s_n\} \mid s_i \in S_i \text{ for } 1 \leq i \leq n\}$$

and $\prod\{S_1, \ldots, S_n\}$ to be the set

$$\{\sqcap\{s_1, \ldots, s_n\} \mid s_i \in S_i \text{ for } 1 \leq i \leq n\}.$$

These operations allow us in definition 15 below to generalize merging in the context of extended literals.

Definition 14 Given two truth values $\mu, \delta \in \mathcal{T}$ such that $\mu \not\geq \delta$.

$$\mathcal{N}(\delta, \mu) = (\{\, \rho \mid \rho \in \mathcal{T},\ \rho \not\geq \delta, \text{ and } \sqcup\{\rho, \mu\} \geq \delta \,\})^*$$

Thus, $\mathcal{N}(\delta, \mu)$ is the set of "smallest" possible elements $\rho \in (\mathcal{T} - \{\delta\})$ such that $\rho \sqcup \mu \geq \delta$.

Definition 15 *(Gp-Resolvent)* Suppose C_1 and C_2 are the em-clauses shown below:

$$C_1 \equiv \quad (B \vee L_1 \vee \ldots \vee L_n)$$
$$C_2 \equiv \quad (B' \vee L'_1 \vee \ldots \ldots \vee L'_m)$$

and $L_i = (A_i : S_i)$, $1 \leq i \leq n$, and $L'_j = \sim (A'_j : S'_j)$, $1 \leq j \leq m$, and $A_1, \ldots, A_n, A'_1, \ldots A'_m$ are unifiable (via mgu θ). Let $\Sigma = \prod\{S_1, \ldots, S_n\}$ and $\Gamma = \coprod\{S'_1, \ldots, S'_m\}$. Obtain a gp-resolvent as follows:

1. For any annotation $\mu \in \Sigma$, if there is a $\rho \in \Gamma$ such that $\mu \geq \rho$, then delete μ from Σ. If the resulting Σ is empty, then a gp-resolvent of C_1 and C_2 is

$$(B \vee B')\theta.$$

2. Σ is not empty after step 1. Now choose a possibly empty $\Sigma_2 \subseteq \Sigma$. Suppose $\Sigma_2 = \{\mu_1, \ldots, \mu_k\}$. Let $\Sigma_3 = \Sigma - \Sigma_2$. Next choose $\rho_1, \ldots, \rho_k$ from Γ. Then there are two subcases.

   (a) If Σ_3 is non-empty, then a gp-resolvent of C_1 and C_2 is

   $$(B \vee B' \vee \sim (A_1 : \mathcal{N}(\rho_1, \mu_1)) \vee \ldots \vee \sim (A_1 : \mathcal{N}(\rho_k, \mu_k)) \vee A_1 : \Sigma_3)\theta.$$

   (b) Otherwise $\Sigma = \Sigma_2$. Then a gp-resolvent of C_1 and C_2 is

   $$(B \vee B' \vee \sim (A_1 : \mathcal{N}(\rho_1, \mu_1)) \vee \ldots \vee \sim (A_1 : \mathcal{N}(\rho_k, \mu_k)))\theta$$

   The em-literals $\sim (A_1 : \mathcal{N}(\rho_1, \mu_1)), \ldots, \sim (A_1 : \mathcal{N}(\rho_k, \mu_k))$ (if they exist) are called *reflections* of the resolution while $A_1 : \Sigma_3$ (if it exists) is called the residue.

The substitution θ is called the *gm-unifier* of the mega-resolution step.

Example 6 Consider the two em-clauses $C_1 = A \vee p : \{\mathbf{lt}, \mathbf{lf}\}$ and $C_2 = B \vee \sim (p : \{\mathbf{t}, \mathbf{f}\})$. Both **lt** and **lf** are not greater than or equal to any of **t** and **f**. There are nine possible gp-resolvents the last of which is subsumed by C_1.

1. $A \vee B \vee \sim (p : \mathcal{N}(\mathbf{t}, \mathbf{lt})) \vee \sim (p : \mathcal{N}(\mathbf{t}, \mathbf{lf}))$
2. $A \vee B \vee \sim (p : \mathcal{N}(\mathbf{t}, \mathbf{lt})) \vee \sim (p : \mathcal{N}(\mathbf{f}, \mathbf{lf}))$
3. $A \vee B \vee \sim (p : \mathcal{N}(\mathbf{f}, \mathbf{lt})) \vee \sim (p : \mathcal{N}(\mathbf{t}, \mathbf{lf}))$
4. $A \vee B \vee \sim (p : \mathcal{N}(\mathbf{f}, \mathbf{lt})) \vee \sim (p : \mathcal{N}(\mathbf{f}, \mathbf{lf}))$
5. $A \vee B \vee \sim (p : \mathcal{N}(\mathbf{t}, \mathbf{lt})) \vee p : \{\mathbf{lf}\}$
6. $A \vee B \vee \sim (p : \mathcal{N}(\mathbf{f}, \mathbf{lt})) \vee p : \{\mathbf{lf}\}$
7. $A \vee B \vee \sim (p : \mathcal{N}(\mathbf{t}, \mathbf{lf})) \vee p : \{\mathbf{lt}\}$
8. $A \vee B \vee \sim (p : \mathcal{N}(\mathbf{f}, \mathbf{lf})) \vee p : \{\mathbf{lt}\}$
9. $A \vee B \vee p : \{\mathbf{lf}, \mathbf{lt}\}$

Note that the first four gp-resolvents contain no residues and that the last one contains no reflections. □

Example 7 Replace C_1 in the above example by the clause $A \vee p : \{\mathbf{lt}, \mathbf{f}\}$. Now **lt** is the only annotation that can not be resolved. There are three possible gp-resolvents.

1. $A \vee B \vee \sim (p : \mathcal{N}(\mathbf{t}, \mathbf{lt}))$
2. $A \vee B \vee \sim (p : \mathcal{N}(\mathbf{f}, \mathbf{lt}))$
3. $A \vee B \vee p : \{\mathbf{lt}\}$ □

It can be seen from definition 14 that if $\mathcal{T}$ is not finitely decomposable, then the set $\mathcal{N}(\rho, \mu)$ annotating a reflection may be infinite. By the definition of em-literals, this is not a syntactically valid em-literal. Semantically, an infinite annotation set causes problems since the completeness of gp-resolution is based on Herbrand's Theorem (see Theorem 2 and Theorem 9) for annotated logics, which holds only for sets of clauses with finite annotations.

Definition 16 *(Gp-Deduction)* A *generalized paraconsistent deduction sequence* from a set S of mega-clauses is a sequence $C_1, \ldots, C_n$ such that:

1. $C_1 \in S$ (or is a renaming of a mega-clause in S) and
2. for $i > 0$, C_{i+1} is a gp-resolvent of C' and C'' where $C', C'' \in S \cup \{C_1, \ldots, C_i\}$.

A *generalized paraconsistent refutation* (gp-refutation) is a generalized paraconsistent deduction sequence whose last em-clause is $\Box$.

9.4.3 Parallel Search Using Gp-resolution

The original motivation for introducing em-clauses and gp-resolution is to eliminate the need for cloning. Intuitively, cloning is useful (e.g., Figure 9.3) in the case when there are two positive literals $A : \mu_1$ and $A : \mu_2$ in two clauses and the least upper bound of μ_1 and μ_2 is greater than or equal to the annotation μ of some negative literal $\sim (A : \mu)$ in a clause, but neither μ_1 nor μ_2 is greater than (or equal to) μ. Now suppose two clauses C_1, C_2 are selected during a deduction for resolution from an unsatisfiable set of clauses S (for simplicity, we assume that S is minimally unsatisfiable). Suppose in addition C_1, C_2 contain $A : \mu_1$ and $\sim (A : \mu)$ respectively but $\mu_1 \not\geq \mu$ so that mega-resolution is not applicable. Then we might postulate the existence of some third clause C_3 that is a logical consequence of S such that C_3 contains $A : \mu_2$ and that $\sqcup\{\mu_1, \mu_2\} \geq \mu$. If such a clause exists, (indeed it must, given the assumption that S is minimally unsatisfiable; otherwise there must be a clause that is a logical consequence of S which contains the literal $A : \rho$ and $\rho \geq \mu$) then we could first "conditionally" resolve $A : \mu_1$ with $\sim (A : \mu)$ and later remove the condition when C_3 is encountered. Of course, we do not know what exact value μ_2 might be. However, it is not difficult to see that μ_2 must be greater than or equal to one of the truth values in the set $\mathcal{N}(\mu, \mu_1)$ (see Lemma 7 in the next subsection). In this sense, a reflection in the resolvent records all possible $A : \mu_2$'s that satisfy the condition $\sqcup\{\mu_1, \mu_2\} \geq \mu$, thus implementing a "target-driven" strategy. This is illustrated in the next example.

Example 8 At the end of Subsection 9.2.3, we showed that for any $\mu \in SEVEN$, $d_2(ann) : \mu$ is a logical consequence of $\{C1, \ldots, C9\}$. In Figure 9.5, we show a gp-refutation of $d_2(ann) : \mathbf{f}$. Note that in the first step when resolving $C4$ with $F1$ on the literals $d_2(X) : \mathbf{h}\,\&\,\mathbf{h}$ and $\sim (d_2(ann) : \mathbf{f})$, we take μ_1, ρ_1 in definition 15 to be $\mathbf{h}\,\&\,\mathbf{h}$ and $\mathbf{f}$ respectively. In this case $\mathcal{N}(\mathbf{f}, \mathbf{h}\,\&\,\mathbf{h}) = \{\mathbf{lt}\}$. $\Box$

On the other hand, as we have already shown, a positive em-literal is equivalent to a disjunction of mega-literals while a negative em-literal is a compact representation for a set (i.e., a conjunction) of mega-literals. So

$F1$: $\sim (d_2(ann) : \mathbf{f})$ (Initial Query)
$F2$: $\sim (s_3(ann) : \mathbf{t}) \vee \sim (s_2(ann) : \mathbf{f}) \vee \sim (d_2(ann) : \mathbf{lt})$ (By $C4$ and $F1$)
$F3$: $\sim (s_3(ann) : \mathbf{t}) \vee \sim (s_2(ann) : \mathbf{f}) \vee \sim (s_1(ann) : \mathbf{t})$ (By $C1$ and $F2$)
Resolving $F3$ with $C6, C7$ and $C8$ yields the empty clause.

Figure 9.5: p-Refutation of $\{C1, \ldots, C9\} \cup \{\sim (d_2(ann) : \mathbf{f}\}$

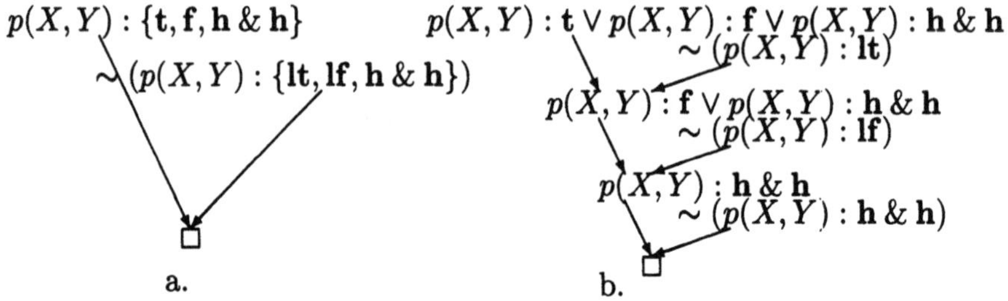

Figure 9.6:

the extended syntax provides naturally a "parallel representation" of regular mega-clauses. Herein lies the true power of em-clauses as it allows for the possibility of an inference rule such as gp-resolution to build-in a parallel search capability.

The next example demonstrates how by using gp-resolution in the extended notation, it is now possible to combine in a single gp-deduction step an inference that would require several deductions in regular mega-clauses using p-resolution.

Example 9 Using the same clauses from Examples 4 and 5, suppose we have the (unsatisfiable) set S of mega-clauses over the lattice $SEVEN$.

$$p(X,Y) : \mathbf{t} \ \vee \ p(X,Y) : \mathbf{f} \ \vee \ p(X,Y) : \mathbf{h} \,\&\, \mathbf{h},$$

$$\sim (p(X,Y) : \mathbf{lt}), \sim (p(X,Y) : \mathbf{lf}), \sim (p(X,Y) : \mathbf{h} \,\&\, \mathbf{h})$$

In the extended notation, S could be represented as the set

$$S' = \{p(X,Y) : \{\mathbf{t}, \mathbf{f}, \mathbf{h} \,\&\, \mathbf{h}\}, \sim (p(X,Y) : \{\mathbf{lt}, \mathbf{lf}, \mathbf{h} \,\&\, \mathbf{h}\})\}.$$

A gp-refutation R of the two em-clauses in S' is shown in Figure 9.6a. R requires a single gp-resolution. On the other hand, a p-refutation of the mega-clauses in S requires three p-deduction steps (see Figure 9.6b). Hence a single gp-resolution corresponds to three simultaneous p-resolution steps. In other words, we would need to perform the three p-resolution steps in parallel in order to obtain a p-refutation of S in as few steps as R. □

To give a quick analysis of a simple case, suppose there is a positive em-literal $A : \{\mu_1, \ldots, \mu_n\}$ and m negative em-literals $\sim (A : T'_1), \ldots, \sim (A : T'_m)$ where for all μ_i, $1 \leq i \leq n$, there exists a β_i in some T'_j, $1 \leq j \leq m$ such that $\mu_i \geq \beta_i$. Then it is possible to gp-resolve away $A : \{\mu_1, \ldots, \mu_n\}$ in $k \leq n$ steps. Now depending on how compact the representation is, k may be as little as 1. This occurs when the annotation set of one of the negative em-literals contains all of $\beta_1, \ldots, \beta_n$. In the worst case when the annotation set of each of the negative em-literals contains at most one of $\beta_1, \ldots, \beta_n$, then $k = n$. On the other hand for p-resolution, the equivalent set of regular mega-clauses for the situation described above will always require n p-deductions.

This illustrates that in general, gp-resolution may require considerably fewer deductions to reach the same conclusion as p-resolution operating on equivalent clauses. An additional gain from the fewer deductions is that the number of unifications is correspondingly reduced.

9.4.4 Soundness and Completeness of Gp-resolution

Lemma 4 Suppose α, β, γ are three truth values such that $\alpha \leq \beta$ and $\beta \not\geq \gamma$. If $\sqcup\{\delta, \alpha\} \geq \gamma$ for some truth value $\delta \not\geq \gamma$, then $\delta \not\geq \beta$ and $\beta \not\geq \delta$.

Proof. Consider one such δ. If $\delta \geq \beta$, then $\delta \geq \alpha$ by the transitivity of $\geq$. Hence $\sqcup\{\delta, \alpha\} = \delta \geq \gamma$. This contradicts the assumption that $\delta \not\geq \gamma$. On the other hand if $\delta \leq \beta$, then $\sqcup\{\delta, \alpha\} \leq \beta$. Since $\gamma \leq \sqcup\{\delta, \alpha\}$ by assumption, it follows that $\gamma \leq \beta$. This contradicts the assumption that $\beta \not\geq \gamma$. □

Lemma 5 *(Soundness of* $\prod$ *and* $\coprod$*)*

1. If $I \models A_1 : S_1 \vee \ldots \vee A_n : S_n$, and $A_1, \ldots, A_n$ are unifiable via mgu θ, then $I \models (A_1 : \prod\{S_1, \ldots, S_n\})\theta$.

2. If $I \models \sim (A_1 : S_1) \vee \ldots \vee \sim (A_n : S_n)$ and $A_1, \ldots, A_n$ are unifiable via mgu θ, then $I \models \sim (A_1 : \coprod\{S_1, \ldots, S_n\})\theta$.

Proof.

1. Clearly, for every $\alpha \in S_1 \cup \ldots \cup S_n$, there is an element $\beta \in \prod\{S_1, \ldots, S_n\}$ such that $\alpha \geq \beta$. Now $I(A_1\theta) \geq \alpha$ for some $\alpha \in S_1 \cup \ldots \cup S_n$. It follows that $I(A_1\theta) \geq \beta$ for some $\beta \in \prod\{S_1, \ldots, S_n\}$ and hence I satisfies $(A_1 : \prod\{S_1, \ldots, S_n\})\theta$.

2. There exists some S_i such that for every $\alpha \in S_i$, $I(A_1\theta) \not\geq \alpha$. Now each $\beta \in \coprod\{S_1, \ldots, S_n\}$ has the form $\sqcup\{\alpha_1, \ldots, \alpha_i, \ldots, \alpha_n\}$ for some $\alpha_i \in S_i$. Clearly, $\alpha_i \leq \beta$. It follows that $I(A_1\theta) \not\geq \beta$. Whence I satisfies $\sim (A_1 : \coprod\{S_1, \ldots, S_n\})\theta$.

□

Theorem 7 *(Soundness of Gp-Resolution)* Gp-resolution is sound.

Proof. Let I be a model of em-clauses C_1, C_2 where C_1, C_2 are as in definition 15. Let θ be the unifier also as in definition 15. Then by the above lemma, I satisfies the following two clauses:

$$E_1 \equiv (B \vee A : \prod\{S_1, \ldots, S_n\})\theta$$

$$E_2 \equiv (B' \vee \sim (A : \coprod\{S'_1, \ldots, S'_m\}))\theta.$$

I satisfies every ground instance E'_1, E'_2 of E_1, E_2. Suppose E'_1, E'_2 have respectively the forms $(B \vee A : S)$ and $(B' \vee \sim (A : S'))$, where $S = \prod\{S_1, \ldots, S_n\}$ and $S' = \coprod\{S'_1, \ldots, S'_m\}$. If $I \not\models A : S$, then I satisfies the resolvent of E'_1, E'_2 as $I \models B$. But if $I \models A : S$, then $I(A) \geq \mu$ for some $\mu \in S$, say μ_i. There are two cases to consider:

1. If in performing step 1 of gp-resolution, μ_i is deleted from Σ, then $\mu_i \geq \rho$ for some $\rho \in S'$. It follows that $I(A) \geq \rho$. Hence I does not satisfy $\sim (A : S')$. As $I \models E'_2$, it follows that $I \models B'$. Hence I satisfies the resolvent $(B \vee B')$.

2. μ_i is in Σ at step 2. So choose $\Sigma_2 \subseteq \Sigma$ and let $\Sigma_3 = \Sigma - \Sigma_2$.

   (a) If $\mu_i \in \Sigma_3$ so that the residue exists, then the result follows immediately as I satisfies the residue.

   (b) Otherwise $\mu_i \in \Sigma_2$. Note that for every $\rho \in S'$, $\mu_i \not\geq \rho$. Now if $I(A) \geq \rho$ for some $\rho \in S'$ so that I does not satisfy $\sim (A : S')$, the result follows by the same argument as in case 1. Otherwise for all $\rho \in S'$, $I(A) \not\geq \rho$. Since for all $\rho \in S'$ and for all $\gamma \in \mathcal{N}(\rho, \mu_i)$, $\sqcup\{\gamma, \mu_i\} \geq \rho$ and $\gamma \not\geq \rho$, hence γ and $I(A)$ are incomparable by Lemma 4. It follows that $I \not\models A : \delta$, for any $\delta \in \mathcal{N}(\rho, \mu_i)$. Thus $I \models \sim (A : \mathcal{N}(\rho, \mu_i))$. Whence I satisfies the resolvent.

As the above holds for any ground instances of E_1 and E_2, the theorem follows. □

We are able to show refutation completeness of gp-resolution at this point for a restricted class of unsatisfiable em-clauses.

Definition 17 An em-literal $A : S$ [resp. $\sim (A : S)$] is simple iff S is a singleton. An em-clause C is said to be simple if every em-literal in C is simple. If only the positive em-literals in C are simple, we say C is positive simple. A negative simple em-clause is similarly defined.

Lemma 6 Suppose C_1 and C_2 are positive simple em-clauses and they are gp-resolvable. Then all gp-resolvents of C_1, C_2 are positive simple.

Proof. The only place where a new positive em-literal is created in definition 15 is in step 2.a when Σ_3 is non-empty. Since C_1, C_2 are positive simple, it follows that Σ in definition 15 is a singleton. Since Σ_3 contains at most Σ, it follows that Σ_3 can contain no more than a single annotation. □

Lemma 7 Let μ_1, μ_2 and β be truth values. If $\mu_1 \not\geq \beta$, $\mu_2, \not\geq \beta$, and $\sqcup\{\mu_1, \mu_2\} \geq \beta$. Then there exists a $\beta_1 \in \mathcal{N}(\beta, \mu_1)$ such that $\mu_2 \geq \beta_1$.

Proof. $\mathcal{N}(\beta, \mu_1)$ contains all the minimal truth values ρ such that $\sqcup\{\rho, \mu_1\} \geq \beta$. Clearly, since $\sqcup\{\mu_1, \mu_2\} \geq \beta$, μ_2 must be greater than or equal to one of the ρ's. □

Lemma 8 Let S be a finite set of ground positive simple em-unit clauses. S is unsatisfiable iff there are positive em-units $A : \mu_1, \ldots, A : \mu_k, k \geq 1$ and a negative em-unit $\sim (A : T)$ such there exists a $\rho \in T$ such that $\rho \leq \sqcup\{\mu_1, \ldots, \mu_k\}$. (Note that in the case $k = 1$, then $\sqcup\{\mu_1, \ldots, \mu_k\} = \mu_1$.)

Proof. The "only if" case is trivial. In the "if" case, suppose the statement of the lemma is false. Let A be a ground atom occurring in S and let $A : \mu_1, \ldots, A : \mu_k$ be all the positive em-units of S with A as literal. Suppose $\sqcup\{\mu_1, \ldots, \mu_k\} = \mu$. Let I be the interpretation that assigns A the truth value μ. $I \models A : \mu_i$ for all $1 \leq i \leq k$ since $\mu_i \leq \mu$. Moreover, $I \models \sim (A : T)$ for any negative em-unit $\sim (A : T)$ since for all $\rho \in T$, $\rho \not\leq \mu$. Hence I is a model of S, contradicting our assumption that S is unsatisfiable. □

Theorem 8 *(Ground Completeness of Gp-resolution)* Suppose S is a finite unsatisfiable set of ground positive simple em-clauses. Then there exists a gp-refutation of S.

Proof Outline. The main task is in establishing the result for the base case. Once more let $k(S)$ be the excess literal number of S.

(Base Case) $k(S) = 0$. S is a set of unit ground em-clauses. By Lemma 8, there are two cases to consider.

1. There are em-units $A : \mu$ and $\sim (A : T)$ such that there exists $\rho \in T$ such that $\mu \geq \rho$. Applying gp-resolution to the two em-units, we obtain the empty clause.

2. There are $k > 1$ positive em-units $A : \mu_1, \ldots, A : \mu_k$ and a negative em-unit $\sim (A : T)$ such that there exists a $\rho \in T$ where $\rho \leq \sqcup\{\mu_1, \ldots, \mu_k\}$ and for all $\mu_i, 1 \leq i \leq k$, $\rho \not\leq \mu_i$. Gp-resolve $A : \mu_1$ with $\sim (A : T)$ by choosing Σ_2 in definition 15 to be μ_1 and ρ_1 to be ρ; we obtain the reflection $\sim (A : \mathcal{N}(\rho, \mu_1))$. Suppose $\sqcup\{\mu_2, \ldots, \mu_k\} = \alpha$. Without loss of generality, assume $\rho \not\leq \alpha$. By Lemma 7, $\mathcal{N}(\rho, \mu_1)$ contains an annotation β such that $\beta \leq \alpha$. So we can again gp-resolve, this time $A : \mu_2$ with $\sim (A : \mathcal{N}(\rho, \mu_1))$. It is easy to see that by repeating this

process $k-1$ times for each $A:\mu_1,\ldots,A:\mu_{k-1}$, we obtain $\sim(A:T')$ where there exists a $\delta\in T'$ such that $\delta\le\mu_k$. Gp-resolving $\sim(A:T')$ with $A:\mu_k$ results in the empty clause.

(Inductive Step) Similar to the proof of ground completeness for p-resolution. □

Lemma 9 *(Lifting Lemma for Gp-Resolution)* If C_1' and C_2' are instances of C_1 and C_2, respectively, then for each gp-resolvent C' of C_1' and C_2', there is a gp-resolvent C of C_1 and C_2 such that C' is an instance of C. □

We extend Herbrand's Theorem (Theorem 2) to em-clauses.

Theorem 9 Suppose T is any set of em-clauses such that the set of annotations, $\mu\in T$, occurring in T is finite. Then T has a model iff every finite subset of the set of ground instances of T has a model.

Proof Outline. The proof follows from observing that any em-clause has a straightforward translation by DeMorgan's Laws to an equivalent set of mega-clauses. □

Theorem 10 *(General Completeness of Gp-resolution)* Suppose S is a finite unsatisfiable set of positive simple em-clauses. Then there exists a gp-refutation of S.

Proof Outline. Using the new Herbrand's Theorem (Theorem 9), the proof is analogous to the proof of Theorem 5. □
Since mega-clauses are special cases of em-clauses, we have the following.

Corollary 1 Suppose S is a finite unsatisfiable set of mega-clauses. Then there exists a gp-refutation of S. □

The following conjecture is currently under investigation.

Conjecture 1 Suppose S is a finite unsatisfiable set of em-clauses. Then there exists a gp-refutation of S. □

9.5 Conclusions

Our principal aim in this paper was to develop a sound and complete proof procedure for annotated logic over an arbitrary complete lattice of truth values. As shown by our examples, these annotated logics provide a framework for formal reasoning about systems exhibiting inconsistency. In this section, we briefly discuss some future research directions.

1. In [9], an inference rule called decomposition was introduced. Even though in [21] we have demonstrated that decomposition is not necessary for completeness in the presence of cloning, it would be interesting to see how decomposition affects the efficiency of p and p* resolutions.

2. To study the behavior of various traditional resolution strategies as they apply to p and p* resolution.
3. To extend TiPPLe to incorporate em-clauses and gp-resolution.
4. To study strategies for restricting the search space of gp-resolution.

Acknowledgement James Lu would like to thank Nasir Khann for useful discussions.

References

[1] R. Anderson and W. W. Bledsoe (1970): A Linear Format for Resolution with Merging and a New Technique for Establishing Completeness. JACM 17(3), 525–534.

[2] N. D. Belnap (1977): A Useful Four-Valued Logic. In: G. Epstein and J. M. Dunn, eds.: Modern Uses of Many-valued Logic. D. Reidel, 8–37.

[3] H. A. Blair and V. S. Subrahmanian (1987): Paraconsistent Logic Programming. Proc. 7th Conference on Foundations of Software Technology and Theoretical Computer Science. Lecture Notes in Computer Science 287, 340–360, Springer Verlag. Extended version in: Theoretical Computer Science 68, 135–154.

[4] H. A. Blair and V. S. Subrahmanian (1988): Paraconsistent Foundations for Logic Programming. Journal of Non-Classical Logic 5(2), 45–73.

[5] C. L. Chang and R. C. T. Lee (1973): Symbolic Logic and Mechanical Theorem Proving. Academic Press.

[6] P. D. Cheng and J. Y. Juang (1987): A Parallel Resolution Procedure Based on Connection Graph. Proc. Sixth AAAI, 13–17.

[7] N. C. A. da Costa (1974): On the Theory of Inconsistent Formal Systems. Notre Dame J. of Formal Logic 15, 497–510.

[8] N. C. A. da Costa and E. H. Alves (1981): Relations between Paraconsistent Logic and Many-valued Logic. Bulletin of the Section of Logic 10, 185–191.

[9] N. C. A. da Costa, J. J. Lu, V. S. Subrahmanian, and L. J. Henschen (1989): Automatic Theorem Proving in Paraconsistent Logics: Foundations and Experimental Results. Proc. 10th CADE, LNCS, Springer Verlag, 72–86.

[10] N. C. A. da Costa, V. S. Subrahmanian, and C. Vago (1989): The Paraconsistent Logics $\mathbf{P}\mathcal{T}$. to appear in: Zeitschrift fur Mathematische Logik und Grundlagen der Mathematik, 37, 1991.

[11] B. A. Davey and H. A. Priestley (1990): Introduction to Lattices and Order. Cambridge University Press.

[12] M. C. Fitting (1989): Bilattices and the Theory of Truth. Journal of Philosophical Logic 18, 225–256.

[13] M. C. Fitting (1988): Logic Programming on a Topological Bilattice. Fundamenta Informatica 11, 209–218.

[14] M. C. Fitting (1988): Bilattices and the Semantics of Logic Programming. To appear in: Journal of Logic Programming.

[15] D. Fishman and J. Minker (1975): Π-Representation: A Clause Representation for Parallel Search. Artificial Intelligence 6, 103–127.

[16] M. Kifer and T. Krishnaprasad (1989): An Evidence Based Framework for a Theory of Inheritance. Proc. 11th International Joint Conf. on Artificial Intelligence, 1093–1098.

[17] M. Kifer and E. Lozinskii (1989): RI: A Logic for Reasoning with Inconsistency. LICS–89.

[18] M. Kifer and J. Wu (1989): A Logic for Object Oriented Logic Programming. Proc. 8th ACM Symp. on Principles of Database Systems, 379–393.

[19] D. W. Loveland (1970): A Linear Format for Resolution. Proc. IRIA Symp. on Automatic Demonstration, Lecture Notes in Mathematics, Springer, 147–162.

[20] D. W. Loveland (1972): A Unifying View of some Linear Herbrand Procedures. JACM 19, 366–384.

[21] J. J. Lu, L. J. Henschen, V. S. Subrahmanian and N. C. A. da Costa (1990): Automated Deduction in Paraconsistent Logics. The full version of this paper. Tech Report, Northwestern University.

[22] J. Roberge (1988): Distributed Resolution-based Theorem Proving Systems. Ph. D. Thesis, Northwestern University.

[23] V. S. Subrahmanian (1987): On the Semantics of Quantitative Logic Programs. Proc. 4th IEEE Symposium on Logic Programming, Computer Society Press, Washington DC, 173–182.

[24] V. S. Subrahmanian (1989): Algebraic Properties of the Space of Multivalued and Paraconsistent Logic Programs. Proc. 9th Intl. Conf. on Foundations of Software Technology and Theoretical Computer Science, LNCS 405, Springer-Verlag, 56–67.

[25] L. Wos, D. Carson, and G. Robinson (1964): The Unit Preference Strategy in Theorem Proving. Proceedings of the AFIPS Conference 26, 615–621.

[26] L. Wos (1988): Automated Reasoning: 33 Research Problems. Prentice Hall.

Chapter 10

Compiling Recursive Functional Prolog Programs with List Structure into Procedural Languages

Young K. Nam
Lawrence J. Henschen

Department of Electrical Engineering and Computer Science
Northwestern University
Evanston, Illinois

Abstract

We describe methods for translating certain classes of recursive Prolog programs representing functions into ordinary, statement-oriented languages like Pascal or C. The techniques of deductive databases are applied to Prolog programs that are functional, recursive, have no side effects (such as printing values in a particular order) and whose defining rules are mutually exclusive.

10.1 Introduction

We are developing a new kind of Prolog compiler which will generate code directly in more traditional languages like Pascal or C and which is thus directly suitable for integration into traditional software libraries. The compiled code is not simply an alternative form of Prolog interpreter or a variation of WAM-type code and does not require a Prolog interpreter to execute. Rather, our proposed compiler will translate sets of rules into ordinary (Pascal, C, etc.) control structures like WHILE loops and uninterpreted function symbols into ordinary data structures like RECORDs. The approach taken is based on the methods previously applied to deductive databases ([3, 7]). The basic ideas are to distinguish primitively evaluable predicates from defined predicates and to identify for recursive predicates the infinite set of derivable clauses from which simple iterative programs can be derived. We assume the reader is familiar with the basic compilation approach in deductive databases, although we will illustrate the process in the next section for some very simple Prolog programs.

In order to extend the database technique to Prolog, there are several issues that must be addressed. First, unlike deductive databases, function symbols must now be handled. There are two kinds of function symbols in Prolog—the symbols representing computable functions (like arithmetic, etc.) and the uninterpreted function symbols. Techniques for computable functions are already known (see [3]). Uninterpreted function symbols will be transformed into data structures, in particular RECORDs. Second, databases normally find all answers to a query at once whereas Prolog systems normally stop after the first answer. Indeed, some Prolog queries have an infinite set of answers unlike the case for databases. A third issue centers around the declarative/procedural question. Prolog programs are often written to take advantage of the order of the rules and even the literals within a rule. The ordinary deductive database techniques disregard the order of the rules. A fourth issue concerns the non-logical operators of Prolog like CUT and ABORT. We do not claim at this point to have solutions for all of these issues. In particular, we have not yet even considered the non-logical operators, nor can we handle very complex, mutually recursive rule sets (as is the case in deductive databases).

In Section 10.2 we present some motivating examples to illustrate the connection between database compilation and the techniques proposed here. We also recall some of the basic results from deductive database compilation. In Section 10.3 we discuss the handling of uninterpreted function symbols. In Section 10.4 we review our earlier work briefly. In Section 10.5 we review the definitions of functionality and mutually exclusiveness. In Section 10.6 we describe the method. Finally in Section 10.7 we discuss some further research problems.

10.2 Motivating Example/Review of Deductive Database Compilation

We consider first a simple example from deductive databases. Let PAR and F be database relations containing parent and father information and let the following rules be given:

```
ANC(X,Y) :- F(X,Y).                    (1)
ANC(X,Y) :- PAR(X,Z),ANC(Z,Y).         (2)
:- ANC(john,?).                        (3)
```

The answers are given by evaluating all expressions of the form

```
:- PAR(john,X1),...,PAR(Xn,Xn+1),F(Xn+1,?)   (4)
```

The compilation method would produce an optimized evaluation method, in this case a simple while loop of the form

```
X := john; (* X is the current working set *)
put JOIN of X and F into answers;
while X not empty
  do JOIN X and PAR(X,Z) putting Z into Y;
       (* Y is the next level of descendants *)
    add JOIN of Y and PAR into answers;
    X := Y;(* for next iteration, the current level
              becomes the new working set *)
  od;
```

For details see [7]. We merely note here that the generation of new variables Xi in (4), the choice of variable Z to put into Y, and the assignment statement X := Y in the program are all done automatically by analyzing the unifiers in the connection graph.

We now consider a very simple Prolog program to print out a list of objects:

```
PRINT([]) :- write("that's all, folks").    (5)
PRINT([U|L]) :- write(U), PRINT(L).         (6)
:- print(A).                                (7)
```

In order to see how the compilation method of deductive databases can be applied, let us first rewrite rules (5) and (6) in a form that appears to be function free:

```
PRINT(X) :- X=NULL, write("that's all, folks").             (8)
PRINT(X) :- head(X,U), tail(X,L), write(U), PRINT(L).       (9)
```

The predicates write, head, and tail are primitively evaluable. (We will discuss in the next section how to automatically derive (8), (9), head and tail from rules (5) and (6).) The compilation technique of [7] classifies this the same as the ANCESTOR example and generates the program:

```
X := A; (*x is the current working list *)
[X=NULL, write("that's all, folks")];
while X not equal NULL
   do JOIN X and
      [head(X,U); tail(X,L); write(U);
      if L=NULL then write("that's all, folks");]
      putting L into Y;
         (* [U,L] is the next level in the list *)
         (* there is no "answer" result *)
      X := Y;(* for next iteration, the current tail
                  becomes the new working list *)
   od;
```

We have kept the exact form of the program except that there is no answer set for this problem. The first bracketed statement would obviously be written normally as an if-then statement, and the statement that JOINs X and the second bracketed expression would normally be written as two assignment statements followed by the write statement. The use of the set notation has been dropped because head and tail are functional. If functionality was not known, then the program would have included sets of (U,L) tuples, and the statement X := Y would have been a set-oriented statement. Except for rewriting the bracketed parts, the above program looks remarkably like what an ordinary Pascal programmer who had never heard of Prolog might have written.

The general case of linear recursion has been described extensively in the literature (see, e.g., [7]). The canonical case has a recursive predicate defined by two rules as follows:

R(X,Y) :- B(X,Y). (10)
R(X,Y) :- A(X,U), R(U,Z), C(Z,Y). (11)

For a query of the form

:- R(a,W). (12)

it easy to see that the set of all answers can be obtained from the set of expressions

B(a,W) (13)
A(a,U1) B(U1,Z1) C(Z1,W) (14)
A(a,U1) A(U1,U2) B(U2,Z2) C(Z2,Z1) C(Z1,W) (15)

.
.
.

A(a,U1) $\cdots$ A(Un-1,Un) B(Un,Zn) C(Zn,Zn-1) $\cdots$ C(Z1,W) (16)

.
.
.

The set of expressions can be determined given just the form of the query without regard to actual values for the instantiated argument. Thus these expressions can be formulated and optimized procedural code using WHILE loops generated for computing answers when the database is being designed, hence the term compilation. Moreover, [7] gives precise termination conditions that can be applied to any actual argument of a real query.

We point out that the A, B, and C in the above rules can themselves consist of several literals. Further, the grouping of the literals into the A- and C-parts is governed by a dataflow analysis based on which arguments in the generic query form represent inputs, as is described in [7]. Because the variable patterns in the solution expressions display such a regular pattern, the set of expressions is often written as

$$A^n \; B \; C^n \tag{17}$$

where it is understood that the various occurrences of A and C are connected in the above pattern. Notice that if the goal (12) is resolved with the recursive rule (11) and then the goal A(a,U) is evaluated next, the variable U would be instantiated. Then the recursive subgoal would also match the generic query form, i.e., it would have its first argument instantiated and return its second argument as answer. This property is called stability in [7], and a modified form of it will be used in Section 10.6.

10.3 Uninterpreted Function Symbols

Uninterpreted symbols like CONS (e.g., []) serve to structure the data, and therefore we transform them into data structures in the target procedural language. We outline a method to automatically infer record structures for the data, leaving only questions about simple data types. (It is unclear whether or not other data structures, like arrays, are relevant and could be automatically inferred from the Prolog program.) Consider rules (5) and (6) from Section 10.2 defining the PRINT predicate and the set of contexts for the various symbols:

```
PRINT([]).
PRINT([X,L]), PRINT(L), write(X).
```

Here we have kept the literals from one clause on a single line; however, for our present purpose distinguishing between antecedent and consequent occurrences is unimportant. By assuming that different occurrences of each predicate (e.g., PRINT) must take the same kind of data, we can draw the following conclusions:

a. because write is primitive and takes primitive type data as argument, the X in rule (6) must represent primitive data;

b. again from rule (6), the second part of [.,.] should be of the same type as the expression [.,.] itself because L occurs as a top-level argument of PRINT and as the second argument of [.,.];

c. the constant NULL ([]) is of the same type as [.,.].

Thus [.,.] is a record structure with two fields—a primitive data type and a field of the same type as the record itself, that is, of type POINTER to the record. The constant NULL occurs in the same context as the record. Thus, PRINT should take as argument a POINTER to this kind of record, and NULL should be a constant of type POINTER.

Note that the type of the primitive data forming the first field cannot be fully determined from rules (5) and (6). If there were other contexts (for example, X occurred in an arithmetic expression), then additional information could be gleaned. However, Prolog (like Lisp and other languages) does not enforce strong typing, so that in general it may be impossible to determine a single type for a simple variable or argument position of a predicate or function. One solution is to mimic the run-time type determination of Prolog by using a variant record for each datum—one field containing a code for the current type and the other containing the actual data. This would require references to be compiled as case statements. This solution is clumsy, and we believe in many cases a more precise type determination can be made at compile time.

Once the basic data structures and types have been determined, it is straightforward to generate predicates like head and tail and to translate their usages to assignment statements like U := X↑.field1 and L := X↑.field2. Of course, a compiler would choose non-mnemonic names, but would still generate references to data and pointer fields of an isomorphic record structure.

10.4 Previous Work

Keep in mind that our main concern is how to handle recursively defined predicates, and we assume that the recursion is in fact linear. In [10, 11] we developed methods to compile general linear recursive Prolog predicates computing with lists into procedural code that directly mimicked the operation of the Prolog program as it would occur with a typical Prolog interpreter, that is, to ensure that the same sequence of operations occurs. To that end we had to distinguish two kinds of Prolog programs—determined and undetermined. As in the deductive-database case, we often assume the recursive part of the definition of the predicate has been reduced (through resolution, for example) to directly recursive rules and base rules of the form:

$$R \text{ :- } A_1,\ldots,A_n,R,C_1,\ldots,C_m. \quad (18)$$
$$R \text{ :- } B. \quad (19)$$

Execution would be invoked by a goal of the form:

:- R. (20)

(As in the preceding section we do not need to show the variables within the various literals explicitly at this point.)

In [10, 11], we called the Aj literals the left part and the Cj literals the right part. Unlike the database case, our goal in [10, 11] was to exactly mimic a Prolog interpreter, and therefore we could not permute the goals in the body according to the dataflow analysis. Because of this, variables that occurred in the right side and in the recursive body literal but nowhere else had to be handled in a special way. Such variables were called tail variables. Programs containing rules with tail variables were called undetermined programs; otherwise they were said to be determined.

We present two simple examples to illustrate the difference between determined and undetermined programs.

Example 10.4.1 *A determined program.*

```
p([]).                                    (21)
p([X|L]) :- write(X),q(L),write(X).       (22)
q([X|L]) :- write(X),r(L),write(X).       (23)
r([X|L]) :- write(X),p(L),write(X).       (24)
```

No rule contains a variable occurring in both a right literal and the recursive body literal only, i.e., a tail variable, so this is a determined program. For goals of the form :- p(A), A is a list; the X's in (21)-(24) are bound variables. This program contains no unbound or answer variables. The database techniques in [7] would mark both occurrences of write in each rule as determined literals and would automatically reorder the goals, for example

```
p([X|L]) :- write(X),write(X),q(L).       (22')
```

because in the database context it doesn't matter how you evaluate the rules and literals as long as the system finds all the answers. Obviously, the reordered rules will not produce the same effect as the original ones; that is why in [10] we modified the notion of determined/undetermined of [7] and referred instead to the left and right parts of rules. The basic analysis of [7] still applies, however, and a goal like :-p(a) has at most one successful evaluation, namely

write,write,write,... (empty),...,write,write,write (25)

where we have again left the individual variables off. Note that if the length of the goal list is not a multiple of 3, the goal will not succeed and the palindromic image will not get printed. Also note that for a list whose length is a multiple of 3, the right parts of the rules cause the reverse image to be printed after the end of the list has been reached (i.e., after rule (21) applies).

Example 10.4.2 *An undetermined program.*

```
p([],0).                                          (26)
p([X|L],Y) :- write(X), q(L,N), Y=N+1.            (27)
q([X|L],Y) :- write(X), r(L,N), Y=N+1.            (28)
r([X|L],Y) :- write(X), p(L,N), Y=N+1.            (29)
```

This is an undetermined program because of the tail variable N in each of (27)-(29). For a goal like :-p(a,Y), the X's are again bound variables. The solution expression is similar to (25) except that 1. the right side contains assignments to a sequence of connected variables and 2. the values of the variables in the right side are not known until the base rule actually fires.

In [10, 11] we were forced to use separate approaches and data structures to handle these two kinds of programs. Again, we remark that the overriding concern in [10, 11] was to exactly mimic an ordinary Prolog compiler, but with compiled, procedurally-oriented code.

Note that if there were base rules for q and r in the above examples, then :- p(a,Y) could succeed for a list of length not a multiple of 3 by using the appropriate other base clause. For the case of Example 10.4.2, depending on the actual base clause used, the processing of the right side could start with different values. This was not relevant in Example 10.4.1 because the right side variables were already determined by the goal and did not depend on the base clause used at the midpoint. (This explains in part why determined and undetermined programs had to be handled differently.) A key point in the method to be presented is that each level of recursion should consume a portion of the input list (otherwise the Prolog program would never terminate). In most cases, even when there are several recursive rules they will all consume the same number of head elements. Thus, it would be quite easy to determine which base rule would be applicable for a given list. For example, if there were a base rule for predicate r in the above examples, then any input list whose length was congruent to 2 modulo 3 would succeed and would use the base rule for r.

10.5 Functional Predicates and Mutually Exclusive Rules

Very often Prolog predicates represent functions instead of general relations; that is, for a given set of values as input for certain arguments, the remaining arguments are uniquely determined. In addition, the definitions are written in such a way that there are no side effects of interest (like the write operations in Examples 10.4.1 and 10.4.2); the object is simply to find the unique answer. In this section we define the functionality and mutually exclusiveness of a program, and in the next section we describe techniques for compiling such programs into much more efficient code than described in [10].

10.5.1 Modes and Functional Dependencies

A mode [6, 9] for a predicate p is a pair (b,u), where b and u are mutually exclusive sets of top-level argument positions in p. A call to predicate p is consistent with mode (b,u) if the set of arguments of p instantiated with ground terms by the call is exactly b. A mode for the head predicate of a rule can be propagated to the body literals in an obvious way (see for example [6] or the dataflow analysis of [7]). Although not generally required in Prolog, very often predicates are used with a single mode. Further, often a Prolog program P and a mode (b,u) for a predicate p yields a kind of functional dependency. That is, given a tuple of values B for the positions b, there is one and only one tuple of values C for positions u such that $P \vdash p(B,C)$. In this case, we say the predicate p is functional relative to mode (b,u); if the mode is clear from context, we also just say p is functional.

10.5.2 Mutually Exclusive Linear Recursive Functional Programs

In [6] the concept of mutually exclusive rules is introduced. Two rules with the same head literal are mutually exclusive for a given call if at most one of them succeeds for that call. Two rules are mutually exclusive relative to a mode M if they are mutually exclusive relative to every call consistent with M. The idea can be extended in an obvious way to sets of more than two rules. We are interested in recursive programs whose sets of recursive rules are mutually exclusive and refer to such programs as mutually exclusive programs.

Example 10.5.1 *A mutually exclusive program for the mode* ({1},{2})

```
p([],0).                                        (30)
p([X|L],Y) :- X>0, p(L,Z), a(Z,Y).              (31)
p([X|L],Y) :- X=0, p(L,Z), b(Z,Y).              (32)
p([X|L],Y) :- X<0, p(L,Z), c(Z,Y).              (33)
```

We will be concerned in the next section with exactly where within the rules the determination can be made as to which rule will succeed. Recall that because we are not concerned with side effects, the techniques of [7] will be used to reorder the literals to be evaluated in a natural order from left to right. Thus, we consider a set of rules forming the recursive part of a program and refer to the set of expressions formed by the bodies of those rules up to the point where it is determined which rule succeeds as the set of mutually exclusive expressions. In Example 10.5.1, the mutually exclusive expressions are $X>0$, $X=0$ and $X<0$. Example 10.5.1 is particularly simple because the mutually exclusive expressions occur as the first literal in each clause. In general this need not be the case. The mutually exclusive expressions may include several literals from each clause and may even include the recursive body literal (i.e., the determination cannot be made until after the recursive call). The max program in the next section is of this latter form. The max program can be handled relatively easily because the recursive rules have no A-part, i.e., no goals to the left of the recursive body literal. The general linear case (with both A- and C-parts) in which the mutually exclusive expressions include the recursive predicate are difficult to handle but fortunately do not seem to occur much in practice.

10.6 Compiling Functional, Mutually Exclusive Programs Having No Side Effects

We now consider how mutually exclusive Prolog rules defining a functional predicate whose computation does not involve side effects can be compiled into procedural code. The method we describe here is much simpler and produces vastly more efficient procedural code than the method given in [10], where our goal was to exactly reproduce all the effects of a Prolog execution. We make heavy use of the techniques and concepts of [7] in this section.

10.6.1 Preliminary remarks

First, because the rules of interest here generate no side effects and compute exactly one result, reordering the literals in the body of a rule is perfectly acceptable. We may therefore adopt the notions of determined/undetermined variables and literals of [7] and may freely reorder the literals to form determined and undetermined parts as in [7].

Second, recall that for a single recursive rule and base rule like

R :- B. (34)
R :- A, R, C. (35)

and a generic goal, if A and C are the determined and undetermined parts for the generic goal form, then the set of expressions of the form

$$A^n \; B \; C^n \quad (36)$$

generates all answers for any actual goal matching the generic format. When there is more than one recursive rule, the general form for answers is slightly more complicated. For example,

R :- A_1, R, C_1. (37)
R :- A_2, R, C_2. (38)

leads to expressions of the form

$$A_{i1} \; A_{i2} \ldots A_{in} \; B \; C_{in} \ldots C_{i2} \; C_{i1} \quad (39)$$

where each ij is 1 or 2. Note that the index sequences on the A- and C-sequences are palindromic. In the database setting, the set of all such expressions must be evaluated in order to find all answers. However, if the predicate R is functional and the rules are mutually exclusive relative to the mode of the generic goal, then for any actual goal G exactly one such sequence for exactly one value of n will succeed and will produce the one and only one answer for G. Of course, the value of n as well as the choice of rule one or two at each point will vary for different actual goals.

Because of the mutual exclusion property, we will write the general solution expression (39) as

$$\begin{pmatrix} A_1 \\ A_2 \end{pmatrix}\begin{pmatrix} A_1 \\ A_2 \end{pmatrix}\cdots\begin{pmatrix} A_1 \\ A_2 \end{pmatrix} B \begin{pmatrix} C_1 \\ C_2 \end{pmatrix}\cdots\begin{pmatrix} C_1 \\ C_2 \end{pmatrix}\begin{pmatrix} C_1 \\ C_2 \end{pmatrix} \quad (40)$$

The intent is that for some level n of recursion and one palindromic threading of the alternatives, the resulting expression will produce the result. We will illustrate this in the next section after which we will describe a modification of the methods of [7] aimed at producing simple and efficient procedural code based on these two remarks.

One might ask at this point why the methods of [10] should not still be used. The answer lies in the fact that even though the predicate of the goal is functional, the Prolog search for the unique answer may still involve a tremendous amount of redundant search. Consider the following example:

Example 10.6.1 *Find the maximum of a list.*

p(X,X). (41)
max([X],X). (42)
max([X|L],X) :- p(X,X), max(L,M), X>M. (43)
max([X|L],M) :- p(X,X), max(L,M), X≤M. (44)

Figure 10.1 shows a resolution tree for Example 10.6.1 with max([1,2,3],L) as goal literal. Notice, for instance, that the children of nodes {2} and {6} are completely the same as are the children of nodes {10} and {13}. Actually, the nodes contained in the threaded line are unnecessary.

Note that in general it is not possible for the Prolog programmer to avoid this problem by cleverly ordering the rules because the order of usage of rules, and therefore potential back-ups and redundancy, depends on the data in the individual goal. Occasionally non-logical operators like CUT and ABORT can be used to improve performance, but these destroy some of the declarative nature of logic programming. By applying the techniques of [7] and the previous two remarks to reorder the rules and goals, we will be able to generate procedural code that does in fact avoid most, and in some cases all, of this kind of redundancy without resorting to non-logical aids.

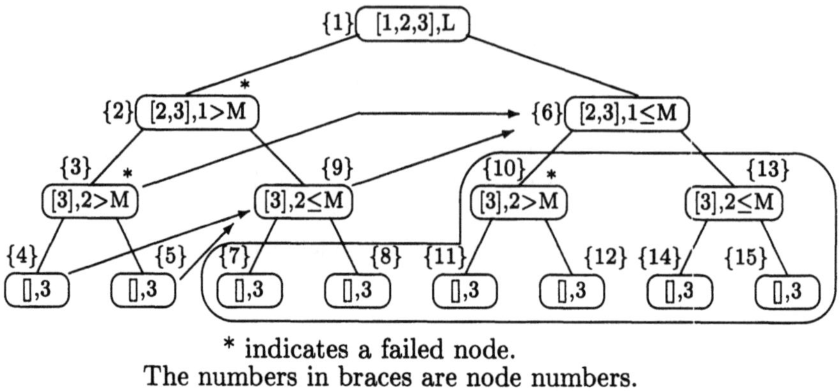

Figure 10.1: Resolution Tree for Example 10.6.1

10.6.2 Examples

We present two admittedly simple examples in this section to illustrate the basic approach. Consider a program to delete an element from a list.

Example 10.6.2 *Delete an element from a list.*

delete(X,[],[]). (45)
delete(X,[Y|L],M) :- X=Y, delete(X,L,M). (46)
delete(X,[Y|L],[Y|M]) :- X≠Y, delete(X,L,M). (47)

and a generic goal of the form

:- delete(a,[y_1,...,y_n],W). (48)

The mode is ({1,2},{3}). Clearly, this program is functional and mutually exclusive relative to this mode. Furthermore, each application of rule (46) or (47) consumes exactly one element from the list in the second argument position as discussed earlier. In order to motivate some of the details to follow, let us rewrite the two recursive rules as

delete(X,[Y|L],M) :- X=Y, delete(X,L,N), M=N. (49)
delete(X,[Y|L],M) :- X≠Y, delete(X,L,N), M=[Y|N]. (50)

We have taken the undetermined terms in the head literals and replaced them with variables, adding explicit assignments in the bodies of the rules. For a goal

:- delete(a,[b_1,b_2,...,b_n],W). (51)

According to the remarks of Section 10.6.1, the solution looks like

$$\begin{pmatrix} a = b_1 \\ a \neq b_1 \end{pmatrix} \begin{pmatrix} a = b_2 \\ a \neq b_2 \end{pmatrix} \cdots \begin{pmatrix} a = b_n \\ a \neq b_n \end{pmatrix}$$

$$N_n = []$$

$$\begin{pmatrix} N_{n-1} = N_n \\ N_{n-1} = [b_n|N_n] \end{pmatrix} \cdots \begin{pmatrix} N_1 = N_2 \\ N_1 = [b_2|N_2] \end{pmatrix} \begin{pmatrix} W = N_1 \\ W = [b_1|N_1] \end{pmatrix} \quad (52)$$

First, according to the remark about consuming one element each time, we can determine that the number of applications of recursive rules is the same as the length of the second argument of the goal, in this case n. Expression (52) is then interpreted as follows: at each point, one and only one of the alternatives $(a = b_i)$ or $(a \neq b_i)$ applies. When the termination condition is reached, assignments to the variables N_i can be made but, of course, in reverse order. At each point, the assignment chosen for N_i is determined by the corresponding successful goal, either $a = b_i$ or $a \neq b_i$. In [7] we analyzed the form of the solution-expression set and generated an obvious procedural program to accomplish the same thing; in that case, the obvious program was a nested loop form. For the present case, the obvious procedural mechanism that corresponds exactly to the solution expression is to traverse the goal list from head to tail saving the list values (in a stack) and marking which ones were equal to the search value a. This should be followed by a loop that performs the assignments to the N_i variables, but in reverse order of course (which is why the first loop saves its results in a stack).

Now consider a Prolog program to find the maximum of a list. Again, we will write the rules so as to make the assignment of the answers explicit.

```
max([X],X).                                  (53)
max([X|L],N) :- max(L,M), X>M, N=X.          (54)
max([X|L],N) :- max(L,M), X≤M, N=M.          (55)
```

For a generic goal of the form

```
:- max([y1,...,yn],W).                       (56)
```

the general solution expression is

$$N_n = y_n \begin{pmatrix} y_{n-1}{>}N_n, N_{n-1}{=}y_{n-1} \\ y_{n-1}{\leq}N_n, N_{n-1}{=}N_n \end{pmatrix} \cdots \begin{pmatrix} y_1{>}N_2, W{=}y_1 \\ y_1{\leq}N_2, W{=}N_2 \end{pmatrix} \quad (57)$$

Notice again that we may immediately infer from the forms of the rules that the level of recursion will be the same as the length of the input list. This is a special case from [7] in which there is no determined part (nothing to the left of max in the rule bodies). Still, the transformation into procedural code is similar to the previous case, namely, traverse to the end of the list (doing nothing) and then back out picking one or the other of the two possibilities at each level. Notice that because there is no determined part, there is no need to stack up the results of the choices for which rule applied during the traversal from head to tail, as there was in the case of delete.

We close this section with two remarks. First, we want to emphasize again that because of the special nature of the rule sets considered here, we are able to analyze the reasoning process at a metalevel in very much the same way as we did in [7], and we are able to do this for generic goals in a compile phase. Again, this kind of compilation is totally different from the standard compilation of Prolog into WAM code, for example. Second, the procedural programs produced for these two examples are still not what an ordinary Pascal programmer would produce; ordinary Pascal programs for these two problems would process lists from head to tail without the second phase of backing out. We discuss the idea of reversing the order of operations for programs in the concluding section and only remark here that it is an open research problem.

10.6.3 Summary of Method

We summarize here the key points and requirements of the method. We assume we are dealing with a predicate P which is functional relative to a specified mode M and whose program is mutually exclusive. As in the case for deductive databases, we assume that the rules have been reduced through resolution to a set of linearly recursive rules, each with P as the head predicate and one body literal over P, and a (set of mutually exclusive) base rule(s). These are standard assumptions and operations from [7]. In addition, for the initial presentation we make the following assumptions.

1. The variables in the recursive rules can be renumbered in such a way that the combination of head/body literal in each rule is the same. This was the case for delete and max; for example, the rules for delete had the form

   delete(X,[Y|L],M) :- ... delete(X,L,N) ... (58)
   delete(X,[Y|L],M) :- ... delete(X,L,N) ... (59)

   This assumption allows us to easily rewrite the solution expression in the form of (40).

2. We assume that at least one of the arguments of the head literal in the recursive rules is a head-tail type term, i.e., of the form [...|...]. Further, the recursive rules should consume at least one element of the chosen list argument. These conditions together guarantee that the recursion will terminate for any goal consistent with M. Note, it is possible to have several input arguments be lists and several be consumed by each level of recursion, as for example in the case of comparing two lists equal.

3. We assume that any argument which is in a position which is undetermined relative to M (i.e., is an answer argument) and which is a head-tail type term has been replaced by a variable in the head literal plus an assignment operation in the body. We did this for delete when we generated rules (49) and (50) from rules (46) and (47).

Recall that the general solution for databases is given by the set of expressions of the form A^n B C^n and that any number of these can produce answers. Thus, in [7] it was necessary to generate a procedural program consisting of an outer while loop plus an embedded inner loop to evaluate the expanding C-side expression at each outer level. In the present case we know that exactly one expression will lead to the answer for any given goal. Applying the motivation of [7] we formulate the basic procedural scheme as a three-step process:

1. process the A-side until the arguments match the base condition, saving any necessary values and the results of the mutually exclusive expressions in a stack; the values to be saved are those where a variable occurs in both an A-part and a C-part of a rule;

2. apply the (appropriate) base case;

3. process the C-side, choosing the appropriate C for the different rules based on the information saved in the stack.

As with the database case, there are special cases (e.g., no A-side) in which a simplified program can be derived. However, unlike the database case, the termination is determined by finding the situation in step 1 that matches the base case of the recursion. (Recall, for databases, several levels of recursion might produce answers, and the base case(s) had to be applied at each of those levels. This led to a termination condition based on failure to expand the A-side expression.)

We describe several cases and the corresponding procedural forms.

The simplest case is when the mutually exclusive expression forms the entire A-side of each recursive rule. This was the case for delete. Step 1 consists of a while loop whose body consists of a case statement which picks out the one case that applies. The appropriate information for the chosen rule is saved in the stack. It is conceivable that some portion of the mutually exclusive expression is itself *not* functional. For example, we might have two rules of the form

...:- parent(X,Y), male(X), ... (60)
...:- parent(X,Y), female(X), ... (61)

Here, parent is not functional, but the test for male or female reduces to one of a person's two parents.

A slight extension to the first case occurs when the A-side contains other goals. These goals are simply evaluated in the appropriate subcases before the necessary information is pushed to the stack.

There are occasionally situations where the recursive rules contain no C-side at all, as was the case for PRINT. In this situation, obviously, there is no need for step 3 and therefore also no need for a stack.

Very often, the recursive rules have no A-sides. The program for max fits this case. Of course, this case does not have a step 1 except for the part that finds the match for the base case. These base cases are typically either match the empty list or match a list of only one element. In either case, the effect is to move to the end of the list and begin "backing out." For this kind of program it is the loop in step 3 that determines which rule should apply at each stage. In actual implementation we could opt to stack up the values of the list or to assume some primitive in the target procedural language for reversing lists.

The last case we shall discuss here occurs when there are both A- and C-sides but the mutually exclusive expressions include the recursive goal. In this case step 1 must, in fact, evaluate and stack the A-sides of all the recursive rules at each level. Step 3 then completes the determination of which single rule holds at each level.

10.7 Future Research

While the basic scheme of the compilation method has been described, there are many aspects that require additional development before a fully automated compiler can be implemented. We have not said yet how the mutually exclusive expressions are found. This is a theorem-proving problem, and we need to develop perhaps special strategies and methodologies for applying automated reasoning programs to this task.

Another interesting theorem-proving task is to automatically determine when step 3 can be eliminated. Both delete and max could be written in Pascal as a single head-to-tail list traversal, whereas the code produced by our method at present generates a head-to-tail traversal followed by a backing-out process. As humans, of course, we can recognize situations where the operations written in the C-sides of the rules can be "reversed." For example, in the max program we know the sequence of comparisons/assignments in formula (57) can be reversed because the tests can be reversed and the assignments made starting from the front of the list. Unfortunately, this involves a considerable amount of high-level knowledge and reasoning about arithmetic. A similar situation occurs for the delete program; our knowledge about lists tells us as humans that the new list can be constructed in a single head-to-tail pass. But the knowledge and reasoning used to reach this conclusion is vastly different than for the max program. Is it possible to determine what kinds of knowledge are required to make such a transformation and develop the corresponding knowledge base and reasoning component of our compiler so that it can automatically make such transformations?

There are a variety of other less interesting problems to work out. Among these are problems like optimizing when there are common subexpressions among the recursive rules (as in (60) and (61)) and organizing cases where the A- and/or C-parts have references to other recursive functions (for example, the function to see if one list contains another list refers to a second function that determines if a single element is in the second list). We believe these will turn out to be relatively straightforward extensions to work already done in deductive databases.

References

[1] A. V. Aho and J. D. Ullman (1979): Universality of Data Retrieval Languages. ACM PODS.

[2] F. Bancilhon and R. Ramakrishnan (1986): An Amateur's Introduction to Recursive Query Processing Strategies. ACM SIGMOD.

[3] M. Chen and L. Henschen (1985): On the Use and Internal Structure of a Logic-based Decision Support System. J. DSS 1(3), 205–219, Sept.

[4] W. Clocksin and C. Mellish (1981): Programming in Prolog. Springer-Verlag.

[5] S. K. Debray and P. Mishra (1988): Denotational and Operational Semantics for Prolog. J. Logic Program. 5(1), 61–91, Mar.

[6] S. K. Debray and D. S. Warren (1990): Towards Banishing the Cut from Prolog. IEEE Trans. Software Eng. 16, 335–349, Mar.

[7] L. Henschen and S. Naqvi (1984): On Compiling Queries in Recursive First-order Databases. J. ACM 31(1) 47–86.

[8] J. Lin (1989): On Transforming of Logic Program via Variable-Dependence Analysis and Fact Propagation. Ph. D. dissertation, Dept. EECS, Northwestern Univ.

[9] C. S. Mellish (1981): The Automatic Generation of Mode Declarations for Prolog Programs. Dep. Artificial Intell., Univ. Edinburgh, DAI Res. Paper 163, Aug.

[10] Y. Nam and L. J. Henschen (1990): Compiling Linear Recursive Programs with List Structure in Prolog into Procedural Languages. Proc. COMPSAC90, Chicago, IL, Nov.

[11] Y. Nam and L. J. Henschen (1991): On Generating Efficient Procedural Codes from Linear Recursive Prolog Programs with List Structure. Proc. of Intl. Phoenix Conf. on Communications and Computers, Phoenix, AZ, March, (to appear).

[12] J. Naughton, R. Ramakrishnan, Y. Sagiv, and J. Ullman (1989): Efficient Evaluation of Right-, Left-, and Multi-linear rules. Proc. ACM SIGMOD Intl. Conf. 18(2), June.

[13] R. Reiter (1978): Deductive Question-Answering on Relational Databases. In: Gallaire and Minker, eds.: Logic and Databases. Plenum Press.

[14] N. Rowe (1988): Artificial Intelligence through Prolog. Englewood Cliffs, N. J.: Prentice-Hall.

[15] D. Troy, C. Yu, and W. Zhang (1989): Linearization of Nonlinear Recursive Rules. IEEE Trans. Software Eng. 15, 1109–1119, Sept.

[16] D. H. D. Warren (1977): Implementing Prolog-Compiling Predicate Logic Programs. Dep. Artificial Intell., Univ. Edinburgh, Scotland, Res. Reps. 39, 40.

Chapter 11

Aligning Multiple RNA Sequences[1]

Ross Overbeek
Ian Foster

Mathematics and Computer Science Division
Argonne National Laboratory
Argonne, Illinois

11.1 Introduction

Each cell in a living organism contains a complete design for all of the mechanisms that sustain its life. This blueprint is encoded in DNA and is called the *genome* of the organism. The basic activity of the cell involves maintaining an environment in which appropriate concentrations of specific proteins drive the reactions that characterize its normal existence. Under changing external conditions, concentrations of some proteins will be increased and others will be decreased. A new protein molecule is produced by copying a section of the DNA blueprint onto a "tape" and then feeding this tape through a "machine" that produces the protein corresponding to the design on the tape. The tape is actually an RNA molecule, and the machine that constructs new proteins under control of such a tape is called the ribosome.

This basic framework has been understood for many years. However, our understanding of the exact details of the process has necessarily developed extremely slowly. Most of what is known about the reactions that sustain life

[1]This work was supported by the Applied Mathematical Sciences subprogram of the Office of Energy Research, U. S. Department of Energy, under Contract W-31-109-Eng-38.

has been gleaned from indirect evidence produced by cleverly constructed and carefully executed experiments. We are currently in the midst of a technological advance that may dramatically accelerate our understanding; it is now possible to read the DNA blueprints, and the costs of doing so are dropping rapidly. The likely impact of this technological advance can be compared with that of VLSI technology on computing since the mid-1970s. In some sense, the basic science is still the same; however, the rapid technological advances allow exploration of questions that were previously beyond our reach.

The accumulation of databases containing sections of DNA (*sequences*) from many organisms creates new research opportunities. For example, it becomes possible to compare the genetic blueprints of different organisms. We are still at the very early stages of this investigation. At this time, complete genetic listings are available only for viruses. As viruses rely on host cells for many biological functions, viral genomes lack the designs for most of the fundamental mechanisms that drive the cell. However, within ten years there may well be hundreds of complete genomes for simple cellular organisms in our databases.

Although complete nonviral genomes are not yet available, comparative analyses of specific sections of genomes is already well under way. Evolutionary relationships between existing organisms imply definite correspondences between the functional components in their genomes. Hence, investigations often focus on similarities and differences in a single functional unit as it occurs in related organisms. This focus on a single functional component permits one to collect versions from many organisms at a fraction of the cost of accumulating even a single complete genome. Comparative analysis applied to such collections has already produced exciting results. Some collections already contain corresponding functional units from hundreds of organisms, and it seems likely that collections of thousands of related sequences will be available within two to three years.

During the past two years, we have worked closely with a team of biologists who have accumulated several such collections of related sequences. This team, headed by Carl Woese of the University of Illinois at Urbana, is currently attempting to understand exactly how the ribosome performs its fundamental role in sustaining life. As mentioned above, the ribosome is the machine that produces protein molecules from instructions encoded in tapes of RNA called "messenger RNA," or mRNA. A functional ribosome in an organism such as *E. coli* will contain three distinct RNA molecules (called rRNA, since these molecules are *ribosomal* RNA), along with over fifty distinct proteins. In simple organisms, the three rRNA molecules are called the 5S rRNA, the 16S rRNA, and the 23S RNA. Woese and his colleagues have collected sequences for these molecules from many hundreds of distinct organisms. In this paper, we focus on the 16S rRNA collection, which comprises over four hundred sequences. The methods that Woese and his colleagues use to extract clues from this collection are of central interest, since this examination of a single functional unit is prototypical—it will be similar in many respects to the attempts of other teams to elucidate other molecular mechanisms.

Extracting Clues from RNA Sequences

The first stage in an investigation of the role of a molecule like 16S rRNA is to identify the molecule's structural units. The *primary structure* of an RNA molecule is expressed as a sequence of characters from the alphabet {A,C,G,U}. For example, the *E. coli* 16S rRNA has a primary structure expressed as a sequence of 1,542 characters. Primary structure is determined by a process called *sequencing.*

Once the primary structure is known, the next step involves determination of how the molecule is folded: its *secondary structure.* The currently hypothesized secondary structure of the *E. coli* 16S rRNA is shown in Figure 11.1. Note that the molecule folds up and that bonds are formed between specific characters in the primary sequence. These bonds will almost always be between A/U, G/C, and G/U pairs. Very occasionally, a G/A interaction occurs (see, for example, the interaction between the G at position 148 and the A at position 174). Lower-level structural units are apparent from the figure. For example, it is natural to think of the sections listed in the following table as distinct structural units.

Positions	Contents
1-8	AAAUUGAA
9-13	GAGUU
14-16	UGA
17-19	UCA
20	U
21-25	GGCUC
⋮	⋮

Given the constraints on the characters that bond, and given the nicely nested pattern formed by the folded molecule, one is tempted to believe that it would be simple to write a program that would take a sequence and produce a "most likely" folded form. Unfortunately, such programs can produce literally thousands of plausible patterns. More accuracy can be gained by looking for foldings that minimize free energy, but these computations are prohibitively expensive for large molecules; furthermore, since the molecule exists in a complex, such energy minimization computations can at best reduce the number of plausible foldings.

In fact, the key to producing an accurate estimate of the secondary structure involves comparative analysis between sequences from several organisms. This process begins by collecting the corresponding sequences from a set of organisms and forming an *alignment.* Such an alignment is used to visually portray the correspondences that exist between versions of the molecule in distinct organisms. For example, here is a small extract from an alignment of the 16S rRNA:

```
                  <<<<   <<<<<    >>>>> >>>>
E. coli        UAUUGCACAAUGGGC-GCAAGCCUGAUGCAGCCAUGCCGCGUGUAUGAAG
F. halmephi    UAUUGGACAAUGGGC-GAAAGCCuGAUCCAGCCAuGCcGngUGUGUGAAG
Thb. thioox    UUUUUCGCaAUGGGG-GCAACCCuGACGAAGCAAUGCCGcGUGAAUGAAG
Fus. nuclea    UAUUGGACAAUGGACCGAGAGUCUGAUCCAGCAAUUCUGUGUGCACGAUG
M. iowae       UUUUUnACAAUGGGC-GAAAGCCUgAUGGAGCAAUCCCGcGUGGAUGAUG
Mb. formici    ACCUCCGCAAUGCAC-GAAAGUGCGACGGGGGAAACCCAAGUGCCA----
```

The scientists who created this alignment took the sequences of the 16S rRNA from the differing organisms and inserted *indels* (i.e., the dashes) to cause corresponding parts of the sequences to align visually. Lowercase letters are meant to convey uncertainty, and the presence of characters such as "n" are used to convey ambiguity ("there is a character, but it could be anything").

To gain some insight into how the secondary structure is deduced, note that the little alignment given above corresponds to positions 365 through 413 in the *E. coli* sequence. If one examines the secondary structure (Figure 11.1), one can see that positions 368–371 bond with positions 390–393 and positions 375–379 bond with 384–388 (to form structures called *helixes*). For convenience, we have included a line just above the alignment to represent the bonded positions. Now, look at the columns that include *E. coli* positions 378 and 385:

Organism	378	385
E. coli	G	C
F. halmephi	G	C
Thb. thioox	G	C
Fus. nuclea	A	U
M. iowae	G	C
Mb. formici	A	U

Note that in each case the positions contain one of the common bonding pairs (G-C for four of the organisms and A-U for the other two). The columns seem to *covary*. It appears as if a mutation in one position produced a second mutation in the corresponding position. In reality, a mutation in one of the bonding columns normally produces a nonbonding pair; if a second mutation does not occur fairly quickly in the corresponding position, the organism disappears through natural selection. This selection against nonbonding pairs produces the detectable phenomenon of covariance.

Covariance is the central tool used in formulating estimates of secondary structure in large RNA molecules. The construction of an alignment that accurately depicts the correspondences between sequences from a variety of organisms is a basic requirement. From an initial alignment, one can construct

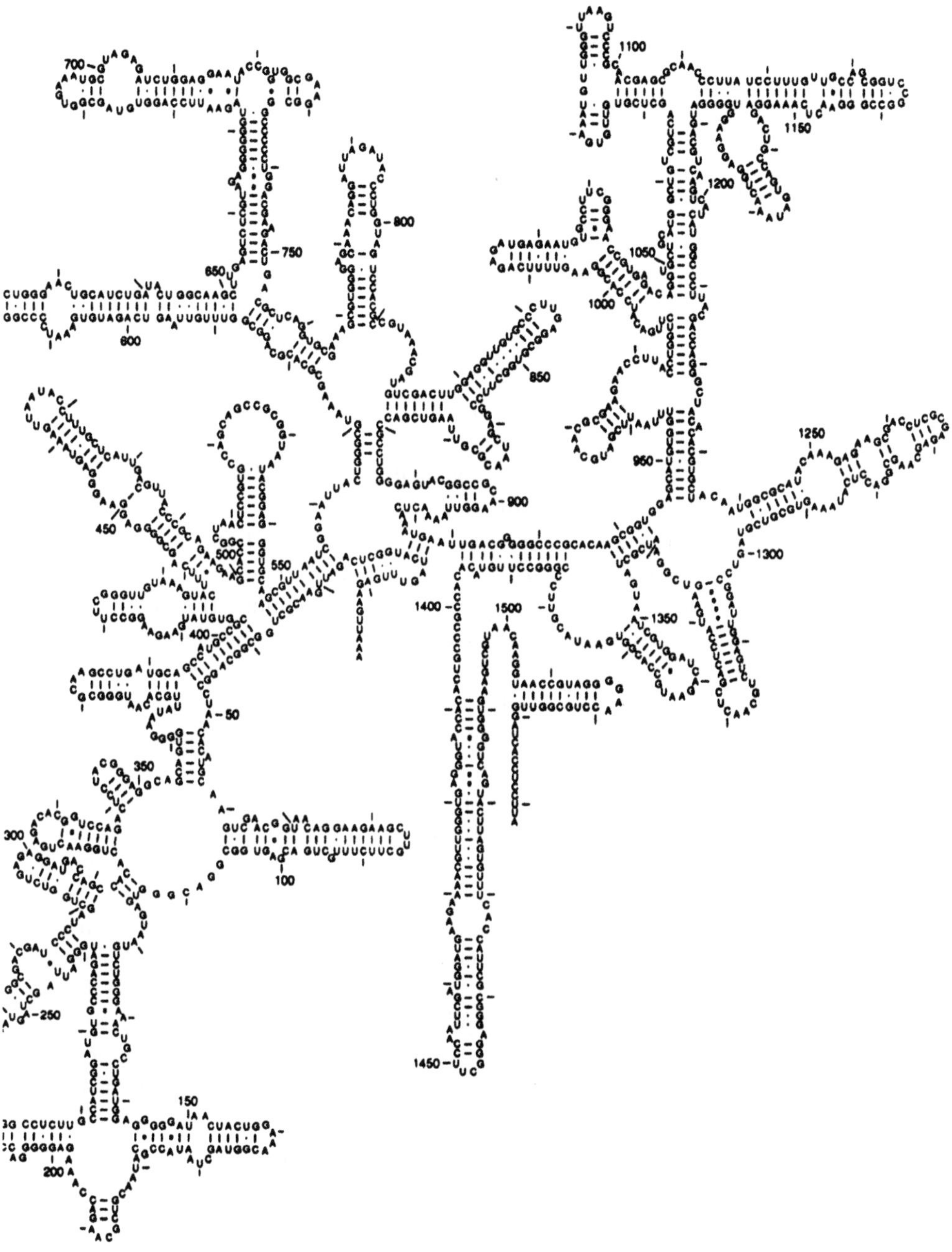

Figure 11.1: Secondary Structure of *E. coli* 16S rRNA

an estimate of secondary structure and phylogenetic relationships. These can be used to improve the alignment, which can then be used to improve computations of secondary structure and phylogeny. That is, the three problems of alignment, secondary structure, and phylogeny all interrelate; a successful exploration begins with the construction of an initial alignment.

11.2 The Alignment Problem

We focus in this paper on the problem of automatically creating an initial alignment of a large collection of related RNA sequences. We shall present algorithms that we believe make far fewer errors than the best alternatives now available. Before we can proceed to a detailed discussion of alignment algorithms, we need to specify what is meant by a "correct alignment" and indicate how we determine which characters in a generated alignment are correct.

Surprisingly, there is no commonly accepted notion of what constitutes a correct alignment. Part of the difficulty comes from the fact that different alignments can be produced to represent different underlying relationships. For example, one alignment might be designed to represent evolutionary origins (i.e., when two characters occur in the same column, this amounts to asserting that they correspond to the same character in the closest common ancestor), while a second alignment might be created to represent correspondence of structural units. Normally, two alignments of this sort would be consistent—two characters occur in the same column in one alignment if and only if they occur in the same column in the other. Unfortunately, there are well-known cases in which this is not true.

The alignments produced by the biologists with whom we work are designed to represent structural correspondence. Hence, we shall cast our problem as: "Automatically generate an alignment of RNA sequences that conforms as closely as possible to what an expert human would produce in his quest to identify the structural components of the RNA sequences." We shall evaluate our progress against some of the more carefully constructed RNA alignments now available. We shall explicitly refrain from attempting to formulate a more precise notion of optimality. We believe that this approach is justified for the following reason. Many man-years of effort went into the construction of the 16S rRNA alignment; a great deal is known about the placement of each character and the distribution of values in each structural unit. While a few of the placements represent essentially arbitrary choices, they represent a small percentage of the overall alignment.

In what follows, we shall refer to three distinct alignments of 16S rRNA sequences. The *large alignment* contains 402 distinct 16S rRNA sequences. The *small alignment* contains 15 sequences selected at random from the large alignment. The *covering alignment* contains 34 sequences that have the property that each column in the large alignment contains at least one nonindel character from these 34 sequences.

11.3 Related Work

We now provide a brief summary of previous attempts to devise algorithms for automatic alignment of RNA sequences.

11.3.1 Aligning Two Sequences

A number of algorithms have been proposed for aligning two sequences. The better algorithms utilize dynamic programming to create alignments that minimize a precisely defined notion of cost. They are sensitive to parameters that characterize the cost function and require time proportional to the product of the lengths of the two sequences being aligned. A number of factors, including computation cost, make dynamic programming-based approaches impractical for large numbers of sequences. However, they prove useful when used as subcomponents in more sophisticated algorithms.

One of the most widely used dynamic programming algorithms is that of Smith and Waterman. We studied the performance of this algorithm on sequences from the covering alignment. After systematically examining a wide range of parameter settings, we picked those that gave the lowest error rates. The algorithm produced error rates ranging from 0.8% to 37%, with an average of 9.4%. Note that these results should not be viewed as either "good" or "bad"; rather, they reflect the degree of similarity between the sequences being aligned. Only 4 of the 561 possible pairings produced an error rate of less than 2% (which corresponds to about 30 distinct positions in the alignment of the pair of sequences). It is this wide diversity among sequences in a collection that makes the task of constructing a good alignment so difficult.

Other algorithms have been proposed that provide more rapid, but somewhat less accurate, alignments of two sequences [14, 1, 11]. We experimented with one of these and achieved error rates in the range 0.7% to 41%, with an average of 9.7%. We chose to use this in our multiple-sequence alignment algorithm, as it greatly improves performance while reducing accuracy only very slightly.

11.3.2 Aligning Multiple Sequences

A number of algorithms for multiple-sequence alignment have been described in the literature [12, 5, 2, 10, 3, 13, 6, 4]. However, most of these can deal only with a small number of sequences. We know of only two that have been used to align large numbers of RNA sequences, although several other programs would also be capable of aligning at least a modest number of such fairly long sequences:

1. *clustal,* a program described in [7, 8]; and
2. *bioserv,* an email service that we established to offer multiple-sequence alignment services to a number of biologists. This uses an alignment algorithm described in [3].

Both of these programs have been used to align as many as 400–500 sequences that range in length from 1400 to 1700 characters. When *bioserv* was used to align the sequences in the covering alignment, it produced an error rate of approximately 17%. *Clustal* produced an error rate of 23% on the same sequences.

The publicly available version of *clustal* can be used to align up to 100 sequences. The *bioserv* system requires rapidly increasing system resources once the set of sequences gets larger than 100–200.

Of the systems cited, only the algorithm described in [3] attempts to reduce the complexity of the overall computation by splitting the problem into independent sections by recognizing conserved regions. In that algorithm, the splitting was achieved by using a technique based on recognition of "pins": unique, conserved subsequences. The basic approach of reducing the problem into subproblems, solving the subproblems, and reconciling the solutions to provide an overall alignment is fundamentally useful (and will be retained in an alignment algorithm based on the techniques described below); however, the approach used for detecting "pins" in *bioserv* is far inferior to the technique for problem reduction described below.

Some of the cited work focuses on generalizing the dynamic programming algorithm that has been so successfully used in aligning pairs of sequences [4]. Indeed, by cleverly constraining the basic dynamic programming algorithm, researchers have been able, in practice, to compute optimal alignments of as many as six protein sequences of 100 characters each. This is well beyond what one would expect to be achievable, given the computational characteristics of the dynamic programming algorithms. One is tempted simply to disregard this result, given our basic objective of producing good alignments of hundreds or even thousands of sequences, where each sequence is often thousands of characters in length. That would be a mistake. One of the contributions of this paper is an algorithm to detect conserved sections of a molecule, reducing the overall alignment problem to a number of far smaller problems that will require a more sophisticated analysis. These problems may well be amenable to solution using multiple-sequence dynamic programming algorithms.

Of the remaining systems, the most successful proceed by forming an estimate of the phylogenetic tree relating the sequences. It is important to note that, since the estimation of phylogenetic relationships is itself a computationally demanding task (at least currently), the estimate is necessarily quite crude in most cases. Once the tree has been computed, one proceeds to perform a sequence of pairwise alignments, first using sequences at the leaves, and then aligning the subalignments. The principal difficulty here is finding an effective method for forming acceptable alignments of subalignments; this problem is almost identical to the problem faced by the system based on "pinning" discussed above. It is worth pointing out one of the common difficulties with this approach.

Consider a section of a large RNA molecule that functions as a "structural unit"—that is, this section of the molecule is under constraints that dramatically limit the set of "acceptable" forms the sequence can take. If the structural unit is one side of a helix, then the constraints are particularly complex; the individual characters must complement characters in the opposing structural unit. To illustrate the basic difficulty, we shall use a contrived example. Consider the case in which the structural unit can be either four or five characters long, and the acceptable values are as follows:

5-Character Values	4-Character Values
ACGUU	AGUU
GUACA	GACA
CACGG	CCGG
UGUAC	UUAC

Presumably, it is obvious how to line up the 4-character values against the 5-character ones. Yet, if one proceeds by forming a consensus of each subalignment and then aligning the result, the obvious relationship is lost (since the consensus simply indicates that any character can occur in each position with equal frequency). We bring up this contrived example only to indicate that any method of aligning consensus sequences or profile analysis will find it difficult to properly handle structural units in which the basic constraints are not on the individual characters, but are somewhat more complex. This problem is particularly acute in sections of RNA in which a helix shows a wide variation in character content in each of the bonding structural units (but the structural units do remain complementary).

11.4 Our Contributions

In the rest of this paper, we present two new algorithms for the automatic construction of initial alignments of related RNA sequences. We demonstrate that these algorithms are significantly more accurate than previous proposals.

In our approach, alignment is viewed as a two-stage process: first one fills in the "solid" sections, and then a completely distinct approach is used to fill in the remaining portions. We provide techniques based on the notions of consistency and verification (defined below) for recognizing and filling in the solid sections. These can be used to create *partial alignments* in which not all characters in the sequence appear and in which each character has an associated "confidence level," which indicates the confidence with which it can be placed in the given column.

The specific contributions of this paper are as follows.

1. We present an efficient algorithm for "splitting" an alignment problem into smaller subproblems. This algorithm requires time proportional to the number of sequences. When this algorithm was run on the three sample collections (recall that these comprise 15, 34, and 402 16S rRNA sequences), it split each set without error. This algorithm is important because it greatly reduces the complexity, and hence the computational requirements, of the alignment problem.

2. We present an algorithm for forming a "partial alignment"—an alignment in which not all characters from the original unaligned sequences are included. This algorithm has two major advantages compared to previous proposals. First, it is substantially more accurate. We have been able to place as many as 74% of the total characters in a large rRNA alignment alignment at error rates of less than 0.3%. In these alignments, the best alternative algorithms make on the order of 15–25% errors on the complete alignment. On the subset of the complete alignment that contains exactly those characters placed in the partial alignment, the alternative algorithms have error rates of approximately 1.5–2.5% (the partial alignment algorithm detects and aligns the conserved sections of the molecule, so it is not surprising that the error rate is relatively low for these sections). Second, the algorithm has a natural mechanism for assigning confidence levels to the placement of specific characters. This permits us to accurately determine the sections of the alignment that are almost completely certain, along with those that are least certain.

We believe that these two algorithms represent appropriate first stages for a complete alignment algorithm that will attain error rates well below the systems that are now commonly in use.

11.5 Basic Concepts

Both our algorithms are based on a single simple idea, namely: *employ consistency checks to produce reliability estimates.* As an example, consider the problem of producing an accurate alignment of two sequences S_1 and S_2. Assume that we have a third, related sequence, S_3. We begin by using a standard binary alignment algorithm to produce an estimate of how S_1 and S_2 should be aligned. We also align the pairs S_1/S_3 and S_2/S_3. We now have three distinct alignments, each of which asserts a correspondence between characters from two sequences. We derive from these correspondences the notions of "verification" and "consistency" and use these notions to arrive at a scheme for computing a "confidence level" in a correspondence.

Verification An alignment of c_1 in S_1 and c_2 in S_2 is *verified* with the alignments S_1/S_3 and S_2/S_3 if there is a character c_4 in S_3 such that S_1/S_3 aligns c_1 and c_4, and S_2/S_3 aligns c_2 and c_4.

Consistency An alignment of characters c_1 in S_1 and c_2 in S_2 is *consistent* with the alignments S_1/S_3 and S_3/S_2 if there is no character c_3 in S_3 such that either

1. S_1/S_3 aligns c_1 with c_3, S_2/S_3 aligns c_4 with c_3, and c_4 and c_2 are distinct characters from S_2; or
2. S_2/S_3 aligns c_2 with c_3, S_1/S_3 aligns c_5 with c_3, and c_5 and c_1 are distinct characters from S_1.

Confidence Level Assume that we have aligned a sequence S with a *target* sequence T and with one or more *verification* sequences, $S_1, \ldots, S_v$. We associate a *confidence level* with each character c_1 that is aligned with a character in c_2 in the S/T alignment. This confidence level is computed as follows. First, we define three quantities:

k1 = the number of verification sequences that have alignments consistent with the alignment of S and T,

k2 = the number of verification sequences that produce inconsistent alignments, and

k3 = the number of verification sequences that produce alignments that verify the alignment of S and T.

Then, we define the confidence level as

$$L = (k_3/v) * (k_1/(k_1 + k_2)).$$

11.6 Partial Alignment Algorithm

The algorithm for constructing a partial alignment from sequences $S_1, S_2, \ldots, S_n$ is summarized in Figure 11.2. Note that this algorithm requires time proportional to the number of sequences to be aligned. The size of the verification set introduces a constant factor into the cost.

11.6.1 Accuracy

A partial alignment produced by this algorithm will contain many erroneous correspondences. (This is clear from the fact cited above that binary alignments of the 16S rRNA sequences produce errors of up to 37%.) However, it turns out that the confidence levels associated with the correspondences allow us to accurately identify which characters are most likely to be misplaced. As

1. Pick a "target sequence" and call it S_T, where S_T is a sequence of length m, t_1,t_2, ..., t_m. The partial alignment will have m columns.

2. Pick v verification sequences distinct from S_T. Call these V_1, V_2, ..., V_v.

3. For each sequence Si of the n sequences, construct the ith row of the partial alignment using these steps:

   (a) Align S_i and S_T.

   (b) For each character C_j in S_i that aligned against a character t_k in S_T, make an entry in position k of the ith row of the partial alignment. The entry should contain a 2-tuple:

   j - the character from S_i that falls at that location

   L - the confidence level at which the placement is being made. The confidence level is computed by using the algorithm described in Section 11.5, with S_T as the target sequence and the V_i as the verification sequences.

Figure 11.2: Partial Alignment Algorithm

an example, we present in Table 11.1 statistics for a partial alignment of the 402 sequences in the large alignment, using 10 verification sequences. The total number of unambiguous characters is 586,227. Note that we can place 80% of the unambiguous characters into the correct column with an error of about 2%, and we can place 62% of the characters making less than 1 error per sequence.

11.6.2 Choosing Target and Verification Sequences

The partial alignment algorithm does not specify how to choose target and verification sequences. Clearly, the columns that get filled in, and the characters that get inserted with confidence, depend critically on the choice of target and verification sequences. We have not yet identified clear-cut rules for making these choices but have instead experimented with a few common-sense approaches. In particular, we have tried as a target the longest sequence (on the grounds that it had the most columns to be filled in) and a sequence of average length (on the grounds that it is is less likely to be genetically distant from most of the other sequences). We have chosen verification sequences by just sorting the sequences on length (i.e., number of nonambiguous characters) and picking evenly distributed sequences from the sorted list. We have also experimented with taking evenly distributed sequences from a set formed

Table 11.1: Partial Alignment of Large Collection

Min. Conf. Level	Chars. Placed	% Placed	Chars. Misplaced	% Error
0.1	469,870	80.1	9,647	2.05
0.2	462,693	78.9	6,507	1.40
0.3	456,297	77.8	5,063	1.10
0.4	448,439	76.4	3,358	0.74
0.5	441,739	75.3	2,402	0.54
0.6	435,791	74.3	1,998	0.45
0.7	420,763	71.7	1,253	0.29
0.8	398,373	67.9	708	0.17
0.9	368,328	62.8	367	0.09
1.0	368,328	62.8	367	0.09

by eliminating the longest 25% and the shortest 25% of the sequences. All of these approaches seem to give good results. We plan to study the exact effects of different choices.

11.7 Splitting Alignments

RNA molecules of the sort we are aligning have sections that are highly conserved. Hence, there are normally at least some columns in which highly reliable placements will occur in each of the rows. This situation leads to a very simple strategy for partitioning the overall alignment problem: create a partial alignment using only a few verification sequences and split the problem on columns that are completely filled with characters above a designated confidence level. We tested this approach on both the covering alignment and the large alignment using both four and ten verification sequences. The results achieved with ten verification sequences for the covering and large alignments are summarized in Tables 11.2 and 11.3, respectively. In each case, we see that the number of columns in which splitting could be achieved, and the size of the largest gap, varied with the confidence level at which splitting was performed.

In each case, splitting at the 1.0 confidence level produced a totally accurate split (i.e., all of the sequences were partitioned accurately). This was true even when only four verification sequences were employed. The implication for constructing highly accurate alignment algorithms is clear: we can split a computationally demanding task using a small verification set before expending large efforts to align the sections that are not highly conserved.

Table 11.2: Splitting the Covering Alignment

Min. Conf. Level	Split. Col.	Largest Gap	Misp. Chars.	Error Rate
0.1	459	315	47	0.30%
0.2	436	315	21	0.14%
0.3	433	315	18	0.12%
0.4	417	315	13	0.09%
0.5	404	315	11	0.08%
0.6	392	315	8	0.06%
0.7	378	315	3	0.02%
0.8	347	316	0	0.00%
0.9	310	316	0	0.00%
1.0	310	316	0	0.00%

Table 11.3: Splitting the Large Alignment

Min. Conf. Level	Split. Cols.	Largest Gap	Misp. Chars.	Error Rate
0.1	319	336	17	0.01%
0.2	313	339	5	0.00%
0.3	309	339	1	0.00%
0.4	306	339	1	0.00%
0.5	291	344	1	0.00%
0.6	278	344	0	0.00%
0.7	254	370	0	0.00%
0.8	245	370	0	0.00%
0.9	213	397	0	0.00%
1.0	213	397	0	0.00%

11.8 Completing the Alignment

Once an overall alignment problem has been partitioned, there are a variety of ways to proceed to complete the alignment. An obvious first step is to form multiple partial alignments and reconcile them. Early experiments on the small alignment indicate that about 10% of the unplaced characters can reliably be assigned columns when two separate partial alignments are created, rather than a single partial alignment. We have not yet pursued the critical issue of how many characters can be assigned by a series of partial alignments that gradually accumulate more evidence. However, the evidence that we have strongly suggests that, following the original partitioning of the problem, the creation of a series of partial alignments that are then merged into a single alignment is one good way to complete an estimate of the "solid" sections of the alignment.

Once the solid sections of an alignment have been filled in, what remains are sections that either

1. have the same secondary structure, but have high variability in the content of the primary structure, or

2. have distinct secondary structure formations within major phylogenetic groupings.

The standard approach to constructing an alignment, used in both *clustal* and *bioserv*, is based on creating an estimate of the phylogenetic tree relating the sequences. Then, sequences that are most closely related are aligned; these small alignments are recursively reconciled to produce a final alignment. This technique might be the best available way to handle these highly variable regions of the alignment. Alternatively, once estimates of major divisions in the phylogenetic tree have been made, the approach of creating partial alignments for subsets of the sequences in a region might offer more accuracy.

11.9 Summary

The automatic generation of alignments of RNA sequences is critical to understanding of basic cellular processes. Unfortunately, most algorithms in current use are extremely limited in the problem size they can deal with, and produce alignments that are far inferior to those produced by skilled humans, even when the human makes the alignment with no hypothesis about secondary structure. In fact, only two programs have been produced that are capable of aligning large number of sequences, and even these perform poorly: they produce error rates of 15-25% on a set of 34 fairly diverse 16S rRNA sequences. These high error rates dramatically curtail the utility of the algorithms.

In this paper, we have proposed a new approach to multiple-sequence alignment based on the notions of consistency and verification. We have described algorithms based on these notions and demonstrated their effectiveness on large test problems.

Our approach allows us to split large alignments into smaller subproblems using an algorithm that takes time proportional to the number of sequences. In addition, it allows us to assign an estimate of confidence to our placement of characters. We have found that we can place 40–65% of the characters in a typical alignment with extremely high confidence (producing error rates of under 0.2%). The abilities to accurately align major sections of the sequences and to recognize exactly which sections are properly aligned are particularly useful to biologists—even in cases where major regions remain essentially unaligned.

Our algorithms have been incorporated in a system that produces a complete alignment in which the accurately aligned sections are indicated by a color map, while confidence levels on the remaining sections of the alignment are illustrated by different degrees of shading. The coupling of this output with a good sequence editor that allows the biologist to manually align the problematic regions represents a useful advance in approaches to aligning large sets of RNA sequences.

We are currently working on algorithms that manipulate multiple partial alignments to reduce the size of unaligned regions. In effect, these algorithms accumulate additional evidence in areas of low confidence.

At some point, however, good algorithms will have to properly handle sections of the molecule that display high variability. The approach that a human uses is based on detection of secondary structure using a limited set of sequences, and then use of the structure to constrain alignment of additional sequences. We are currently building in a limited ability to automatically detect secondary structure and to use it to align variable regions. We expect, however, that it will take a considerable amount of effort before we can actually match the performance of expert humans.

References

[1] S. Altschul, W. Gish, W. Miller, W. Myers, and D. Lipman (1990): Basic Local Alignment Search Tool. J. Mol. Biol., 214.

[2] G. Barton and M. Sternberg (1987): A Strategy for the Rapid Multiple Alignment of Protein Sequences. J. Mol. Biol. 198, 327–337.

[3] R. Butler et al. (1989): Aligning Genetic Sequences. In: Strand: New Concepts in Parallel Programming. Englewood Cliffs, N. J.: Englewood Cliffs.

[4] H. Carrillo and D. Lipman (1988): The Multiple Sequence Alignment Problem in Biology. SIAM J. Appl. Math. 48, 1073–1082.

[5] D.-F. Feng and R. Doolittle (1987): Progressive Sequence Alignment as a Prerequisite to Correct Phylogenetic Trees. J. Mol. Evol. 23, 351–360.

[6] C. Henneke (1989): A Multiple Sequence Alignment Algorithm for Homologous Proteins Using Secondary Structure Information and Optionally Keying Alignments to Functionally Important Sites. CABIOS 5, 141–150.

[7] D. Higgins and P. Sharp (1988): Gene 73, 237–244.

[8] D. Higgins and P. Sharp (1989): CABIOS 5, 151–153.

[9] D. Lipman, S. Altschul, and J. Kececioglu (1989): A Tool for Multiple Sequence Alignment. Proc. Natl. Acad. Sci. 86.

[10] H. Martinez (1988): A Flexible Multiple Sequence Alignment Program. Nucleic Acids Research 16, 1683–1691.

[11] P. Powell (1989): FASTSIM: A New Algorithm for Rapid Sequence Similarity Determination, Technical Report. TR 89–94, Computer Science Department, University of Minnesota, Minneapolis, Minnesota.

[12] W. Taylor (1988): Pattern Matching Methods in Protein Sequence Comparison and Structure Prediction. Protein Engineering 2, 77–86.

[13] M. Vingron and P. Argos (1989): A Fast and Sensitive Multiple Sequence Alignment Algorithm. CABIOS 5, 115–121.

[14] W. Wilbur and D. Lipman (1983): Rapid Similarity Searches of Nucleic Acid and Protein Data Banks. Proc. Natl. Acad. Sci. 80, 726–730.

Chapter 12

Similarity, Uncertainty and Case-Based Reasoning in Patdex

Michael M. Richter
Stefan Wess

Department of Computer Science
University of Kaiserslautern
Kaiserslautern, Germany

12.1 General Considerations

PATDEX[1] is an expert system that uses case-based reasoning in the diagnosis of faults of complex machines. It is integrated in the MOLTKE workbench[2] for technical diagnosis, which was developed at the University of Kaiserslautern in recent years (cf., e.g., [4, 5, 23]); MOLTKE contains other parts as well (cf., e.g., [16]), in particular a model-based approach (cf. [21, 22]); it is essentially in PATDEX [3] that the heuristic features are located. The use of cases also plays an important role in knowledge acquisition. In this paper we present the underlying principles of PATDEX and embed its main concepts into a theoretical framework. This research has a number of mainly indirect connections to the work of Woody Bledsoe. We mention his interest in analogy, his early connectionist work, and his influence in merging mathematics and artificial

[1] PATtern Directed EXpert Systems
[2] MOdels, Learning and Temporal Knowledge in Expert Systems for Technical Domains

intelligence. For the first author the main point to note is that Woody Bledsoe brought him into contact with AI at an early stage. More than twenty years ago we started a lively discussion that still goes on and that will, it is hoped, last for many more years.

12.1.1 Similarity

Similarity and uncertainty have in common that both can be described by measures of values, e.g., in the real interval $[0, 1]$. At first glance it seems that here the analogy between these concepts comes to an end; we will, however, discuss some more connections later on.

A similarity measure $sim(x, y)$ can be defined on arbitrary objects of interest as physical objects, situations, problems or formulae; let U be the (finite) universe of these objects. The basic axioms for sim are:

1. $sim(x, x) = 1$ (reflexivity)
2. $sim(x, y) = sim(y, x)$ (symmetry)

The dual notion is that of a *distance measure* $d(x, y)$, which may attain arbitrary nonnegative values. In the corresponding axioms reflexivity reads as $d(x, x) = 0$. One does not require, however, the triangle inequality, and one allows $d(x, y) = 0$ for $x \neq y$, which means that d is neither a metric nor even a pseudo-metric.

One says that d and sim correspond to each other iff there is an order reversing one-one mapping

$$f : range(d) \rightarrow range(sim)$$

such that $f(0) = 1$ and $sim(x, y) = f(d(x, y))$; we denote this by $d \equiv_f sim$. Popular candidates for f are: $f(z) = 1 - \frac{z}{1+z}$ for unbounded $range(d)$ or $f(z) = 1 - \frac{z}{max}$, if $range(d)$ has a greatest element max.

Usually individual values of sim or d are not so much of interest as certain relations between them. For use in analogical reasoning the following relations are basic. If d is a distance measure and sim a similarity measure then we define

$$R_d(x, y, u, v) \quad :\Longleftrightarrow \quad d(x, y) \leq d(u, v) \tag{12.1}$$
$$R_{sim}(x, y, u, v) \quad :\Longleftrightarrow \quad sim(x, y) \geq sim(u, v) \tag{12.2}$$

and

$$S_d(x, y, z) \quad :\Longleftrightarrow \quad R_d(x, y, x, z) \tag{12.3}$$
$$S_{sim}(x, y, z) \quad :\Longleftrightarrow \quad R_{sim}(x, y, x, z) \tag{12.4}$$

It is very often easier to determine the relation $S_d(x, y, z)$ than the distance measure d itself and it is also sufficient for many applications.

We say that d and sim are *compatible* iff

$$R_d(x, y, u, v) \iff R_{sim}(x, y, u, v) \tag{12.5}$$

Compatibility is ensured by $d \equiv_f sim$ for some f.

If $M \subseteq U$, then a $y \in M$ is called *most similar* to x with respect to M iff

$$\forall z \in M : S_d(x, y, z)$$

holds. As usual in topology the measures also define a neighborhood concept. For $\epsilon > 0$ we put

$$V_\epsilon(x) := V_{d,\epsilon}(x) := \{y \mid d(x, y) \leq \epsilon\}$$

and analogously $V_{sim,\epsilon}(x)$ is defined; if d is a metric then these sets are ordinary closed neighborhoods. $S_d(x, y, z)$ expresses the fact that each neighborhood of x that contains z also contains y.

There are two uses of similarity we have in mind that are related to each other and play a role in PATDEX. In both cases we regard the objects x as incomplete descriptions of other unknown objects x_0 in which we are really interested.

In the first case $sim(x, y)$ is simply interpreted as the degree of certainty that $x_0 = y_0$ holds. For the other extended intention we regard the objects as problems where some problems are associated with solutions. Then $sim(x, y)$ shall express the degree of certainty that the solution for x is also a solution for y (or, more generally, can easily be modified to a solution of y).

We will generally assume that objects are described by attributes that may take on certain values. There is a fixed number of attributes but the objects of the universe U are identified with tuples of attribute values for some set A of attributes where certain attributes may be missing. This gives rise to many similarity measures and we will discuss one in Section 12.1.3.

12.1.2 Uncertainty

When dealing with incomplete information a certainty measure or factor $\mu(x)$ expresses the degree of certainty that the event x has taken place. Here $\mu(x)$ is, without loss of generality, a real number in $[0, 1]$; it may be interpreted in various ways, e.g., as a probability or some kind of evidence. When dealing with problems and solutions uncertainty applies to two kinds of objects. One was mentioned above, namely the uncertainty of two descriptions describing the same problem. In addition we have the uncertainty of having the correct solution. In case-based reasoning (see 12.1.4 below) both aspects are intimately connected. The problem that in the presence of incomplete information different objects sometimes cannot be distinguished with certainty was treated in the investigations of rough sets and we will introduce the basic notions of this field.

The approach introduced in [17] defines *rough sets* as pairs $(U, \approx)$ where $\approx$ is a binary relation on the universe U called the *indiscernability relation.* If $x \approx y$ holds then x and y are called indistinguishable. With each subset $M \subseteq U$ two approximations are associated:

$$M_l := \{x \in U \mid y \in M \text{ for all } y \text{ with } x \approx y\} \text{ is the lower approximation.}$$
$$M_u := \{x \in U \mid y \in M \text{ for some } y \text{ with } x \approx y\} \text{ is the upper approximation.}$$

The difference $\partial M = M_u/M_l$ is called the *boundary* of M. If $x \in M_l$, then x certainly belongs to M; if $x \in M_u$, then x is surely not in M; ∂M is the area of uncertainty. The uncertainty of $M \subseteq U$ can be measured by the correlated accuracy measure

$$\mu_{\approx}(M) := \frac{card(M_l)}{card(M_u)}; \text{ clearly } 0 \leq \mu_{\approx}(M) \leq 1 \text{ holds.}$$

Rough sets occur in various ways. There are essentially two different types that are both connected with diagnostic problems:

- The indiscernability relation is transitive.
- The indiscernability relation is not transitive.

In the first case $\approx$ defines a partition of U into blocks of indistinguishable elements. The typical example for this arises as above when the objects in U are described by attributes that may take on certain values. Each set A of attributes defines an indiscernability relation $\approx_A$, where $x \approx_A y$ holds iff the values of all attributes in A for x and y are identical.

Because of the finiteness of the universe U the upper and lower approximations of sets as well as the correlated accuracy measure can in principle be computed. In [17] it is shown how this can be expressed in terms of rules and how one can apply it to classification problems; an extension to a decision logic is presented in [18]. For larger numbers these computations become very time-consuming, however.

A prototype for the second case arises from a distance measure d. For each $\epsilon > 0$ there is an indiscernability relation $\approx_{d,\epsilon}$ defined by

$$x \approx_{d,\epsilon} y \text{ iff } y \in V_{d,\epsilon}(x)$$

In case d is a metric, $\approx_{d,\epsilon}$ is transitive and the blocks are of the form $V_{d,0}(x)$. The intention is, of course, that elements in these blocks are in some sense absolutely indistinguishable, i.e., there is no further information available to separate these elements.

12.1.3 Diagnosis

Our area of interest is fault diagnosis and we need to introduce the basic notions.

We assume a fixed number N of *symptoms* $S_1, \ldots, S_N$. With each symptom S_i a range R_i is associated; in principle symptoms are nothing other than attributes. Typically R_i is either a real interval $[a, b]$ or the Boolean domain $\{0, 1\}$ or some other finite set. Symptoms may take on values in their range and these values are assumed to be the only source of information.

Values of symptoms are obtained by carrying out a *test*. A test can be an observation, a measurement, or simply the answer to a question. In some situations certain tests may not be allowed.

The information at some stage of the diagnostic process is usually incomplete and is expressed in the form of an *information vector* or a *situation*. A situation is a vector $Sit = (a_{i1} \ldots, a_{ik})$, $1 \leq i, j \leq N$, such that each $a_{ij} \in R_{ij}$. The components of Sit are the known symptom values whereas the values of the remaining symptoms are unknown. A situation is *complete* if every symptom has a value.

Situations are arranged in the information graph. Its nodes are labelled with situations, and an edge goes from Sit_1 to Sit_2 if Sit_2 has one more component then Sit_1 and there is a test t_j available that can provide the value necessary to extend Sit_1 to Sit_2; in this case t_j is a label of the edge.

A *diagnosis* (or fault description) is a formula of the first order predicate calculus using constants and relations over the ranges R_i; the precise form of these formulae is not of interest here. To avoid technical difficulties we assume always a single fault. This means that the set of complete situations is partitioned into sets representing these faults; a special set is *"no fault"* and, if wanted, another one is *"unknown fault."* The applicability of this approach relies on the fact that at least the "interesting" faults can be fully described. In the diagnosis of even complex machines this assumption is usually satisfied; in medical diagnostics this sometimes may be doubtful.

In a diagnostic problem some complete situations have occurred but are only partially known, i.e., one is confronted with some incomplete situation Sit. The task is to determine the diagnosis of the unknown complete situation (at least with some certainty). At first glance this seems to be a pure classification problem. For each set A of symptoms and each fault diagnosis ϕ we can compute the accuracy introduced above as

$$\mu(\phi) = \mu_{\approx}(\{Sit \mid Sit \text{ is a complete situation that satisfies } \phi\})$$

where $\approx = \approx_A$ as introduced in Section 12.1.2. If $\mu(\phi) = 1$, then ϕ is decided by the symptoms in A and the symptoms not in A are redundant. To recognize redundant symptoms some of the computations mentioned in 12.1.2 are useful; in PATDEX they are complemented using the case-base.

With equal right one can say, however, that the real problem is to find an optimal way to complete incomplete situations sufficiently enough so that a diagnosis with a high degree of certainty can be established. This task has been attacked less successfully in the literature. Hence, when given a situation Sit one may proceed in two ways:

1. Select a test t_k in order to reach a node in the information graph that allows a better diagnosis than in Sit.

2. Present candidates for a diagnosis, possibly accompanied by a certainty factor.

In PATDEX (as well as in the general MOLTKE approach) these two steps play the central role. The second step contains mainly a classification problem. The first step is more complex and can be discussed from different points of view. Above we said that we want an optimal way to complete the information; the term *optimal* is not clearly defined here, however. The main point is that we want an optimization with respect to an unknown target, namely the true diagnosis. We believe that purely information theoretic based approaches like the Top-Down-Inductive-Decision-Trees (ID3, cf. [20]) are insufficient for our purposes, so PATDEX chooses another way.

12.1.4 Case-Based Reasoning

In case-based reasoning (cf., e.g., [14, 13, 25, 8]) one has a base of cases where a case is an ordered pair $c = (\text{problem}, \text{solution})$. The cases are stored in a case base. Instead of solving a new problem directly the case base is employed in order to use solutions from earlier problems.

Hence the heart of case-based reasoning is analogical reasoning (cf., e.g., [11, 9, 10, 12, 27]). The various definitions of case-based reasoning differ in the way that the case base is involved, see, e.g., [28]. All descriptions have in common that the case base has a structure that enables the system to find suitable cases that help in solving an actual problem. This structure plays an important role and therefore case-based reasoning has two ingredients:

- Use analogy to infer solutions from solved problems for an actual problem.

- Add each new case to the case base and update its internal structure.

We will first concentrate on analogy. Given an actual problem p two steps have to be carried out:

1. Select a suitable case $c' = (p', s')$ in the case base.

2. Transfer the solution s' of c' in order to get a solution s of p.

The selection of a suitable case uses a similarity measure. Given p one selects $c' = (p', s')$ such that p' is most similar to p with respect to the case base. In our diagnostic context the problem is a situation and the solution is the diagnosis; hence a case has the syntactic form $c = (Sit, \phi)$. We assume that the diagnosis was correct for the recorded case. There are three basic possibilities for this:

1. Sit did not have sufficiently many symptom values to determine ϕ but one had a good guess.

2. Sit did determine ϕ but contained redundant information, i.e., unnecessary symptom values.

3. Sit did determine ϕ and no smaller situation would do so.

Since one does not know which possibility has occurred one has to take care of all of them.[3] This is a requirement not only for the structure of the case base but also for the similarity measure. A very general type of appropriate similarity measures for situations, which are also used in PATDEX, is of the form (cf. [3, 2]):

$$sim(Sit_1, Sit_2) = \frac{\alpha\, card(E) + \beta\, card(C) + \eta\, card(U_1) + \gamma\, card(U_2)}{card(E \cup C \cup U_1 \cup U_2)} \quad (12.6)$$

where α, β, γ and η are real numbers and

- E := set of attributes with the same values for Sit_1 and Sit_2
- C := set of attributes with different values for Sit_1 and Sit_2
- U_1 := set of attributes with values for Sit_1 but not for Sit_2
- U_2 := set of attributes with values for Sit_2 but not for Sit_1

When a new problem is solved successfully then this gives rise to a new case, which is added to the case base. The experience gained in the solution process can, however, also be used to eliminate cases. The situations of the cases are organized in the information graph; in PATDEX this graph will be labelled in a particular way in order to use it for test selection.

[3] We emphasize that occurrences of incompleteness (1) and redundancy (2) are independent of one another.

12.2 The PATDEX-System

12.2.1 Motivation and Overview

PATDEX[4] is a part of the MOLTKE-System, which was developed in recent years at the University of Kaiserslautern. PATDEX arises from the considerations of the previous section. PATDEX employs the techniques developed in these contexts both in explicit and implicit ways; we will not discuss these details here. Instead we are interested in some shortcomings of these approaches in real world applications that we will point out next.

- The problem with the use of rough sets based on the relations $\approx_A$ for sets A of symptoms consists mainly of the fact that the diagnosis problem is considered only as a classification problem, in particular the indiscernability relations $\approx_A$ for different A are totally unrelated to each other. If $A \subseteq B$ then $\approx_B$ is finer and therefore more informative than $\approx_A$; this should be reflected in the system. Also, one would like to get hints concerning which B for a given A provides the most suitable new information.

- The difficulty with the similarity measure is that its quality is related to the final success of the whole reasoning procedure; this is an *a posteriori* criterion. *A priori* it is not clear what the criteria for similarity of objects should be; they do not only depend on the objects themselves but also on the pragmatics of reasoning. In case-based reasoning it is usually clear whether a solution for a given problem is correct but it is far from clear what it means to say that two problems are similar enough that the solution for one problem also works for the other one. An even more serious difficulty arises when the world of problems is continuously changing.

All this suggests that the similarity should not be defined in some fixed way but instead should be the result of an adaptive learning process. This will be carried out later on.

12.2.2 The PATDEX/1 Prototype

The first version of PATDEX is PATDEX/1. This prototype contains the basic structures that have been extended later on as described in Section 12.2.3.

In this section we will briefly describe this prototype. As basic techniques, PATDEX/1 applies learning by memory adaptation and analogical reasoning. The system has capabilities to memorize and utilize both its individual experiences and its statistical information. The reasoning process that uses this

[4]Actually, there are two systems, PATDEX/1 and PATDEX/2. By PATDEX (or the Patdex approach) we denote all the information that relates to both systems.

experience knowledge is combined with another one that focuses on similarities. The overall process of diagnosis is based on the analogical problem solving algorithm (APS) proposed by [11]. The process is started by the user giving some observed symptom values as input to the system.

The toplevel algorithm of PATDEX reads as follows:

Input : The actual situation Sit
Output : diagnosis ϕ or failure

1. Find a case in the case base with a situation Sit' most similar to Sit. If there is no case with a situation at least "minimally similar" to Sit then stop with failure.
2. If Sit and Sit' are "sufficiently similar" then accept the diagnosis ϕ of Sit' also for Sit and goto 4.
3. Otherwise select a test in order to obtain an improved situation and goto 1.
4. If the diagnosis is correct then add the case (Sit, ϕ) to the case base and stop with success.
5. If the diagnosis is not correct then cancel temporarily (i.e., for the actual problem) all cases with diagnosis ϕ and goto 3.

Here we need an external teacher who says whether a diagnosis is correct or not. We also have to explain *"minimally similar"* and *"sufficiently similar."* For this we need a partition of the case base, which is given after the introduction of the similarity measure. Finally, we have to describe the selection of the next test.

Therefore PATDEX has two main features: similarity and the experience net. Both make use of the case base but are independent and could work in parallel.

For situations PATDEX uses as a first proposal the similarity measure from equation 12.6 with parameters $\alpha = 1$, $\beta = -2$, $\gamma = \eta = -1/2$. It should be remarked that this measure is normalized to $[-2, 1]$ and reads as

$$sim(Sit_1, Sit_2) = \frac{card(E) - 2card(C) - 1/2(card(U_1) + card(U_2))}{card(E \cup C \cup U_1 \cup U_2)} \quad (12.7)$$

This special choice of the parameters is at the moment mainly motivated by experimental results. It has a defensive, pessimistic character. A high negative contribution to the measure is given for conflicting symptom values, i.e., we strongly wish to avoid false diagnoses.

If the value assigned to a given case by the similarity measure exceeds a lower bound ϵ (hypothesis-threshold), this case is said to be qualified for further processing. If the value exceeds an upper bound δ it is even qualified as diagnosis (diagnosis-threshold). Both thresholds are locally defined for each case of the case base. If, for a given case, the similarity value equals 1 this case is said to be proven.

A case becomes disqualified for further use in a particular diagnosis session as soon as all symptoms contained in that case do not hold, given a situation encountered during diagnosis, or if there are no unknown symptom values any more and the specified case does not exceed the diagnosis-threshold. Another reason for disqualification is given if the case the system chooses as its hypothesis is refused by the user. For the use of cases in the top-level algorithm we will define different similarity classes. For this we choose real numbers ϵ and δ such that $0 < \epsilon < \delta < 1$ and define:
Sit_1 and Sit_2 are called

indistinguishable	$sim(Sit_1, Sit_2) = 1$
sufficiently similar	$\delta \leq sim(Sit_1, Sit_2) < 1$
probably similar	$\epsilon \leq sim(Sit_1, Sit_2) < \delta$
at least minimally similar	$0 \leq sim(Sit_1, Sit2) < \epsilon$
not minimally similar	$sim(Sit_1, Sit_2) < 0$

We note that we here have the possibility to make the numbers ϵ and δ dependent on the respective cases.

Test Selection—the Planning of the Diagnostic Process

The analogical problem solving mechanisms of PATDEX/1 are adjusted to the needs of the given domain within the field of technical diagnosis. The basic hypothesis is that the observable similarities concerning the fault behavior of the technical system under consideration normally have similar causes. Therefore the description of the situation of the known case serves as a guideline for the completion of the given partial description of the target case. Thus, analogical transfer for technical diagnosis in PATDEX/1 means: eventually completing a partial description using the most similar case and the experience graph for the guidance of this process. That is, the target situation is completed upon suspicion, so to speak, and then evaluated with respect to new ascertained symptoms and to the relation between the similarity value of the target situation and the given thresholds of the actual most similar case. Successes and failures of this process have their effect in an improvement of the underlying thresholds, whereas the typicalness or frequency of cases has its effect in an improvement of the weights in the experience graph.

With each situation Sit we can associate the set

$$C(Sit) = \{Sit' \mid Sit' \text{ is a complete extension of } Sit\}.$$

When Sit is presented in principle every element of $C(Sit)$ could have occurred. One can take advantage of the fact that not all these elements are equally likely. The experience net will record the information for this purpose.

The information net is an information graph (see 12.1.3) where the nodes represent situations from the case base only. A directed edge leads from Sit_1 to Sit_2 iff Sit_2 contains one more symptom value than Sit_1. In addition, the edges are labelled with determination factors.

Let C be the case base. The determination factor $\Delta(Sit_2, Sit_1)$ is the estimated conditional probability of Sit_2 given Sit_1, i.e.,

$$\Delta(Sit_1, Sit_2) = \frac{card(\{c \in C \mid c = (Sit, \phi), Sit \text{ extends } Sit_2\})}{card(\{c \in C \mid c = (Sit, \phi), Sit \text{ extends } Sit_1\})} \quad (12.8)$$

The quality of the determination factor is a function of the quality of the case base and it is clear that the determination factors should be adapted continuously.

Evaluation of PATDEX/1

PATDEX/1 is a stand-alone prototype. It was completely implemented before the completion of the MOLTKE workbench. It served for modeling the given facts of case-based knowledge processing using the diagnostic problem solving of an expert service technician as a guideline. Important features of this approach are the combination of similarity and experience for the diagnosis of technical systems and the differentiation between classification and test selection. This has to be seen as the fulfillment of a requirement of the underlying real world application. Particularly *derivational analogy* [10] can be elegantly applied to the field of technical diagnosis. Compared with a human engineer PATDEX/1 came off very well, in particular with respect to the similarity measure that has been defined in equation 12.7.

Shortcomings of PATDEX/1 are the difficulty of generalizing the similarity measure and the fact that the case-focussing test selection is not necessarily globally optimal. Usually the complexity of handling the experience graph is super-exponential in space and time because in the worst case all sequences of symptom values have to be represented. Furthermore, PATDEX/1 takes no advantage of causal or functional background knowledge. This increases the possibilities of faulty diagnoses when too many redundant symptom values are presented or some relevant ones are missing.

12.2.3 PATDEX/2

PATDEX/2 (cf. [6, 28]) is an integral part of the MOLTKE workbench that allows the utilization of all its qualities (proposed in [26]). Therefore it is possible to switch between case-based reasoning and the interpretation of a

MOLTKE knowledge base during problem solving. The use of causal knowledge enables PATDEX/2 to identify pathologic symptom values. Thus, redundant information can be filtered off and cannot be the cause for a false diagnosis any more. By the exploitation of functional background knowledge additional symptom values can be derived from the known ones. In this manner the selection of the most similar case is considerably speeded up. The overall case-based reasoning approach that is used by PATDEX/2 is comparable to the memory-based reasoning approach proposed by Stanfill and Waltz in [25].

An important aspect of our PATDEX/2 approach is to view the relevance of certain symptom values for special situations as a part of the empirical knowledge that shall be learned. In PATDEX/2 we combine the case-based reasoning approach for diagnosis with a connectionist approach for learning this empirical knowledge. These relevance measures $w_{ij} \in [0,1]$ are represented by means of a relevance matrix R= $[w_{ij}]$ where the symptoms S_i and diagnoses ϕ_j occur as indices of the rows and columns, respectively. In the course of time the weights of the symptoms, i.e., the elements of the relevance matrix, are learned by PATDEX/2. The strategy for learning the entries of the relevance matrix is similar to the competitive learning mechanism proposed in [24]. The matrix R reads as follows:

$$\mathrm{R} = \begin{array}{ccccc} & \phi_1 & \phi_2 & \cdots & \phi_m \\ S_1 & w_{11} & w_{12} & \cdots & w_{1m} \\ S_2 & w_{21} & w_{22} & \cdots & w_{2m} \\ \vdots & \vdots & \vdots & \ddots & \vdots \\ S_n & w_{n1} & w_{n2} & \cdots & .w_{nm} \end{array}$$

For the degree of relevance of a certain symptom value it is important whether it is a consequence of the normal functioning of the technical system or of a fault—e.g., **`relay 21K3 switched`** is of the first kind while **`voltage 214 too high`** is a pathological symptom value. To identify certain pathological symptom values PATDEX/2 can use the functional background knowledge that is represented in the MOLTKE workbench.

Since PATDEX/1 uses its similarity measure *sim* only for the comparison of two cases, it is not necessary to define relations between symptom values. PATDEX/2 extends this view of similarity by the additional use of local similarity measures $\omega_i(a_{ik}, a_{il})$ that determine the similarity between possible symptom values $a_{ik}, a_{il}, \in R_i$ of a symptom S_i. If one of the symptom values is unknown then the similarity ω_i evaluates to zero.

The introduction of R and ω_i leads to the definition of a new similarity measure, which is normalized to $[0,1]$ and matches the definition made in Section 12.1.1:

$$sim(Sit_1, Sit_2) = \frac{\alpha\mathcal{E}}{(\alpha\mathcal{E} + \beta\mathcal{C} + \eta\mathcal{U}_1 + \gamma\mathcal{U}_2)} \tag{12.9}$$

The attribute sets get replaced by the sums of their relevances. Based on the sets defined in Section 12.1.4, those sums are defined by multiplying the relevance of a specific symptom S_i (represented by the relevance matrix R) with the similarity of the observed symptom value a_{ik} in Sit_1 and the defined symptom value a_{il} in the actual case $c = (Sit_2, \phi_j)$:

$$\mathcal{E} := \sum_{S_i \in E} w_{ij}\omega_i(a_{ik}, a_{il}) \tag{12.10}$$

$$\mathcal{C} := \sum_{S_i \in C} w_{ij}(1 - \omega_i(a_{ik}, a_{il})) \tag{12.11}$$

$$\mathcal{U}_1 := \sum_{S_i \in U_1} v_{ij}(1 - \omega_i(a_{ik}, a_{il})) = card(U_1) \tag{12.12}$$

$$\mathcal{U}_2 := \sum_{S_i \in U_2} w_{ij}(1 - \omega_i(a_{ik}, a_{il})) = \sum_{S_i \in U_2} w_{ij} \tag{12.13}$$

We point out here that ω_i is zero for symptoms S_i that belong to one of the attribute sets U_1 or U_2, because the corresponding symptom values are unknown (cf. Section 12.1.4).

We also restrict the representation of redundant symptoms (i.e., $S_i \in U_1$) to pathological ones. Thus, observed redundant symptom values representing the normal behavior of the underlying technical system cannot decrease the value of *sim* any more. Since PATDEX/2 focusses on the learning of symptom relevance only for the respective diagnosis, no entries for redundant symptoms S_i can be created. Here we need an alternative weighting v_{ij}. In PATDEX/2 we define $\forall i, j : v_{ij} = 1$, which is motivated by the above mentioned restriction of U_1.

By the use of these definitions we get a similarity measure *sim* that is dependent on the values represented in the relevance matrix. After each erroneous diagnosis the weights of the relevance matrix are changed. Thus, the similarity measure *sim* is the result of an adaptive learning process.

Test Selection

As opposed to other known case-based systems, which concentrate on the aspect of classification, PATDEX/2 uses case-based mechanisms for classification as well as for test selection. In PATDEX/2 the case-focussing test selection procedure is extended by a case-based one.[5] This is globally optimal as compared with the already known (strategy) cases. In PATDEX/2 a fixed limit

[5] This subcomponent of PATDEX/2 is a case-based reasoning system of its own, where strategy cases are used that can be automatically generated out of the known diagnostic cases. As it is an improvement of the experience graph and, beyond that, the cost estimation procedure can be viewed as a kind of graph interpretation, we maintain the denotation experience graph for PATDEX/2 for reasons of simplicity.

exists concerning the number of representable strategy cases. This helps to deal with the exponential complexity of the procedure (for the worst case all possible subsets of symptom values have to be represented).[6] If the limit is reached the more typical cases will displace the less typical ones. PATDEX/2 uses an A*-like cost estimation algorithm to choose from among several similar strategy cases. If PATDEX/2 cannot find a sufficiently similar case, a case-focusing test selection procedure, such as in PATDEX/1, will be applied.

12.3 Discussion and Evaluation

Well-known case-based reasoning approaches that can be compared to PATDEX are, among others, the PROTOS ([19, 7, 8]), CASEY [15], CREEK [1] and memory-based reasoning, e.g., MBRTALK [25], approaches. In applying case- and analogy-based reasoning techniques to the problematic nature of test selection and in integrating connectionist methods for adaptive learning, PATDEX goes beyond the state of the art defined by these (and similar) systems, respectively.

PATDEX is similar to the memory-based reasoning approach but, it combines it with the use of additional background knowledge. PATDEX not only exploits causal background knowledge, as CASEY does, but also functional knowledge. PATDEX is a computational model that is fully implemented, unlike CREEK. PATDEX/1, in fact, has been tested since early 1989. Because of its adaptive capabilities PATDEX can be easily used for other domains within the field of technical diagnosis. This is not so easily done for PROTOS because many numerical values and the relational structure needed for the explanation facility have to be transferred to the new domain.

12.4 Conclusion

PATDEX has its well-defined role within the MOLTKE workbench. Thus, PATDEX is fully integrated in an overall view of knowledge acquisition and integration as well as learning. From this view point, PATDEX defines a new state of the art for case-based reasoning systems (with the restriction to the field of technical diagnosis).

In summary, PATDEX closes the gap between the general theoretical basis provided by similarity, uncertainty, rough sets and the associated techniques on the one hand and the requirements of real applications on the other.

[6]In practice, only a small subset of the possible strategy cases occur. Thus, in spite of the limitation of the number of strategy cases, good test selections can be achieved.

12.5 Acknowledgments

The authors are indebted to Klaus Althoff for various discussions. He also was involved in the integration of PATDEX into the MOLTKE projects (cf. [2]), of which he created major parts. We thank S. Nökel and Mike Stadler for correcting an earlier version of this paper.

References

[1] A. Aamodt (1990): A Computational Model of Knowledge-Intensive Learning and Problem Solving. Proc. EKAW–90, 1–20.

[2] K.-D. Althoff (1991): Lernen aus Fallbeispielen zur Diagnose Technischer Systeme. Doctoral dissertation, University of Kaiserslautern, (forthcoming).

[3] K.-D. Althoff, A. De la Ossa, F. Maurer, M. Stadler, and S. Wess (1989): Adaptive Learning in the Domain of Technical Diagnosis. Proc. of Workshop on Adaptive Learning, FAW Ulm, July.

[4] K.-D. Althoff, B. Faupel, S. Kockskämper, R. Traphöner, and W. Wernicke (1989): Knowledge Acquisition in the Domain of CNC Machining Centers: the MOLTKE Approach. Proc. of EKAW–89, 180–195.

[5] K.-D. Althoff, F. Maurer, and R. Rehbold (1990): Multiple Knowledge Acquisition Strategies in MOLTKE. Proc. EKAW–90, 21–40.

[6] K.-D. Althoff and S. Wess (1991): PATDEX/2: Case-Based Knowledge Acquisition in Moltke, Technical Report, University of Kaiserslautern, (forthcoming).

[7] R. Bareiss, K. Branting, and B. Porter (1988): The Role of Explanation in Exemplar-based Classification and Learning. Proc. Case-Based Reasoning, AAAI.

[8] R. Bareiss (1989): Exemplar-Based Knowledge Acquisition. Academic Press.

[9] J. G. Carbonell (1983): Learning by Analogy: Formulating and Generalizing Plans from Experience. In: R. S. Michalski, J. G. Carbonell, and T. M. Mitchell, eds.: Machine Learning. Palo Alto: Tioga Publishing Co.

[10] J. G. Carbonell (1986): Derivational Analogy in Problem Solving and Knowledge Acquisition. In: R. S. Michalski, J. G. Carbonell, and T. M. Mitchell, eds.: Machine Learning II. Morgan Kaufmann.

[11] M. L. Gick and K. J. Holyoak (1980): Analogical Problem Solving. Cognitive Psychology 12, 306–355.

[12] R. Hall (1989): Computational Approaches to Analogical Reasoning: A Comparative Analysis. Artificial Intelligence 39.

[13] K. Hammond (1989): Case-Based Planning. Academic Press.

[14] J. Kolodner (1983): Maintaining Organization in a Dynamic Long-term Memory. Cognitive Science 7, 243–280.

[15] P. Koton (1988): Reasoning About Evidence in Causal Explanations. Proc. AAAI–88, 256–261.

[16] K. Nökel (1989): Temporal Matching: Recognizing Dynamic Situations from Discrete Measurements. Proc. IJCAI–89.

[17] Z. Pawlak (1984): Rough Classification. Int. J. of Man-Machine Studies 20, 469–483.

[18] Z. Pawlak (1990): Decision Logic. Preprint Warszawa.

[19] B. Porter (1986): PROTOS: An Experiment in Knowledge Acquisition for Heuristic Classification Tasks. Proc. First Intern. Meeting on Advances in Learning, Les Arcs, France.

[20] J. R. Quinlan (1986): Induction of Decision Trees. Machine Learning 1, 81–106.

[21] R. Rehbold (1989): Model-Based Knowledge Acquisition from Structure Descriptions in a Technical Diagnosis Domain. Proc. Avignon–89.

[22] R. Rehbold (1991): Integration Modellbasierten Wissens in Technische Diagnostik–Expertensysteme. Dissertation, University of Kaiserslautern, (forthcoming).

[23] M. M. Richter, T. Pfeifer, K.-D. Althoff, B. Faupel, K. Nökel, and R. Rehbold (1990): Final Report, Project X6, SFB RAI—Knowledge-Based Systems, Kaiserslautern.

[24] D. E. Rumelhart and D. Zipser (1985): Feature Discovery by Competitive Learning. Cognitive Science 9, 75–112.

[25] C. Stanfill and D. Waltz (1988): The Memory-based Reasoning Paradigm. Proc. DARPA Workshop on Case-Based Reasoning, Morgan Kaufmann.

[26] M. W. van Someren, L. L. Zheng, and W. Post (1990): Cases, Models or Compiled Knowledge: a Comparative Analysis and Proposed Integration. Proc. EKAW–90.

[27] J. Wolstencroft (1989): Restructuring, Reminding and Repair: What's Missing from Models of Analogy. AI Communications 2(2).

[28] S. Wess (1991): PATDEX/2: Ein System Zum Adaptiven, Fallfokussierenden Lernen in Technischen Diagnosesituationen. SEKI–Working-Paper, SWP–1–91, Kaiserslautern 1991 (in German).

Chapter 13

Formal and Informal Proofs

J. A. Robinson

Syracuse University
Syracuse, New York

Pay attention to the patter by means of which we convince someone of the truth of a mathematical proposition. I mean the patter which awakens intuition.

Ludwig Wittgenstein

It is only the very unsophisticated outsider who imagines that mathematicians make discoveries by turning the handle of some miraculous machine.

G. H. Hardy

When you are satisfied that the theorem is true, you start proving it.

G. Polya

Proofs are not only a means to certainty, but also a means to understanding. Behind each formal proof there lies an idea, or perhaps several ideas.

Saunders MacLane

Proof, in its best instances, increases understanding by revealing the heart of the matter.

Philip Davis and Reuben Hersh

It is a great pleasure to take part in this tribute to Woody Bledsoe in celebration of his seventieth birthday. In his 1985 Presidential Address to the American Association for Artificial Intelligence [2], Bledsoe describes the vision which came to him in a dream a quarter of a century earlier, in 1960, a vision in which he saw the future wonders which AI could make happen. I for one am very glad that he reacted so boldly and courageously to that dream by deciding

> then and there to quit my job and set about spending the rest of my life helping bring this dream to reality ... to make machines act in some fundamental ways like people.

Ever since then, his leadership and enthusiasm have been major assets not only in the relatively small research community devoted to mechanical theorem-proving and automated deduction, but also in the artificial intelligence community at large. His words eloquently express what many of us feel:

> I *do* want to see *my* computer prove difficult new theorems in mathematics, reason by analogy, engage in intelligent discourse, use common sense ... that's what *really* drives me.

I did not meet him until several years later. I too had quit my job in 1960, for similar reasons, although my own vision of the possible future of artificial intelligence was not then as clear as his (and, I am sure, still is not). We first met at the Sandia Corporation in April, 1964, when I went there to give a talk on the resolution principle, which was at that time still quite a recent development. He was on that occasion, as he has been ever since, a source not only of expert, friendly encouragement, but also of constructive, far-sighted, and entirely helpful criticism. Thank you, Woody!

It seems appropriate on this occasion to set down a few reflections on a theme in which Woody has long been interested, and which he has often touched on in his writing and lecturing: the enormous contrast between formal ('machine') proofs and informal ('people') proofs. As the quotations at the head of this piece illustrate, the contrast between formal and informal reasoning concerns almost everyone who has ever had to deal in the business of mathematical proof. Some professionals are plainly uneasy about the inability of formal proofs to fulfill the main purpose for which proofs are supposed to exist: to provide understanding, and to create conviction, in the 'thinking subject'.

For example, the logician J. R. Shoenfield reveals some of this uneasiness in the Introduction to his *Degrees of Unsolvability* [6]. His frank comments are almost confessional and apologetic in tone. They are made in the light of his decision to expound his reasoning in an expository style markedly different from that of his previous writings. He was clearly quite aware that in opting for this easier and more intelligible manner of communicating whatever it is that proofs are supposed to communicate, he was departing not only from his own earlier practice but also from the conventions of austere formality prevailing both in his own special field (the theory of recursive functions) and in much of the rest of modern mathematics:

> My main purpose here is to show that proofs in degree theory need not be as complicated and incomprehensible as they appear to be in the literature ... The main difference between the proofs here and those in the literature is that I have directly given the idea behind the proof, instead of translating it into an obscure language from which the reader must translate back.

I believe that all of us in the field of mechanical theorem-proving share Woody's feeling of frustration that our present limited logical knowledge and the rigid algorithmic technology of today's computer science have so far forced us to work with *formal* proofs (often expressed, as Shoenfield laments, in an 'obscure language'). We would all be much happier if the proofs constructed by our theorem-proving programs were the same kind of informal but clear and rigorous arguments that we ourselves naturally prefer to use in our actual thinking.

An informal proof need not necessarily be lacking in rigor. Its informality comes not so much from its being less rigorous than a formal proof, but rather from its being a kind of meaningful narrative. It is more like a story, or even a drama, conveyed to us in language calling on our semantic and intuitive understanding. It is not simply a computation which we see to be 'correct' because of our perception of its syntactic structure. To follow an informal proof as it unfolds in time is to understand the story as it develops. This requires us to grasp the *meaning* of the words and diagrams, and this in turn calls for extensive 'cognitive processing' in which we mentally represent the entities, structures, relationships and events which are the *dramatis personae* of the proof and the episodes of its plot development.

Informality, of course, does not entail simplicity. Nor are informal arguments entirely immune from the banes of complexity and incomprehensibility. What is important is that they can at their best be simple, clear, elegant and compelling. Formal proofs, on the other hand, seem always to be essentially hard to understand, and inevitably complicated. Why is this? Is it somehow part of the nature of a formal proof?

We have probably all suffered what might be called 'post-formalization blues'. These often descend on us after we have taken the (sometimes considerable) trouble to formalize an informal proof whose idea we understand, only to find that in the resulting formal proof the idea has somehow become quite difficult or even impossible to spot. Formalization seems to hide the ideas or even to destroy them.

Is this an unavoidable side effect of formalization? Can it be that a formal proof always, by its very nature, will be uninformative and hard to understand? Are we always going to have to find out what a formal proof really means by 'informalizing' it—by (in effect) translating it back again into the informal proof from which it originally came?

I think that the answer to these questions is that, unfortunately, it is no accident that formal proofs are deficient in this way. The reason for this is that, at bottom, all systems of formal proof are essentially similar in that they are really no more than notational systems for providing a display or trace of a systematic virtual search computation. Indeed in some formal proof systems, for example in the semantic tableau system and the resolution system, the formal proofs can even be generated as traces of *actual* search computations. What one of these virtual or actual search computations does is to carry out a systematic search of the set of interpretations of the formal language in which the formal proof is written. In any one formal language the search is made through the same set in every case. However, each different proof problem—determined by the conclusion to be proved and the premisses on which the proof may be based—provides its own particular criterion for ending the search. For the goal of the search itself is always this: to find *an interpretation which will make all the premisses come out true while at the same time making the conclusion come out false.* In other words, what is sought is a *counterexample* to the claim that the conclusion is a logical consequence of the premisses. It is the *absence* of such a counterexample which establishes the relationship of logical consequence between the premisses and the conclusion. So, provided that the systematic search is, and is seen to be, guaranteed to find a counterexample if one exists, it will be the *failure* of the search to find a counterexample which will represent the *success* of the attempt to prove the theorem. This 'no counterexample' view of formal proofs has been forcefully advocated by, among others, the logicians Georg Kreisel and Jaakko Hintikka.

A formal proof is then, in this view, simply a suitably organized *display of the reasons why the search for a counterexample to the theorem fails.*

From our standpoint the problem is that what we get from such a 'no counterexample display' proof is only part of what we look for in a proof. We expect two things: firstly, a *guarantee* (of the conclusion's truth, given that of the premisses) and, secondly, an *explanation* (of why the conclusion must be true, if the premisses are). But what we learn from the bare information that a systematic search has failed to turn up a counterexample is simply that there is no interpretation which meets the requirements for being one. No light is shed upon *why* the search *had to* come out the way it did. The formal proof leaves us knowing only that no counterexample exists, without understanding why. It gives us the guarantee but not the explanation.

If we cannot have both, is it not better to have the explanation?

13.1 The Mutilated Chess Board Problem

Before we pursue this point of view further, let us look at an example of the way in which a 'proof by mere search' differs from a 'proof by explanation'. The well-known problem of the mutilated chess board illustrates this distinction very nicely.

This problem seems to have been first posed by the philosopher and logician Max Black, as one of a series of 'exercises in reasoning' in his philosophy textbook *Critical Thinking* (Prentice Hall, 1946; second edition 1952). In his words (p. 157):

> An ordinary chess board has had two squares—one at each end of a diagonal—removed. There is on hand a supply of 31 dominos, each of which is large enough to cover exactly two adjacent squares of the board. Is it possible to lay the dominos on the mutilated chess board in such a manner as to cover it completely?

Professor Black offers no solution as such. However, in the back of the book he gives a list of 'hints for the solution of exercises in reasoning' among which we find the following hint for the mutilated chess board problem (p. 433)

> What are the colors of the squares that the domino covers when placed on the chess board?

The problem next appears in Martin Gardner's *Mathematical Games* column in the February 1957 issue of the magazine *Scientific American*. He acknowledges Max Black's book as the source. Whereas Black had asked merely whether a covering is possible (thus inviting a one-word answer, yes or no) Gardner asks for the *reasoning* justifying the yes or the no:

> If so, show how it can be done. If not, prove it impossible.

Gardner published the solution in the next issue (March 1957) of the magazine :

> It is impossible ... and the proof is easy. The two diagonally opposite corners are the same color. Therefore their removal leaves a board with two more squares of one color than of the other. Each domino covers two squares of opposite color, since only opposite colors are adjacent. After you have covered 60 squares with 30 dominos, you are left with two uncovered squares of the same color. These two cannot be adjacent, therefore they cannot be covered by the last domino.

It is not easy to think of an improvement on Gardner's statement of the impossibility proof. It is simple, clear, elegant and convincing. And informal!

This proof is presumably the one which Max Black had in mind when he gave the hint involving the colors of adjacent squares.

The problem appears again in Claude Berge's *Theorie des Graphes et ses Applications* (Dunod, Paris 1958, pp. 175–176). His informal proof is not quite as satisfying as Gardner's more direct argument. Berge says:

> There is ... a simple, very elegant, proof: let us color the squares alternately white and black (as on the usual chess board), and remark that the two missing squares have the same color; each domino necessarily covers a white square and a black square, so that if the covering were possible with 31 dominos, the truncated chess board would have 31 white squares and 31 black squares, which contradicts our remark.

Interestingly, Black did not provide, along with his statement of the problem, a diagram of the mutilated chess board, whereas Gardner did. Moreover Gardner's diagram showed the usual pattern of alternating colors on which the 'proof by explanation' of the impossibility of a covering essentially rests. It should perhaps be emphasized that (as Black's wording and hint suggest) the characteristic two-color pattern of the chess board is not a given of the problem. A purer form of the problem would be in terms of a plain array of squares, with eight rows and eight columns, removing the first square in the top row and the last square in the bottom row. Let the trick of coloring alternate squares be discovered as part of the problem-solving process!

Of course, it is clear (and here is the point of the example) that the impossibility of a covering can be proved without bringing in this idea at all. We can simply set up a brute-force search which would find a covering if one existed, and record the fact that it fails to find one. This is the 'no counterexample' approach.

There are many ways to organize such a search, but just to fix ideas consider the following one. So as to make the search surveyable, let us consider a truncated 4×4 board. Let the 14 squares of the mutilated 4×4 board be numbered consecutively, starting at 1, from left to right and from top to bottom (so that adjacent squares have numbers differing either by 1 or by 4). The mutilated board will therefore be numbered like this:

	1	2	3
4	5	6	7
8	9	10	11
12	13	14	

Let a *horizontal placement of a domino* be a pair $(a, a+1)$ of numbers in which $a+1$ is not a multiple of 4 and where

$$1 \leq a \leq 13,$$

and a *vertical placement of a domino* be a pair $(a, a+4)$ of numbers satisfying

$$1 \leq a \leq 10.$$

A *placement of a single domino* is either a horizontal placement of a domino or a vertical placement of a domino, and a *placement of k dominos* is then a sequence of $2k$ numbers in which no number occurs more than once and in which successive pairs of numbers are placements of dominos, in ascending order of their first number. Thus (on the 4×4 board) the pairs $(1,2)$ and $(1,5)$ are placements of a single domino, and the sequence $(1,2,3,7,5,6,10,11)$ is a placement of four dominos. With this encoding, we can present a display of a systematic search for a placement of 7 dominos in the form of a tree figure whose nodes, at the *kth* level, are placements of k dominos in which each successive placement of a domino covers the earliest square not yet covered. In this way we systematically explore all the different ways of laying dominos on the board.

Thus for the truncated 4×4 board, the root of the search tree (the *0th* level) is the empty sequence, or the placement of 0 dominos. At the *1st* level we have 2 different placements of 1 domino

$$(1,2) \quad (1,5),$$

at the *2nd* level, 3 different placements of 2 dominos

$$(1,2,3,7) \quad (1,5,2,3) \quad (1,5,2,6),$$

at the *3rd* level, 4 different placements of 3 dominos

$$(1,2,3,7,4,5) \quad (1,2,3,7,4,8) \quad (1,5,2,3,4,8) \quad (1,5,2,6,3,7),$$

at the *4th* level, 6 different placements of 4 dominos

$$(1,2,3,7,4,5,6,10) \quad (1,2,3,7,4,8,5,6) \quad (1,2,3,7,4,8,5,9)$$

$$(1,5,2,3,4,8,6,7) \quad (1,5,2,3,4,8,6,10) \quad (1,5,2,6,3,7,4,8),$$

at the *5th* level, 9 different placements of 5 dominos

$$(1,2,3,7,4,5,6,10,8,12) \quad (1,2,3,7,4,8,5,6,9,10)$$

$$(1,2,3,7,4,8,5,6,9,13) \quad (1,2,3,7,4,8,5,9,6,10)$$

$$(1,5,2,3,4,8,6,7,9,10) \quad (1,5,2,3,4,8,6,7,9,13)$$

$$(1,5,2,3,4,8,6,10,7,11) \quad (1,5,2,6,3,7,4,8,9,10)$$

$$(1,5,2,6,3,7,4,8,9,13),$$

and at the 6*th* level, 8 different placements of 6 dominos

$$(1,2,3,7,4,5,6,10,8,12,9,13) \quad (1,2,3,7,4,8,5,6,9,13,10,11)$$

$$(1,2,3,7,4,8,5,6,9,13,10,14) \quad (1,5,2,3,4,8,6,7,9,13,10,11)$$

$$(1,5,2,3,4,8,6,7,9,13,10,14) \quad (1,5,2,3,4,8,6,10,7,11,9,13)$$

$$(1,5,2,6,3,7,4,8,9,13,10,11) \quad (1,5,2,6,3,7,4,8,9,13,10,14).$$

At this point in the construction it becomes apparent that the search for a placement of 7 dominos must fail, since after each placement of 6 dominos the two remaining uncovered squares are not adjacent.

Provided that we are convinced that this tree of domino placements is the record of a systematic, complete search through all possibilities, it is, as it stands, completely satisfying evidence that *there is no covering* of (that is, no placement of 7 dominos on) the truncated 4×4 board. It is clear that the same method will produce a proof (or a covering) in the case of the truncated 8×8 board.

Yet the impossibility of covering the truncated board remains unexplained even after it has been established, by failure of this search, beyond doubt. The parity argument, however, proves the nonexistence of a covering not just by designing a search which fails to find one, but *by revealing why no such search could ever be successful.* What this example illustrates is the difference between a proof which merely certifies that there is no counterexample to some proposition and thereby establishes its truth, and a proof which not only demonstrates the nonexistence of a counterexample but also supplies some kind of reason why not.

13.2 The 27-subcube Problem

In the same issue (*Scientific American*, February 1957) as that containing the chess board problem Martin Gardner also posed the following problem:

> A carpenter, working with a buzz saw, wishes to cut a wooden cube, three inches on a side, into 27 one-inch cubes. He can do this easily by making six cuts through the cube, keeping the pieces together in the cube shape. Can he reduce the number of necessary cuts by rearranging the pieces after each cut?

This interesting problem is similar to the chess board problem in that it can be solved informally but elegantly, in this case by an appeal to our intuitions about 3-dimensional space, as well as more formally by recourse to a systematic search method. However, it is not at all clear how to organize the search, so that the explanatory proof may be the only practical way to solve the problem. In a letter to the magazine (*Scientific American*, March

1957) it was claimed by Frank Hawthorne that it was he who had first posed this problem in the September-October 1950 issue of *Mathematics Magazine*. The March 1957 issue of *Scientific American* also contains Gardner's version of the solution:

> There is no way to reduce the cuts to fewer than six. This is at once apparent when you focus on the fact that a cube has six sides. The saw cuts straight—one side at a time. To cut the one-inch cube at the center (the one which has no exposed surfaces to start with) must take six passes of the saw.

To appreciate the compelling force of the key idea of this proof, one must imagine the six faces of that central cube and 'feel' the impossibility of creating them by fewer than six acts of cutting. Most people to whom I have shown this example find that mentally viewing the image of the central cube evokes a very strong three-dimensional intuition, partly visual, partly tactile, partly kinesthetic. However, because of the possibility of stacking some or all of the pieces in some clever manner before each act of sawing, it seems to be quite a difficult matter to make a formal survey of all possible sequences of actions.

It is amusing to note that if in this problem the carpenter is not allowed to saw through more than one piece of wood at a time, the minimum number of sawings required to separate out all 27 subcubes is the same as the maximum number, and that these consecutive sawings can be made in any manner at all! For consider: each sawing separates a single piece of wood into two pieces, thereby increasing by 1 the total number of pieces of wood on the work bench. Since the carpenter starts out with 1 piece of wood and finishes with 27 pieces, he must at the end have made exactly 26 sawings, no matter how cleverly he chose to make them. This elegant observation is due to Dr. Chris Witzgall, and is in itself an excellent example of an informal intuitive argument which is entirely convincing.

The philosopher Kant and the mathematician Brouwer both believed that it is our intuitions of space and time which explain the characteristic certainty of our mathematical knowledge. They simply accepted these intuitions as being phenomenologically 'there' as part of the way our conscious thinking occurs. Today we are in no better position than were they to account for these intuitions solely by introspection, but it begins to seem possible that advances in our knowledge of the neural architecture of the brain might throw some light on such things. For example, it is now much clearer how the brain automatically processes visual information so as to make sense of what is projected onto the retinas of our eyes. Our neural computer simply *responds* to this information after only a short delay during which there is time for the propagation of nerve impulses through only a few dozen synapses. Objects and features of objects 'spring to the eye' spontaneously. Most of the time we do not choose (consciously and deliberately) to see what we see. We just see it. Professor Hardy [3] felt that it is only by deliberately engineering just such spontaneous, immediate perceptions in the minds of their readers and listeners that the makers of mathematical proofs obtain their effects. He evidently considered mathematical cognition to be analogous to vision. The thinking subject either grasps what is before it, or does not. This led him to the somewhat paradoxical conclusion that

> there is, strictly, no such thing as mathematical proof; ... we can, in the last analysis, do nothing but *point*.

It seems clear that one cannot escape this conclusion. Any argument, however rigorous, is a collection of *immediate* judgments whose compelling nature rests on the direct intuition of the necessity of the perceived (or conceived) state of affairs. If one were to analyze such an immediate judgment into two or more subordinate judgments, then these in turn would have to supply the immediacy of conviction: the regress cannot be infinite. Judgments must ultimately be grounded and not require further support. For example, it would be absurd to demand proof that, when 'a' is substituted for 'b' in the expression '$2 + b$', the resulting expression is '$2 + a$'. We simply *see that it is so.*

13.3 Resolution Proofs as Displays of Failed Searches

Elsewhere [5] I have argued that the resolution principle can best be viewed as arising naturally out of an analysis of a uniform systematic search scheme for finding a satisfying interpretation of a set of clauses. A resolution proof, on this view, is essentially just what we have left to show for our trouble when we have completed such a search for a satisfying interpretation, but which has failed because the set of clauses is not in fact satisfiable. This general way of looking at formal proofs has been forcefully advocated (but with respect to other formal systems) by Hintikka [4], Beth [1], and other logicians. It has the additional advantage of supplying an elegant *completeness proof* for the formal system of deduction. Here, however, it is not the completeness aspect of the 'no counterexample theory' of resolution which I want to dwell on, but rather the way in which it reveals what the cognitive content, or meaning, of a resolution proof actually is.

In the case of resolution, then, the systematic search associated with each problem is made through the set of all interpretations of the so-called *Herbrand Base* associated with that problem. In the present discussion, we can simply take it that the Herbrand Base of a problem is a denumerable set of atomic sentences or *atoms*, and that moreover there is a particular fixed enumeration

$$A_1, A_2, \ldots$$

of the atoms in terms of which all searches are uniformly organized. The various interpretations that we consider in the search are represented by the corresponding sequences

$$v_1, v_2, \ldots$$

of *truth values*, each v_i being one of f, t. An atom A_i is true (false) in such an interpretation if v_i is t (f).

Now a *ground clause* is a formal sentence

$$P \rightarrow Q$$

where P and Q are finite sets of atoms. This notation is just a shorthand way of representing a conditional sentence of the form

$$(P_1 \wedge \ldots \wedge P_m) \rightarrow (Q_1 \vee \ldots \vee Q_n)$$

in which either or both of m and n can be 0. When both m and n are 0 we have the so-called *empty clause*, which is false in every interpretation.

Regarding P and Q as sets of atoms allows us to abstract from the order and multiplicity of conjuncts and disjuncts, and loses no information since the atoms in P are always made into a conjunction and those in Q are always made into a disjunction. Since the empty conjunction is equivalent to t and the empty disjunction is equivalent to f, the empty clause is equivalent to the conditional sentence $\mathtt{t} \rightarrow \mathtt{f}$, which is equivalent to f.

So, in general, a ground clause $P \rightarrow Q$ is true in an interpretation V if there is an atom in Q which is true in V, or an atom in P which is false in V. Otherwise $P \rightarrow Q$ is false in V.

Let S be a set (finite or infinite) of ground clauses. An interpretation V *satisfies* S if and only if each clause in S is true in V. S is *unsatisfiable* just in case *no* interpretation satisfies it. In other words, an unsatisfiable set of clauses is one for which, no matter what interpretation we consider, at least one of its clauses is false in that interpretation.

Consider then the problem of showing that an unsatisfiable set S of ground clauses *is* unsatisfiable. One way to do this is to organize a search through the interpretations which will certainly detect one which satisfies S, if any such interpretations exist. The following tree method meets this requirement. We construct a *search tree*, whose nodes are the finite words (including the empty word) over the alphabet f, t of truth values. The root node of the search tree is the empty word. Every node V has exactly two immediate successor nodes, namely the words $V\mathtt{f}$ and $V\mathtt{t}$. We think of an n-letter word $v_1 v_2 \ldots v_n$ as giving the first n elements of an interpretation, thus providing truth values for the first n atoms.

It is clear that each branch of the search tree corresponds to an interpretation V; the truth value v_i of the atom A_i in V is the last (rightmost) letter in the *ith* node along the branch corresponding to V.

A node $v_1 \ldots v_n$ of the search tree is said to be a *failure node* for S if at least one clause in S is false when its atoms are given truth values as prescribed in

$v_1 \ldots v_n$. For example the clause

$$(A_2 \wedge A_4) \rightarrow (A_1 \vee A_3 \vee A_5)$$

is false with respect to the node `ftftf`.

If S is an unsatisfiable set, then every branch of the search tree will consist entirely of failure nodes from some node onwards. Let us then prune away all failure nodes on each branch after the earliest one, and label that failure node (which is now a leaf of the tree) with one of the clauses which is false with respect to the truth values specified by it. The resulting tree, with each of its leaves, by construction, a failure node labelled by a clause in S which it falsifies, is called a *failure tree* for S. An appeal to König's Lemma assures us that a failure tree for an unsatisfiable S is a finite tree, even when S is an infinite set.

Just as it stands, a failure tree for S can serve as a formal proof of the unsatisfiability of S. It is a display of the failure of the search (represented by the search tree) to find an interpretation which satisfies S.

However, even though a failure tree for S obviously guarantees that S is unsatisfiable, it sheds little or no light on why this is so. Presumably in any given case the reason why there is no way to satisfy the set of clauses is entirely bound up with what those clauses are saying. In other words, we would have to take into account the *meanings* of the predicate symbols and function symbols in the problem, and not depend just on the logical form exhibited in the syntactic structure of the clauses.

A failure tree does tell us something, though. It tells us (of course) that there is no satisfying interpretation for the set of clauses, but it also tells us how 'difficult' it is to determine that this is so. The number of nodes in a failure tree for S which are *not* failure nodes is a measure of how much searching has to be done before the failure of the search is completely established. Expressed in more cognitive terms, this number is a measure of how difficult it is to 'see' that S is unsatisfiable. We call it the *degree of difficulty* of the set of clauses. Resolution now enters the picture, as a device for reducing the degree of difficulty of an unsatisfiable set of clauses.

Note that if S explicitly contains the empty clause, its degree of difficulty will be zero, since the root of the search tree will already be a failure node, so that a failure tree will consist simply of this single node, labelled by the empty clause.

Now if a set of ground clauses contains two ground clauses of the form $M \rightarrow N \cup \{A\}$ and $\{A\} \cup P \rightarrow Q$ then we say that the ground clause $M \cup P \rightarrow N \cup Q$ is a *(ground) resolvent* of S.

We note that a failure tree for an unsatisfiable set S of ground clauses must either be the trivial tree of only one node (because S contains the empty clause), or else must contain a pair of leaves both of which are immediate descendents of the same node V. Suppose that the word V is of length n. Then V is an *inference node* of the failure tree, and its *depth* is n. The reason for this terminology is the fact that the two failure nodes $V\mathtt{f}$ and $V\mathtt{t}$ are labelled respectively by clauses of the form

$$M \rightarrow N \cup \{A_{n+1}\} \qquad \{A_{n+1}\} \cup P \rightarrow Q$$

in which the sets P, Q, M and N are all included in the set $\{A_1, \ldots, A_n\}$. Thus the resolvent $M \cup P \rightarrow N \cup Q$ can be inferred from, and added to, S, in the knowledge that it is not already a member of S.

The resolvent clause $M \cup P \rightarrow N \cup Q$ cannot be a member of S already, since, if it were, the node V *would be a failure node*: all atoms in M and P are true in V, and all atoms in N and Q are false in V. Therefore when we add this clause to S we obtain an unsatisfiable set with a strictly lower degree of difficulty, since V is a failure node for the new set but not for the old set.

What this discussion shows is that if the empty clause is not a member of an unsatisfiable set S of ground clauses, we can always add a resolvent clause to S, by one step of resolution, thereby producing a set with a lower degree of difficulty. The added clause is of course a logical consequence of S. This resolution process can be repeated until at some stage it is the empty clause which is added.

Thus, overall, we started with a finite subset of S consisting of those clauses which were selected by the search tree and displayed as the labels of the leaves of the failure tree. These clauses serve as the premisses of a formal resolution proof whose inferences are then chosen by successive steps of the procedure outlined above. The result is a resolution proof containing no more inference steps than the number of times that we iterated the procedure (and possibly fewer, since some of the resolvents which are generated may not be used in subsequent steps).

The formal proof, however, has not improved our *understanding* of the unsatisfiability of the set S one iota. All that it has really accomplished for us is to *speed up the process of detecting the unsatisfiability* which could already be detected by the brute-force technique of constructing a failure tree for S. Indeed, after each successive resolution step a set of clauses is obtained which the search tree would find to be unsatisfiable in a shorter time. The final set, containing as it does the empty clause which is false already at the root node, is *immediately recognizable as unsatisfiable* (its degree of difficulty being zero).

If we add the extra feature of unification, and go over the story again so as to deal with resolution in general instead of merely with ground resolution, we will learn nothing new. The ground resolution case is enough to make the point, that a formal resolution proof is no more than a display of an unsuccessful search for a counterexample, and sheds no light on *why* there is no counterexample.

I believe that this is typical of all formal proofs, no matter what their particular flavor may be. There is nothing special about resolution beyond its being an unusually simple formal system. The semantic tableau method of

Beth and Hintikka (described in references [1] and [4]) is a much more elaborate formal system in which there are many different patterns of inference; this fact complicates the story without essentially changing its moral.

The moral is that *informal proofs are the real stuff of mathematical understanding.* Therefore to understand how informal proofs work, we have to understand mathematical understanding, and to understand *mathematical* understanding we have to understand *understanding* in general. But this is just what the general mission of artificial intelligence has always been.

We cannot really separate out a mechanical theorem-proving mission from the general artificial intelligence mission without narrowing the goal down to the point where it loses much of its attraction. By concentrating, as perhaps we have tended to do, on formal systems of proof with their powerful, attractive property of completeness, we may have been only kidding ourselves that we were making our machines be 'more like people'. The example of A. P. Morse to the contrary notwithstanding, people just don't, in fact, work well with formal proofs, just as they don't, in fact, play chess very well by using the minimax method, an evaluation function, and alpha-beta pruning. Real chess is played 'informally' by people, who at their best bring to this task an overwhelming arsenal of intuitions and direct immediate judgments based on vast experience as well as rapid responses.

13.4 Conclusion

We are today still, alas, a very long way from knowing how informal proofs work. But surely this is what we have to achieve if we are to come closer to the ultimate goal of making our machines be more like people, at least in the domain of mathematical cognition. I just cannot see how we shall make progress in this direction until we know a great deal more than we now do about the way the brain works.

Our mechanizable formal systems are no more than (sometimes quite sophisticated) automatic methods for constructing displays which reveal the impossibility of counterexamples to the logical entailments which we feed to them as inputs. This is of course an extremely valuable technique: it is certainly no part of my purpose here to downgrade the usefulness of well-designed mechanical theorem-proving programs, the best of which today are quite remarkably effective. What I have tried to do is to suggest why it is that 'real' informal-but-rigorous proofs are so different from the formal ones which our theorem-provers construct, and why I fear they must always remain so. Nor does my point have anything to do with how proofs are *discovered.* That is an extremely mysterious business, to be sure; but it can be so even for formal proofs or for the purely combinatorial search trees of the game of chess. It is not how proofs, informal or formal, are discovered, but how they are understood, and how they affect our knowledge, that matters.

If there ever comes a time when Woody's dream is fulfilled, one of the ways we shall know it will be not only that our machines will explain elegant informal proofs to us, but also that we will explain elegant informal proofs to our machines, and be confident that they can understand and appreciate what we are telling them.

References

[1] E. W. Beth (1955): Semantic Entailment and Formal Derivability. Originally published in 1955, reprinted in: Jaakko Hintikka, ed.: Philosophy of Mathematics. Oxford University Press, 1969. 9–41.

[2] W. W. Bledsoe (1985): I Had a Dream, AAAI Presidential Address, 19 August 1985, The AI Magazine 7, 57–61.

[3] G. H. Hardy (1929): Mathematical Proof. Mind 38, 1–25.

[4] J. Hintikka (1955): Form and Content in Quantification Theory. In: Two Papers on Symbolic Logic, Acta Philosophica Fennica 8, Helsinki.

[5] J. A. Robinson (1979): Logic: Form and Function. Edinburgh University Press and Elsevier North Holland.

[6] J. R. Shoenfield (1971): Degrees of Unsolvability. Elsevier North Holland.

Chapter 14

PTTP and Linked Inference[1]

Mark E. Stickel

Artificial Intelligence Center
SRI International
Menlo Park, California

Institute for New Generation
Computer Technology
Tokyo, Japan

14.1 Introduction

This paper reports on a combination of two research efforts, the Prolog Technology Theorem Prover (PTTP), and theory resolution in the guise of linked inference. Both received a great deal of encouragement and support over the years from Woody Bledsoe, by personal discussions and by his inviting me to discuss these topics with his students and colleagues at the University of Texas and MCC. He also invited me to cochair with him an AAAI workshop on the encompassing topic of high-performance theorem proving at Stanford in 1989.

This paper is also somewhat concerned with the larger issue of the tradeoffs between large and small inference operations, between keeping or discarding and rederiving intermediate results, and between controlling or ignoring redundancy. Woody has always been interested in these tradeoffs and, for example, was fascinated by the effectiveness of hyperresolution despite the extra unification operations it requires by failing to keep intermediate binary resolvents. Linked inference operations can increase the size of inference operations

[1]This research was supported by the National Science Foundation under Grant CCR-8922330. The views and conclusions contained herein are those of the author and should not be interpreted as necessarily representing the official policies, either expressed or implied, of the National Science Foundation or the United States government.

above that of hyperresolution operations even more than hyperresolution does above binary resolution, so these tradeoffs must be reexamined in this new context. It is too early for us to have definite answers or sometimes even opinions on these matters, but we describe PTTP, its features and some of the deficiencies that we seek to overcome by this work. We then discuss linked inference and how PTTP can be used to efficiently implement linked inference.

14.2 The Prolog Technology Theorem Prover

Prolog is a logic programming language that has the standard theorem-proving functions of resolution and unification, in restricted form, as its basic operations. Early compiled implementations of Prolog executed thousands of resolution operations per second, while conventional theorem-proving programs based on resolution and unification operated at the much slower rate of tens of inferences per second. This performance difference, along with predictions now being achieved of Prolog systems capable of executing millions of inferences per second, makes it worthwhile to examine how Prolog implementation techniques can be used to improve the performance of theorem-proving programs.

Prolog systems offer high-speed execution of logic programs written in the language of first-order predicate calculus clauses, a language in which theorem-proving problems are also often expressed. Can a Prolog system be used as a theorem-proving program by simply giving it the clauses of a problem? It is obvious that, Prolog being a general-purpose symbolic programming language, it could be used to write theorem-proving programs, but this does not imply that their performance would be any higher than programs written in other programming languages. However, the direct execution of the clauses of a theorem-proving problem by a Prolog system, if this were possible, could offer a substantial performance benefit.

Unfortunately, direct execution by a Prolog system of clauses of theorem-proving problems is rarely successful. From the standpoint of executing such clauses, Prolog has three large deficiencies:

- Its unification algorithm is unsound and possibly nonterminating.
- Its search strategy is incomplete.
- Its inference system is incomplete for non-Horn clauses.

The Prolog Technology Theorem Prover (PTTP) [7] represents a particular approach to preserving as much as possible the character and performance of an implementation of a Prolog system while overcoming these deficiencies. PTTP basically implements Loveland's model elimination theorem-proving procedure [2, 3] in the same manner as Prolog.

A simple and effective means of doing this is to transform clauses of theorem-proving problems to executable Prolog clauses [9]; three transformations are used:

- A transformation for sound unification renames variables in clause heads so that the "occur check" problem cannot arise while unifying a clause head and a goal. Remaining unification steps are placed in the body of the clause where they are performed by a new built-in predicate that does sound unification with the occur check.
- A transformation for complete depth-bounded search adds depth-bound arguments to each predicate and depth-bound test and decrement operations to the bodies of nonunit clauses. A driver predicate can then be used to produce a sequence of bounded depth-first searches that simulates complete breadth-first search.
- A transformation for complete model elimination inference adds an argument for the list of ancestor goals to each predicate and adds ancestor-list update operations to the bodies of nonunit clauses; additional clauses are added to perform the model elimination reduction and pruning operations. Prolog inference plus the model elimination reduction operation, which succeeds if a goal is unifiable with the complement of an ancestor goal, is complete for non-Horn clauses.

PTTP is one of the fastest theorem-proving programs in existence when evaluated by its inference rate and performance on easy problems. At SRI, it has been used to solve reasoning problems in planning and natural-language-understanding systems effectively. However, PTTP's high inference rate is achieved at the cost of allowing neither more flexible search strategies nor elimination of redundancy in the search space by subsumption, and its high inference rate can be overwhelmed by its exponential search space. It therefore appears unsuitable for proving many difficult theorems for which conventional resolution theorem-proving programs such as McCune's OTTER system [4] have demonstrated some success.

An advantage of PTTP is that the model elimination procedure is a highly restricted form of resolution that is compatible with the set of support strategy. Just as with Prolog, PTTP's backward chaining from the theorem to be proved is important for applications in which there are many irrelevant axioms, such as planning, natural-language understanding, and deductive databases. Systems like OTTER generally function by attempting to create the deductive closure of the axioms; this is inappropriate in the presence of many irrelevant axioms. If the set of support strategy is used in an attempt to restrict inferences to be relevant to the theorem, completeness may not be preserved, since the hyperresolution and unit-resulting resolution (UR-resolution) procedures customarily used are not always compatible with the set of support strategy.

PTTP has the further advantage of completeness of its inference system and search strategy. Although being able to prove your theorems is more important than any claim of completeness, completeness still has a practical as well as theoretical advantage. With completeness, failure to find a proof of a theorem can always be ascribed to not running the theorem-proving program long enough. Without completeness, there may be doubt about whether the failure is the result of infelicitous use of incompatible restrictions, such as hyperresolution and the set of support strategy, or the absence of costly but occasionally needed operations such as factoring.

Now, what does PTTP do wrongly that limits its ability to prove many theorems that OTTER can prove? Overbeek and Wos [12] stress the importance of subsumption to control redundancy in the search space. Prolog and PTTP lack subsumption, and their implementations largely preclude it. Subsumption requires storage of derived clauses and comparison of old and new derived clauses to see if one subsumes another. Derived clauses are not stored in Prolog and PTTP. Instead, they are implicitly represented, one at a time on the stack, and are destroyed on backtracking.

PTTP's lack of subsumption is the result of a deliberate trade-off between high speed inference and architectural simplicity versus greater control of redundancy and more complex implementation. Storing derived clauses for subsumption costs time and space. The model elimination procedure allows inferences only between a derived clause and an axiom or ancestor clause. The small number of clauses that can be used in an inference operation makes term indexing largely unnecessary. Performing subsumption without indexing is usually prohibitively slow, but indexing itself can be an expensive operation that slows the rate of inference. If the search space is small enough, the higher inference rate of a system that lacks subsumption will cause it to prove the theorem more quickly. Some problems once thought to be unsolvable without subsumption have been solved by PTTP [12]. But, for larger search spaces, the reduction in the size of the search space possible through subsumption more than compensates for the extra time subsumption requires. Although the search spaces for model elimination and, say, hyperresolution are both potentially highly redundant, they are not redundant in precisely the same way, and the caching of solutions to previously seen goals [5] is probably a more appropriate redundancy control mechanism for PTTP than subsumption, though caching imposes similar storage and indexing requirements.

A not so deliberate but crucial deficiency of PTTP is the relative inflexibility of its search strategy. Like Prolog, PTTP is compelled to explore all the consequences of the current derived clause before backtracking to examine alternative clauses. To do otherwise, e.g., to place derived clauses on an agenda and operate on them in order of apparent merit, would again require the storage of derived clauses instead of allowing their highly efficient implicit representation on the stack. But in systems like OTTER, even a default merit-ordered search strategy that orders clauses by their symbol count often (but not always) works much better than a first-in-first-out level saturation strategy. Weighting, basically a more refined symbol counting technique, allows finer control over which clause to choose next and produces even better performance if used well.

Finally, another problem with PTTP is the absence of demodulation, i.e., simplification by equalities. Actually, it would be easy to add demodulation to PTTP, but it would not generally be useful. Although demodulation sometimes shortens a proof, its principal benefit is that its absence can allow an enormous number of variants of clauses that are equationally equal. For example, in group-theory problems where e is the identity, g the inverse operation, and f the multiplication operation, clauses that are otherwise identical may contain the semantically equivalent terms a, $f(e,a)$, $f(e,f(a,e))$, $g(g(a))$, $g(f(g(a),e))$, $f(a,f(b,g(b)))$, and so on. With a suitable set of demodulators, these terms could all be simplified to a and the resulting duplicate clauses could be eliminated by subsumption. However, if subsumption is absent, as in PTTP, there appears to be little benefit to the simplification.

Can PTTP still be used fruitfully in the proof of difficult theorems despite these problems? Perhaps not as a stand-alone theorem-proving program, but as a component of a larger reasoning system. Tarver [10] has had some success with decomposing set-theory problems into subproblems of manageable size for PTTP. Here, we discuss a different approach in which PTTP is used to implement some inference operations of an OTTER-like theorem-proving program. The general idea is to use PTTP instead of ordinary inference operations to derive clauses. These derived clauses are then demodulated, subsumed, stored, and placed on an agenda for future use just as in OTTER. The benefits of rapid inference by PTTP and the redundancy control and flexible search strategy of OTTER could be combined thereby in a single system.

The principal characteristic of PTTP that results in its high performance is the Prolog-style compilation of clauses. The objective of combining PTTP and OTTER-like systems could be achieved by simply compiling clauses for use as nuclei in hyperresolution or UR-resolution operations; such inferences are virtually a degenerate case of PTTP inference. This has been suggested previously [1], but it is likely to be of only marginal utility. The reason is that these operations do not consume a high enough proportion of the total time needed to prove a theorem. A greater benefit could be achieved by compiling the kind of inference operations for which execution consumes a higher proportion of the total time. Linked inference appears to be a good candidate.

14.3 Linked Inference

Many advances in automated theorem-proving research have consisted of replacing sequences of small inference steps by a single larger inference step.

For example, a hyperresolution operation may replace a sequence of P_1-resolution operations, a UR-resolution operation may replace a sequence of unit-resolution operations, a paramodulation operation may replace a sequence of resolution operations with substitutivity axioms, and an associative-commutative unification operation may replace a sequence of resolution or paramodulation inferences with associativity and commutativity axioms. Theory resolution [6] is a general approach to the creation of large inference operations that can shorten proofs and reduce the size of the search space.

Binary resolution allows the derivation of the clause $A \vee B$ from the clauses $\neg P \vee A$ and $P \vee B$. This is justified since $\neg P$ and P are syntactically contradictory. Theory resolution extends resolution to resolve on sets of literals that are semantically contradictory. For example, when "$<$" is interpreted as a total ordering relation, the clause $(x < y) \vee (x = y) \vee (x > y)$ is valid and justifies the inference of $A \vee B \vee C$ from $\neg(a < b) \vee A$, $\neg(a = b) \vee B$, and $\neg(a > b) \vee C$. The result can be obtained in three binary-resolution steps or, alternatively, in a single theory-resolution step if the theory of a total ordering is built in. Clauses of the built-in theory need not be present. Theory resolution is arguably better than binary resolution because its inference steps are larger and the clauses that define the theory can be omitted.

The way in which a theory is to be implemented is unspecified. A theory can be implemented as a "black box" recognizing or suggesting theory-resolution operations, with knowledge of its internal structure unnecessary to the system that actually performs them. The lack of specification of the means for implementing a theory is advantageous because there is freedom to implement it in the most efficient manner, with few constraints on the strategies and data structures to be used.

Still, a theory has to be implemented somehow; one method that minimizes the effort needed is to use the clauses of the theory in implicit or hidden resolution operations.

This is similar to the linked inference principle of Veroff and Wos [11, 13]. They define linked variants of ordinary inference operations such as binary resolution, hyperresolution, and UR-resolution. Ordinary UR-resolution is defined to operate on n unit clauses (called electrons) and a single nonunit clause (called the nucleus) that has $n+1$ literals, and to produce a unit-clause result; it is often useful, though, to extend its definition to allow a nucleus with only n literals and thereby allow derivation of the empty clause as well as unit clauses. If each of n literals of the nucleus is unifiable with the complement of one of the n electrons, the instantiated remaining literal of the nucleus can be derived. For example, $eats(a, b)$ can be derived in a single UR-resolution step from $\neg carnivore(x) \vee \neg meat(y) \vee eats(x, y)$ and $carnivore(a)$ and $meat(b)$.

In linked UR-resolution, the electrons are not required to match the literals in the nucleus directly, but can be matched indirectly via linking clauses. For example, $eats(a, b)$ can be derived in a single linked UR-resolution step from $\neg carnivore(x) \vee \neg meat(y) \vee eats(x, y)$ and $lion(a)$ and $veal(b)$ in which $lion(a)$

is linked to $\neg carnivore(x)$ by the linking clauses $\neg lion(u) \vee feline(u)$ and $\neg feline(v) \vee carnivore(v)$ and $veal(a)$ is linked to $\neg meat(y)$ by the linking clause $\neg veal(w) \vee meat(w)$. The complexity of allowable linked inference operations can be restricted by parameters such as a limit on the number of linking clauses that can be used.

Despite the many examples of larger inference steps having the potential of making theorem-proving programs more effective, the larger linked inference operations are not certain to be beneficial. The permission to perform large linked inference operations does not exclude the possibility of also performing ordinary unlinked operations. In particular, any ordinary UR-resolution operation is defined to also be a linked UR-resolution operation, which happens to use no linking clauses. In the example above, $feline(a)$, $carnivore(a)$, and $meat(b)$ could be derived as well as $eats(a, b)$. From these, $eats(a, b)$ could be derived many times: by ordinary UR-resolution from $carnivore(a)$ and $meat(b)$, by linked UR-resolution from $lion(a)$ and $veal(b)$, by linked UR-resolution from $feline(a)$ and $meat(b)$, and so on. Deep inferences are performed in parallel with shallow inferences that ultimately lead to the same results. Using linked inference instead of ordinary inference does not confer the same nearly automatic benefit one gets from using hyperresolution or UR-resolution instead of binary resolution. Some of the shallow inferences should be prohibited to make linked inference effective. In theory resolution, a theory is built into the resolution operation and clauses of the theory do not appear; this prevents the derivation of some of the shallow inferences. For example, the taxonomic theory that includes the facts that lions are felines, felines are carnivores, and veal is meat could be built in using theory resolution. Then $eats(a, b)$ could be derived from $lion(a)$ and $veal(b)$ but $feline(a)$, $carnivore(a)$, and $meat(b)$ could not be. In linked inference, the clauses of a theory are not omitted, so their usage must be restricted.

The partitioning and target strategies of the linked inference principle restrict how clauses are used and thus restrict which linked inference operations are performed; in particular, they often eliminate many of the shallow inferences in favor of deep inferences. The partitioning strategy requires the user to designate for each clause whether it can be used as an electron, nucleus, or linking clause or some combination thereof. The target strategy requires the user to designate which literals can appear in the result of a linked inference operation. To simulate theory resolution's building in the clauses $\neg lion(u) \vee feline(u)$, $\neg feline(v) \vee carnivore(v)$, and $\neg veal(w) \vee meat(w)$, in linked inference, the partitioning strategy could be used to restrict these clauses for use only as linking clauses and the target strategy could prevent their literals from appearing in the result.

Computing linked inference operations can be very costly compared to computing ordinary unlinked inference operations, because the addition of linking clauses creates so many extra alternative inference operations. Intuition and experience with OTTER's implementation of linked UR-resolution suggest that making inferences in a linked inference system is proportionately much more costly than making inferences in an unlinked one. While the costs of demodulation, subsumption, and indexing of newly derived results may exceed the cost of deriving them in an unlinked system, the majority of the time may be spent deriving new results in a linked system. Thus, it is beneficial to seek the most efficient methods of computing linked inference operations and, in particular, to explore the efficient implementation of the model elimination theorem-proving procedure by PTTP as the basis for an efficient implementation of linked inference with compiled nucleus and linking clauses.

14.4 The Unit-Resulting PTTP

Linked inference and the Prolog Technology Theorem Prover explore similar search spaces: linked inference constructs the graph of resolution operations that comprise a single linked inference operation, while PTTP constructs a similar graph that constitutes a model elimination refutation. The essential difference in their operation is that linked inference is meant to derive new clauses from old and use them in future inference operations after they are demodulated, pass subsumption tests, are stored, and reach the top of the agenda; PTTP tries to find a complete refutation from the input clauses without producing any intermediate derived clauses.

The Unit-Resulting Prolog Technology Theorem Prover (UR-PTTP) is an extension of PTTP that derives new clauses. UR-PTTP's behavior is similar to that of linked UR-resolution, but it operates by constructing a graph of a model elimination refutation that is complete with the exception of a single literal. If that literal could also be refuted, the refutation would be complete.

It can also be insightful to regard UR-PTTP as performing a limited form of abductive reasoning. In abductive reasoning, if Q is true and we want an explanation for its being true, and $P \supset Q$ is known, then P can be hypothesized as a possible reason. PTTP tries to prove a query Q from axioms by means of Prolog-style backward chaining and the model elimination reduction operation. UR-PTTP has an extra inference rule that permits it to assume a literal instead of proving it. When UR-PTTP is asked to prove a query Q, it can prove it completely as PTTP does, or it can return a literal P assumed in the course of searching for a proof; all other literals in the refutation graph are proved in the usual way. UR-PTTP thereby demonstrates that Q can be proved from the assumption P and that, if P is proved later, Q is a consequence. Of course, UR-PTTP may return zero or more single-literal assumptions P, any one of which, if true, would suffice to prove Q.

It is certainly possible to extend PTTP to assume any number of literals instead of a single one. Then Q could be proved to be a consequence of the conjunction of the assumed literals. We have implemented procedures for doing this and applied them to the task of natural-language interpretation [8]. However, we will restrict our attention here to a unit-resulting extension of PTTP that assumes only single literals, so that it behaves similarly to the UR-resolution and linked UR-resolution inference rules, which derive unit clauses from sets of unit clauses. UR-PTTP is meant to be used instead of these inference rules in an OTTER-like theorem-proving program.

We describe the basic operation of an OTTER-like system that uses UR-PTTP. It computes the deductive closure of a set of unit clauses of a problem by inferring new unit clauses by means of UR-PTTP, which has compiled all the nonunit clauses of the problem. It can be used as a refutation system by showing that the empty clause is in the deductive closure. Two lists of unit clauses are maintained: the has-been-given list, which is initially empty, and the to-be-given list, which initially contains all the unit clauses of the problem. The has-been-given list contains those unit clauses for which all immediate UR-PTTP consequences have already been derived. The to-be-given list is an agenda of unit clauses that have not yet been used as input to UR-PTTP. The following actions are performed repeatedly. A highest-rated unit clause Q is selected and deleted from the to-be-given-list, and added to the has-been-given list. UR-PTTP attempts to prove $\neg Q$ from the unit clauses on the has-been-given list. If $\neg Q$ is proved without any assumption, the refutation is complete, since $\neg Q$ contradicts Q. If $\neg Q$ is proved subject to the assumption P, then $\neg P$ is demodulated and tested for forward subsumption. If it is not forward subsumed, it is added to the to-be-given list and used in backward subsumption and, if it is an equality, backward demodulation operations. By proving that $\neg Q$ can be proved if P is assumed, we will have demonstrated that Q can be refuted if $\neg P$ can be, so the demodulated form of $\neg P$ is added to the to-be-given list if it is not redundant.

UR-PTTP makes deep inferences, as linked UR-resolution does, but is a system in which clauses are compiled for efficient execution in the same manner as Prolog. It inherits PTTP's methods for compiling sound unification with the occur check, depth-bounded search, and model elimination reduction and pruning operations. It is complete: it can prove any unit-clause theorem without an assumption if the depth bound is high enough, but is intended to be used with lower bounds to produce proofs with assumptions that are used to solve the problem incrementally.

Like linked UR-resolution, inferences made by UR-PTTP must be restricted to limit the size of the search space UR-PTTP explores and to eliminate some of the shallow inference while retaining the deep ones. Our restrictions differ somewhat from those of linked UR-resolution. We believe neither set of restrictions is powerful enough, and would like to be able to specify arbitrary criteria for accepting or rejecting a partial linked inference operation or model elimination refutation. Providing such power in a usable way is likely to be very difficult, though UR-PTTP's reliance on the Prolog logic programming language may facilitate this.

UR-PTTP compiles all nonunit clauses in the same way and does not make the linked UR-resolution type of distinction between nucleus and linking clauses. Inferences can be restricted by specifying which predicates with which polarity are not allowed to be assumed. In the earlier example, taxonomic literals headed by *lion*, $\neg$*lion*, *feline*, $\neg$*feline*, and so on, might not be allowed to be assumed, but those headed by *eats* and $\neg$*eats* might be. The default is to allow any literal to be assumed unless the user specifically designates it as nonassumable. Thus, the most logically complete usage of UR-PTTP is available by default, and restrictions that might limit completeness must be stated explicitly by the user.

Linked UR-resolution restricts the size of potential inference operations by the link-depth and deduction-size parameters. The link depth of an inference is the maximum number of linking clauses that link an electron to the nucleus. Any linked inference operation alternatively could be performed by a sequence of binary-resolution steps. The deduction size is the number of binary-resolution steps that a linked inference operation represents. A finite value of either parameter can be used to make the search space of linked inference operations finite. UR-PTTP does not use the former, since it does not distinguish between nucleus and linking clauses, but its cost parameter, described below, is quite similar to the deduction size for linked UR-resolution. A nice feature of a bound on the deduction size or cost is that small increases in the bound generally result in the search space growing relatively slowly. This is used to advantage in PTTP's iterative deepening search to produce relatively slow, smooth growth in the size of the search space as the depth bound increases, though the UR-PTTP search space may grow more quickly, since there may be a large number of clauses on the has-been-given list available for use. Increasing the allowable link depth will almost certainly greatly increase the size of the search space, since a larger link depth generally permits inferences with much larger deduction size, unless this is also bounded.

UR-PTTP, like PTTP, normally evaluates the cost of a deduction by counting the number of subgoals that are added to the graph. Backward chaining with an n-literal clause increases the cost by $n-1$, the number of literals in the body of the clause. Solving a goal by assuming it or unifying it with a unit clause or the complement of an ancestor goal does not add to the cost of the deduction. Nonunit clauses can have user-assigned nonnegative costs larger or smaller than the number of literals. It is very much in the spirit of linked inference for the user to be able to assign costs to axioms explicitly to encourage or discourage their use in cost-bounded linked inference operations, and to control the behavior of the system more precisely. For example, it might be useful to assign permutative axioms like $\neg p(x, y, z) \vee p(y, x, z)$ zero cost so that these axioms can be used freely; this has the effect of building in permutative unification. Zero-cost axioms must be used carefully as they can result in infinite branches in the search space; in this case that problem is solved by means of the model elimination pruning rule, which rejects goals that are identical to an ancestor goal and prevents the sequence of goals $p(a,b,c), p(b,a,c), p(a,b,c), \ldots$. In UR-PTTP, the cost of a deduction is taken to be the sum of the costs of the clauses used minus the cost of the most expensive clause. Thus, a bound of zero would result in derivation of all ordinary UR-resolvents regardless of the length of the nucleus. Unless zero-cost axioms are used, only ordinary UR-resolvents could be derived with a bound of zero.

The Prolog version of PTTP [9] has been modified to create UR-PTTP, and a prototype OTTER-like system that uses UR-PTTP as described above has been written in Prolog.

14.5 Conclusion

The Prolog Technology Theorem Prover has a very high inference rate, but its lack of more flexible search strategies and redundancy control impairs its ability to prove difficult theorems. However, an extension of PTTP can be used as a component of a larger theorem-proving program that has the desired search strategy and redundancy control mechanisms. The extension efficiently computes large, potentially slow inference operations, such as theory resolution operations in which the theory is represented by clauses. Results of such inference operations can be checked for redundancy and ordered in an agenda by the larger theorem-proving program.

The Unit-Resulting Prolog Technology Theorem Prover we developed performs roughly the same inferences as Veroff and Wos's linked UR-resolution operation, but is based on PTTP and shares its high-speed compiled inferences.

Limited experimentation with UR-PTTP suggests that the idea has merit, and the positive results of more extensive experiments with linked UR-resolution [11] should apply to UR-PTTP as well, given the similar behavior of the systems. Should linked inference prove to be a viable inference procedure, the methods of UR-PTTP are virtually certain to be useful, since they employ PTTP/Prolog-style clause-compilation methods for high performance. Linked inference procedures and UR-PTTP are tools that can be used to control the inference process to find proofs more quickly. Like any complex tool, they are not suitable for all problems and require a great deal of care in their use, but can be very helpful if used skillfully. Further research should investigate which inferences should be allowed by these large-inference-step procedures and how to make it easy to program the behavior of the procedures.

References

[1] R. Butler, E. Lusk, W. McCune, and R. Overbeek (1986): Paths to High-Performance Automated Theorem Proving. 8th International Conference on Automated Deduction, Oxford, England, 588–597.

[2] D. W. Loveland (1969): A Simplified Format for the Model Elimination Procedure. Journal of the ACM 16(3), 349–363.

[3] D. W. Loveland (1978): Automated Theorem Proving: A Logical Basis. Amsterdam: North-Holland.

[4] W. W. McCune (1990): OTTER 2.0 Users Guide, Technical Report ANL-90/9, Mathematics and Computer Science Division, Argonne National Laboratory, Argonne, IL.

[5] D. A. Plaisted (1988): Non-Horn Clause Logic Programming Without Contrapositives. Journal of Automated Reasoning 4(3), 287–325.

[6] M. E. Stickel (1985): Automated Deduction by Theory Resolution. Journal of Automated Reasoning 1(4), 333–355.

[7] M. E. Stickel (1988a): A Prolog Technology Theorem Prover: Implementation by an Extended Prolog Compiler. Journal of Automated Reasoning 4(4), 353–380.

[8] M. E. Stickel (1988b): A Prolog-Like Inference System for Computing Minimum-Cost Abductive Explanations in Natural-Language Interpretation. International Computer Science Conference '88, Hong Kong, 343–350.

[9] M. E. Stickel (1990): A Prolog Technology Theorem Prover: a New Exposition and Implementation in Prolog. International Symposium on Design and Implementation of Symbolic Computation Systems, Capri, Italy, 154–163.

[10] M. Tarver (1990): An Examination of the Prolog Technology Theorem Prover. 10th International Conference on Automated Deduction, Kaiserslautern, W. Germany, 322–335.

[11] R. Veroff and L. Wos (to appear): The Linked Inference Principle, I: The Formal Treatment. Journal of Automated Reasoning.

[12] L. Wos, R. Overbeek, and R. Lusk (1991): Subsumption, A Sometimes Undervalued Procedure. In: J.-L. Lassez and G. Plotkin, eds.: Computational Logic, Essays in Honor of Alan Robinson. Cambridge, MA: MIT Press.

[13] L. Wos, R. Veroff, B. Smith, and W. McCune (1984): The Linked Inference Principle, II: The User's Viewpoint. 7th International Conference on Automated Deduction, Napa, CA, 316–332.

Chapter 15

Automated Reasoning and Bledsoe's Dream for the Field[1]

Larry Wos

Mathematics and Computer Science Division
Argonne National Laboratory
Argonne, Illinois

Abstract

In one sense, this article is a personal tribute to Woody Bledsoe. As such, the style will in general be that of private correspondence. However, since this article is also a compendium of experiments with an automated reasoning program, researchers interested in automated reasoning, mathematics, and logic will find pertinent material here. The results of those experiments strongly suggest that research frequently benefits greatly from the use of an automated reasoning program. As evidence, I select from those results some proofs that are better than one can find in the literature, and focus on some theorems that, until now, had never been proved with an automated reasoning program, theorems that Hilbert, Church, and various logicians thought significant. To add spice to the article, I present challenges for reasoning programs, including questions that are still open.

[1]This work was supported by the Applied Mathematical Sciences subprogram of the Office of Energy Research, U. S. Department of Energy, under Contract W-31-109-Eng-38.

15.1 Coming Attractions

Woody, before getting deeply into the material, I feel certain that you would like to have a brief review of what is to come in this letter/article. I think I know you fairly well, and I think you know me fairly well. But, in the lattice of knowledge that one person has about another, greater than my knowledge of you and your knowledge of me is my knowledge of me—or, with regard to the focus of this article, of my research. Mathematics does indeed offer the use of some appropriate—or farfetched—metaphors.

I think of this document as a letter/article because it is partly a personal tribute to you—and hence, the style is often that of private correspondence—and partly a compendium of experiments with an automated reasoning program. At least to me, the results of those experiments strongly suggest that research frequently benefits greatly from the use of an automated reasoning program. Even the most skeptical should be rather startled at what can now be done with a reasoning program. Especially for you, I promise to select from those results some proofs that are better than one can find in the literature. I also promise to focus on some theorems that, until now, had never been proved with an automated reasoning program—theorems that Hilbert, Church, and various logicians thought worth their attention. Since—as you so well know—I also wish to interest others in what the field has to offer, some of my remarks and observations are directed to researchers in general.

The focus of this article is a significant fraction of your grand dream for automated reasoning. From your entrance into the field of automated reasoning until the present, you have daydreamed about a computer friend. Rather than the whole of your grand dream—which will take many years to realize, if ever—I shall concentrate on those aspects that are now a reality, reached far sooner than anyone could have guessed but two decades ago. Although these aspects—also present from the beginning in my dream for the field of automated reasoning—focus on but a fraction of your grander dream, nevertheless the fraction is itself a significant dream to have.

You envision an automated reasoning program that will, when instructed for the first time to prove one or more theorems, frequently succeed, and succeed without the researcher playing a key role. You dream of reasoning programs that contribute to mathematics or logic—by finding a better proof than expected, or producing a shorter proof than previously known, or answering an open question. And you hope to see mathematicians and logicians using reasoning programs as assistants in their research. In this article—although you will be hard to convince—I show that this significant fraction of your dream is in fact a reality. As evidence, I outline a fragment of the history of automated reasoning from its birth (approximately 1960) to the present, drawing mainly from experiences with automated reasoning programs designed by colleagues at Argonne National Laboratory. You and I both know that I shall focus mainly on recent results obtained with McCune's powerful program OTTER [17]; I reserve my comments concerning other programs for Note 6. Even in your terms, I may succeed in staying with the truth—we shall see, yes?

Rather than a detailed account of people and places, the presentation consists of a sequence of snapshots showing how the effectiveness of automated reasoning gradually increased. The snapshots focus on individual questions, problems, and theorems, and the successes and failures resulting from their consideration by some automated reasoning program. By discussing some of the recent advances in theory, implementation, and application, I answer those who assert the impossibility of the effective automation of reasoning. How can such skeptics still exist?

To address the three cited objectives of your and my dream for an automated reasoning program and to illustrate how far automated reasoning has advanced, I give examples, examples of successfully proving theorems on the first try—and without much guidance from the user. A second set of examples focuses on finding better proofs than were expected, producing shorter proofs than were known, and answering open questions. The final examples concern the current use of an automated reasoning program by mathematicians and logicians.

The rate of recent advances suggests—certainly to me—that still more of your dream is within reach, as the field continues to offer intriguing challenges. Presented here are some of those challenges, including questions—still open—that are amenable to attack with an automated reasoning program. Although the progress is accelerating, there will always exist challenges for research in various areas of theory, implementation, and application.

Now at this point I can hear you saying, "Larry, I hear you; my students hear you; the world hears you. But, although I shouldn't be giving advice, I will. Tell me more, and tell the others who might be curious about the new results—expand on what you have just said, and don't mind a little repetition. And, oh yes, let me see in writing some of your opinions and biases, so we can all have a crack at them."

15.2 Motivation, Dedication, and Organization

As in any course, article, book, or meal, certain preliminaries are in order before the entree is served. However, if you prefer to feast immediately, you might simply fast-forward to the next section.

The choice of subject for this letter/article is based in part on many stimulating and enlightening conversations I have had with you, Woody, and is based in part on some recent successful experiments with McCune's automated reasoning program OTTER. By anticipating what you might say, let me now expand somewhat on the three cited goals that are part of your dream for automated reasoning. As will be shown with various examples, regarding those important goals, your dream is now a reality.

First, you and I envision an automated reasoning program that will, when instructed for the first time to prove one or more theorems, frequently succeed, and succeed without the researcher playing a key role. Yes, I agree with the thought I am certain you have at this moment, the key point concerns the degree to which the success depends on the researcher's guidance. After all, you and I aim at a program that can function as a colleague—for example, finding a proof with little more than a hint from us. As evidence that this goal has been reached, I shall discuss (in Section 15.4) a successful attempt to use OTTER to prove 68 theorems suggested by Dana Scott, and (in Section 15.5) I shall include a new axiom system and a shorter proof. If you were focusing on your grand dream, you would at this point comment that we still do poorly with a randomly selected theorem from a typical advanced mathematics text, and you would be right. But think of it: 68 theorems in one run with OTTER, with no knowledge of any of the corresponding proofs and—of at least fair significance—theorems suggested by Scott.

You and I also dream of reasoning programs that contribute to mathematics or logic—by finding a better proof than expected, or producing a shorter proof than previously known, or—best of all—answering an open question. I shall give you examples of each of the three types of contribution. When you read the full story—which I include in this article—of how that already-mentioned new axiom system was actually found, I suspect you will consider the episode as evidence of contributing a bit to mathematics and logic. On the other hand, I suppose the story could also be used as evidence that at least one person interested in automated reasoning conducts research in an odd way. Some gambles win, and some lose—and, sometimes, it is hard to tell which has occurred.

And of course you and I hope to see mathematicians and logicians using reasoning programs as assistants in their research. Well, they now do, occasionally, use such programs, and some occurrences are briefly discussed.

Since I cannot with certainty rank the given three goals, I shall organize this article by simply considering them in the order cited. The promised outline of a fragment of the history of automated reasoning from its birth (approximately 1960) to the present will provide an appropriate perspective for a full appreciation of the significance of reaching the three objectives. Although from today's perspective the beginning was not particularly impressive or promising—I think you share this view—you may find that the evidence offered here shows that we have made impressive advances.

The examples have been chosen to reflect the state of the art at various points between 1960 and 1990, beginning with an example that today's researcher would correctly classify as offering no challenge, and ending with examples that—at least some—mathematicians and logicians consider interesting theorems. Far more interesting to me than history is the current state of the art, for it demonstrates that research does benefit greatly from the use of an automated reasoning program. The examples will occasionally be accompanied by personal commentary concerning their significance. To answer those who assert the impossibility of the effective automation of reasoning, I also consider some of the recent advances in theory, implementation, and application.

For many of the examples, the corresponding input clauses are included and often even the clauses that illustrate a successful completion of the problem under study. To aid you in judging whether the dice were loaded or the deck stacked—even you might be a bit skeptical—I occasionally discuss the particular approach that was used to succeed, the parameter settings, and the reasons for choosing that approach. Again, occasional commentary is given. Finally, to stimulate research, I include problems that I consider challenging—or even surpassing—the limits of current automated reasoning programs, and offer questions—still open—that are amenable to attack with some existing program.

Throughout this article, in addition to proving theorems, I also discuss diverse ways in which a general-purpose, automated reasoning program can aid research; I focus mainly on McCune's marvelous and portable program OTTER. As you know, this program is based on the paradigm common to the programs designed by colleagues at Argonne. The lesser-known uses include the checking of given proofs, the discovery of problems for testing programs and ideas, the systematic search for shorter proofs, and the identification of errors in the chosen axiom system. Appropriate examples will be included here. For the more familiar use of proving theorems, examples will be given to show how a researcher can move from one attack to another in search of that which is sufficiently effective, thus illustrating the versatility offered by some automated reasoning programs.

For the final item of this section, I note that, throughout this article, I express various unsupported opinions, opinions based mostly on experimentation from 1964 to the present. I do so at your repeated encouragement. Indeed, when in conversation I have commented that I can give neither proof nor overwhelming data as the basis for certain utterances, you have replied that I need only issue an appropriate warning. So, Woody, here goes: This article contains commentary that is little more than opinion, and perhaps even bias. The commentary might add spice—for some, the taste of jalapeno peppers—to the presentation (in the following sections) of shorter proofs, new proofs, a new axiom system, new results, new techniques, and new uses of an automated reasoning program.

15.3 A Fragment of Automated Reasoning from Birth through Adolescence

This small taste of history provides a perspective for the following sections. In those sections I focus on the new material, and in them I give the hard evidence that three goals of your grander dream have been reached, at least to an important extent. The history presented in this section will also permit others to see how far away those goals were—even a few years ago—how far we have come, and (to some extent) how we got where we are.

The first problem (the Davis-Putnam example [2]) I identify with the field is, by today's standards, not very impressive. The Davis-Putnam example asks for a proof of the unsatisfiability of the following set of clauses. (For this and succeeding examples, the notation for clauses is that used to present a problem to OTTER, where "-" means **not** and "|" means **or**.)

```
P(x,y).
-P(y,f(x,y)) | -P(f(x,y),f(x,y)) | Q(x,y).
-P(y,f(x,y)) | -P(f(x,y),f(x,y)) | -Q(x,f(x,y) |
     -Q(f(x,y),f(x,y)).
```

Before the introduction of binary resolution [21], this problem was accepted as a challenge for reasoning programs. On the other hand, with the use of binary resolution, a proof was obtainable in 1963 in essentially no CPU time. Of course, rather than deprecating the quality of the early research, the simplicity of the Davis-Putnam example illustrates what little power was offered by the programs existing before 1963. The problem did serve nicely as a beginning, and—more important to me—the ease with which it was solved with a reasoning program motivated my attempt to obtain computer proofs of theorems from the mathematics literature.

The first theorem that was tried is a simple classroom exercise: If in a group the square of every element x is the identity e, the group is commutative. The attempt to prove the theorem failed; after 2,000 clauses were retained, memory was exhausted (on an IBM 704 computer). Since the failure led to the formulation of the set of support strategy [31], I am still fond of the theorem, even though I recognize its utter simplicity. Since the importance of strategy in general and of the set of support strategy in particular has been repeatedly proven [34], I am continually disappointed at the lack of research directed to that aspect of automated reasoning. To stimulate such research—which was, I admit, one reason I wrote my second book [36]—you and I could pool our personal resources and offer a \$20 prize for a significant contribution in the area of strategy.

The first attempt to prove that, in a ring, the product of $-x$ and $-y$ is xy also failed. The failure led to the use of lemma adjunction, specifically, the lemmas that assert that the product in either order of 0 and x is 0.

The first attempt to prove that subgroups of index 2 are normal failed. To succeed, a primitive form of case analysis was used; one case focused on an element in the subgroup, the other case on an element outside the subgroup. I find this theorem appealing, for it offers some difficulty for a person to prove, and it is somewhat significant from the viewpoint of mathematics. The index 2 problem has added appeal for me, for the attack that succeeded is representative of the way I prefer case analysis to be treated, rather than, say, by automatically splitting every ground clause into its individual literals.

Even at the end of the sixties—because of the lack of power offered by the then-current reasoning programs—I knew it was futile to attempt to seek a computer proof of the "commutator theorem": If in a group the cube of every x is the identity e, then $[[x,y],y] = e$ for all x and y, where the commutator $[x,y]$ of any two elements x and y of a group is the product of x, y, the inverse of x, and the inverse of y. Indeed, although I presented this theorem as an example of the potential use of paramodulation [20], its included clause-notation proof was obtained by hand. As it turned out, the first computer proof of the theorem was obtained with hyperresolution [22]; the first paramodulation proof by computer was not obtained until approximately six years after the problem was suggested. This theorem is now, and will always be, important to me, for its study solidified my interest in paramodulation [33], an interest that has never diminished. The theorem also holds special meaning for me, for Ross Overbeek's reading of the corresponding paper prompted him to enter the field; any estimate of the value of that occurrence falls short of the mark.

The four theorems and the results were typical of the 1960s, at least of the research I shared with colleagues. Failure was the expected outcome, and we were seldom disappointed. In other words, the first of your goals for the field of automated reasoning—succeeding moderately often on the first attempt to prove a theorem not tried before—was far in the future, if reachable at all. As it turned out, and as I shall exemplify in the next section, two decades after the close of the 1960s would see the attainment of this goal. Examples of success on the first attempt include theorems of interest to Hilbert, Church, Frege, Bernays, Łukasiewicz, and Tarski.

The second goal—a reasoning program making contributions to mathematics and logic—brings me even more excitement than does the first goal. However, except for a brief consideration in the late 1960s of attempting to answer an open question concerning the independence of the first three axioms of the five for a ternary Boolean algebra [30], I was not involved in such an activity until the late 1970s. In fact, except for the discovery by Guard and associates of SAM's lemma [4], I was almost certain such contributions would not occur in my lifetime. Clearly—and most fortunately—I was in error, for the period from 1978 to the present witnessed the answering of open questions from various fields of mathematics and logic, where the answers were obtained with substantial assistance from an automated reasoning program. The fields include combinatory logic, finite semigroups, ternary Boolean algebra, Robbins algebra, equivalential calculus, and various areas of logic related to implicational calculus.

As for your third goal—the actual use by mathematicians and logicians of automated reasoning programs—until the early 1980s, the only item that occurs to me concerns Paul Halmos and his visit to the University of Texas. On that visit—and I checked this with Halmos—you and Halmos, using a theorem prover, were able to prove a theorem from *Naive Set Theory* [5]. Unfortunately, my best recollection says that the theorem occurs between pages 43 and 45; perhaps you can supply the needed details. I am not sure this early 1970s incident actually qualifies for "use of an automated reasoning program by a mathematician", but it is in the vicinity.

On the other hand, without doubt, Kalman's use at Argonne of the program ITP [12, 13] during his stay in the early 1980s clearly qualifies; Kalman was studying various aspects of equivalential calculus, in part in the context of group theory. Kalman continues to conduct research with the aid of a reasoning program; he is now using OTTER for additional studies in group theory. For a second example, Scott has just successfully used OTTER (on his Macintosh) to obtain a proof of the completeness of a Łukasiewicz axiom system for sentential calculus.

History is treating our goals well—or do you remain understandably skeptical? Resist if you can, but I suggest such skepticism may be difficult to maintain when you see what happened here in August 1990, research prompted by Scott's stimulating communication by email. Of course, there will always exist research problems to solve in various areas of automated reasoning. We would not wish it any other way, for we have each spent a major fraction of our lives in this field. We have certainly shown that the automation of careful reasoning is far more than feasible, contrary to those who assert its impossibility. Although the type of reasoning found in mathematics is indeed deep and beyond complete capturing, we have come much closer than I would have guessed when I entered the field in 1963.

Clearly—and this remark is particularly appropriate when addressed to me—care must be exercised to control the obvious enthusiasm; the last time I exercised such care occurred in approximately 1802. Yes, I am aware that your grand dream is far from realized and that many graduate students could outperform the best of the existing automated reasoning programs if the competition centered on randomly selected theorems from mathematics texts. But how many of those students have answered an open question—even further, a question posed by Kaplansky? As you well know, such a question—as well as other open questions—has been answered with crucial assistance from one of Argonne's reasoning programs.

Exercising my usual conservative approach, which brings to eye, ear, and mind a herd of stampeding elephants—just between you and me—the results of the next two sections are truly exciting and beautiful!

15.4 Successes on the First Try: Goal 1

In this and succeeding sections—finally getting to new results and new uses—I am going to emphasize the most recent experiences I have had with OTTER, experiences involving theorems suggested by Dana Scott. The theorems are from various logical calculi, such as sentential calculus. Beginning in 1879, such theorems have occupied the attention of Frege, Russell, Hilbert, Tarski, Bernays, Łukasiewicz, Church, and others of their stature; two books that contain pertinent material are [10, 11]. I shall try hard to tell the truth about how the successes occurred, under what conditions, in what order, and how easily for my colleague McCune and me. However, when I am experimenting and not in the presence of one of my colleagues, I can easily lose the fine details. But, you need not worry; most, if not all, of what I present can be independently verified, so the loss of some of the minute history is not devastating.

When Scott returned home after a visit to Argonne in August 1990, he sent by email 68 theorems for OTTER's attempt at proof. The theorems (whose negations are given shortly) are numbered 4 through 71 in a study by Łukasiewicz; they are provable from the following three axioms expressed in clause notation. The first of the three can be thought of as transitivity, $(x \to y) \to ((y \to z) \to (x \to z))$. The second can be read as $(\neg x \to x) \to x$, and the third as $x \to (\neg x \to y)$.

```
(L1)  P(i(i(x,y),i(i(y,z),i(x,z)))).
(L2)  P(i(i(n(x),x),x)).
(L3)  P(i(x,i(n(x),y))).
```

The rule of inference is condensed detachment, captured with the following clause when hyperresolution is the chosen inference rule.

```
-P(i(x,y)) | -P(x) | P(y).
```

Łukasiewicz accurately claims that the given three axioms are complete for sentential calculus. An axiom set is complete for sentential calculus if the set of formulas that can be deduced from it with the use of condensed detachment and instantiation is precisely the set of formulas (in effect, unit clauses) that are true under all assignments of true and false, with *implication* and *negation* used as logicians use these terms. Therefore, the following 68 theorems, presented as negations to seek a contradiction, must hold. Frequently, throughout the remainder of this article, the theorems in the unnegated form will be referred to as theses. Frequently also, the ANSWER literal (which is ignored by the program when applying an inference rule) is appended to a clause to indicate the role of the clause. The ANSWER literal will usually be abbreviated with "$ANS", the word "thesis" with "th", the word "negated" with "neg", and the word "binary" with "bin".

```
-P(i(i(i(i(q,r),i(p,r)),s),i(i(p,q),s))) | $ANS(neg_th_04).
-P(i(i(p,i(q,r)),i(i(s,q),i(p,i(s,r))))) | $ANS(neg_th_05).
-P(i(i(p,q),i(i(i(p,r),s),i(i(q,r),s)))) | $ANS(neg_th_06).
-P(i(i(t,i(i(p,r),s)),i(i(p,q),i(t,i(i(q,r),s)))))
  | $ANS(neg_th_07).
-P(i(i(q,r),i(i(p,q),i(i(r,s),i(p,s))))) | $ANS(neg_th_08).
-P(i(i(i(n(p),q),r),i(p,r))) | $ANS(neg_th_09).
-P(i(p,i(i(i(n(p),p),p),i(i(q,p),p)))) | $ANS(neg_th_10).
-P(i(i(q,i(i(n(p),p),p)),i(i(n(p),p),p))) | $ANS(neg_th_11).
-P(i(t,i(i(n(p),p),p))) | $ANS(neg_th_12).
-P(i(i(n(p),q),i(t,i(i(q,p),p)))) | $ANS(neg_th_13).
-P(i(i(i(t,i(i(q,p),p)),r),i(i(n(p),q),r))) | $ANS(neg_th_14).
-P(i(i(n(p),q),i(i(q,p),p))) | $ANS(neg_th_15).
-P(i(p,p)) | $ANS(neg_th_16).
-P(i(p,i(i(q,p),p))) | $ANS(neg_th_17).
-P(i(q,i(p,q))) | $ANS(neg_th_18).
-P(i(i(i(p,q),r),i(q,r))) | $ANS(neg_th_19).
-P(i(p,i(i(p,q),q))) | $ANS(neg_th_20).
-P(i(i(p,i(q,r)),i(q,i(p,r)))) | $ANS(neg_th_21).
-P(i(i(q,r),i(i(p,q),i(p,r)))) | $ANS(neg_th_22).
-P(i(i(i(q,i(p,r)),s),i(i(p,i(q,r)),s))) | $ANS(neg_th_23).
-P(i(i(i(p,q),p),p)) | $ANS(neg_th_24).
-P(i(i(i(p,r),s),i(i(p,q),i(i(q,r),s)))) | $ANS(neg_th_25).
-P(i(i(i(p,q),r),i(i(r,p),p))) | $ANS(neg_th_26).
-P(i(i(i(p,q),q),i(i(q,p),p))) | $ANS(neg_th_27).
-P(i(i(i(i(r,p),p),s),i(i(i(p,q),r),s))) | $ANS(neg_th_28).
-P(i(i(i(p,q),r),i(i(p,r),r))) | $ANS(neg_th_29).
-P(i(i(p,i(p,q)),i(p,q))) | $ANS(neg_th_30).
-P(i(i(p,s),i(i(i(p,q),r),i(i(s,r),r)))) | $ANS(neg_th_31).
-P(i(i(i(p,q),r),i(i(p,s),i(i(s,r),r)))) | $ANS(neg_th_32).
-P(i(i(p,s),i(i(s,i(q,i(p,r))),i(q,i(p,r))))) | $ANS(neg_th_33).
-P(i(i(s,i(q,i(p,r))),i(i(p,s),i(q,i(p,r))))) | $ANS(neg_th_34).
-P(i(i(p,i(q,r)),i(i(p,q),i(p,r)))) | $ANS(neg_th_35).
-P(i(n(p),i(p,q))) | $ANS(neg_th_36).
-P(i(i(i(p,q),r),i(n(p),r))) | $ANS(neg_th_37).
-P(i(i(p,n(p)),n(p))) | $ANS(neg_th_38).
-P(i(n(n(p)),p)) | $ANS(neg_th_39).
-P(i(p,n(n(p)))) | $ANS(neg_th_40).
-P(i(i(p,q),i(n(n(p)),q))) | $ANS(neg_th_41).
-P(i(i(i(n(n(p)),q),r),i(i(p,q),r))) | $ANS(neg_th_42).
-P(i(i(p,q),i(i(q,n(p)),n(p)))) | $ANS(neg_th_43).
-P(i(i(s,i(q,n(p))),i(i(p,q),i(s,n(p))))) | $ANS(neg_th_44).
-P(i(i(s,i(q,p)),i(i(n(p),q),i(s,p)))) | $ANS(neg_th_45).
-P(i(i(p,q),i(n(q),n(p)))) | $ANS(neg_th_46).
```

```
-P(i(i(p,n(q)),i(q,n(p)))) | $ANS(neg_th_47).
-P(i(i(n(p),q),i(n(q),p))) | $ANS(neg_th_48).
-P(i(i(n(p),n(q)),i(q,p))) | $ANS(neg_th_49).
-P(i(i(i(n(q),p),r),i(i(n(p),q),r))) | $ANS(neg_th_50).
-P(i(i(p,i(q,r)),i(p,i(n(r),n(q))))) | $ANS(neg_th_51).
-P(i(i(p,i(q,n(r))),i(p,i(r,n(q))))) | $ANS(neg_th_52).
-P(i(i(n(p),q),i(i(p,q),q))) | $ANS(neg_th_53).
-P(i(i(p,q),i(i(n(p),q),q))) | $ANS(neg_th_54).
-P(i(i(p,q),i(i(p,n(q)),n(p)))) | $ANS(neg_th_55).
-P(i(i(i(i(p,q),q),r),i(i(n(p),q),r))) | $ANS(neg_th_56).
-P(i(i(n(p),r),i(i(p,q),i(i(q,r),r)))) | $ANS(neg_th_57).
-P(i(i(i(i(p,q),i(i(q,r),r)),s),i(i(n(p),r),s))) |
      $ANS(neg_th_58).
-P(i(i(n(p),r),i(i(q,r),i(i(p,q),r)))) | $ANS(neg_th_59).
-P(i(i(s,i(n(p),r)),i(s,i(i(q,r),i(i(p,q),r))))) |
      $ANS(neg_th_60).
-P(i(i(p,r),i(i(q,r),i(i(n(p),q),r)))) | $ANS(neg_th_61).
-P(i(i(n(n(p)),q),i(p,q))) | $ANS(neg_th_62).
-P(i(q,i(p,p))) | $ANS(neg_th_63).
-P(i(n(i(p,p)),q)) | $ANS(neg_th_64).
-P(i(i(n(q),n(i(p,p))),q)) | $ANS(neg_th_65).
-P(i(n(i(p,q)),p)) | $ANS(neg_th_66).
-P(i(n(i(p,q)),n(q))) | $ANS(neg_th_67).
-P(i(n(i(p,n(q))),q)) | $ANS(neg_th_68).
-P(i(p,i(n(q),n(i(p,q))))) | $ANS(neg_th_69).
-P(i(p,i(q,n(i(p,n(q)))))) | $ANS(neg_th_70).
-P(n(i(i(p,p),n(i(q,q))))) | $ANS(neg_th_71).
```

My esteemed colleague McCune, with whom I have conducted most of the experiments reported here, immediately submitted the 68 theorems to OTTER. He chose hyperresolution as the only inference rule; he placed the clause for condensed detachment in the axioms, the three Łukasiewicz axioms in the set of support, and the 68 negations in the passive list—a list that is inactive except for both testing for the completion of a proof by detecting unit conflict and checking for forward subsumption [21]. For the weighting strategy [15, 34, 36]—the strategy used by OTTER to decide where next to focus its attention—he chose to use symbol count. The motivation was to approach the theorems as simply as possible, keying on the three Łukasiewicz axioms, using the 68 negations only to detect the corresponding proofs (if found), and causing OTTER's search for proofs to be directed by focusing at any given point on the shortest formula available. OTTER proved 33 of the theorems before it was decided that more CPU time would yield little additional. A similar attempt with ROO, a parallel version of OTTER, produced 48 proofs.

Motivated by the strong urge to demonstrate OTTER's power by sending proofs of all 68 theorems at once to Scott, I decided to make one additional attempt. The attempt was based on replacing the use of symbol count (to direct OTTER in its choice of where next to focus its attention) by a set of 68 weight templates, each matching the corresponding pattern of occurrences of the function i in one of the theorems under attack. I reasoned—hazarded is probably more accurate—that, if the 68 theorems were the steps in a proof of a desired result, then each merited emphasis if and when deduced.

It worked, and better than I would have guessed; OTTER proved all 68 theorems in less than 16 CPU minutes on a SPARCstation. As it turned out, Scott was interested mainly in eight of the theses: 16, 18, 24, 21, 35, 39, 40, and 49. He wished to have a single proof of all eight, rather than piecing eight separate results together and tediously removing duplicate steps and such, which he could compare to the Łukasiewicz proof for style differences. A ploy that succeeded focuses on taking the disjunction of the corresponding eight negations and placing it among the axioms; the desired proof is found when the empty clause is generated with the disjunction playing the role of nucleus. He informed me that the resulting 46-step proof more or less matches the style of that given by Łukasiewicz.

But the story does not end there. Its ending is given in the next section, where it properly belongs, as an example of a small contribution to logic made by an automated reasoning program.

Although three attempts were required to obtain all 68 proofs, the study led to the formulation of an approach—not scientifically justified—that was used on the following example. The theorem in question asks for a proof that the third axiom (FL3) of the Frege/Łukasiewicz axiom system *FL* (presented by Church) follows from the six axioms (F1) through (F6), Frege's original axiom system *F* for sentential calculus. (The first two of the three axioms—theses 18, 35, and 49—of which *FL* consists are, respectively, the first and second of *F*.) When OTTER considered the following seven clauses, coupled with the clause for condensed detachment, it found the desired proof on the first try.

```
(F1 th_18)  P(i(x,i(y,x))).
(F2 th_35)  P(i(i(x,i(y,z)),i(i(x,y),i(x,z)))).
(F3 th_21)  P(i(i(x,i(y,z)),i(y,i(x,z)))).
(F4 th_39)  P(i(i(x,y),i(n(y),n(x)))).
(F5 th_40)  P(i(n(n(x)),x)).
(F6 th_46)  P(i(x,n(n(x)))).
(FL3 neg_th_49)  -P(i(i(n(p),n(q)),i(q,p))) | $ANS(step_FL3).
```

The approach was again to rely on the weight templates whose pattern of the function i matches the 68 theses just discussed.

At this point, you might properly wonder what possible explanation I can give for such an action, especially since I was seeking but one of the 68 proofs, that of 49. Explanation: if the use of those 68 patterns as weight templates enabled McCune and me to send to Scott the proofs of the test problems he suggested, then those patterns must be good ones, good ones for other theorems in this area of logic—even if the axioms (hypotheses) were changed. Reasoning of this type accounts for the fact that some gamblers win great sums; it also accounts for the fact that most gamblers lose their entire stake. As it turned out, the sought-after proof was obtained in less than 1 CPU second, a proof of 14 steps. Since the proof was obtained on the first attempt, do we not have overwhelming evidence of the realization of the first goal, or is the evidence—at least to this point—underwhelming?

I then took for weight templates the patterns suggested by the 14 steps of the proof, set the appropriate flags to instruct OTTER to seek possibly shorter proofs of (FL3), and obtained in one run the following 11-step proof in 6 CPU seconds. (In Note 5, I have more to say concerning the importance of seeking shorter proofs and the use of weight templates.) The number in the first field gives the corresponding position of the clause among those retained during OTTER's attempt. The expression *hyper,x,y,z* typically denotes that x is the clause in focus, y is the nucleus, z is the other satellite (for condensed detachment), and hyperresolution is the inference rule.

```
1 [] -P(i(x,y)) | -P(x) | P(y).
2 [] P(i(x,i(y,x))).
3 [] P(i(i(x,i(y,z)),i(i(x,y),i(x,z)))).
4 [] P(i(i(x,i(y,z)),i(y,i(x,z)))).
5 [] P(i(i(x,y),i(n(y),n(x)))).
6 [] P(i(n(n(x)),x)).
7 [] P(i(x,n(n(x)))).
8 [] -P(i(i(n(p),n(q)),i(q,p))) | $ANS(step_FL3).
--------------------
22 [hyper,4,1,3] P(i(i(x,y),i(i(x,i(y,z)),i(x,z)))).
25 [hyper,5,1,4] P(i(n(x),i(i(y,x),n(y)))).
33 [hyper,6,1,2] P(i(x,i(n(n(y)),y))).
50 [hyper,22,1,7] P(i(i(x,i(n(n(x)),y)),i(x,y))).
56 [hyper,25,1,2] P(i(x,i(n(y),i(i(z,y),n(z))))).
61 [hyper,33,1,3] P(i(i(x,n(n(y))),i(x,y))).
86 [hyper,56,1,50] P(i(x,i(i(y,n(x)),n(y)))).
523 [hyper,61,1,2] P(i(x,i(i(y,n(n(z))),i(y,z)))).
541 [hyper,523,1,3] P(i(i(x,i(y,n(n(z)))),i(x,i(y,z)))).
675 [hyper,86,1,4] P(i(i(x,n(y)),i(y,n(x)))).
708 [hyper,675,1,541] P(i(i(n(x),n(y)),i(y,x))).
```

Clause (708) contradicts clause (8), and the proof is complete.

I have no idea whether this proof is the same as, equivalent to, or better than that to be found in the literature.

The motivation for studying the next theorem—another example of succeeding on the first try—was simple curiosity on my part. If Scott was satisfied that OTTER proved the completeness of the three-axiom system of

Łukasiewicz by proving the eight theses, 16, 18, 24, 21, 35, 39, 40, and 49, then an obvious theorem to attempt to prove is the converse: from these eight axioms, one can deduce each of (L1), (L2), and (L3). (As we later learned, for a complete system, one need consider only the second, fifth, and eighth; in other words, the other five are dependent on these three.) The same approach was used again, directing OTTER's search for the desired three proofs by using the 68 weight templates. The theses were proved in a single run, in the order (L3), (L1), and (L2). The proofs were obtained in the interval of CPU seconds between 97 and 101.

You might find the following example of particular interest, for it dramatically demonstrates how far we have come and exhibits OTTER's nearly linear performance with respect to clause generation. The theorem to prove asserts that the third of the following six axioms, Frege's system for sentential calculus, is dependent on the remaining five.

```
(F1 th_18)  P(i(x,i(y,x))).
(F2 th_35)  P(i(i(x,i(y,z)),i(i(x,y),i(x,z)))).
(F3 th_21)  P(i(i(x,i(y,z)),i(y,i(x,z)))).
(F4 th_39)  P(i(i(x,y),i(n(y),n(x)))).
(F5 th_40)  P(i(n(n(x)),x)).
(F6 th_46)  P(i(x,n(n(x)))).
```

Using the approach described in this section, but with a smaller set of weight templates, OTTER succeeded in proving—on the first attempt—that the third axiom is indeed dependent on the other five.

For the following reasons, this first-attempt success is (to me) both stimulating and enlightening. The proof was obtained (on a SPARCstation) in just under 19 CPU hours. The proof was completed upon retention of a clause numbered 11331, after generating over 30,000,000 clauses. The performance of OTTER was nearly linear; during the first few CPU seconds, clauses were generated at the rate of 550 per CPU second, and during the last few CPU seconds, clauses were generated at the rate of 460 per CPU second. I could not resist letting OTTER run for additional time to see what would happen to the generation rate. At the 43 CPU hour mark, after generating just under 66,000,000 clauses, the rate had finally dropped to 300 clauses per CPU second—pretty impressive to see such small degradation over such a long period of time. A fast, tireless, and *accurate* assistant, OTTER provides a nice complement to the researcher with insight. The mathematician or logician might make marvelous discoveries, after mastering and then using this program. The proof OTTER found consists of 74 steps, not counting, of course, the five (input) axioms, all of which are used.

I have a tag or postscript to the discussion of this Frege dependence theorem. As OTTER was searching for a proof that the third axiom is dependent on the other five, my colleague McCune informed me (after browsing in one of Łukasiewicz's books) that the dependence could be proved from the first two axioms alone. On a second computer, OTTER was asked to search for such a proof. We were motivated, of course, by curiosity about which run would finish first—the long run just discussed, or this second attempt—and by the wish to compare the two proofs, should they be found. The second attempt succeeded first, in under 52 CPU seconds, finding a proof consisting of 11 steps. Yes, for this newer theorem, one attempt was again sufficient. Considered together, the two results suggest that the presence of unneeded but usable axioms can sharply detract from a program's effectiveness.

The final detailed example of success on the first try concerns proving the completeness of the axiom system consisting of (the unnegated form of) theses 19, 37, and 59. McCune informed me of this system from his reading of Łukasiewicz. Scott has commented to me that the system is indeed interesting—and he can explain why—but that Łukasiewicz does not explain why nor does he give the corresponding proofs. OTTER's proof that theses 19, 37, and 59 form a complete system rests on the deduction of (FL1), (FL2), and (FL3), mentioned earlier as a complete system. Since these proofs may not exist in the literature, OTTER's proofs are included here. The following three proofs are, respectively, for the deduction of (FL1), (FL2), and (FL3); the third required approximately 406 CPU seconds, and the first two required less than 1 CPU second.

```
1 [] -P(i(x,y)) | -P(x) | P(y).
2 [] P(i(i(i(x,y),z),i(y,z))).
21 [] -P(i(q,i(p,q))) | $ANS(step_th_18).
--------------------
35 [hyper,2,1,2] P(i(x,i(y,x))).
```

Clause (35) contradicts clause (21), and the proof is complete.

```
1 [] -P(i(x,y)) | -P(x) | P(y).
2 [] P(i(i(i(x,y),z),i(y,z))).
3 [] P(i(i(i(x,y),z),i(n(x),z))).
4 [] P(i(i(n(x),z),i(i(y,z),i(i(x,y),z)))).
24 [] -P(i(i(p,i(q,r)),i(i(p,q),i(p,r)))) | $ANS(step_th_35).
--------------------
35 [hyper,2,1,2] P(i(x,i(y,x))).
43 [hyper,35,1,3] P(i(n(x),i(y,i(x,z)))).
44 [hyper,35,1,2] P(i(x,i(y,i(z,x)))).
68 [hyper,43,1,4] P(i(i(x,i(y,i(z,u))),i(i(z,x),i(y,i(z,u))))).
72 [hyper,68,1,35] P(i(i(x,i(x,y)),i(z,i(x,y)))).
79 [hyper,72,1,43] P(i(x,i(n(y),i(y,z)))).
81 [hyper,79,1,79] P(i(n(x),i(x,y))).
87 [hyper,81,1,4] P(i(i(x,i(y,z)),i(i(y,x),i(y,z)))).
98 [hyper,87,1,44] P(i(i(x,y),i(x,i(z,y)))).
299 [hyper,98,1,68] P(i(i(x,i(y,z)),i(y,i(x,z)))).
```

```
338 [hyper,299,1,98] P(i(x,i(i(x,y),i(z,y)))).
1127 [hyper,338,1,68] P(i(i(x,y),i(i(y,z),i(x,z)))).
1208 [hyper,1127,1,299] P(i(i(x,y),i(i(z,x),i(z,y)))).
1285 [hyper,1208,1,68] P(i(i(x,i(y,z)),i(i(x,y),i(x,z)))).
```

Clause (1285) contradicts clause (24), and the proof is complete.

```
1 [] -P(i(x,y)) | -P(x) | P(y).
2 [] P(i(i(i(x,y),z),i(y,z))).
3 [] P(i(i(i(x,y),z),i(n(x),z))).
4 [] P(i(i(n(x),z),i(i(y,z),i(i(x,y),z)))).
27 [] -P(i(i(n(p),n(q)),i(q,p))) | $ANS(step_th_49).
--------------------
35 [hyper,2,1,2] P(i(x,i(y,x))).
42 [hyper,35,1,4] P(i(i(x,i(y,n(z))),i(i(z,x),i(y,n(z))))).
43 [hyper,35,1,3] P(i(n(x),i(y,i(x,z)))).
44 [hyper,35,1,2] P(i(x,i(y,i(z,x)))).
48 [hyper,42,1,35] P(i(i(x,n(x)),i(y,n(x)))).
68 [hyper,43,1,4] P(i(i(x,i(y,i(z,u))),i(i(z,x),i(y,i(z,u))))).
72 [hyper,68,1,35] P(i(i(x,i(x,y)),i(z,i(x,y)))).
79 [hyper,72,1,43] P(i(x,i(n(y),i(y,z)))).
81 [hyper,79,1,79] P(i(n(x),i(x,y))).
87 [hyper,81,1,4] P(i(i(x,i(y,z)),i(i(y,x),i(y,z)))).
98 [hyper,87,1,44] P(i(i(x,y),i(x,i(z,y)))).
101 [hyper,87,1,35] P(i(i(x,y),i(x,y))).
104 [hyper,101,1,87] P(i(i(x,i(x,y)),i(x,y))).
111 [hyper,104,1,48] P(i(i(x,n(x)),n(x))).
112 [hyper,104,1,35] P(i(x,x)).
299 [hyper,98,1,68] P(i(i(x,i(y,z)),i(y,i(x,z)))).
338 [hyper,299,1,98] P(i(x,i(i(x,y),i(z,y)))).
1127 [hyper,338,1,68] P(i(i(x,y),i(i(y,z),i(x,z)))).
1203 [hyper,1127,1,1127] P(i(i(i(i(x,y),i(z,y)),u),i(i(z,x),u))).
1208 [hyper,1127,1,299] P(i(i(x,y),i(i(z,x),i(z,y)))).
1234 [hyper,1127,1,299] P(i(i(i(x,i(y,z)),u),i(i(y,i(x,z)),u))).
1257 [hyper,1203,1,1203] P(i(i(x,i(y,z)),i(i(u,y),i(x,i(u,z))))).
1260 [hyper,1203,1,1127] P(i(i(x,y),i(i(i(x,z),u),i(i(y,z),u)))).
1321 [hyper,1208,1,4] P(i(i(x,i(n(y),z)),
                           i(x,i(i(u,z),i(i(y,u),z))))).
1492 [hyper,1257,1,1234] P(i(i(x,i(y,z)),i(i(u,x),i(y,i(u,z))))).
1517 [hyper,1260,1,299] P(i(i(i(x,y),z),i(i(x,u),i(i(u,y),z)))).
1640 [hyper,1321,1,3] P(i(i(i(x,y),z),i(i(u,z),i(i(x,u),z)))).
2266 [hyper,1640,1,1492] P(i(i(x,i(i(y,z),u)),
                              i(i(v,u), i(x,i(i(y,v),u))))).
2269 [hyper,1640,1,299] P(i(i(x,y),i(i(i(z,u),y),i(i(z,x),y)))).
2751 [hyper,2269,1,112] P(i(i(i(x,y),z),i(i(x,z),z))).
```

```
3002 [hyper,2751,1,1517] P(i(i(i(x,y),z),i(i(z,u),i(i(x,u),u)))).
3204 [hyper,3002,1,111] P(i(i(n(x),y),i(i(x,y),y))).
3331 [hyper,3204,1,299] P(i(i(x,y),i(i(n(x),y),y))).
3544 [hyper,3331,1,2266] P(i(i(x,y),i(i(z,y),i(i(n(z),x),y)))).
3879 [hyper,3544,1,299] P(i(i(x,y),i(i(z,y),i(i(n(x),z),y)))).
3982 [hyper,3879,1,35] P(i(i(x,i(y,z)),i(i(n(z),x),i(y,z)))).
4126 [hyper,3982,1,81] P(i(i(n(x),n(y)),i(y,x))).
```

Clause (4126) contradicts clause (27), and the proof is complete.

In 1923, Hilbert offered as a complete axiom system for sentential calculus that consisting of theses 18, 21, 22, 30, 3, and 54. Later, it was proved that 30 is dependent on the remaining axioms in Hilbert's system. On the first try, OTTER proved (with a 3-step proof) the dependence of 30 in less than 2 CPU seconds and, in that same run, proved the completeness of Hilbert's system. The first completeness proof proceeds by deducing the system consisting of theses 19, 37, and 59, completing a 9-step proof in just over 7 CPU seconds. In just under 17 CPU seconds, OTTER found a different 28-step proof of completeness by deducing the system *FL* consisting of theses 18, 35, and 49. However, later in the same run, a far shorter proof was found in approximately 1,733 CPU seconds, a proof consisting of 12 steps. Still in the same run, a 30-step proof of the completeness of Hilbert's system was found by deducing Frege's system *F* (excluding the dependent axiom) in approximately 143 CPU seconds. A shorter 16-step proof was then found in approximately 3,706 CPU seconds. A 6-step proof of the Łukasiewicz system was obtained in approximately 961 CPU seconds, and a 5-step proof in approximately 1,848 CPU seconds.

A few observations concerning these results are now in order. The way OTTER can be used is strikingly different from the approach typically taken with a colleague; Notes 4 and 7 elaborate on this point. In the latter case, a decision is made concerning which single path is best to pursue; for example, if the completeness of the Hilbert axiom system is the objective, the decision might be to attempt to prove the Lukasiewicz system or, instead, to attempt to prove Frege's. In contrast, with OTTER as your assistant, a good approach consists of asking this program to consider simultaneously a number of possible paths to a completeness proof; more comments are given in Note 5. Indeed, as evidenced in the preceding paragraph, that approach was markedly successful—one run produced a number of different completeness proofs for the axiom set under study. Among those many proofs, some differ in the intermediate steps but are alike in that each completes by deducing all the members of a given complete system. In contrast, other proofs obtained in that single run complete by deducing the members of a totally different complete system.

To me it seems clear that, at least in some areas, all has changed from the early 1960s, from the mid-1970s, and even from five years ago. Indeed, as the following summary suggests—and it may not convince you, for you can drive a hard bargain—certain important aspects of your dream for the field are now a reality. An unsophisticated approach in which one chooses the obvious inference rule and obvious weights to direct the search (by symbol count) yields many proofs on the first try. With a little bit of sophistication, an approach can be formulated, based on the study of one set of theorems, and then used to obtain on the first try proofs of theorems that merited the attention of great minds. You and I know that in no way do the results of the cited experiments lessen the achievements of the various logicians involved in the studies beginning with Frege in 1879. Instead—at least I would like to find that this is the case—the successes with OTTER mark a singular achievement for automated reasoning and suggest that the field has indeed advanced significantly.

15.5 Contributions to Mathematics and Logic: Goal 2

Since this article is in part written for you, to add a special touch, I present new material and, further, material obtained during the writing of this article. The new results include an axiom system that may not have been seen before, and what appears to be a proof far shorter than is found in the literature; each was obtained with the aid of the program OTTER. To complement the new material and to give additional evidence of the possible value of automated reasoning to mathematics and logic, I shall begin by briefly reviewing some open questions that were answered with the assistance of a reasoning program. Some of these questions might prove challenging and useful for those who wish to compare the effectiveness of various programs, to evaluate the potential of a new approach, or—by attempting to answer the questions unaided by a computer friend—to gain an appreciation for how far the field has come.

Of course, I can now hear you asking one of your provocative questions. "Larry, before you give your review of the past ten or so years, how far *has* the field come?" The accurate answer to your question, it seems to me, depends on whether the focus is on how far we *have* traveled or on how far we have *to* travel. To graphically see what I mean, we can look to the space program. In particular, the man in the moon is no longer a myth; indeed a person has walked on it. But the distance to the moon is dwarfed by the distance to the outermost planet, which is in turn miniscule when compared to the distance to another galaxy. So, how far has the space program advanced? Four decades ago, I would have doubted that we would reach the moon; two decades ago, I did doubt that an automated reasoning program would play a key role in answering open questions. Those two misplaced doubts have not shaken me; I still have doubts. For one example, if Texas is larger than Illinois—and, mind you, I said if—it is to make room for taller people, for tall Texans.

As you know, beginning in 1978, automated reasoning programs designed by members of the Argonne group have been used to answer various open questions, in finite semigroups posed by Kaplansky [29], in equivalential calculus posed by Kalman [35], in ternary Boolean algebra posed by Grau [30], in finite semigroups posed by McFadden [14], and in combinatory logic posed by Smullyan [24, 25, 37]. In my opinion, the successes in combinatory logic provide the most powerful evidence that automated reasoning has made contributions of note.

Barendregt [1] defines combinatory logic as an equational system satisfying the combinators S and K, where the respective actions of the constants S and K are given by the following two clauses.

```
EQUAL(a(a(a(S,x),y),z),a(a(x,z),a(y,z))).
EQUAL(a(a(K,x),y),x).
```

The Smullyan questions (answered by my colleague McCune and me) focus on what are called *fragments* of combinatory logic, and not the logic in its entirety. One of Smullyan's questions concerns finding, if such exists, an appropriate combinator to be used with the combinator B (defined with the following clause) to construct a fixed point combinator.

```
EQUAL(a(a(a(B,x),y),z),a(x,a(y,z))).
```

The combinator Θ is a *fixed point combinator* if and only if, for all combinators x, the equation $\Theta x = x(\Theta x)$ holds. The *strong fixed point property* holds for the fragment with basis $\mathbf{B}$ if and only if there exists a fixed point combinator Θ such that Θ is expressed purely in terms of the elements of B.

Statman answered Smullyan's question in the affirmative [26] by using the combinator W, defined by the following clause.

```
EQUAL(a(a(W,x),y),a(a(x,y),y)).
```

In addition to Statman's combinator, the fourth in the following list of five, McCune and I found four others by using an automated reasoning program [16].

$$\begin{aligned}
\Theta_1 &= B(B(B(WW)W)B)B \\
\Theta_2 &= B(B(WW)W)(BBB) \\
\Theta_3 &= B(B(WW)(BWB))B \\
\Theta_4 &= B(WW)(BW(BBB)) \\
\Theta_5 &= B(WW)(B(BWB)B)
\end{aligned}$$

Smullyan did indeed seem impressed by the discovery of the additional four. Later, by using the *kernel strategy* [37] applied by OTTER, we found an infinite class of infinite sets of such combinators. That result is still extremely gratifying to me, especially in my role as mathematician. Indeed, to find such richness where, just one year earlier, it was not known whether any appropriate combinators existed, and then to make the discovery because of formulating a new strategy that could be effectively applied by an automated reasoning program, signaled to me a significant advance in the field.

If you are curious and immediately wonder whether we simply built on Statman's combinator, I can accurately report that we did not; indeed, we found other combinators to act in the place of *W*. In particular, recalling that expressions in combinatory logic are assumed to be left associated unless otherwise indicated, our combinator *N* and the combinator *H* serve well.

$$Nxyz = xzyz$$
$$Hxyz = xyzy$$

The proofs—obtained with OTTER applying the kernel strategy—that the fragments with respective bases of combinators *B* and *N* and of *B* and *H* satisfy the strong fixed point property were completed by finding the following [37].

$$B(B(N(BB(N(BBN)N))N)B)B$$
$$H(B(H(HB))B)B(HH)$$

The combinator *N* may merit further study, for–in addition to offering as much (or more) power as *W* for constructing fixed point combinators–$BN(BB)$ acts as *S* does. The significance of this comment rests with the fact that *S* plays a vital role in the study of combinatory logic as a whole—*S* and *K* form a complete axiom system for the logic. Without the Argonne program OTTER and the kernel strategy, the combinator *N* might have remained undiscovered.

Smullyan then posed corresponding questions about the presence of the strong fixed point property for the fragments with respective bases of combinators *B* and *L* and of *Q* and *L*, defined by the following equations.

$$Lxy = x(yy)$$
$$Qxyz = y(xz)$$

McCune and I proved that neither fragment could satisfy the strong fixed point property [18]; OTTER again played a key role.

Smullyan also posed questions concerning the existence of *combinations*, an expression involving some given set of combinators that behaves as some other given combinator does. When an appropriate combination exists, he wished also to know the length (in symbol count) of the smallest and how many of such there are. The questions McCune and I answered focus on the following combinators.

(B)	$Bxyz = x(yz)$	(Q)	$Qxyz = y(xz)$
(C)	$Cxyz = xzy$	(Q_1)	$Q_1xyz = x(zy)$
(F)	$Fxyz = zyx$	(T)	$Txy = yx$
(G)	$Gxyzw = xw(yz)$	(V)	$Vxyz = zxy$

From the theorems proved in the study, the following summarizes the important results [39]. The combinators to be used for attempting to find an appropriate combination are B and T. Two combinations exist that act as Q does, each of length 6, and no shorter exists. Two combinations exist that act as Q_1 does, each of length 6, and no shorter exists. Two combinations exist that act as C does, each of length 8, and no shorter exists. Five combinations exist that act as F does, each of length 8, and no shorter exists. Ten combinations exist that act as V does, each of length 10, and no shorter exists. Five combinations exist that act as G does, each of length 10, and no shorter exists.

Questions of either type—those focusing on fixed point combinators and those focusing on combinations—might prove challenging and useful for evaluating a new reasoning program or a new idea. A full appreciation of the power offered by today's automated reasoning program can be gained by accepting the challenge of attempting to answer the questions just discussed, and to do so unaided by a computer program. What satisfaction you and I might derive—especially me with my experiences at the poker table—from witnessing a person's reaction to accepting the challenge, when said person is one of those who still considers automated reasoning at best to have the significance of a mouse eating a grain of wheat! Since you might be curious, paramodulation was the chosen inference rule for both studies of combinatory logic, and the set of support strategy played a key role. The time required to find fixed point combinators is frequently less than 5 CPU seconds, when the kernel strategy is used. Except for the combinator G, the respective times in CPU seconds for the studies of combinations are 6, 14, 19, 50, and 248. For G, the results were obtained from a Prolog program used in a generate-and-test mode; that Prolog program is *not* a reasoning program. In addition to finding the appropriate combinations for the combinator G, the Prolog program was used to show that no smaller combinations exist. Since attempts with OTTER did not succeed, we have an opportunity for the big stakes players.

And now for the new material gathered during the writing of this article: a possibly new axiom system for sentential calculus, a sharply shorter proof of the completeness of the axiom system proposed by Łukasiewicz (consisting of L1, L2, and L3), and some related items. The following story, showing precisely how the axiom system was discovered, provides powerful evidence of the value of having the program OTTER as a member of a research team. The use of OTTER dramatically reduced the turnaround time and facilitated the immediate verification of a conjecture. The story centers on an episode of temporary miscommunication that was the primary force in the discovery of the axiom set to be introduced. At the same time, the story provides another example of the serendipity of research.

After McCune and I dispatched the 68 theorems suggested by Dana Scott in an email, Scott and I began an email exchange of ideas and results. He played the role of logician, bringing knowledge of various calculi into play; I played the role of computer scientist, supplying knowledge relevant to the varied uses of OTTER. Scott's contributions included suggestions of theorems for OTTER to try to prove and suggestions of where to find other theorems for OTTER to try to prove. My contributions were to respond to Scott's suggestions by including input files, designed to enable OTTER to perform effectively, and proofs, obtained from OTTER using the various input files. Because the third attempt (at proving the 68 theses) succeeded in proving all 68 and in less than 16 CPU minutes on a SPARCstation, we were off and running.

As indicated earlier, among those 68 theses numbered 4 through 71 (with 1 through 3 being the complete axiom system proposed by Łukasiewicz), the set consisting of theses 18, 35, and 49 can be proved to be another complete axiom system, namely, *FL*. In other words, our success with Scott's original suggestion included a proof of the completeness of the Łukasiewicz system. Therefore, an obvious challenge was to attempt to have OTTER prove the converse, that the system consisting of theses 18, 35, and 49 is complete, where the proof would rest on deducing theses 1, 2, and 3.

It worked, and on the first try; theses 2 and 3 were proved in approximately 2 and 4 CPU seconds, respectively. However, as it turned out, the deduction of the remaining thesis, 1, proved more difficult. In fact, were it not for the ability to run OTTER in the background—permitting me to write this article and simultaneously conduct research—and were it not for successes in the preceding years reached only after substantial CPU time had been used, the attempt to complete the desired proof might have been abandoned. Instead, thesis 1 was proved after 26,600 CPU seconds.

The natural move next was to follow another of Scott's suggestions by studying other known axiom systems, which, as it turned out, led to the earlier-mentioned episode of miscommunication. Buried in the text of one of the papers by Łukasiewicz is the mention of the complete system consisting of theses 19, 37, and 59; no proof of its completeness is given, nor does he give any discussion of the importance of this system. Again, on the first try, OTTER proved it complete in less than 7 CPU seconds, with a 29-step proof in which the known system consisting of theses 18, 35, and 49 is deduced. Flushed with pleasure at this new success, I immediately sent the proof to Scott by email. However, as became clear from Scott's return email, I had miscommunicated—I had left the impression that this system, consisting of theses 19, 37, and 59, was a new axiom system, whose discovery belonged to me and, of course, to OTTER. What to do?—instant notification of my error, which I sent, certainly seemed hardly enough. After all, I had in effect been congratulated—which I obviously did not deserve—and, perhaps equally disturbing, Scott had expressed interest in the axiom system.

What was called for—if it could be found—was a genuinely new axiom system. The preferred way to rectify my communication error was to present—if at all possible—Scott with an axiom system that was genuinely new, *not* already in the literature. Unfortunately, I had no intuitive grasp of the elements of sentential calculus, and no knowledge of how its study had proceeded.

Therefore, on the surface, the idea of making such a search was out of the question—unless a guess and reliance on the ever-present and valued assistant OTTER could save the day.

The morning after the episode of miscommunication, noting that various complete systems existed each consisting of three theses—1 and 2 and 3, 18 and 35 and 49, and 19 and 37 and 59—I read through the 68 originally suggested by Scott. My objective was to replace one of theses 19, 37, and 59, for Scott had liked that system. Influenced by the apparent relation of 37 to 59, the placement and occurrence of their variables and the function n, the thesis that appealed to me above the rest was 60. Therefore, the system to try to prove complete was that consisting of theses 19, 37, and 60. I selected the naive approach of seeking a deduction of thesis 59, since both systems share 19 and 37. Such an approach is naive, for it might actually be easier to deduce another system entirely, say that consisting of theses 18, 35, and 49; I must admit that I did not even consider that possibility.

Again OTTER succeeded on the first try; an 11-step deduction of thesis 59 was obtained in less than 3 CPU seconds. Rather than sending that proof to Scott, I then had OTTER seek a shorter proof, which it found. Here is the 8-step proof that I sent to Scott, found by OTTER in just under 5,000 CPU seconds; a shorter proof exists, which will also be given. As observed earlier, the number in the first field indicates the corresponding position among the retained clauses.

```
1 [] -P(i(x,y)) | -P(x) | P(y).
8 [] P(i(i(i(x,y),z),i(y,z))).
9 [] P(i(i(i(x,y),z),i(n(x),z))).
10 [] P(i(i(u,i(n(x),z)),i(u,i(i(y,z),i(i(x,y),z))))).
25 [] -P(i(i(n(p),r),i(i(q,r),i(i(p,q),r)))) | $ANS(step_th_59).
--------------------
34 [hyper,8,1,8] P(i(x,i(y,x))).
40 [hyper,10,1,9] P(i(i(i(x,y),z),i(i(u,z),i(i(x,u),z)))).
59 [hyper,40,1,34] P(i(i(x,i(y,i(z,u))),i(i(z,x),i(y,i(z,u))))).
63 [hyper,59,1,34] P(i(i(x,i(x,y)),i(z,i(x,y)))).
68 [hyper,63,1,63] P(i(x,i(i(y,i(y,z)),i(y,z)))).
30752 [hyper,68,1,68] P(i(i(x,i(x,y)),i(x,y))).
30758 [hyper,30752,1,34] P(i(x,x)).
30954 [hyper,30758,1,10] P(i(i(n(x),y),i(i(z,y),i(i(x,z),y)))).
```

Clause (30954) contradicts clause (25), and the proof is complete.

Scott noted that this new axiom system, in which 59 is replaced by 60, offers less appeal, but the proof of 59 is interesting. He then suggested a way to shorten the proof, which OTTER succeeded in doing. Where the preceding 8-step proof is *organic*, meaning that none of the steps (theses) of the proof contains a subthesis that is true under all assignments of true and false, the following 7-step proof is inorganic.

```
1 [] -P(i(x,y)) | -P(x) | P(y).
2 [] -P(i(i(n(p),r),i(i(q,r),i(i(p,q),r)))) | $ANS(step_th_59).
3 [] P(i(i(i(x,y),z),i(y,z))).
4 [] P(i(i(i(x,y),z),i(n(x),z))).
5 [] P(i(i(u,i(n(x),z)),i(u,i(i(y,z),i(i(x,y),z))))).
--------------------
6 [hyper,3,1,3] P(i(x,i(y,x))).
10 [hyper,5,1,4] P(i(i(i(x,y),z),i(i(u,z),i(i(x,u),z)))).
18 [hyper,10,1,6] P(i(i(x,i(y,i(z,u))),i(i(z,x),i(y,i(z,u))))).
21 [hyper,18,1,6] P(i(i(x,i(x,y)),i(z,i(x,y)))).
26 [hyper,21,1,6] P(i(x,i(y,y))).
27 [hyper,26,1,26] P(i(x,x)).
33 [hyper,27,1,5] P(i(i(n(x),y),i(i(z,y),i(i(x,z),y)))).
```

Clause (33) contradicts clause (2), and the proof is complete.

It appears that the two proofs are the shortest that exist among, respectively, the organic and inorganic. A review of this anecdote suggests correctly how a dialogue between researchers can be sharply enhanced by access to an automated reasoning program of OTTER's type.

The preceding story has an interesting postscript, showing, among other things, that wild guesses coupled with OTTER's power and rapid turnaround can lead to additional advances. Roughly a week after the discovery of the system consisting of theses 19, 37, and 60, Scott suggested yet another possible axiom system. He suggested replacing thesis 60 by the following thesis, to be called 60a.

```
P(i(i(i(i(y,z),i(i(x,y),z)),u),i(i(n(x),z),u))).
```

In approximately 47 CPU seconds, OTTER proved (on the first try) the completeness of Scott's system by deducing thesis 59 with the following 8-step proof.

```
1 [] -P(i(x,y)) | -P(x) | P(y).
8 [] P(i(i(i(x,y),z),i(y,z))).
9 [] P(i(i(i(x,y),z),i(n(x),z))).
10 [] P(i(i(i(i(y,z),i(i(x,y),z)),u),i(i(n(x),z),u))).
25 [] -P(i(i(n(p),r),i(i(q,r),i(i(p,q),r)))) | $ANS(step_th_59).
--------------------
27 [hyper,8,1,8] P(i(x,i(y,x))).
30 [hyper,9,1,8] P(i(n(i(x,y)),i(y,z))).
35 [hyper,27,1,10] P(i(i(n(x),y),i(z,i(i(u,y),i(i(x,u),y))))).
176 [hyper,35,1,30] P(i(x,i(i(y,i(z,u)),i(i(i(v,z),y),i(z,u))))).
193 [hyper,176,1,176] P(i(i(x,i(y,z)),i(i(i(u,y),x),i(y,z)))).
213 [hyper,193,1,8] P(i(i(i(x,y),i(i(z,y),u)),i(y,u))).
253 [hyper,213,1,10] P(i(x,x)).
273 [hyper,253,1,10] P(i(i(n(x),y),i(i(z,y),i(i(x,z),y)))).
```

Clause (273) contradicts clause (25), and the proof is complete.

OTTER also obtained a completeness proof of the Scott system by deducing the three axioms of *FL*; the 33-step proof was obtained in approximately 89 CPU seconds. In approximately 94 CPU seconds, an 11-step deduction of thesis 60 was obtained. Of the completeness proofs focusing on the (independent) Frege system, a 31-step proof was obtained in approximately 2,912 CPU seconds. Hilbert's system was deduced with a 24-step proof in approximately 921 CPU seconds. In approximately 2,677 CPU seconds, the Łukasiewicz axiom system was obtained with a 24-step proof. As is so typical of the experiments presented in this article, all of the completeness proofs for the Scott system were obtained in a single run.

A post postscript: I have just found a 4-step, organic proof of thesis 59 from theses 19, 37, and 60; therefore, earlier, I should have said "it appeared that ..."—this article does indeed have the character of a letter. When I notified Scott of this 4-step proof, his reaction, given by email, was that it might indeed be "a very neat proof that would not be obvious to a human investigator". He explained that it is not particularly easy to do unification in one's head—and is he ever right! (Note 1 is relevant to the preceding.) Here is the proof.

```
1 [] -P(i(x,y)) | -P(x) | P(y).
8 [] P(i(i(i(x,y),z),i(y,z))).
9 [] P(i(i(i(x,y),z),i(n(x),z))).
10 [] P(i(i(u,i(n(x),z)),i(u,i(i(y,z),i(i(x,y),z))))).
25 [] -P(i(i(n(p),r),i(i(q,r),i(i(p,q),r)))) | $ANS(step_th_59).
--------------------
33 [hyper,10,1,9] P(i(i(i(x,y),z),i(i(u,z),i(i(x,u),z)))).
44 [hyper,33,1,8] P(i(i(x,i(y,z)),i(i(i(u,y),x),i(y,z)))).
68 [hyper,44,1,8] P(i(i(i(x,y),i(i(z,y),u)),i(y,u))).
99 [hyper,68,1,10] P(i(i(n(x),y),i(i(z,y),i(i(x,z),y)))).
```

Clause (99) contradicts clause (25), and the proof is complete.

Of course—in keeping with my personality with which you are so familiar, and because I simply delight in proofs and in successes with OTTER—I would like to include virtually everything learned because of Scott's prompting; I clearly thrive on successful proof finding with OTTER, especially when the impetus is provided by a great scholar. However, since I can hear you giving me the appropriate warning, I shall postpone some items to a later section (Random Notes), omit many, and conclude this section with the promised shorter proof.

The theorem in question asserts that the Łukasiewicz system, consisting of theses 1, 2, and 3, is complete for sentential calculus. Scott informs me that the Łukasiewicz proof, deducing the three axioms of *FL*, is 46 steps in length; using OTTER on the Macintosh, Scott obtained essentially the same proof. The following 30-step proof, deducing the three theses of *FL*, is the shortest so far obtained; a somewhat different 30-step proof that is not organic has also been found with OTTER.

```
1 [] -P(i(x,y)) | -P(x) | P(y).
4 [] -P(i(q,i(p,q))) | -P(i(i(p,i(q,r)),i(i(p,q),i(p,r)))) |
        -P(i(i(n(p),n(q)),i(q,p))) |
        $ANS(step_all_Church_FL_18_35_49).
8 [] P(i(i(x,y),i(i(y,z),i(x,z)))).
9 [] P(i(i(n(x),x),x)).
10 [] P(i(x,i(n(x),y))).
--------------------
34 [hyper,8,1,8] P(i(i(i(i(x,y),i(z,y)),u),i(i(z,x),u))).
35 [hyper,9,1,8] P(i(i(x,y),i(i(n(x),x),y))).
38 [hyper,10,1,8] P(i(i(i(n(x),y),z),i(x,z))).
41 [hyper,34,1,34] P(i(i(x,i(y,z)),i(i(u,y),i(x,i(u,z))))).
43 [hyper,34,1,8] P(i(i(x,y),i(i(i(x,z),u),i(i(y,z),u)))).
52 [hyper,38,1,34] P(i(i(x,n(y)),i(y,i(x,z)))).
57 [hyper,38,1,9] P(i(x,x)).
59 [hyper,52,1,38] P(i(x,i(y,i(n(x),z)))).
66 [hyper,41,1,35] P(i(i(x,i(n(y),y)),i(i(y,z),i(x,z)))).
67 [hyper,43,1,41] P(i(i(x,i(i(y,z),u)),i(i(y,v),
                        i(x,i(i(v,z),u))))).
103 [hyper,59,1,9] P(i(x,i(n(i(i(n(y),y),y)),z))).
272 [hyper,66,1,103] P(i(i(i(i(n(x),x),x),y),i(z,y))).
303 [hyper,272,1,57] P(i(x,i(i(n(y),y),y))).
310 [hyper,303,1,67] P(i(i(n(x),y),i(z,i(i(y,x),x)))).
317 [hyper,310,1,66] P(i(i(i(i(x,y),y),z),i(i(n(y),x),z))).
335 [hyper,317,1,34] P(i(i(x,i(y,z)),i(i(n(z),y),i(x,z)))).
346 [hyper,317,1,57] P(i(i(n(x),y),i(i(y,x),x))).
350 [hyper,317,1,38] P(i(i(n(x),n(y)),i(y,x))).
377 [hyper,350,1,38] P(i(x,i(y,x))).
416 [hyper,377,1,8] P(i(i(i(x,y),z),i(y,z))).
508 [hyper,416,1,350] P(i(n(x),i(x,y))).
509 [hyper,416,1,346] P(i(x,i(i(x,y),y))).
593 [hyper,508,1,335] P(i(i(n(x),y),i(n(y),x))).
644 [hyper,593,1,508] P(i(n(i(x,y)),x)).
714 [hyper,509,1,41] P(i(i(x,i(y,z)),i(y,i(x,z)))).
765 [hyper,714,1,8] P(i(i(i(x,i(y,z)),u),i(i(y,i(x,z)),u))).
771 [hyper,714,1,346] P(i(i(x,y),i(i(n(y),x),y))).
1755 [hyper,771,1,67] P(i(i(n(x),y),i(i(z,x),i(i(y,z),x)))).
1809 [hyper,1755,1,644] P(i(i(x,i(y,z)),i(i(y,x),i(y,z)))).
1870 [hyper,1809,1,765] P(i(i(x,i(y,z)),i(i(x,y),i(x,z)))).
```

Clauses (377), (1870), and (350), respectively, contradict the literals of clause (4), and the proof is complete.

15.6 Mathematicians and Logicians as Users of a Program: Goal 3

Although a precise estimation of the value of having many mathematicians and logicians as users of automated reasoning programs is essentially impossible, their use would clearly advance the field markedly. A master of some area, group theory for example, who was also quite familiar with the intricacies of the use of OTTER would almost certainly make significant contributions to that area. I must admit that this observation seems difficult to doubt when you review what has occurred in the past decade regarding the answering of open questions from a variety of fields of mathematics and logic. Granted that none of the long-standing deep problems have been solved; however, the investigators could hardly be classed as masters of any of the disciplines from which the questions were taken. I also agree with you that failure would be the likely outcome if an automated reasoning program were asked to prove a randomly selected theorem from a standard textbook. Given these pluses and minuses, the goal of having mathematicians and logicians counted among the users of such programs is indeed an important goal and—how can it be otherwise!—an important component of your and my dream for automated reasoning. Finally, I can report that the current set of those using an automated reasoning program does now include such researchers.

Although I have heard rumors of such uses at the University of Texas and in England—rumors that I have not verified—I prefer to confine my comments to what I know personally. Also, recognizing the lack of a clear distinction between that which is strictly computation and that which contains reasoning steps, I shall, nevertheless, pretend otherwise and cite as my first example John Kalman's research. Kalman is a logician at the University of Auckland in New Zealand. In the late 1970s, he had his own theorem-proving program which he heavily used for his research in equivalential calculus [8]. Prompted by his desire to become familiar with our efforts in the field, Kalman visited Argonne and suggested leaving with us seven theorems to attempt to prove with our program. He did not have to wait too long before seeing results. Indeed, ten minutes after he presented the seven theorems, he had in his hands five proofs obtained with a version of the program that would eventually be called AURA [23]. After tasting this finest of brandies, he was converted, shortly thereafter becoming a constant and active user of our programs—and still is today. So convinced of their value, he has written an extensive workbook for the use of ITP, a book that contains many gems that merit study [9]. He is currently using OTTER to aid his research in group theory, but in the context of divisional calculus.

For some time, copies of OTTER have also been in the hands of Corrado Boehm (combinatory logic) University of Rome, Robert Meyer (relevance logic) Australian National University, R. Goldblatt (mathematics) Wellington University in New Zealand, and Roger Hindley (logic) University College in Wales. Most recently, we sent a copy of OTTER to Dana Scott (logic) Carnegie Mellon University, which he has successfully used on his Macintosh. Barely a trickle, indeed; but, compared to 1975, a veritable torrent!

On the other hand, recognizing that in some sense we have hardly moved from zero, you might find interest in the following excerpts from two phone conversations I had with Irving Kaplansky. Approximately a decade ago, I called him at the University of Chicago to obtain a possible open question to attack with our program. He supplied such a question, the question mentioned earlier regarding the possible existence of certain finite semigroups; Note 3 might be of interest. During that conversation, when automated theorem proving in general was broached, Kaplansky surprised me with the following comments, which I have just checked in a second phone conversation. "You have a friend in court. Should you continue in your research, the mathematics papers written in 2060 or thereabouts will be at a new level of reading, with much of the current level within reach of a theorem-proving program." In our second conversation, prompted by my wish to clear with him my inclusion of his earlier comments, he and I agreed that progress is indeed painfully slow, as it is with programs to automate translation from one language to another.

The snail travels fast, thinks the defenseless plant with fear; the snail travels slow, observes the cat eager to play. I believe—of course, you know I actually mean that I am certain—the significant point is that, finally, we *are* traveling, making recognizable progress. What will you and I witness before we move elsewhere? I wonder, I await, and, most truthfully—how many know that truth takes on numerous possible values?—I thrive on the expectation.

15.7 New and Old Uses for an Automated Reasoning Program

At this point—taking you up on your comment that you do not mind a little repetition—I shall present a general review of what is possible when using an automated reasoning program, in particular, when using OTTER. At the simplest level, in addition to the familiar use of a program like OTTER for finding proofs, such a program offers unexpected power for proof checking, as the following example shows. If a proof has no steps missing, you can take each of its steps and use weight templates to instruct OTTER to prefer expressions that match the pattern of function symbols in the steps of the proof.

An example that I just ran focuses on the completeness proof for the system consisting of theses 19, 37, and 60 in which an 8-step deduction of thesis 59 is found. The goal is to have OTTER proof check the given proof. Therefore, OTTER was given the instruction to use symbol count to assign priority (by weight) to all clauses other than those that match one of the following eight weight templates.

```
weight_list(pick_and_purge).
weight(P(i(x,i(y,x))), 2).
weight(P(i(i(i(x,y),z),i(i(u,z),i(i(x,u),z)))), 2).
weight(P(i(i(x,i(y,i(z,u))),i(i(z,x),i(y,i(z,u))))), 2).
weight(P(i(i(x,i(x,y)),i(z,i(x,y)))), 2).
weight(P(i(x,i(i(y,i(y,z)),i(y,z)))), 1).
weight(P(i(i(x,i(x,y)),i(x,y))), 1).
weight(P(i(x,x)), 2).
weight(P(i(i(n(x),y),i(i(z,y),i(i(x,z),y)))), 2).
end_of_list.
```

The flags for seeking shorter proofs were turned off. And here comes an illustration of the unexpected power when using OTTER in a proof-checking mode; Note 3 is relevant.

Instead of verifying by "finding" the expected 8-step proof, OTTER found the 7-step deduction of thesis 59, the same 7-step proof given earlier. The explanation rests in part with the fact that weight templates of the kind just given are treated as if all variables are indistinguishable. Since step five, just as step one, of the shorter proof matches in its pattern of the function i the first weight template, it is preferred over clauses that do not match one of the templates. OTTER happened to find the shorter proof first and was prevented from completing the longer proof because the two proofs share the same last step and forward subsumption was in use. Disappointment seems misplaced; after all, the shorter proof is often the more elegant.

On the other hand, were you to insist upon verifying the given 8-step proof, or were you to prefer to avoid proofs that are not organic, OTTER offers most of what is needed. In the first case, if the weight of the fifth and sixth steps is set to 1 rather than to 2, OTTER finds the 8-step proof. If, in addition, back subsumption (which discards any retained clause that is subsumed by a clause retained later) is turned off, and if you instruct OTTER to seek more than one proof, then OTTER finds both proofs.

In the second case, the avoidance of proofs that are less likely to be organic is achieved with demodulation [32]. Specifically, if the following demodulator list is included in the input, where "$T" is interpreted as "true", then all clauses containing as a proper subexpression $i(t,t)$ for some term t are immediately purged.

```
list(demodulators).
(i(i(x,x),y) = discard).
(i(y,i(x,x)) = discard).
(i(discard,x) = discard).
(i(x,discard) = discard).
(n(discard) = discard).
(P(discard) = $T).
end_of_list.
```

An analysis of these demodulators shows that the thesis $i(x,x)$ is not discarded, an important observation, for this thesis often plays a key role in a completeness proof for an axiom system for sentential calculus as well as for many other proofs.

Of course, if you do not have a proof in its entirety to check, but have instead some steps that are conjectured to be useful (as when hints are given to a starred exercise), then the corresponding weight templates can be included. In other words, for those points between the ends of the spectrum focusing, respectively, on proof checking and proof finding, OTTER offers much of what is needed.

Somewhat related is the use of OTTER to monitor progress, its own and that of the researcher; I recommend Note 4. For a sample of this use, you can imagine wishing to prove some specific theorem, but finding it difficult to choose between various possible attacks. With one choice of attack, the conjecture asserts that, if lemmas A, B, and C could be proved, then the theorem would be within reach. With a second choice, the key lemmas appear to be D and E. With a third choice, lemmas F, G, and H are thought to offer what is needed. With OTTER and without much effort, the researcher can pursue simultaneously all three lines of attack. A way to proceed is to represent each of the eight lemmas as a unit clause, negate each, include with each the ANSWER literal with an argument designating which lemma is represented, place the negations in the passive list, and start the run. With Unix, the run can be placed in the background. Then, one simply glances occasionally at the output file to read the occurrences of UNIT CONFLICT.

The ANSWER literal, which is ignored by OTTER when testing for UNIT CONFLICT, is included to facilitate monitoring of the program's progress and of the researcher's chosen attacks. In addition, you can include in the axioms list the disjunction of the appropriate negated lemmas disjoined with an appropriate ANSWER literal and occasionally glance at the output file in search of a desired occurrence of the EMPTY CLAUSE; if the disjunction is placed in the passive list, it will be ignored, thus defeating the purpose of its inclusion. Using a disjunction approach is a means for avoiding the tedious piecing together of the various parts of a proof, and is useful for automatically removing duplicate steps from the various parts; in other words, this ploy returns the union of the various parts.

In contrast to the preceding uses, you can even use OTTER for the mundane task of detecting errors in the input or errors in the axioms, errors such as a simple typo or an incorrect interchange of symbols. The means for such detection is that of reading the retained clauses resulting from using the input that might contain an error—perhaps unsuspected at that. In fact, I recently had such an experience. The run was motivated by yet another study of a possible axiom system. A proof was found more rapidly than I expected. However, rather than satisfaction, you can imagine my reaction when I found that the proof contained the following undesirable clause.

```
P(x).
```

In the study in question, the clause amounts to asserting that everything is true—obvious nonsense! Fortunately, as exhibited in the proofs given earlier, OTTER supplies, for each deduction, the immediate ancestors from which the deduction is obtained. By tracing backward through the ancestry tree, the flaw in the input was easily found.

Even when you are certain that the input is flawless, you might wish to examine OTTER's output to decide whether the chosen approach is satisfactory. In particular, one of OTTER's flags can be used to cause the output file to contain all of the generated clauses, including those that are discarded for some reason. An inspection of the generated clauses can lead to the discovery that much of the CPU time is being consumed on what is clearly fruitless consideration. Especially for the expert in the field from which the problem under study is chosen, such an inspection can prove of immense value; Note 4 is perhaps of interest here. Indeed, it seems to me that, if so many successes can be so readily obtained by someone who is hardly more than conversant with a field, then the possibilities are most promising if the user is instead a master of the discipline under attack.

You can also use OTTER to produce test problems of varying difficulty, problems to be used to evaluate the relative strength of a new program or the potential of a new approach. A graduated set of such problems certainly appears to be one of the useful items to have for research in automated reasoning. Such a set should also prove useful as exercises for a course, a book, or an exam. One means for obtaining the desired set of problems, or at least obtaining some of them, is to have OTTER run well past the CPU time needed to complete some simple task. For example, let the task be to prove theses 39 and 40 from the members of *FL*, 18, 35, and 49. Taken together, theses 39 and 40 can be viewed as asserting, for all x, the equality of $n(n(x))$ and x. OTTER completes the given assignment in less than 4 CPU seconds, after retaining 156 clauses. If you then take advantage of the fact that OTTER's performance is nearly linear with regard to the deduction of conclusions and allow the program to run for a few CPU hours, a proof of the dependence of thesis 21 is also obtained. In approximately 7.4 CPU hours, OTTER produces a 55-step proof of the dependence of thesis 21 on theses 18, 35, and 49. For those who prefer an approach to automated reasoning based on Prolog technology, this problem appears to offer a small challenge; in particular, a level saturation approach with OTTER obtained an 11-step proof in approximately 9.1 CPU hours. The explanation for finding an 11-step proof rather than the 55-step proof rests with the fact that thesis 21 is dependent on theses 18 and 35 alone; the longer proof relies on the use of thesis 49, thus admitting to the search clauses in which the function n occurs. In fact, 28 of the 55 clauses of the longer proof contain at least one occurrence of n. A bigger challenge for the approach based on Prolog technology focuses on proving thesis 1, the first of the three axioms Łukasiewicz used for sentential calculus, from theses 18,

35, and 49.

Clearly, a quick review of the various uses covered here—from those directly related to research to those that might well be considered mundane—correctly suggests that I am captivated by experiments with an automated reasoning program. That I find the program OTTER to be a great companion and a valuable assistant is manifestly obvious. Finally, you are definitely warranted in concluding that—rather than the impossibility of an effective automation of logical reasoning, as has too frequently been suggested in the literature—I am building a case for my view that the field has demonstrated in various ways that we have arrived.

15.8 Challenges and Open Questions

My friend and colleague George Robinson asked me on September 15, 1990, whether the field was still as prone to bs as it was when he and I studied automated theorem proving. I told him that all was sharply changed; many researchers now welcomed challenges and the opportunity to attack open questions with an existing program. Therefore, in addition to the challenges presented by the theorems discussed so far in this article, others seem in order, including questions that remain open here in 1990.

Open Question in Robbins Algebra

The following question has been open for decades, receiving attention even from Tarski and his students [6, 27]. Is every Robbins algebra a Boolean algebra? A Robbins algebra is a nonempty set satisfying the following three axioms, expressed in clause notation, in which the function o can be interpreted as plus and the function n as negation.

```
(R1)  EQUAL(o(x,y),o(y,x)).
(R2)  EQUAL(o(o(x,y),z),o(x,o(y,z))).
(R3)  EQUAL(n(o(n(o(x,y)),n(o(x,n(y))))),x).
```

A Boolean algebra is a nonempty set S with two operations, plus and times, and a 0 and a 1. Each operation is commutative, and each distributes over the other. The 1 is a multiplicative identity, and the 0 is an additive identity. In addition, for every x, the negation of x exists with x plus its negation equal to 1 and x times its negation equal to 0. An alternative axiomatization of Boolean algebra consists of (R1), (R2), and Huntington's axiom (H3) [7].

```
(H3)  EQUAL(o(n(o(n(x),y)),n(o(n(x),n(y)))),x).
```

Since it is known that every finite Robbins algebra is Boolean, two choices exist. You can attempt to find an infinite model that satisfies the three axioms for a Robbins algebra but violates a property for a Boolean algebra, or you can attempt to prove that Robbins implies Boolean. If you rely on the assistance of an automated reasoning program, with the intention of proving that Robbins does *not* imply Boolean, an added challenge is presented by the need to cope with infinite sets. If instead you rely on such a program with the objective of proving that Robbins *does* imply Boolean, then an added challenge focuses on coping with the extremely large number of clauses that will ordinarily be encountered, even if paramodulation is the chosen inference rule.

Open Questions in Combinatory Logic

Combinatory logic presents a number of open questions of which the following offer intrigue. Does the fragment with a basis consisting of the combinators B and S alone satisfy the strong fixed point property? The first two of the following three clauses specify the actions of B and S, respectively; the third clause corresponds to the denial of the strong fixed point property.

```
EQUAL(a(a(a(B,x),y),z),a(x,a(y,z))).
EQUAL(a(a(a(S,x),y),z),a(a(x,z),a(y,z))).
-EQUAL(a(y,f(y)),a(f(y),a(y,f(y)))).
```

If paramodulation is the inference rule of choice, then, of course, you must add a clause for reflexivity of equality, $x = x$. If the inference rule of choice is one of the resolution-based rules, hyperresolution for example, then in addition you must add clauses for symmetry, for transitivity, and for substitutivity in each of the arguments of the function a.

For a similar question, does the fragment with a basis consisting of the combinators B and N_1 satisfy the strong fixed point property? The behavior of the combinator N_1 is given by the following clause.

```
EQUAL(a(a(a(N1,x),y),z),a(a(a(x,y),y),z)).
```

To further encourage research focusing on the automation of model generation, here are two open questions. Does there exist a finite model satisfying the combinators B and L in which the strong fixed point property fails to hold? Does there exist a finite model satisfying the combinators Q and L in which the strong fixed point property fails to hold? That some (possibly infinite) model of the type under discussion in each of the two questions exists follows from results quoted earlier from the research with my colleague McCune [18].

Because of the lack of formal criteria to determine precisely which questions merit the classification of open, I point out that the preceding questions from combinatory logic are my questions. I do not mean to imply that I know the answers, for I do not; nor, as far as I can ascertain, does anyone else.

Challenge Problem from Many-Valued Sentential Calculus

As for a new problem (from among the theorems whose proof is known) to take the place of the problems previously considered tough for an automated reasoning program to solve, the following theorem from many-valued sentential calculus provides an excellent target. The theorem, sometimes known as the fifth conjecture of Łukasiewicz, asserts that, from the first four of the following five clauses, the fifth can be deduced by using condensed detachment.

```
P(i(x,i(y,x))).
P(i(i(x,y),i(i(y,z),i(x,z)))).
P(i(i(i(x,y),y),i(i(y,x),x))).
P(i(i(n(x),n(y)),i(y,x))).
P(i(i(i(x,y),i(y,x)),i(y,x))).
```

The problem asks for an approach that enables an automated reasoning program to prove the Łukasiewicz conjecture, using condensed detachment. Two points to keep in mind are (1) a proof of the conjecture has been obtained where the problem was rephrased as one for equality, and (2) McCune and I have a 63-step proof using condensed detachment, a proof obtained by using OTTER in a mode that is in part proof checking and in part proof finding. Our solution, obtained in a style somewhat reminiscent of that discussed earlier in which you have an outline and OTTER fills in the holes, does *not* count, for we relied heavily on the equality-oriented proof that was obtained with OTTER.

Challenge Problems from Sentential Calculus

The last two challenges in this section come from Dana Scott. Both focus on a slightly different notation for sentential calculus, namely, the replacement of $n(x)$ by $i(x, F)$, where F is thought of as meaning false. Scott asks whether, among the following five axioms given in clause form, axioms 1, 2, and 4 are complete for sentential calculus, and whether axioms 1, 4, and 5 are.

```
(axiom 1)  P(i(i(x,y),i(i(y,z),i(x,z)))).
(axiom 2)  P(i(i(i(x,F),x),x)).
(axiom 3)  P(i(x,i(i(x,F),y))).
(axiom 4)  P(i(F,x)).
(axiom 5)  P(i(i(i(x,F),F),x)).
```

Axiom 3 is included because its deduction from either set will establish the corresponding completeness result. As it turns out, the following axiom, 2a, is provable from axioms 1, 2, and 3; it is obviously more general than axiom 2.

```
(axiom 2a)  P(i(i(i(x,y),x),x)).
```

Therefore, the complete axiom system consisting of 1, 2, and 3 is equivalent to that consisting of 1, 2a, and 3.

15.9 Random Notes

For your enjoyment and curiosity, I decided to include the following notes. Some of them express the views we have exchanged; some are simply my unsupported opinions (which you have encouraged me to express). Some of the notes are directly related to the earlier material, some are distantly related, and some are perhaps unrelated.

Note 1. That a number of individuals still maintain the impossibility of the effective automation of logical reasoning attests to resistance—and, perhaps, to fear. No automated reasoning program will ever replace the skilled mathematician or logician. But, in view of the frequency, variety, and ease of success reported here and elsewhere, how can one doubt the value of such a program as an automated reasoning assistant?

As my esteemed colleague Bob Boyer has been heard to say, there is no way to cheat when answering an open question. Since a number of such questions have been answered with the claim that an automated reasoning program played a key role, then the only apparent complaint must rest with the belief that the program did very little. If that is the objection, I recommend that the skeptic review the 4-step proof deducing thesis 59 from theses 19, 37, and 60, coupled with Scott's comment. That proof certainly appears to be an example of a proof that a person might not find.

For a more impressive result, I recommend that the curious examine Steve Winker's original 159-step proof that establishes that the formula *XHN* provides a complete axiomatization of equivalential calculus [38]. Winker's result refuted a long-standing conjecture, and it was obtained with substantial assistance from an automated reasoning program (AURA) designed by colleagues at Argonne National Laboratory.

For equally impressive results, but ones that may offer even more depth, I recommend that the difficult-to-persuade examine the research (with my colleague McCune) focusing on fixed point problems in combinatory logic [37, 18]. In that context, the kernel strategy, used to find fixed point combinators, is of particular satisfaction; its use provides the most uniform success with randomly selected problems from some area of mathematics or logic.

Note 2. Automated reasoning programs can find proofs that a person, even with training and knowledge, might not find; again, the earlier cited 4-step proof of thesis 59 provides an appropriate example. More obvious is the case in which an open question was answered, where an automated reasoning program played a key role. Such a program makes no tacit assumptions. Such a program nicely complements the expertise that can be brought to a study by the researcher. The objection that one cannot know *a priori* which inference rule to use seems spurious; in many computing environments, one can simply try a number of runs, simultaneously. When a person, rather than a program, is one's colleague, the preferred attack is often unclear.

On the other hand, the desire for choosing quickly and wisely the best inference rule and strategy provides an excellent motivation for researchers in our field to propose and test metarules for making such choices. Only, let the testing at least include hard problems and new problems, instead of relying on those that have been repeatedly quoted and that are now easy to solve. Of course, I recommend an occasional attempt to answer an open question; but the tests must focus mainly on known results, for, otherwise, progress cannot be measured. I also give the obvious recommendation that data must be given to support claims, CPU time compared to other programs or approaches, number of conclusions drawn, number retained, and the like.

Note 3. To me, it is clear that the field has matured, at least passing through adolescence. Only a few years ago, what chance would we have had with the 68 theorems suggested by Scott? Before OTTER's arrival, what program could perform almost linearly with respect to drawing conclusions? Who would have guessed that, by 1990, a program would exist that could be effectively used (at one end of the spectrum) for proof checking and (at the other end) for finding a new axiom system?

If the field moves forward just a bit more, such a program might well ease the burden of refereeing for journals of various types—give such a future program the outline supplied in the paper to be refereed, and instruct it to fill in the gaps. For a distantly related example, OTTER found that, among the 68 theses of Łukasiewicz sent by Scott, 12 properly subsumes 11, 18 properly subsumes 17, and 26 properly subsumes 27; a pair of this type might occur in a proof to be refereed. Also, when OTTER was given the obvious eight weight templates and asked recently to proof check the 8-step proof of thesis 59 from theses 19, 37, and 60, the 7-step proof was found first. In other words, OTTER sometimes finds a better proof than the one under study, and sometimes finds a theorem more general than either of two theorems—neither of which subsumes the other—whose negations are included to permit the program to prove both theorems in one run.

Note 4. Conducting research with OTTER is strikingly different from conducting research with a person. With a person, one is ordinarily forced to choose, from among the possible attacks, the one that seems most promising. It is simply too confusing and too arduous to sit back and try to consider two or more paths of inquiry simultaneously. However, with OTTER, you can pursue many attacks in a single run. I certainly took this approach in my studies of the completeness proofs for various axiom systems.

It was simple; you choose the axiom system to be proved complete, place its members in the set of support, and, for each other known complete axiom system, place in the axioms the disjunction of the negations of its members. I think it helps to monitor OTTER's progress: you can place the individual negations—each disjoined with an ANSWER literal designating its role—in the passive list, preventing them from participating except for unit conflict and subsumption. Occasional glances at the output file show how things are progressing. Asking about UNIT CONFLICT shows which members of which axiom system have been proved, which is where the ANSWER literal comes into play. Asking about EMPTY CLAUSE shows which systems have been deduced or, equivalently, which completeness proofs have been obtained.

The other striking difference focuses on the occurrences of success measured in CPU time or measured in the number of retained clauses. Specifically, proofs often occur in bunches. OTTER often proves some of the theorems under consideration in the first few CPU seconds and with a small number of retained clauses. Then, often, an isolated proof is found after say 20 CPU minutes and the retention of 3,000 clauses. At that point, sometimes, OTTER seems to go off and hide, for glances at the output file in search of new successes from among the targets yield nothing new. However—and I am just beginning to accept this—after a long gap in time and retained clauses, another desired result is often obtained.

For such an example, it suffices to glance at the following results of the successful attempt to prove the dependence of theses 1, 2, 3, 16, 21, 24, 39, and 40 on the set consisting of 18, 35, and 49.

```
UNIT CONFLICT at  2.02 sec  104 [bin,103,13] $ANS(step_th_2).
UNIT CONFLICT at  3.69 sec  144 [bin,143,18] $ANS(step_th_39).
UNIT CONFLICT at  3.76 sec  149 [bin,148,15] $ANS(step_th_16).
UNIT CONFLICT at  3.87 sec  157 [bin,156,19] $ANS(step_th_40).
UNIT CONFLICT at  4.04 sec  162 [bin,161,14] $ANS(step_th_3).
UNIT CONFLICT at 1288.96 sec
                           3211 [bin,3210,16] $ANS(step_th_24).
UNIT CONFLICT at 26600.33 sec
                           8089 [bin,8088,12] $ANS(step_th_1).
UNIT CONFLICT at 26635.09 sec
                           8205 [bin,8204,17] $ANS(step_th_21).
```

Note 5. Using OTTER is, to me, often reminiscent of collaborating with a brilliant colleague whose mental processes are complex and sometimes mysterious. Rather than simple poetry or worse, I am instead referring to the intricacy and interplay of OTTER procedures and options.

For example, let the objective be to find one or more shorter proofs. The obvious approach is to set the flags instructing OTTER to compare derivation lengths and to use back subsumption and thus seek, in a straightforward manner, to measure the lengths of two deductions of the same conclusion, preferring the shorter. The approach is a good one, but far more can be done, and certain hidden complications exist. Perhaps not so obvious, the shorter proof terminating in clause D may be missed because of encountering on the way two deductions of clause C, where the first deduction has shorter length than the second, but where the second deduction contains clauses that can be used repeatedly on the path from C to D. In other words, the shorter subpath on the way to proving a theorem may lead to pursuing a longer path to the finish.

Since the topic of shorter proofs is important—such proofs are often more elegant—I shall simply touch on additional options and other complications that are in part hidden. Rather than instructing OTTER to direct its search based on choosing as the focus of attention the "simplest" clause not yet used, you can instead use level saturation, a breadth-first search. In the context of seeking shorter proofs, the appeal of an approach based on level saturation rests with the correlation between proofs of lower level and shorter proofs. However, you can quickly see that there can exist two proofs with the first having lower level than the second, but with the second being shorter. The level of input clauses is 0, and the level of a deduced clause is one greater than the maximum of the levels of the immediate ancestors used to deduce the clause. A proof can be found at level 4 with a length of 7, in contrast to a second proof of the same result having level 5 and length 5. Therefore, an approach based on level saturation is not guaranteed to find the shortest proof, for, if forward subsumption is used—which is almost always recommended—the second proof can have its last step subsumed, preventing completion. In fact, any form of subsumption, backward or forward, can interfere with finding the shortest proof, for the subsumed clause—although less general than its subsumer—may be the key step in a shorter proof.

Nevertheless, when using the ancestor_subsume flag, to avoid retaining many copies of the same clause, the use of back subsumption is recommended. The clause A ancestor-subsumes the clause B if and only if (1) A properly subsumes B or (2) A and B are identical and the derivation length of A is less than or equal to that of B. Just as there can exist clauses A and B that subsume each other, there can also exist two clauses that ancestor-subsume each other. In either case, if forward subsumption is in use, OTTER retains the earlier copy. Rather than asking OTTER to directly seek shorter proofs by comparing derivation lengths, you might instead prefer to use a level-saturation approach with the ancestor_subsume flag turned off. With that choice, you might well decide to use back subsumption to purge less general clauses in favor of newer and more general clauses. However, you should be warned that the use of back subsumption can, unfortunately, interfere with finding a shorter proof. To see how such interference can occur, you can imagine the deduction of two clauses A and B such that B is deduced after A is deduced and such that B back subsumes A. If in addition the path to A is shorter than the path to B, and if the use of A or B could lead in one step to the completion of the desired proof, then the actions of back subsumption can prevent you from finding the shorter proof.

The use of weighting—which is inescapable with OTTER, for the focus of attention will be based on symbol count or on the input weight templates—can block the discovery of a shorter proof. Indeed, the use of so-called less complex expressions in preference to more complex can force the program to miss a proof that might be shorter though relying on more complex clauses.

Perhaps the word "miss" is not the best word in a technical sense; the word "delay" might be better, delay for an inordinate amount of CPU time, for example, so that the proof is in fact missed.

In a similar way, the inclusion of dependent clauses—although promoting naturalness and perhaps the imitation of the approach a person might take—can also interfere with finding a shorter proof. For example, the successful attempt to prove thesis 21 dependent on theses 18, 35, and 49 produced a 55-step proof in which all three theses were used; in contrast, a second attempt in which thesis 49 was omitted produced an 11-step proof.

Although you might strongly and understandably object, I think we have yet more evidence to support the position that the best approach to the automation of logical reasoning does *not* emphasize the imitation of the way a person works. Naturalness of proof and of approach is fine; but its absence to some degree or to a total degree may be the key to obtaining the desired result. All is okay; better to increase the effectiveness of research by complementing the researcher, without the concern of imitation. I was startled by the obvious appeal—to a surprising number of people—coupled with the misleading nature of a recent talk in which the speaker asserted that the successful dispatching of a challenge problem proposed by Frank Pfennig, though laudable, suffered from the fact that Łukasiewicz (the author of the theorem in question) would never have in effect generated the 6,000,000 conclusions that the program did. I am certain that Łukasiewicz did not and would never take such an approach; but the observation is at best irrelevant, if the object is to evaluate the merit of the attack taken by an automated reasoning program.

Finally, intuition can get in the way of finding the desired shorter proof. For example, intuition might suggest that the thesis $i(x, i(y, x))$ is a key to finding a shorter proof. To impose that intuition on OTTER's search, you can use a weight template to give a low weight (high preference) to the following clause.

```
P(i(x,i(y,x))).
```

OTTER would accept the instruction, emphasizing the role of the preceding clause, which might in turn send the program down a long, long path culminating in a proof. The resulting proof might be far longer than that produced if the clause were simply given a weight based on symbol count or, even further, given a high weight to prevent or delay its use.

Quite different from the direct approach to seeking a shorter proof with OTTER is that in which a number of targets are given simultaneously. For example, if the objective is to prove that some given set of axioms is complete for some calculus, by including in the axioms clauses each of which is the disjunction of the negations of a known set of axioms, OTTER can in effect attempt to use each disjunction as a target. If you also set the proofs limit flag to 0, giving OTTER permission to continue to seek proofs until time or memory is exhausted, then the program may in fact succeed in completing a number of different and sometimes unrelated proofs. Then, rather than OTTER comparing proof lengths, the user makes the comparison. This approach to seeking a shorter proof can, of course, be combined with the direct approach in which OTTER compares derivation lengths of the same deduction.

Of course, the beauty—some might say difficulty—of using OTTER is captured by the converse; such an emphasis based on the researcher's intuition might be just what is needed to find the shorter proof. Indeed, when Scott suggested that the 8-step proof of thesis 59 from theses 19, 37, and 60 could be shortened and also stated how that could be accomplished, I simply added weight templates to encourage the use of one clause and discourage the use of two others; it did work, as evidenced by the resulting 7-step proof, as he predicted. I view such occurrences as proof that OTTER is an assistant whose value is great—and growing. Conducting research with a person is, in reality, no less complicated.

Note 6. Just as different mathematicians and different logicians have different strengths, so also is it true for automated reasoning programs. No program compares to the Boyer-Moore program for verifying computer code or verifying algorithms. The Kapur-Zhang program appears to be the best for solving problems in which associative/commutative unification can be used. As for the approach based on Prolog technology, Stickel's program sets the standard. However, for my type of experimentation, no program offers nearly as much as OTTER does. Even further, for serious research in some area of mathematics or logic, I strongly conjecture that nothing that currently exists compares to OTTER. The fact that, with respect to drawing conclusions, this program performs almost linearly gives it a distinct edge. Even after 19 CPU hours on a SPARCstation, the rate of deductions had changed only from 550 clauses per CPU second to 460. One of McCune's masterful moves was to adapt the use of discrimination trees [19] to the application of forward subsumption, a procedure that is indispensable for serious research [40].

Sometimes, the objective is to find the "best" proof, which may mean the shortest, the most elegant, the most natural, or simply one that the researcher particularly likes. With the use of weighting and the ancestor flag and demodulation, OTTER often enables one to take a proof and polish it to produce a better proof. A case in point: when Scott suggested that the 8-step proof of thesis 59 from theses 19, 37, and 60 could be shortened to a 7-step proof and showed how, it was trivial to instruct OTTER to follow Scott's suggestions. Later, when the assignment for OTTER was to simply verify that proof to check that no shorter proof existed, the use of level saturation failed in an odd way—a 4-step proof was found instead.

Note 7. Clearly, the use of OTTER as a team member requires reorientation. For one of the more striking examples—when compared to working with another person—no means exists to ask OTTER how it is doing, whether it is close to a proof, or whether it is stuck in a trough. Instead, one notices that the proofs often come in clusters and that the gaps between successes can vary widely. I have often had the urge to truncate a run because of noting that no new proofs had been found in the preceding 3 CPU hours, only to discover, after resisting the urge, that more proofs would be found. For example, in a run being made as I write this section, in which the object is to find a proof of the completeness of a single axiom (found by Meredith) for sentential calculus, one of the desired results was obtained at the 15 CPU-second mark, and the next result was not obtained until the 5,606 CPU-second mark.

Also, I find it much easier to wander through OTTER's results than through those of a person; the explanation rests in part with no missing steps, precise history of derivation, and the like. Such wanderings have produced unexpected treasure. For example, before I was told that theses 18, 35, and 49 form a complete axiom system for sentential calculus, I was using OTTER to attempt to prove that the system consisting of theses 16, 18, 21, 24, 35, 39, 40, and 49 was complete. The approach focused on deducing theses 1, 2, and 3, the Łukasiewicz axiom system. In that study, I had also instructed OTTER to seek, where possible, shorter proofs. You can imagine the delight and excitement I experienced at discovering, during my examination of OTTER's success with proving theses 1, 2, and 3, that the shorter proofs did not use theses 24, 39, and 40. In other words, I had a proof that theses 24, 39, and 40 were dependent on theses 18, 35, and 49—possibly, although unlikely, a new result. Even though the result was already known, as it turned out, the example clearly illustrates—at least to me—the potential value of OTTER as a research assistant. The fact that theses 16 and 21 are also dependent—although not discovered in my cited experiment—detracts for me in no way. On the contrary, the experiment now gives me yet another way to seek the existence of dependent axioms.

15.10 A Sincere Debate

At this point, before giving a summary and conclusions, an exchange of views is in order. I move that you and I more or less debate. Since you have lost your vote, for you are not present, the motion carries. However, fear not, for I shall also take your part—at least to the extent of guessing what you might say at various points. In fact, I shall give you first say.

"Larry, Larry, you are really confused. There is no way that anyone has come near to fulfilling my dream, not even you, for whom I have the greatest respect. You and your colleagues at Argonne are pretty good at solving some mathematics problems, that's for sure. But so is Macsyma and Wu's algorithm. In all three cases, the problem is that the mainstream of mathematics is just not being addressed. Who cares about little problems like centuries-old geometry theorems, propositional calculus independence results, or tricky little integration problems? Who cares? Not any significant portion of the working mathematicians of this world."

You raise a number of excellent points in your first salvo. I am certain that your sole motive is to get at the truth, so let me address two of those points. You are indeed correct that your grand dream has hardly been touched, but what about those aspects addressed in this article/letter? We can prove theorems suggested by a well-respected researcher, such as Dana Scott, and do so on the first try. A few mathematicians and logicians are now experimenting with an automated reasoning program, OTTER in particular. And—most significant, at least to me, and I hope to others—we have used our programs to answer open questions, and from a variety of fields.

The kernel strategy that McCune and I formulated for answering questions concerning fixed point combinators appears to offer far more power in that area than any person can offer—no disrespect intended to anyone. Who would have thought that by the mid-1980s there would exist an area in which the probability of success was higher than one half when a problem from that area was submitted to an automated reasoning program? And, please note, we usually have the answer in less than 5 CPU seconds. So, we have at least fulfilled one of Ross Overbeek's dreams. Second, as for the mainstream not caring, at least (in alphabetical order) Kalman, Kaplansky, and Scott care some. In fact, Kalman, on a visit to Argonne, told me that the work that Winker and I did added stature to equivalential calculus. I am getting more space than you are.

"Larry, you were a mathematics graduate student at two great schools, the University of Chicago and the University of Illinois. Take down the fifteen or so graduate-level mathematics texts that you studied and mastered. Open those 7,000 pages to any page you want. Is there a program on earth that can check that page, even if it is recast in a formalism of your choice? How many of those pages could be processed? I bet not 10. Quaife's splendid work with OTTER shows that to check regular math (for example, set theory or number theory), you have to put in dozens of lemmas to get through a regular math proof with OTTER, say one lemma per line of proof! That is proof checking, not automated theorem proving. So we are probably less than 10/7000th of the way towards fulfilling my dream. And I will bet you any amount of money that my dream will be fulfilled only when machines do as people do, as I have been saying and trying to do for two decades: use analogy, use a huge database of known mathematics, use examples, use domain-specific heuristics."

You again score points, and in abundance. You see, I focus on what has been accomplished, rather than on what has not. Indeed, an overwhelming majority of what is found in mathematics books is outside our power—now. But, we cannot begin in the middle; mathematics did not begin there. We have come so far from those early days in the 1960s, but the journey has only begun. Only in the past few years have there existed more than a scant number of researchers trying to do mathematics with a reasoning program. Too many people in the field wasted time early—in my judgment—by focusing exclusively on theory rather than experimenting heavily. Granted that much of Quaife's excellent effort is proof checking; however, a study of it might unearth new ideas to replace much of the checking with proof finding.

I have the greatest respect for your analysis; but, nevertheless, I do not advocate having computers approaching mathematics as people do. Instead, I favor a team consisting of a mathematician and an automated reasoning program—a team that may eventually be far more powerful than either could be! In that regard, although our dreams—yours and mine—diverge sharply, we must each believe that the field clearly merits total devotion to it. Proof: the sum of the years you and I have given to the corresponding research approaches half a century!

"Larry, I hope I have not been too harsh!"

Just harsh enough—not too much, but, more important, not too little. Questioning from those within the field is required—if the field is to advance. However—and I also hope you find my remarks not too harsh—I think that the disappointment and sometimes disapproval regarding the automation of reasoning, especially in the context of mathematics, misses the following fundamental and essential point. In my view, to evaluate the effectiveness of the better automated reasoning programs by comparing them to the best of mathematicians and logicians—for example, those who currently hold the position of full professor and those who in the past have written some of the standard texts—is the wrong comparison to make. If instead the comparison is with people in general, or even with randomly chosen college students, the more effective reasoning programs now available will receive excellent grades, scoring much higher than the vast majority of people. Indeed, given a set of problems from mathematics and logic, the best of such programs—OTTER is my choice—reason far more effectively than almost all people do. For an example, the ability of OTTER to simplify and canonicalize expressions in most cases dwarfs that of any individual I have encountered. OTTER can apply as many as 4,000 demodulators per CPU second; obviously, no person can compete with such performance.

Of course, let us not ignore the long-term goal of using an automated reasoning program that *does* compare favorably with the better mathematicians and logicians. When and if the field has advanced that far, we shall be savoring the finest brandy and perfect raspberries, listening to Mozart and Bix Beiderbecke reborn, and—perhaps more relevant—answering some of the deepest open questions that exist.

Were I not familiar with the twists and turns of the mind—learned mainly at the poker table—I would be sitting at this keyboard experiencing amazement at the lack of use, by mathematicians and logicians, of some automated reasoning program. What fun they are missing by not using this powerful, portable program! How sad, for the use of a program like OTTER by even a handful of mathematicians and logicians would materially increase the rate of advance of automated reasoning and—from what I can tell—also of mathematics and logic. Could it possibly be that they simply do not know what a valuable assistant for research OTTER is? With OTTER's availability and with access to today's powerful computers, we may have a mystery here, one focusing on how researchers can resist the temptation to add this program to the team. Such users are, in my opinion, one of the key missing pieces to the rapid advance of our field.

The grand objectives are indeed stunning and beautiful—and worth the gamble.

15.11 Summary, Conclusions, and the Future

My congratulations to you, Woody, for winning the 1991 prize for Milestones in Automated Theorem Proving. How poetic it would be if, many years from now, the winner of that prize had found the basis for the corresponding research by reading this article/letter, stimulated by one of the experiments, challenge problems, or open questions presented here! Even further, perhaps this future winner, excited by the demonstration of the marked advance of automated reasoning over the past decade and intrigued by the apparent power of McCune's program OTTER, will have conducted the prizewinning research on a copy of this very program—a copy obtained by anonymous FTP, for example. After all, the challenge problems do provide a wide spectrum of difficulty and are, therefore, useful for testing new programs and evaluating new ideas. Might the impetus come from one of the included problems in which equality plays no role, or might it come from one of the problems in which equality plays the dominant role? I prefer that the impetus be one of the open questions posed here to further stimulate research in automated reasoning and to promote its application to mathematics and logic.

Perhaps, at this very moment, our future prize winner is pursuing one of the three goals on which the organization of this article is based. The pursuit might be of the first goal, which focuses on the ability of an automated reasoning program, when given a theorem to consider, to obtain a proof on the first attempt and, of equal importance, without much guidance from the user. The proviso about the lack of guidance addresses the understandable concern of those who wish or might wish to use such a program, but without the requirement of mastering the various intricacies and subtleties. Perhaps the needed stimulus will be one of the numerous examples given to show that first-attempt success in fact occurs frequently here in 1990.

Instead, the target might be the second goal, which concerns contributions to mathematics and logic resulting from the use of an automated reasoning program. In that case, the impetus for research might be one of the contributions discussed here—a shorter proof than can be found in the literature, an answer to an open question, or the new axiom system. How satisfying that the theorems that are the focus of these contributions and those relevant to the first goal include ones meriting the attention (in alphabetic order) of Bernays, Church, Frege, Hilbert, Kalman, Kaplansky, Łukasiewicz, Scott, and Smullyan.

Unfortunately, compared to the first two goals, less evidence exists for the realization of the third goal, the use by mathematicians and logicians of an automated reasoning program. However, as such use increases—as it assuredly must—automated reasoning will make impressive advances. I eagerly await the succession of open questions that will be answered with the assistance of some automated reasoning program. Will one of those questions be so significant that answering it will merit the awarding of the $100,000 prize you had established for the "first monumental contribution to mathematics made by an automated reasoning program"?

Has the evidence cited here begun to persuade you that the three goals that are an important part of your grand dream for automated reasoning are in fact finally a reality? The case is further strengthened if you also consider the recent advances in theory, implementation, and application. With regard to theory, the formulation of the kernel strategy [37] for attacking fixed point problems in combinatory logic and the full development of the formal treatment for linked inference [28] in which equality plays no essential role fit nicely into the picture of a field making important advances. As for implementation, the nearly linear performance of OTTER coupled with its power marks a singular development. In the area of application, among the uses for an automated reasoning program discussed in this article are those of systematically seeking shorter proofs, identifying dependent axioms, checking proofs, finding new axiom systems, and the familiar proving theorems. These varied uses make an automated reasoning program—especially the program OTTER—an excellent research assistant. Its power reduces the turnaround time dramatically, producing a qualitative improvement in the study of the chosen area of mathematics or logic. The added power is in part derived from McCune's adaptation of the use of discrimination trees [19] for the application of forward subsumption [21] and demodulation [32].

Of course, as you know so well, I am somewhat puzzled at the lack of use of an automated reasoning program by mathematicians and logicians. Could the explanation be that people simply are not familiar with the recent developments? Is it not known that a program like OTTER can apply as many as 4,000 rewrites rules per CPU second on a SPARCstation to simplify and canonicalize rapidly and accurately? Do few know that, by using weighting [15], the expert's knowledge and intuition can be used to direct such a program in its search for the desired information? More generally, is there little awareness of how well the use of an automated reasoning program complements the way a person works?

In unification alone, this complementary aspect is nicely exemplified. Indeed, as Dana Scott observes, unification is neither the most natural nor the easiest process for a person to apply, which explains in part why the use of a reasoning program can lead to finding unexpected and elegant proofs that a person might not discover. An excellent example is provided (in Section 15.5) by the 4-step proof of thesis 59 from theses 19, 37, and 60 (the new axiom system). Dijkstra has commented on the importance of finding such short and elegant proofs [3].

In summary, to me—and I hope to you—the evidence presented in this article strongly suggests that automated reasoning has made significant quantitative and qualitative advances in the decade from 1980 to 1990. Further, this evidence refutes the position that still exists asserting the impossibility of an effective automation of reasoning. Of course, no program will ever replace the mathematician and logician; but programs *do* exist that nicely complement such a researcher. That we have barely begun is clear; that challenges will always exist is preferred; that automated reasoning offers intrigue, excitement, and beauty will continually be true.

Acknowledgments

My gratitude to William McCune for his impressive program OTTER and for being a splendid colleague, and to Dana Scott for his intriguing and excellent suggestions regarding research.

References

[1] H. P. Barendregt (1981): The Lambda Calculus: Its Syntax and Semantics. Amsterdam: North-Holland.

[2] M. Davis and H. Putnam (1960): A Computing Procedure for Quantification Theory. J. of the ACM 7, 201–215.

[3] E. Dijkstra (1989): Dijkstra's Reply to Comments. In: A Debate on Teaching Computing Science. C. of the ACM 32(12), 1397–1414.

[4] J. Guard, F. Oglesby, J. Bennett, and L. Settle (1969): Semi-automated Mathematics. J. of the ACM 16, 49–62.

[5] P. R. Halmos (1960): Naive Set Theory. Princeton: D. Van Nostrand Co., Inc.

[6] L. Henkin, J. Monk, and A. Tarski (1971): Cylindric Algebras, Part I. Amsterdam: North-Holland.

[7] E. Huntington (1933): New Sets of Independent Postulates for the Algebra of Logic, with Special Reference to Whitehead and Russell's Principia Mathematica. Trans. of the AMS 35, 274–304.

[8] J. Kalman (1978): A Shortest Single Axiom for the Classical Equivalential Calculus. Notre Dame J. Formal Logic 19(1), 141–144.

[9] J. Kalman (1986): An ITP Workbook, Technical Report. ANL–86–56, Argonne National Laboratory, Argonne, Illinois.

[10] J. Łukasiewicz (1963): Elements of Mathematical Logic. Oxford: Pergamon Press.

[11] J. Łukasiewicz (1970): The Equivalential Calculus. In: L. Borkowski, ed.: Jan Łukasiewicz: Selected Works. Amsterdam: North-Holland.

[12] E. Lusk, W. McCune, and R. Overbeek (1982): Logic Machine Architecture: Kernel Functions. In: D. W. Loveland, ed.: Lecture Notes in Computer Science, Vol. 138. New York: Springer-Verlag.

[13] E. Lusk, W. McCune, and R. Overbeek (1982): Logic Machine Architecture: Inference Mechanisms. In: D. W. Loveland, ed.: Lecture Notes in Computer Science, Vol. 138. New York: Springer-Verlag.

[14] E. Lusk and R. McFadden (1987): Using Automated Reasoning Tools: A Study of the Semigroup F2B2. Semigroup Forum 36(1), 75–88.

[15] J. McCharen, R. Overbeek, and L. Wos (1976): Complexity and Related Enhancements for Automated Theorem-Proving Programs. Computers and Mathematics with Applications 2, 1–16.

[16] W. McCune and L. Wos (1987): A Case Study in Automated Theorem Proving: Finding Sages in Combinatory Logic. J. Automated Reasoning 3(1), 91–108.

[17] W. McCune (1990): OTTER 2.0 Users Guide, Technical Report. ANL–90/9, Argonne National Laboratory, Argonne, Illinois.

[18] W. McCune and L. Wos (1989): The Absence and the Presence of Fixed Point Combinators, Preprint MCS–P87–0689, Mathematics and Computer Science Division, Argonne National Laboratory, Argonne, Illinois.

[19] W. McCune (1990): Experiments with Discrimination-Tree Indexing and Path Indexing for Term Retrieval. J. Automated Reasoning (to appear).

[20] G. Robinson and L. Wos (1969): Paramodulation and Theorem-Proving in First-Order Theories with Equality. In: B. Meltzer and D. Michie, eds.: Machine Intelligence 4. Edinburgh: Edinburgh University Press.

[21] J. Robinson (1965): A Machine-oriented Logic Based on the Resolution Principle. J. of the ACM 12, 23–41.

[22] J. Robinson (1988): Automatic Deduction with Hyper-Resolution. International J. Computer Mathematics 1, 227–234.

[23] B. Smith (1988): Reference Manual for the Environmental Theorem Prover, An Incarnation of AURA. Technical Report ANL–88–2, Argonne National Laboratory, Argonne, Illinois.

[24] R. Smullyan (1985): To Mock a Mockingbird. New York: Alfred A. Knopf.

[25] R. Smullyan (1987): Private Correspondence.

[26] R. Statman (1986): Private Correspondence with Raymond Smullyan.

[27] A. Tarski (1980): Private Communication.

[28] R. Veroff and L. Wos (1990): The Linked Inference Principle, I: The Formal Treatment. J. Automated Reasoning (to appear).

[29] S. Winker, L. Wos, and E. Lusk (1981): Semigroups, Antiautomorphisms, and Involutions: A Computer Solution to an Open Problem, I. Mathematics of Computation 37, 533–545.

[30] S. Winker (1982): Generation and Verification of Finite Models and Counterexamples Using an Automated Theorem Prover Answering Two Open Questions. J. of the ACM 29, 273–284.

[31] L. Wos, G. Robinson, and D. Carson (1965): Efficiency and Completeness of the Set of Support Strategy in Theorem Proving. J. of the ACM 12, 536–541.

[32] L. Wos, G. Robinson, D. Carson, and L. Shalla (1967): The Concept of Demodulation in Theorem Proving. J. of the ACM 14, 698–709.

[33] L. Wos and G. Robinson (1973): Maximal Models and Refutation Completeness: Semidecision Procedures in Automatic Theorem Proving. In: W. Boone, F. Cannonito, and R. Lyndon, eds.: Word Problems: Decision Problems and the Burnside Problem in Group Theory. New York: North-Holland.

[34] L. Wos, R. Overbeek, E. Lusk, and J. Boyle (1984): Automated Reasoning: Introduction and Applications. Englewood Cliffs, New Jersey: Prentice-Hall.

[35] L. Wos, S. Winker, R. Veroff, B. Smith, and L. Henschen (1984): A New Use of an Automated Reasoning Assistant: Open Questions in Equivalential Calculus and the Study of Infinite Domains. Artificial Intelligence 22, 303–356.

[36] L. Wos (1987): Automated Reasoning: 33 Basic Research Problems. Englewood Cliffs, New Jersey: Prentice-Hall.

[37] L. Wos and W. McCune (1988): Searching for Fixed Point Combinators by Using Automated Theorem Proving: A Preliminary Report. Technical Report ANL–88–10, Argonne National Laboratory, Argonne, Illinois.

[38] L. Wos, S. Winker, W. McCune, R. Overbeek, E. Lusk, R. Stevens, and R. Butler (1991): OTTER Experiments Pertinent to CADE–10. Technical Report ANL–89/36, Argonne National Laboratory, Argonne, Illinois.

[39] L. Wos, S. Winker, W. McCune, R. Overbeek, E. Lusk, R. Stevens, and R. Butler (1990): Automated Reasoning Contributes to Mathematics and Logic. In: Lecture Notes in Artificial Intelligence, Vol. 449. New York: Springer-Verlag.

[40] L. Wos, R. Overbeek, and E. Lusk (1990): Subsumption, A Sometimes Undervalued Procedure. In: J.-L. Lassez, ed.: Festschrift for J. A. Robinson (to appear).

Contributors

Russell W. Anderson: *Graduate Group in Bioengineering, University of California at Berkeley and San Francisco, Berkeley, California 94720, U. S. A.*

O. L. Astrachan: *Computer Science Department, Duke University, Durham, North Carolina 27706, U. S. A.*

A. M. Ballantyne: *Austin Lab for Software Engineering and Computer Science, EDS Research and Development, 1601 Rio Grande, Suite 451, Austin, Texas, U. S. A.*

Wolfgang Bibel: *FG Intellektik, Technische Hochschule, Alexanderstr. 10, D–6100 Darmstadt, Germany*

Anne Olivia Boyer: *614 W. 32nd St., Austin, Texas 78705, U. S. A.*

Robert S. Boyer: *Computer Sciences Department, University of Texas at Austin, Austin, Texas 78712–1188, U. S. A., and 614 W. 32nd St., Austin, Texas 78705, U. S. A.*

Hans J. Bremermann: *Division of Biophysics, Department of Molecular and Cell Biology and Department of Mathematics University of California at Berkeley, 102 Donner Lab, Berkeley, California 94720, U. S. A.*

Alan Bundy: *Department of Artificial Intelligence, University of Edinburgh, Edinburgh, Scotland*

Newton C. A. da Costa: *University of Sao Paulo, Brazil*

Ian Foster: *Mathematics and Computer Science Division, Argonne National Laboratory, 9700 South Cass Avenue, Argonne, Illinois 60439, U. S. A.*

M. Gelfond: *Department of Computer Science, University of Texas at El Paso, El Paso, Texas 79999, U. S. A.*

Lawrence J. Henschen: *Computer Science Department, Northwestern University, Evanston, Illinois 60201, U. S. A.*

V. Lifschitz: *Department of Computer Sciences and Department of Philosophy, University of Texas at Austin, Austin, Texas 78712, U. S. A.*

D. W. Loveland: *Computer Science Department, Duke University, Durham, North Carolina 27706, U. S. A.*

J Strother Moore: *Computational Logic, Inc., 1717 West Sixth Street, Suite 290, Austin, Texas 78703, U. S. A.*

James J. Lu: *Computer Science Department, Northwestern University, Evanston, Illinois 60201, U. S. A.*

Young K. Nam: *Computer Science Department, Northwestern University, Evanston, Illinois 60201, U. S. A.*

Ross Overbeek: *Mathematics and Computer Science Division, Argonne National Laboratory, 9700 South Cass Avenue, Argonne, Illinois 60439, U. S. A.*

A. Rabinov: *Apple Computer, Inc., 20525 Mariani Ave., MS: 37DS, Cupertino, California 95014, U. S. A.*

Michael M. Richter: *Department of Computer Science, University of Kaiserslautern, D–6750 Kaiserslautern, Germany*

J. A. Robinson: *607 Bird Library, Syracuse University, Syracuse, New York 13502, U. S. A.*

Mark E. Stickel: *Artificial Intelligence Center, SRI International, Menlo Park, California 94025, U. S. A., and Institute for New Generation Computer Technology, Tokyo, Japan*

V. S. Subrahmanian: *Computer Science Department, University of Maryland, College Park, Maryland 20742, U. S. A.*

Stefan Wess: *Department of Computer Science, University of Kaiserslautern, D–6750 Kaiserslautern, Germany*

Larry Wos: *Mathematics and Computer Science Division, Argonne National Laboratory, 9700 South Cass Avenue, Argonne, Illinois 60439, U .S. A.*

Index

MIX
Papier aus verantwortungsvollen Quellen
Paper from responsible sources
FSC® C105338

If you have any concerns about our products,
you can contact us on
ProductSafety@springernature.com

In case Publisher is established outside the EU,
the EU authorized representative is:
Springer Nature Customer Service Center GmbH
Europaplatz 3, 69115 Heidelberg, Germany

Printed by Libri Plureos GmbH
in Hamburg, Germany

Arctic Marine Governance

Elizabeth Tedsen · Sandra Cavalieri
R. Andreas Kraemer
Editors

Arctic Marine Governance

Opportunities for Transatlantic Cooperation

Springer

Editors
Elizabeth Tedsen
Ecologic Institute Washington DC
Washington DC
USA

Sandra Cavalieri
R. Andreas Kraemer
Ecologic Institute
Berlin
Germany

ISBN 978-3-662-52222-6 ISBN 978-3-642-38595-7 (eBook)
DOI 10.1007/978-3-642-38595-7
Springer Heidelberg New York Dordrecht London

Softcover reprint of the hardcover 1st edition 2014

Printed on acid-free paper

Springer is part of Springer Science+Business Media (www.springer.com)

Foreword

Nowhere is climate change more visible than in the Arctic, which is a vital and vulnerable component of the Earth's environment and climate system. The melting of Arctic sea ice is progressing rapidly and affecting ecosystems as well as the traditional livelihoods of indigenous peoples. As climate change and economic development accelerate in the Arctic region, the European Union is stepping up its engagement with its Arctic partners to jointly meet the challenge of safeguarding the environment while ensuring the sustainable development of the Arctic region.

The EU sponsored the Arctic TRANSFORM project (<http://arctic-transform.org>), the results of which form a basis for this volume. The goal of Arctic TRANSFORM was to develop transatlantic policy options for supporting adaptation in the marine Arctic environment, placing a special emphasis on involving a broad range of stakeholders to address the major climate issues facing the region. Key project objectives included:

- To promote mutual exchange among EU and US policymakers and stakeholders on policies and approaches in the Arctic in stakeholder working groups;
- To provide a comparative analysis of existing policies and make recommendations with substantial buy-in as to how to strengthen cooperation between the EU and US; and
- To encourage dialogue and thus improve conditions for further transatlantic policy development and more effective protection of the Arctic marine environment.

Today, as the impacts of climate change become apparent, the need for good collaboration on preserving the Arctic marine environment is even more pressing than it was at the time of the project. Enhanced transatlantic cooperation can take advantage of emerging opportunities for improving protection of the Arctic marine environment. Through formal cooperation on policy strategies as well as informal channels of information sharing, research exchanges, and stakeholder networks, a strong transatlantic partnership can create a foundation for improved knowledge and efficient and timely addressing of the challenges ahead.

Helga Schmid
Deputy Secretary General
European External Action Service

Preface

The Arctic region plays an important role in regulating the world's climate and also is highly impacted by climate change, with average temperatures rising almost twice as fast as the rest of the world and sea ice melting much faster than previously predicted. These rapid changes will have significant impacts on human activity in the region and on the Arctic marine environment.

Recognizing the importance of these changes, the Arctic TRANSFORM project was developed to explore the roles of the EU and US in light of the changing climate and the region's political and legal complexities. The project sought to promote exchange between EU and US policymakers and stakeholders on approaches in the Arctic, provide comparative analyses of existing policies, make recommendations as how to strengthen cooperation between the EU and US, encourage dialogue, and thus improve conditions for further transatlantic policy development and more effective protection of the Arctic marine environment. The project was funded by the European Commission (formerly DG External Relations, now European External Action Service) and led by Ecologic Institute, along with the Arctic Center, the Netherlands Institute for the Law of the Sea, and the Heinz Center.

Four years later, this book begins where Arctic TRANSFORM left off, addressing the new and significant changes and developments in the marine Arctic with updates, new policy recommendations, and additional topics to better reflect the current governance environment.

Within four years, the Arctic region has undergone substantial changes. The impacts of climate change are swiftly altering the landscape and ushering in new activity and actors. Likewise, the policy environment has also undergone transformations, including modifications to adapt existing frameworks to the changed and changing Arctic conditions, as described in detail throughout this book.

A look to the future leaves little doubt that this rapid pace of change will continue—for the environment and policy. For instance, in May 2013—after the time of writing of this volume—the Arctic Council Ministerial Meeting in Kiruna is slated to take up a wide range of critical governance issues, including a legally binding agreement on marine oil preparedness and response, updating PAME's Arctic Marine Strategic Plan, and releasing reports for the Arctic Ocean Review and Arctic Biodiversity Assessment. Most important in the transatlantic context, the meeting will also address the EU's application to become a permanent observer of the Arctic Council. In addition, in 2015, the US will assume chairmanship of the Arctic Council.

While such developments may continue to further alter the Arctic governance landscape, we believe that the background, principles, and discussion presented herein, do and will continue to promote better understanding of marine Arctic governance needs and options and help to lay a foundation for future effective environmental governance.

March 2013

E. Tedsen
S. Cavalieri
R. Andreas Kraemer

Acknowledgments

The editors thank and recognize the contributions of the Arctic TRANSFORM project to this book. The Arctic TRANSFORM project was funded by the **European Commission** (formerly DG External Relations, now European External Action Service) and led by **Ecologic Institute**, along with the **Arctic Centre**, the **Netherlands Institute for the Law of the Sea**, and the **Heinz Center**. The project was enriched by discussions and outcomes from five thematic working groups, which engaged more than 50 experts from multiple disciplines on Arctic indigenous peoples, environmental governance, fisheries, offshore hydrocarbon activities, and shipping. In particular, the authors recognize the contribution of the Co-Chairs of each working group as follows:

- **Environmental Governance Working Group**: Dr. Stuart Chapin (University of Alaska, Fairbanks) and Dr. Neil Hamilton (WWF Arctic International Programme).
- **Fisheries Working Group**: Ambassador David Balton (US Department of State, Bureau of Oceans, Environment and Science) and Kjartan Hoydal (North East Atlantic Fisheries Commission).
- **Indigenous Peoples Working Group**: Patricia Cochran (Inuit Circumpolar Conference) and Dr. Mark Nuttall (University of Alberta, University of Oulu).
- **Offshore Hydrocarbon Working Group**: Dr. Cutler Cleveland (Boston University) and Kevin O'Carroll (OSPAR Offshore Industry Committee).
- **Shipping Working Group**: Dr. Lawson Brigham (US Arctic Research Commission) and Rene Piil Pedersen (Danish Shipowners' Association).

The editors thank Lucy Smith and Erika Sulik at Ecologic Institute for their careful attention and support in finalizing this book.

Contents

Abbreviations

ABA	Arctic Biodiversity Assessment
ACAP	Arctic Contaminants Action Program (working group)
ACIA	Arctic Climate Impact Assessment
ACS	Arctic Council System
ADHR	Arctic Human Development Report
AECO	Association of Arctic Expedition Cruise Operators
AEPS	Arctic Environmental Protection Strategy
AMAP	Arctic Monitoring and Assessment Programme (working group)
AMSA	Arctic Marine Shipping Assessment
AOR	Arctic Ocean Review
APM	Associated Protective Measure
ARR	Arctic Resilience Report
ATS	Antarctic Treaty System
AWPPA	Arctic Waters Pollution Prevention Act
BEAC	Barents Euro-Arctic Cooperation
BEAR	Barents Euro-Arctic Region
BePOMAr	Best Practices in Ecosystem-Based Oceans Management in the Arctic
BFR	Brominated Flame Retardant
BWM	Convention for the Control and Management of Ships' Ballast Water and Sediments
CAFF	Conservation of Arctic Flora and Fauna (working group)
CBD	Convention on Biological Diversity
CBMP	Circumpolar Biodiversity Monitoring Programme
CCAC	Climate and Clean Air Coalition
CDEM	Construction, Design, Equipment, and Manning
CFP	Common Fisheries Policy
CITES	Convention on International Trade in Endangered Species
CLCS	Commission on the Limits of the Continental Shelf
CMS	Convention on the Conservation of Migratory Species
COP	Conference of Parties
CPAN	Circumpolar Protected Areas Network
DDE	Dichlorodiphenyldichloroethylene
DDT	Dichlorodiphenyltrichloroethane

DE	Maritime Safety Committee Sub-Committee on Design and Equipment
DOI	United States Department of the Interior
EA	Environmental Assessment
EAF	Ecosystem Approach to Fisheries
EBM	Ecosystem-Based Management
EBSA	Ecologically or Biologically Significant Area
EC	European Community
EEA	European Environment Agency
EEAS	European External Action Service
EEC	European Economic Community
EEZ	Exclusive Economic Zone
EFTA	European Free Trade Agreement
EIA	Environmental Impact Assessment
EPPR	Emergency Prevention, Preparedness, and Response (working group)
ESA	Endangered Species Act
EU	European Union
FAO	Food and Agriculture Organization
FMP	Fishery Management Plan
FPZ	Fisheries Protection Zone
GAIRAS	Generally Accepted International Rules and Standards
HFC	Hydrofluorocarbon
HFO	Heavy Fuel Oil
IACS	International Association of Classification Societies
IAEA	International Atomic Energy Agency
IAIA	International Association for Impact Assessment
IASC	International Arctic Science Committee
ICAO	International Civil Aviation Organization
ICC	Inuit Circumpolar Council
ICCAT	International Commission for the Conservation of Atlantic Tunas
ICES	International Council for the Exploration of the Sea
ICJ	International Court of Justice
ICRW	International Convention on the Regulation of Whaling
ICS	International Chamber of Shipping
IEA	International Energy Agency
IHO	International Hydrographic Organization
ILO	International Labour Organization
IMO	International Maritime Organization
IMP	Integrated Maritime Policy
IPCC	Intergovernmental Panel on Climate Change
IPHC	International Pacific Halibut Commission
IRF	International Regulators Forum
ISA	International Seabed Authority
IUCN	International Union for Conservation of Nature

IUU	Illegal, Unreported, and Unregulated
JAMP	Joint Assessment Monitoring Programme
JIP	Joint Industry Programme
LME	Large Marine Ecosystem
LNG	Liquid Natural Gas
LOS	United Nations Convention on the Law of the Sea
LRTAP	Convention on Long-range Transboundary Air Pollution
MAP	Marine Advisory Programme
MARPOL	International Convention for the Prevention of Pollution from Ships
MEPC	Marine Environment Protection Committee
MOPPR	Marine Oil Pollution Preparedness and Response
MOU	Memorandum of Understanding
MPA	Marine Protected Area
MSC	Maritime Safety Committee
MSFD	Marine Strategy Framework Directive
MSP	Marine Spatial Planning
MSY	Maximum Sustainable Yield
NAFO	Northwest Atlantic Fisheries Organization
NAMMCO	North Atlantic Marine Mammal Commission
NASCO	North Atlantic Salmon Conservation Organization
NAV	Maritime Safety Committee Sub-Committee on Navigation
NDP	Northern Dimension Policy
NEAFC	North East Atlantic Fisheries Commission
NEPA	National Environmental Policy Act
NGO	Non-Governmental Organization
NIRB	Nunavut Impact Review Board
NM	Nautical Mile
NMFS	United States National Marine Fisheries Service
NOAA	United States National Oceanic and Atmospheric Administration
NORDREG	Northern Canada Vessel Traffic Services Regulations
NPAFC	North Pacific Anadromous Fish Commission
NPFMC	North Pacific Fishery Management Council
NTNU	Norwegian University of Science and Technology
OATRC	Oden Arctic Technology Research Cruise
OIC	Offshore Industry Committee
OPRC	International Convention on Oil Pollution Preparedness, Response, and Cooperation
OSPAR	Convention for the Protection of the Marine Environment of the North-East Atlantic
PAME	Protection of the Arctic Marine Environment (working group)
PBSG	Polar Bear Specialist Group
PCB	Polychlorinated Biphenyl
PFO	Perfluorooctane Sulfonate
PICES	North Pacific Marine Science Organization
PM	Particulate Matter

POP	Persistent Organic Pollutant
PSC	Pacific Salmon Commission
PSC	Port State Control
PSSA	Particularly Sensitive Sea Area
RAIPON	Russian Association of Indigenous Peoples of the North
REIO	Regional Economic Integration Organization
RFMO	Regional Fisheries Management Organization
SAMCoT	Sustainable Arctic Marine and Coastal Technology
SAO	Senior Arctic Official
SAR	Search And Rescue
SDWG	Sustainable Development Working Group
SEA	Strategic Environmental Assessment
SLCP	Short-Lived Climate Pollutant
SOLAS	International Convention for the Safety of Life at Sea
SPRS	Swedish Polar Research Secretariat
SRS	Ship Reporting System
STCW	Standards of Training, Certification, and Watchkeeping
TAC	Total Allowable Catch
TEA	Transboundary Environmental Assessment
TEK	Traditional Ecological Knowledge
TFEU	Treaty on the Functioning of the European Union
UN	United Nations
UNECE	United Nations Economic Commission for Europe
UNGA	United Nations General Assembly
UNFCCC	United Nations Framework Convention on Climate Change
UNTS	United Nations Treaty Series
US	United States of America
USGS	United States Geological Survey
VOC	Volatile Organic Compound
VTS	Vessel Tracking Service
WCPFC	Western and Central Pacific Fisheries Commission
WSSD	Word Summit on Sustainable Development

Part I
The Arctic Environment

Chapter 1
Introduction to the Arctic

Erik J. Molenaar, Timo Koivurova, Elizabeth Tedsen, Andrew Reid, and Kamrul Hossain

Abstract Climate change is occurring rapidly in the Arctic, bringing new economic opportunities alongside challenges for environmental governance in the region. Evaluating these changes, and options for effectively addressing them, requires an understanding of existing institutions and frameworks. This chapter provides a foundation for the book and an introduction to Arctic marine governance and transatlantic cooperation, setting the scene with sections on the spatial scope of the Arctic marine area, the law of the sea in the Arctic marine area, the Arctic Council, and the respective Arctic and marine policies of the European Union and the United States.

Based on Molenaa EJ., Corell R, Koivurova T, Cavalieri S. (2008) Introduction to the background papers. Arctic TRANSFORM. 8 Sept 2008.

E. J. Molenaar (✉)
Netherlands Institute for the Law of the Sea (NILOS), Utrecht University, and University of Tromsø, Tromsø, Norway
e-mail: E.J.Molenaar@uu.nl

T. Koivurova · K. Hossain
Northern Institute for Environmental and Minority Law, Arctic Centre, University of Lapland, Rovaniemi, Finland
e-mail: timo.koivurova@ulapland.fi

K. Hossain
e-mail: kamrul.hossain@ulapland.fi

E. Tedsen
Ecologic Institute, 1600 Connecticut, Ave NW, Suite 300, Washington D.C 20009, USA
e-mail: elizabeth.tedsen@ecologic.eu

A. Reid
Ecologic Institute, Pfalzburger Strasse 43/44, Berlin 10717, Germany
e-mail: andrew.reid@ecologic.eu

E. Tedsen et al. (eds.), *Arctic Marine Governance*,
DOI: 10.1007/978-3-642-38595-7_1, © Springer-Verlag Berlin Heidelberg 2014

1.1 Arctic Marine Area

As there is no generally accepted definition of either the 'Arctic' or the 'marine Arctic', the spatial scope of this book has been determined as the marine areas included within the area agreed by the Arctic Monitoring and Assessment Programme (AMAP) working group of the Arctic Council. These are the marine areas north of the Arctic Circle (66°32′N), and north of 62°N in Asia and 60°N in North America, modified to include the marine areas north of the Aleutian chain, Hudson Bay, and parts of the North Atlantic Ocean, including the Labrador Sea. For the purpose of this book, these marine areas are referred to as the 'Arctic marine area' or 'marine Arctic'.

Figure 1.1 shows the AMAP area, as well as the borders of the Arctic according to the Arctic Circle (the parallel of latitude that runs approximately 66.56083° north of the Equator), and to certain scientific parameters (10 °C July isotherm, treeline, marine, and vegetation).

Similarly, there is no universally accepted definition for the 'Arctic Ocean'. However, it is generally accepted that the five coastal states to the Arctic Ocean are Canada, Denmark (in relation to Greenland and the Faroe Islands), Norway, the Russian Federation, and the United States (US).

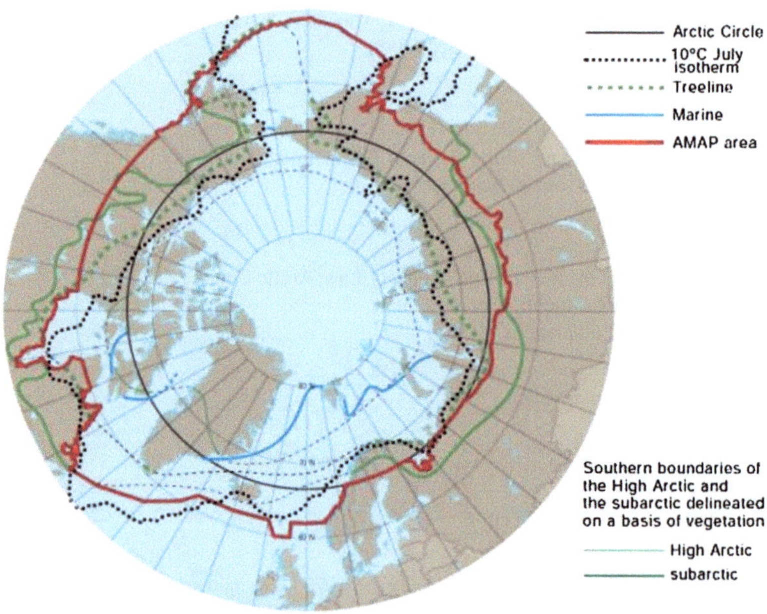

Fig. 1.1 Arctic monitoring and assessment programme (AMAP) boundary (*Source* AMAP (1997))

1.2 Law of the Sea in the Arctic Marine Area

The international law of the sea is made up of a multitude of global, regional, and bilateral instruments, decisions by international (intergovernmental) organizations, and international rules from other sources, including customary international law. The cornerstones of the current international law of the sea are the United Nations (UN) Law of the Sea (LOS) Convention (1982) and its two implementation agreements, the Part XI Deep-Sea Mining Agreement (1994) and the Fish Stocks Agreement (1995). The LOS Convention's overarching objective is to establish a universally accepted, just, and equitable legal order—or 'Constitution'—for the oceans that lessens the risk of international conflict and enhances stability and peace in the international community. As of January 2013, the LOS Convention has 165 parties, the Part XI Deep-Sea Mining Agreement 144 parties, and the Fish Stocks Agreement 80 parties.

All eight Arctic states (Canada, Denmark, Finland, Iceland, Norway, the Russian Federation, Sweden, and the US) are parties to these three treaties, except the US is not a party to either the LOS Convention or the Part XI Deep-Sea Mining Agreement (Division for Ocean Affairs and Law of the Sea 2013). The US is therefore, among other things, not subject to the LOS Convention's Part XV on dispute settlement.[1] The European Union (EU) is party to all three treaties. This is important in view of the fact that Denmark, Finland, and Sweden are Member States of the EU and Iceland and Norway are parties to the EEA Agreement (1993).

All of the global instruments that are part of the law of the sea apply to the marine environment of the entire globe, including therefore the entire marine Arctic, however defined. The mandate of the global bodies associated with these instruments has the same geographical scope. The perception that there is an international law vacuum in the Arctic, which only became a matter of attention following the melting of Arctic ice and the planting of a Russian flag on the seabed of the North Pole in August 2007, is therefore incorrect.

The most basic distinction between marine areas made by the LOS Convention is between the maritime zones of coastal states—also referred to as 'areas within national jurisdiction'—and the commons seaward thereof—also referred to as 'areas beyond national jurisdiction'. The maritime zones of coastal states can consist of: internal waters, archipelagic waters, territorial sea, contiguous zone, exclusive economic zone (EEZ), and the continental shelf. As clarified below, the EEZ includes the continental shelf, but in some cases, there is also an 'outer' continental shelf that extends seaward of the EEZ. The two marine commons are the high seas—usually seaward of the EEZ (where established)—and the so-called 'Area'—seaward of the EEZ or outer continental shelf. The Area is defined as "the sea-bed and ocean floor and subsoil thereof, beyond the limits of national jurisdiction" (LOS Convention 1982, art. 1(1)(1)).

---

[1] In the domain of straddling and highly migratory fish stocks, however, the US is subject to Part XV of the LOS Convention due to its being a party to the Fish Stocks Agreement. See also Reagan (1983).

Except for archipelagic waters,[2] all of the maritime zones recognized in the LOS Convention also exist in the marine Arctic: internal (marine) waters, territorial seas, contiguous zones, EEZs, (outer) continental shelves, the so-called 'Area' (the deep sea-bed beyond continental shelves) and the high seas. There are four high seas pockets (or enclaves) in the marine Arctic, namely the so-called 'Banana Hole' in the Norwegian Sea, the so-called 'Loop Hole' in the Barents Sea, the so-called 'Donut Hole' in the central Bering Sea, and the so-called 'Central Arctic Ocean'. There may be two or more pockets of the Area that could remain in the Arctic Ocean. Some region-specific maritime zones exist as well, as Norway has—instead of a regular EEZ—established a Fishery Zone around Jan Mayen and a Fisheries Protection Zone around Svalbard.

The outer limits of the maritime zones of coastal states are measured from baselines drawn in accordance with several provisions of the LOS Convention. The normal baseline is the low-water line along the coast (*inter alia* LOS Convention 1982, arts. 5–7, 9–14). It should be noted here that sea level rise, a consequence of climate change, could in many situations mean that new baselines may have to be drawn landward of the older ones and, as a consequence, the high seas and the Area will increase in size. In certain situations, the LOS Convention also allows coastal states to draw straight baselines. The straight baselines drawn by Canada around its Arctic Archipelago are regarded by the US and certain EU Member States as inconsistent with international law (Roach and Smith 1996) (Fig. 1.2).

Internal waters lie landward of the baselines. The maximum breadth of the territorial sea is 12 nautical miles (nm; 1 nm = 1,852 m) measured from the baselines, 24 nm is the maximum breadth for the contiguous zone, and 200 nm for the EEZ. Article 76 of the LOS Convention also recognizes that in certain circumstances, the continental shelf extends beyond 200 nm from the baselines. This is the so-called 'outer continental shelf'. Coastal states that take the view that they have an outer continental shelf must submit information on its outer limits on the basis of the criteria in Article 76 to the Commission on the Limits of the Continental Shelf (CLCS). The planting of a Russian flag on the seabed of the North Pole in August 2007 actually took place during the gathering of scientific data for this process. In 2002, the CLCS recommended that the Russian Federation gather and submit new data to complement the data submitted in 2001, which the Russian Federation is still expected to do.

The limits of the outer continental shelf established by the coastal state "on the basis of" the recommendations of the CLCS "shall be final and binding" (LOS Convention 1982, art. 76(8)). So far, only the Russian Federation and Norway have made submissions to the CLCS in relation to their outer continental shelves that lie within the marine Arctic. The CLCS has issued recommendations in relation to both submissions. Canada, Denmark (in relation to Greenland), and the

[2] Terms such as the 'Canadian Arctic Archipelago' and the 'Spitsbergen Archipelago', even if used consistently by Canada and Norway, do not imply that these states qualify—or claim to qualify—as archipelagic states in the domain of the international law of the sea.

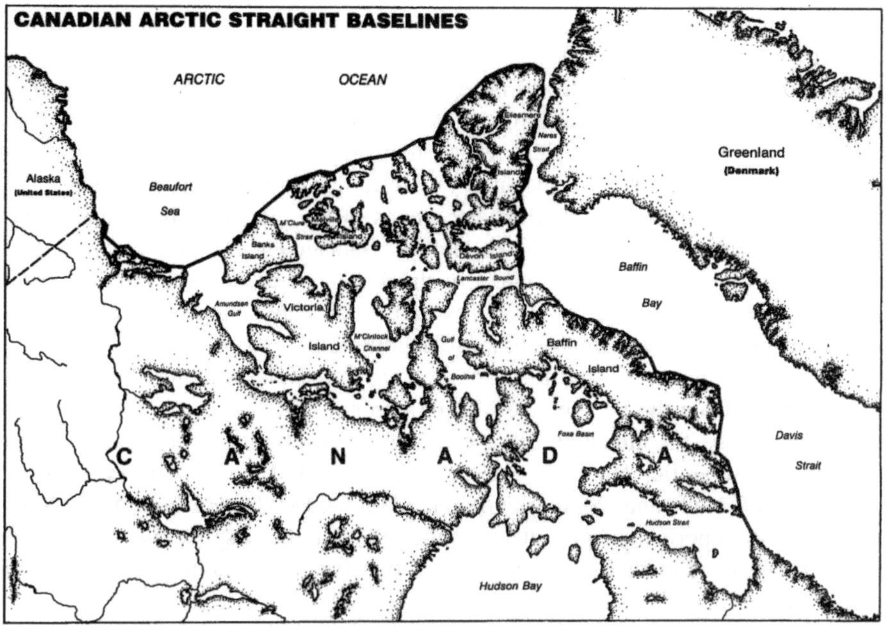

Fig. 1.2 Canadian Arctic straight baselines (*Source* United States Department of State (1992))

US are all engaged in activities to enable them to make submissions to the CLCS, despite the fact that the US is not yet a party to the LOS Convention. Canada has to make its submission before November 2013 and Denmark before November 2014 (Cf. LOS Convention 192, art. 4, Annex II).

The LOS Convention recognizes the sovereignty of a coastal state over its internal waters, archipelagic waters, and territorial sea, the airspace above, and its bed and subsoil. Sovereignty entails exclusive access to and jurisdiction over all resources (living and non-living; e.g., fish and oil) as well as full jurisdiction over all human activities, unless states have in one way or another consented to restrictions on these. The LOS Convention also recognizes that a coastal state has economic and resource-related sovereign rights and jurisdiction in its EEZ and outer continental shelf (if present). These sovereign rights give coastal states exclusive access to and jurisdiction over all resources in the EEZ and over all non-living and some living resources of the continental shelf. Nevertheless, other states have navigational rights or freedoms within the maritime zones of coastal states and, with respect to their EEZ and—where present—outer continental shelf, also the freedoms of overflight and laying of submarine cables and pipelines.

In the high seas, all states have the freedoms already mentioned above as well as the freedom to construct artificial islands and other installations, the freedom of fishing, and the freedom of scientific research. These freedoms are all subject to certain conditions and obligations. The Area and its resources are considered the common heritage of mankind and the International Seabed Authority (ISA) is charged with organizing and controlling all activities of exploration for, and exploitation of, the resources of the Area.

Maritime delimitation is always necessary for adjacent coastal states, but opposite coastal states are only required to do this if the maximum widths of their maritime zones cannot be reached. Many maritime boundaries between adjacent and opposite coastal states in the marine Arctic have already been agreed upon, but some are still under negotiation. By means of the bilateral Murmansk Treaty (2010), Norway and the Russian Federation managed to finalize their lengthy negotiations on a maritime boundary in the Barents Sea and the Arctic Ocean. Enhanced access to the marine Arctic caused by climate change is likely to have been an incentive towards finalizing these negotiations.

Even though the Arctic and the Antarctic are both polar regions, they are radically different from the perspective of international law. The Arctic is not subject to a comparably fundamental disagreement on sovereignty over territory that exists in the Antarctic. The only dispute on title to land territory in the Arctic is that between Canada and Denmark/Greenland on the tiny Hans Island in the Nares Strait.[3] News reports in April 2012 suggested that the dispute would soon be resolved by means of dividing the island in two (see e.g., Humphreys 2012), but by mid-September 2012 this had not (yet) happened.

It is true that there are several unresolved maritime boundaries in the marine Arctic and that most of the outer limits of the continental shelves still have yet to be established. But that is true for most regions in the world. There are nevertheless two international law of the sea disputes that are Arctic-specific. First, the disagreement between Canada and the Russian Federation on the one hand and the US and other states—including several EU Member States—on the other hand, on the legal status of the Northwest Passage and other waters within the Canadian Arctic Archipelago and the waters within the Northern Sea Route, respectively (Bartenstein 2011). As regards Canada, this disagreement is related to Canada's straight baselines around the Canadian Arctic Archipelago (see above). Second is the disagreement between Norway and several other parties to the Spitsbergen Treaty (1920) as to whether or not the treaty also applies seaward of the territorial sea (Molenaar 2012).

Apart from these disputes however, the marine Arctic is not fundamentally different from most other marine areas or oceans; except, of course, the waters adjacent to the Antarctic continent. As all other states have rights under the international law of the sea in all coastal state's maritime zones, the coastal states to the marine Arctic do not have full jurisdiction and control over these areas. In other words: they cannot just do as they please because they have to respect the rights of others.

---

[3] See the Joint Statement by the then Canadian Minister of Foreign Affairs Pettigrew and then Danish Minister of Foreign Affairs Møller made in New York on 19 Sept 2005 and the short article by P.E.D. Kristensen, then Ambassador of Denmark to Canada, published in the *Ottawa Citizen* on 28 July 2005, which place the dispute in the proper perspective of the good and ongoing cooperation between the two states.

1.3 Arctic Council

The first stage of Arctic-wide cooperation started with the 1991 Arctic Environmental Protection Strategy (AEPS), which was adopted in Rovaniemi by the eight Arctic States (AEPS 1991). In the Strategy, six high-priority environmental problems facing the Arctic were first identified (persistent organic contaminants, radioactivity, heavy metals, noise, acidification, and oil pollution) as well as international environmental protection treaties that apply in the region, and specific actions to counter the threats. Interestingly, the AEPS stated that "The implementation of the Strategy will be carried out through national legislation and in accordance with international law, including customary international law as reflected in the [LOS Convention]" (AEPS 1991). As part of the environmental protection action by the eight Arctic States, four environmental protection working groups were established: Conservation of Arctic Flora and Fauna (CAFF), Protection of the Arctic Marine Environment (PAME), Emergency Prevention, Preparedness, and Response (EPPR), and AMAP. Three ministerial meetings (following the signing of the Rovaniemi Declaration (1991) and the AEPS) were held in this first phase of Arctic cooperation, generally referred to as 'AEPS cooperation' (Koivurova and VanderZwaag 2007).

The Arctic Council was established between the eight Arctic states through the Ottawa Declaration in 1996 to enhance Arctic cooperation. The establishment of the Arctic Council broadened the mandate of cooperation to all common issues facing the Arctic (excluding matters related to military security), especially those relating to environmental protection and sustainable development. The four environmental protection working groups of the Strategy were integrated into the structure of the Council, and one new working group was established (the Sustainable Development Working Group (SDWG)). In the absence of a permanent secretariat—although by means of the Nuuk Declaration in 2011 the Minsters agreed to establish a permanent secretariat in Tromsø, Norway—the work of the Arctic Council has been heavily influenced by the priorities that the chair-state lays out for its two-year chair period, at the end of which a ministerial meeting is organized. Senior Arctic Officials (SAOs), a group of high-level officials, guide the work of the Council in between the ministerial meetings. The Arctic Council has also adopted new programmes related to environmental protection, such as the Arctic Council Action Plan to Eliminate Pollution in the Arctic (ACAP), which was recently turned into a sixth working group (re-titled the Arctic Contaminants Action Program), and commissioned the Arctic Climate Impact Assessment (ACIA; ACIA 2005).

One unique aspect of the Arctic Council is the role it gives to the region's indigenous peoples: They are normally accorded the status of non-governmental organizations (NGOs) in different intergovernmental organizations and forums, but the Arctic Council defines them as 'permanent participants', a distinct category of membership between members proper and observers, whom the Arctic Council member states must consult prior to any consensus decision making. The group of observers is large, and consists of intergovernmental and non-governmental organizations as well as states that are active in the Arctic region (Koivurova and VanderZwaag 2007).

The Arctic Council is engaged in various kinds of activities related to the Arctic marine area, especially through its AMAP and PAME working groups, but to some extent also through CAFF, which has marine projects. The Council has produced many important scientific assessments following the dramatic findings of the ACIA, which have played important roles in governing the Arctic marine area. It conducted a comprehensive assessment of Arctic marine shipping, which led to the Arctic Marine Shipping Assessment (AMSA) that was adopted in the 2009 Ministerial Meeting and contains policy recommendations. One of these recommendations urged the Council members to act in concert to push for comprehensive, stringent, and mandatory rules on shipping in extreme polar conditions. The work to convert the non-binding Polar Shipping Guidelines (2009) into mandatory measures is now in progress under the aegis of the International Maritime Organization (IMO).[4] The Council has also sponsored the making of an Oil and Gas Assessment and is in the process of completing the Arctic Biodiversity Assessment, which will play a role in evaluating the effectiveness of conservation policies. Moreover, an instrument that was adopted during the AEPS, the Arctic Council's Offshore Oil & Gas Guidelines (PAME 2009), has been updated already two times and the most recent version endorsed at the 2009 Ministerial Meeting. Currently, the Arctic Ocean Review process is examining gaps in Arctic marine governance and will make recommendations on routes to be followed by member states and other stakeholders in their future Arctic marine policy and law.

Until recently, Arctic cooperation functioned for over fifteen years in a fairly similar and consistent mode of operation. Yet in response to alarming climate change, the Council has recently strengthened the way it functions. In May 2011, the ministers decided to establish a permanent secretariat and adopted the first ever legally binding instrument, the Arctic Search and Rescue (SAR) Agreement (2011), which marks a change in the Council, using treaties as ways of reaching policy goals in marine areas. The agreement is meant to strengthen search and rescue coordination and cooperation efforts in the Arctic by allocating responsibilities to each Arctic state and by establishing procedures for states to cooperate in cases of emergency. There is also an ongoing process to conclude an Agreement on Cooperation on Marine Oil Pollution Preparedness and Response (MOPPR), which is scheduled to be signed during the May 2013 Ministerial Meeting. Like the SAR Agreement, the special scope of the agreement goes beyond the Arctic

[4] To be more specific, AMSA's recommendation was for "updating and mandatory application of relevant parts of the Guidelines for Ships Operating in Arctic Ice-covered Waters (Arctic Shipping Guidelines)". Those were adopted in 2002, and since then the AMSA report was published in April 2009 and the 'Guidelines for ships operating in polar waters' were adopted by the 26th IMO Assembly in November–December 2009); the AMSA recommendation referred to the earlier 2002 Guidelines.

Ocean; the MOPPR Agreement will likely apply not only in Arctic Ocean waters, but also in the Baltic Sea (Gulf of Bothnia), and may also have a few legally non-binding appendices (e.g., a manual on emergency response). Both treaties are firmly anchored in broader agreements already in existence.

Although the Arctic Council is not a treaty-based organization, it seems to have gradually strengthened its ways of conducting policy, which many describe as move from a policy *shaping* body to policy *making* one. The Council continues to do important scientific work also in relation to marine areas, but is getting stronger in terms of its institutional structure and ways of doing policy via legally binding agreements, all adding to the Council's capacity to respond effectively to the challenges of climate change in the region.

1.4 Arctic Policies of the EU and US

1.4.1 EU Arctic Policy and Competences

As global attention has turned towards the environmental and geopolitical changes in the Arctic, the EU has taken an increasing interest and more active role in developing its own Arctic policy. Beginning in 2007 with its Integrated Maritime Policy (European Commission 2007a) and Action Plan for Integrated Marine Policy (European Commission 2007b), the European Commission drew attention to Arctic issues and called for preparation of a report on Arctic Ocean strategic issues. In March 2008, a paper from the European Commission and the High Representative called for the development of an EU Arctic policy, highlighting the increasing geopolitical importance of the Arctic resulting from the melting of the Arctic sea ice and increased accessibility of Arctic waters (High Representative and European Commission 2008).

More concrete Arctic policy development was initiated with the European Parliament's resolution in October 2008 on Arctic governance (European Parliament 2008) and the November 2008 European Commission Communication on 'The European Union and the Arctic Region', which set out proposals for a coordinated Arctic approach for the EU (European Commission 2008). In 2009, the Council of the European Union adopted 'Council conclusions on Arctic issues' which welcomed the gradual formulation of an EU Arctic policy based upon effective mitigation of climate change, multilateral governance, the LOS Convention, maintaining the Arctic as an area of peace and stability, and formulating EU actions and policies that respect the sensitivities of Arctic ecosystems and biodiversity and the rights of indigenous peoples (Council of the European Union 2009). The January 2011 European Parliament's resolution on 'Sustainable EU policy for the High North' affirmed the EU's Arctic interests and commitment to developing policies based on best scientific knowledge, and stressed a need for a coordinated EU policy (European Parliament 2011).

Most recently, in June 2012, the EU Commission and the High Representative for Foreign Affairs and Security Policy adopted a Joint Communication on 'Developing a European Union Policy towards the Arctic Region: progress since 2008 and next steps'(European Commission and High Representative 2012). The Joint Communication sets out the case for a refined policy and increased EU engagement in Arctic issues based on knowledge, responsibility to achieve sustainable development, and engagement with Arctic states, indigenous peoples, and other partners. The Communication provides a continuation of the 2008 Communication, emphasizing the importance of combating climate change, funding research, supporting indigenous peoples, maritime safety, sustainable economic development, and multilateral cooperation. The EU considers the Arctic Council to be the region's primary forum for international cooperation and considers the LOS Convention to be a basis for the management of the marine Arctic.

As the development of an Arctic EU policy has gradually unfolded, certain areas have been consistently highlighted:

- Promoting sustainable development and resource use,
- Protecting and preserving the Arctic in unison with its population,
- Contributing to enhanced Arctic multilateral governance and international cooperation,
- Supporting research and knowledge, and
- Commitment to combating climate change

In December 2008, the European Commission applied to become a permanent observer to the Arctic Council. The decision was postponed in 2009—a move attributed to Canada in response to the EU's ban on seal products, as well as reluctance by Russia. Following the adoption of new criteria for the admission of observers in May 2011, updated information was submitted by the Commission in December 2011 and a decision will be made regarding the EU's status as an observer at the 2013 Ministerial Meeting in Kiruna.

The fact that none of the current EU Member States are coastal states with respect to the Arctic marine area, as defined in this book, is clearly a major feature and constraint of EU policy regarding the Arctic marine area. Currently, three Arctic states—Denmark, Finland, and Sweden—are members of the EU. Iceland applied to become a member of the EU in June 2009 and is now a candidate country. Greenland and the Faroe Islands are not a part of the EU, although Greenland is a member of the Overseas Countries and Territories Association (Koivurova et al. 2010). It should nevertheless be noted that EU law has considerable impact on contracting parties to the EEA Agreement (1993), which includes Iceland and Norway and requires implementation of certain EU legislation related to the common market. Up until now, Norway has decided not to extend the applicability of the EEA Agreement to Svalbard (Koivurova et al. 2010).

While neither the EU nor its Member States can act as coastal states with respect to the Arctic marine area, they can still act in a wide range of other capacities: for instance as flag states, port states, market states, or with respect to their natural and legal persons. In a flag state capacity, EU Member States are able to exercise their rights and discharge their obligations with respect to the Arctic marine area, most notably the freedoms of the high seas in the high seas pockets in the Arctic marine area, the navigational rights and freedoms in the maritime zones of coastal states to the Arctic marine area, and obligations relating to marine living resources and the marine environment connected to these rights and freedoms.

The competence of the EU and its Member States regarding the Arctic marine area is determined by general international law as well as by EU law. It goes without saying that EU Member States cannot confer more extensive competence to the EU then they themselves possess in accordance with international law. Competence between the EU and its Member States is distributed based upon first the Treaty on the Functioning of the European Union (TFEU 2008), the EU Treaty (2010), and other treaties concluded within the framework of the EU.

Most of the EU's sectoral competences that are relevant to the marine Arctic fall under the shared competence between the EU and its Member States. 'Transport' and 'environment' are among the areas listed in Article 4(2) of the TFEU where the EU and its Member States share competence. One of the clearest changes brought by the TFEU is that energy is now also explicitly listed as a shared competence between the EU and Member States. The main exception to these shared competences is the conservation of marine biological resources under the common fisheries policy, an exclusive competence for the EU under Article 3(1) of the TFEU.

1.4.2 US Arctic Policy

For the US, like the EU, the loss of sea ice and prospects of increased activity in the Arctic have heightened interest in the region's future. Unlike the EU, however, the US, by virtue of the state of Alaska, is an Arctic Ocean coastal state.

The US's current Arctic policy was adopted in January 2009 (NSPD-66 2009). The 2009 Presidential Directive, under President George W. Bush, was the US's first official statement on the Arctic since 1994. The 2009 Arctic Policy highlights the national and homeland security interests of the US foremost, but also awards significant attention to protection of the Arctic environment, sustainable development, and regional cooperation. It sets forth the interests of the US in the region and a six-point policy to:

- Meet national security and homeland security needs relevant to the Arctic,
- Protect the Arctic environment and biological resources,
- Ensure sustainable natural resource and economic development,
- Strengthen institutions for cooperating among the Arctic nations,
- Involve indigenous communities in decision making, and
- Enhance scientific monitoring and research into local, regional, and global environmental issues.

Other issues such as international governance, extended continental shelf and boundary issues, scientific cooperation, maritime transportation, and economic and energy issues are also covered. In its Arctic Policy, the US articulates a commitment to cooperation with other Arctic actors through bilateral and other institutional mechanisms. The Arctic Council is viewed as an important institution for Arctic governance, albeit with the intention that the Council remain largely within its current mandate rather than be transformed into an international organization.

The US is not a party to the LOS Convention, thus limiting its participation under the treaty. The 2009 Arctic Policy recommends Senate ratification of the LOS Convention to serve US security and environmental interests. Despite not being a party to the LOS Convention, in a May 2008 meeting of the 'Arctic five' coastal states (Canada, Denmark/Greenland, Norway, the Russian Federation, and the US) in Ilulissat, Greenland, the US, along with the other states, affirmed a commitment to the law of the sea framework for determining continental shelf claims (Ilulissat Declaration 2008). The US has also taken steps to determine the outer limits of its extended continental shelf in accordance with the LOS Convention in preparation for future accession.

Although the US has not formally updated its Arctic policy since 2009, it is currently developing an implementation plan for the National Ocean Policy that pays special attention to the Arctic as a priority region (see Sect. 1.5.2).

1.5 EU and US Marine Policy

1.5.1 EU Maritime Policy

Ocean and coastal management is an active area of EU policy development. While the economic, social, and cultural importance of marine waters to the EU is well established, there is also an understanding that the intensity and scope of its maritime activities is expanding due to advances in technology and increased demand. In recent years, the EU has clearly identified that it is "at a crossroads in our relationship with the ocean" (European Commission 2007a). Simultaneously, the last several decades have witnessed a growing recognition of the limitations of sector-based maritime policies, owing to the strong interlinkages of matters related to the marine environment. In this context, the EU is implementing a new system of integrated and holistic maritime policies, most notably through the Integrated Maritime Policy (IMP; European Commission 2007a) and its associated environmental pillar, the Marine Strategy Framework Directive (MSFD 2008).

Adopted in 2007, the IMP aims to improve coordination of marine policies among EU Member States so as to overcome the compartmentalization and incoherence that occur across sector-based policies. Its objective is to produce a more integrated and holistic approach to governing EU marine waters that will "enhance Europe's capacity to face the challenges of globalization and competitiveness, climate change, degradation of the marine environment, maritime safety and security, and energy security and sustainability" (European Commission 2007a). As the IMP encompasses a broad spectrum of policy areas, it has been called "the most comprehensive policy ever adopted by the EU" (Koivurova 2009). The IMP has a dual focus on economic development while maintaining environmental sustainability.

Perhaps the most significant instrument implemented under the IMP is the MSFD. The MSFD, adopted in 2008, is a legally binding instrument designed to establish a policy framework within which EU Member States will maintain or achieve "good environmental status" of their marine environment by 2020. The MSFD states that marine policies will use an ecosystem-based approach to the management of human activities, with Member States formulating their own national marine strategies with regards to distinct geographic regions, such as the Baltic Sea. It is important to note that the MSFD is not a harmonizing measure intended to produce a uniform set of standards across all Member States with regards to what constitutes good environmental status. Rather, Member States are required to establish their own marine strategies for their own marine waters. Member States determine what constitutes good environmental status, as well as the optimal methods for achieving that target.

1.5.2 US Ocean Policy

Oceans and coasts play a critical role for the US economy and the quality of life of its citizens. Increased attention is being paid to the need for ocean and coastal management to be integrated and scientifically-based, including using an ecosystem approach, and with the involvement of stakeholders at all stages of development and implementation. In light of this, the US has recently adopted a new National Ocean Policy, an initiative aiming to implement an integrated and holistic approach to maritime policy (Exec. Order No. 13547 2000).

Many of the challenges the US faces in sustainably developing ocean and coastal ecosystems and economies stem from mismatches between the way natural systems work and the way the activities affecting them are managed. Management has been fragmented by an outdated and disjointed collection of laws, institutions, and jurisdictions: At the federal level alone, oceans and coasts are managed under more than 140 different federal laws implemented by a wide range of agencies.

Federal law generally applies to areas of the ocean beyond the 3 nm jurisdiction from shore that most states and territories hold within the 200 nm US EEZ. Coastal lands generally fall under the jurisdiction of states. An important foundation for federal management of oceans and coasts in the US is the public trust doctrine. Under the public trust doctrine, the bottom and water column resources seaward of the land are held in trust by the government, which has a duty to ensure that public interests in those lands are protected.

Attempts to improve US ocean governance coordinating structures have met with varying successes. Both President George W. Bush and President Barack Obama have made concerted efforts to address the need for an overarching national ocean policy. Those efforts recently culminated on July 2010 when the Interagency Ocean Policy Task Force released its final recommendations on a new ocean policy (Council on Environmental Quality 2010). On the same day, President Obama signed Executive Order 13547 establishing a National Policy for the Stewardship of the Ocean, Our Coasts, and the Great Lakes (Exec. Order No. 13547 2010). The Executive Order adopts most of the final recommendations in the Interagency Task Force report and directs executive agencies to implement those recommendations under the guidance of a National Ocean Council. It establishes a national policy to ensure the protection, management, and conservation of US ocean and coastal ecosystems and resources, to respond to climate change and ocean acidification through adaptive management, and to coordinate with national security and foreign policy interests. The Order also provides for the development of coastal and marine spatial plans that build upon existing federal, state, tribal, local, and regional decision making and planning processes, which are intended to pave the way for a more integrated, ecosystem-based, flexible, and proactive approach to managing sustainable multiple use of the oceans and coasts.

Subsequently, in 2012, a draft implementation plan for the National Ocean Policy was released and opened to public comment (National Ocean Council 2012). The implementation plan lays out initial steps required to achieve the Policy's objectives, focusing on specific actions and nine priority objectives. One of these priority areas is on "Changing Conditions in the Arctic" and "addressing environmental stewardship needs in the Arctic Ocean and adjacent coastal areas in the face of climate-induced and other environmental changes". The draft plan focuses on the impacts of climate change in the Arctic region, highlighting opportunities and challenges presented by rapidly diminishing sea ice and resulting human and environmental changes and needs. The plan outlines potential actions and outcomes for improving Arctic environmental response management, observing and forecasting sea ice changes, establishing a network of biological observatories, improving Arctic communication networks and architecture, advancing Arctic marine mapping and charging, and improving (national and international) coordination on Arctic issues.

Submerged lands extending beyond three miles from the US coast are subject to federal government jurisdiction, led by the US Department of the Interior (DOI). DOI develops five-year framework leasing programmes for the outer continental shelf (Outer Continental Shelf Lands Act 1975, sec. 18). Marine resources and oceans are governed by a variety of other laws including the Coastal Zone Management Act (CZMA 1972), the Marine Protection, Research, and Sanctuaries Act (1972), and the Clean Water Act (1972). The Alaska Coastal Management Act (ACMA 1977) was passed pursuant to the federal Coastal Zone Management Act and created the Alaska Coastal Management Program. Prior to the passage of ACMA, more than 60 % of Alaska's coastal area was controlled by federal agencies.

Compared to many other countries, state governments in the US play prominent roles. State and regional governance are critical elements in managing sectors like marine fisheries and areas such as coastal zones. Localized efforts to improve ocean and coastal management and coordination have developed in a number of

US coastal states as well as multi-state initiatives in regions that share important ocean and coastal ecosystems.[5]

1.6 Conclusion

Climate change is occurring rapidly in the Arctic region, bringing new economic opportunities alongside challenges for environmental governance in the region. Evaluating these changes, and options for effectively addressing them, requires an understanding of existing institutions and frameworks. From the foundation established here, this book further explores the Arctic policy landscape, changing governance needs, and ways to promote a sustainable future.

References

ACIA (2005) Arctic Climate Impact Assessment. Cambridge University Press, New York

ACMA (1977) Alaska Coastal Management Act. AS 46.40

AEPS (1991) Arctic Environmental Protection Strategy, 14 Jan 1991, 30 I.L.M. 1624

AMAP (1997) Arctic Pollution Issues: A State of the Arctic Environment Report. Arctic Monitoring and Assessment Programme, Oslo

AMSA (2009) Arctic Marine Shipping Assessment 2009 Report. Arctic Council, Apr 2009

Arctic SAR Agreement (2011) Agreement on Cooperation on Aeronautical and Maritime Search and Rescue in the Arctic, 12 May 2011, 50 I.L.M. 1119 (2011). Entered into force on 19 Jan 2013

Arctic Shipping Guidelines (2002) Guidelines for Ships Operating in Arctic Ice-covered Waters, IMO MSC/Circ. 1056, MEPC/Circ. 399, 23 Dec 2002

Bartenstein K (2011) The "Arctic Exception" in the Law of the Sea Convention: A Contribution to Safer Navigation in the Northwest Passage? Ocean Dev Int Law 42:22–52

Clean Water Act (1972) 33 U.S.C. 1251 et seq

Coastal Zone Management Act (1972) 16 U.S.C. 1451 et seq

Council of the European Union (2009) Council conclusions on Arctic issues, 2985th Foreign Affairs Council Meeting (8 Dec 2009)

Council on Environmental Quality (2010) Final Recommendations of the Interagency Ocean Policy Task Force. 19 July 2010

Division for Ocean Affairs and Law of the Sea (2013) United Nations. <http://www.un.org/Depts/los/index.htm>. Accessed 31 Jan 2013

EEA Agreement (1993) Agreement on the European Economic Area, 17 Mar 1993. Entered into force 1 Jan 1994

European Commission (2007a) Communication from the Commission to the European Parliament, The Council, the European Economic and Social Committee and the Committee of the Regions: an Integrated Maritime Policy for the European Union, 10 Oct 2007, COM (2007) 575 final

[5] Notable state initiatives include: Marine Life Protection Act (California), Territorial Sea Plan (Oregon), Puget Sound Partnership and Marine Waters Planning and Management Act (Washington), Ocean and Great Lakes Ecosystem Conservation Act (New York), Massachusetts Ocean Act (Massachusetts), and the Rhode Island Ocean Special Area Management Plan (Rhode Island).

European Commission (2007b) Accompanying Document to the Integrated Maritime Policy, 10 Oct 2007, SEC (2007) 1278

European Commission (2008) Communication from the Commission to the European Parliament and the Council: the European Union and the Arctic Region, 20 Nov 2008, COM (2008) 763

European Commission and High Representative (2012) Joint Communication to the European Parliament and the Council. Developing a European Union Policy towards the Arctic Region: progress since 2008 and next steps, European Commission and High Representative of the European Union for Foreign Affairs and Security Policy, 26 June 2012, JOIN (2012) 19 final

European Parliament (2008) Arctic Governance. European Parliament resolution of 9 Oct 2008 on Arctic governance, P6_TA(2008)0474

European Parliament (2011) A sustainable EU policy for the High North. European Parliament resolution of 20 Jan 2011 on a sustainable EU policy for the High North (2009/2214(INI)), 20 Jan 2011, P7_TA(2011)0024

Exec. Order No. 13547 (2010) Stewardship of the Ocean, Our Coasts, and the Great Lakes. 10 July 2010, 75 Fed. Reg. 43,023

Fish Stocks Agreement (1995) Agreement for the Implementation of the Provisions of the United Nations Convention on the Law of the Sea of 10 December 1982 relating to the Conservation and Management of Straddling Fish Stocks and Highly Migratory Fish Stocks, 4 Aug 1995, 2167 U.N.T.S. 3. Entered into force 11 Dec 2001

High Representative and European Commission (2008) Paper from the High Representative and the European Commission to the European Council. Climate Change and International Security, 14 Mar 2008, S113/08

Humphreys A (2012) New proposal would see Hans Island split equally between Canada and Denmark. National Post, 11 Apr 2012. Available at: <http://news.nationalpost.com/2012/04/11/new-proposal-would-see-hans-island-split-equally-between-canada-and-denmark/>

Ilulissat Declaration (2008) Arctic Ocean Conference. Ilulissat, Greenland. 27 May 2008, 48 I.L.M. 382 (2009)

Koivurova T (2009) A note on the European Union's integrated maritime policy. Ocean Dev Int Law 40:171–183

Koivurova T, VanderZwaag DL (2007) The Arctic Council at 10 Years: Retrospect and Prospects. Univ British Columbia Law Rev 40:121–194

Koivurova T, Kokko K, Duyck S, Sellheim N, Stepien A (2010) EU Competencies Affecting the Arctic: Directorate General for External Policies of the Union. Directorate B Policy Department Study

LOS Convention (1982) United Nations Convention on the Law of the Sea, 10 Dec 1982, 1833 U.N.T.S. 396. Entered into force 16 Nov 1994

Marine Protection, Research, and Sanctuaries Act (1972) 16 U.S.C. 1401 et seq

Marine Strategy Framework Directive (2008) Directive 2008/56/EC of the European Parliament and the Council of 17 June 2009 establishing a framework for community action in the field of marine environmental policy, June 17 2009, 2008 O.J. (L 164)

Molenaar EJ (2012) Fisheries Regulation in the Maritime Zones of Svalbard. Int J Mar Coast Law 27:3–58

Murmansk Treaty (2010) Treaty between the Kingdom of Norway and the Russian Federation concerning Maritime Delimitation and Cooperation in the Barents Sea and the Arctic Ocean, 15 Sept 2010, U.N.T.S. Reg. No. 49095. Entered into force 7 July 2011

National Ocean Council (2012) Draft National Ocean Policy Implementation Plan, 12 Jan 2012. <http://www.whitehouse.gov/sites/default/files/microsites/ceq/national_ocean_policy_draft_implementation_plan_01-12-12.pdf>. Accessed 5 Feb 2013

NSPD-66 (2009) National Security Presidential Directive and Homeland Security Presidential Directive. Arctic Region Policy, 9 Jan 2009, NSPD-66/HSPD-25

Nuuk Declaration (2011) Nuuk Declaration on the occasion of the Seventh Ministerial Meeting of the Arctic Council. 12 May 2011

Ottawa Declaration (1996) Declaration on the Establishment of the Arctic Council, 19 Sept 1996, 35 I.L.M. 1382 (1996)

Outer Continental Shelf Lands Act (1975) 43 U.S.C. 1331 et seq

PAME (2009) Arctic Offshore Oil and Gas Guidelines. Last updated 29 Apr 2009

Part XI Deep-Sea Mining Agreement (1994) Agreement relating to the Implementation of Part XI of the United Nations Convention on the Law of the Sea of 10 Dec 1982, 28 July 1994. 1836 U.N.T.S. 3. Entered into force 28 July 1996

Polar Shipping Guidelines (2009) Guidelines for Ships Operating in Polar Waters, IMO Assembly Resolution A.1024(26), 2 Dec 2009

Reagan PR (1983) President Ronald Reagan, 'Statement on United States Ocean Policy' 19 Weekly Comp Pres Doc 383. 10 Mar 1983

Roach JA, Smith RW (1996) United States Responses to Excessive Maritime Claims, 2nd edn. Martinus Nijhoff Publishers, The Hague

Rovaniemi Declaration (1991) Declaration on the Protection of the Arctic Environment, 14 June 1991, 30 I.L.M. 1624

Spitsbergen Treaty (1920) Treaty concerning the Archipelago of Spitsbergen, 9 Feb 1920, 2 L.N.T.S. 7. Entered into force 14 Aug 1925

TFEU (2008) Treaty on the Functioning of the European Union. Consolidated Version. O.J. C 115/47

Treaty on European Union (2010) Consolidated Version of the Treaty on European Union, 2010 O.J. C 83/01

United States Department of State (1992) Limits in the Seas. No. 112. United States Responses to Excessive Maritime Claims. Available at <http://www.state.gov/documents/organization/58381.pdf>. Accessed 5 Feb 2013

Chapter 2
The Arctic Marine Environment

Arne Riedel

Abstract The following sets the scene for subsequent chapters of this book: It presents a descriptive overview of the Arctic environment, demonstrating the scope of this term and highlighting unique environmental features of the circumpolar region, while focusing on the marine environment in general and the transatlantic region in particular. The global and regional threats for the Arctic environment are presented to establish a basic understanding of the evolving and increasing risks that this relatively pristine area encounters already, and those yet to come. Next to the primary global threat of climate change—bringing with it increasing sea ice loss, ocean acidification, thawing permafrost, and melting glaciers—developments in the areas of pollutants and chemicals, natural resources, shipping, fisheries, tourism, and military activities show increasing impacts on the Arctic environment.

2.1 Introduction

> **The Arctic is a vast, ice-covered ocean**, surrounded by tree-less, frozen ground, that teems with life, including organisms living in the ice, fish and marine mammals, birds, land animals and human societies (NOAA 2012).

This quote by the United States (US) National Administration for Oceanic and Atmospheric Research (NOAA), describes the Arctic environment succinctly: The Arctic environment is unique. Additionally, distinctive and fragile Arctic ecosystems are increasingly threatened, but what exactly is at risk? What are these risks? And why does it matter?

This chapter is based on previous publications by Ecologic Institute within the Arctic TRANSFORM project and EU Arctic Footprint and Policy Assessment (Cavalieri et al. 2010).

A. Riedel (✉)
Ecologic Institute, Pfalzburger Strasse 43/44, Berlin 10717, Germany
e-mail: arne.riedel@ecologic.eu

E. Tedsen et al. (eds.), *Arctic Marine Governance*,
DOI: 10.1007/978-3-642-38595-7_2, © Springer-Verlag Berlin Heidelberg 2014

The Arctic marine area is a place where local communities fight to maintain traditional livelihoods, as well as where the adjacent circumpolar states—and now others—have vested interests. Although the environmental conditions of the Arctic still exhibit extreme, and often challenging, variations in light and temperature, the quickly diminishing ice cover in the polar region is opening up new and increased opportunities for economic activities, particularly during extended periods of daylight in the long northern summers. This includes activities such as hydrocarbon exploration and development, shipping across the (at this time) two possible shipping transport routes (the Northern Sea Route and Northwest Passage), and the possibility of increased fisheries activity. As considered throughout this book, these developments should be tackled with governance approaches tailored to the region's particular requirements and environmental conditions.

Recent scientific assessments describing the regional Arctic landscapes and Arctic ecosystems[1] cover a level of detail that can be summarized only briefly within the limited scope of this chapter. Instead, the chapter's main focus is on setting the scene for the following discussions on governance by providing an overview of aspects to be taken into consideration when considering specific measures and programs in the Arctic. This includes a presentation of the Arctic environment that highlights circumpolar environmental features and the threats they are facing today, focusing on the marine environment, with its links to coastal areas and the mainland where relevant. The impacts on indigenous and other local communities are elaborated on in detail in Chap. 4.

This chapter first gives a brief overview of relevant environmental aspects, with a focus on marine habitats, and highlights distinct geographical particularities within the Arctic region where appropriate (Sect. 2.2). It then turns towards the environmental challenges and threats the region face (Sect. 2.3). Finally, the conclusion (Sect. 2.4) stresses the most critical aspects presented.

2.2 The Arctic Environment

The marine and terrestrial area covered by the term 'Arctic' has been defined in several different ways. The Arctic Monitoring and Assessment Programme (AMAP) working group of the Arctic Council has conducted elaborate studies in the Arctic region and listed existing definitions of the Arctic using geographical, geophysical, and political definitions (AMAP 1998; see also Chap. 1, Sect. 1.2). AMAP's definition of the Arctic is based on the terrestrial and marine areas north of the Arctic Circle (at 66°32′N), with certain exceptions, namely the widening

[1] Including, *inter alia*, AOR 2011, summarizing existing Arctic Council assessments. See also AMAP 2011b and IASC et al. 2011. The most comprehensive overview yet will be provided in the upcoming Arctic Biodiversity Assessment (ABA) scientific report, to be completed in 2013, online at <http://www.caff.is/aba >. Accessed 31 Jan 2013.

of the area to 62°N in Asia and to 60°N in North America, thereby including the marine areas north of the Aleutian islands between Russia and the US, and the Canadian Hudson Bay, as well as parts of the North Atlantic Ocean, including the Labrador Sea.

The Arctic region therefore includes the Arctic Ocean basin, which is surrounded almost entirely by land masses, with straits towards the Atlantic and Pacific Oceans. Arctic coastal land is split within five states' jurisdictions. The largest Arctic Ocean coastal states are Russia and Canada, covering together around 9.5 million km^2 (about 70 %) of the Arctic land mass (AMAP 1998). The other regions, from east to west, are the Euro-Barents region (Norway), Greenland, Northern Canada, and the Bering region, including Alaska and the Pacific coast of Russia.

Biodiversity in the Arctic experiences threats from both inside and outside the region. Bird species, fish stocks, and marine mammals in the Arctic environment are not only linked with external systems by their migration patterns, but Arctic habitats are also affected by global ocean currents and pollution transport into the region (CAFF 2010; see also below Sect. 2.3.2). In return, the Arctic environment plays an important part in the biological, chemical, and physical balance of the planet (CAFF 2010, see also below Sect. 2.3.1).

2.2.1 Marine Environment

Nearly two thirds of the Arctic region is covered by ocean waters. These waters are an important part of the global climate system, due to their influence on deep ocean currents and global circulation of the oceans (ACIA 2005). Winds and precipitation also play an important role in mixing warmer waters from the south with colder Arctic waters; this may be subject to change in a warming climate. Arctic ecosystems depend on these interactions and can be highly vulnerable, although it has not been determined with full certainty how potential climatic changes will affect the Arctic environment or if they will lead to a net warming or cooling (ACIA 2005).

The Arctic Ocean provides for diverse habitats, on the surface as well as in the water column, and in open waters as well as in coastal areas. The last ice age and its glacier formations led to a loss in biodiversity that—in combination with extreme climate variability between seasons, sea ice coverage, little solar radiation, and constantly low water temperatures—has resulted in a unique and varied maritime ecosystem. The Arctic Council's PAME (Protection of the Arctic Marine Environment) working group has identified as many as 17 Large Marine Ecosystems (LMEs), described as large ocean areas sharing fundamental oceanographic characteristics (AOR 2011). From a geological point of view, marine ecosystems in the Arctic are rather young, and Arctic food webs can be characterized in a relatively simple structure: The marine food web in the Arctic is based on primary production of algae that is consumed by zooplankton, which is in turn eaten by fish, which are consumed by seabirds and mammals, including humans (ACIA 2005). Despite its simplicity, the functioning of the food web is critically linked to timing. For instance, algal blooms are sensitive to temperature and sea ice retreat, with implications for the entire food web (ACIA 2005).

Box 2.1 The Arctic Ocean and global oceanic currents

The Arctic marine region includes the Arctic Ocean and the surrounding regional and shelf seas, thereby representing an area of approximately 20 million km^2 (AMAP 1998). The Arctic Ocean consists of two basins—Eurasian and Canadian, divided by the transpolar Lomonosov Ridge. The Arctic Ocean is connected to the Pacific Ocean via the Bering Strait between the US and Russia, and to the Atlantic Ocean via the Nordic Seas (Greenlandic Sea, Icelandic Sea, Norwegian Sea) to the East of Greenland and through Davis Strait and Hudson Strait to the West of Greenland.
Currents from both the Atlantic and Pacific Ocean bring water into Arctic waters-warmer waters from the North Atlantic Current via the Fram Strait and the Barents Sea, and comparatively cooler water via the Bering Strait. Due to the narrow and shallow access of the Pacific through the Bering Strait, most of the water in the Arctic ocean originates from the Atlantic, in a ratio of about 80:20 (AMAP 1998).

The Arctic Ocean's vertical water structure is also influenced by these different influxes: Its cold surface waters are divided into a Polar Mixed Layer with low salinity (down to 30–50 m of depth) and a water column of increasing temperature and rising salinity (halocline, down to 200 m of depth) which differs for the incoming Pacific and Atlantic waters. The halocline generally insulates the upper layers from the warmth stored in Atlantic waters in intermediate depths (200–900 m) and thereby also influences the sea ice cover. When the density of the incoming water masses increases (by temperature cooling and rising salinity), water sinks and flows back via the East Greenland Current and the Canadian Straits, in a ratio of about 75:25 (AMAP 1998).

On the ocean surface, sea ice is a dominant feature in the Arctic marine area (CAFF 2001). Almost half of the Arctic Ocean is covered by a permanent ice cap, which grows and shrinks seasonally with maximum cover in March and minimum cover in September (on changes and threats, see Sect. 2.3.1.1). Sea ice determines physical properties, such as exchange of heat between the atmosphere and ocean and light availability, and provides a unique habitat for Arctic species (ACIA 2005). Sympagic organisms live on or immediately below the sea ice and include primary and secondary species dependent on sea ice, with thicker sea ice supporting more complex sympagic communities. They support pelagic ecosystems in the water column in the open ocean, as well as benthic ecosystems on the ocean floor (Molenaar et al. 2008).

Ice algae develop during spring and throughout summer as light becomes available in the polar region. Polar cod, which provide a key link between zooplankton and marine mammals, live in both sea ice and pelagic environments. Nesting seabirds, such as ivory gulls, feed on polar cod and other small fish and zooplankton at the ice edge. Marine mammals, such as the polar bear, walrus, ringed and bearded seals among others, and also whales, such as narwhales, belugas, and bowhead whales, depend on the sea ice for food and survival (CAFF 2010).[2] In the water column and on the seabed, fish, crustaceans and—again—marine mammals also find their habitats, including Atlantic cod, haddock, Alaska pollock, Pacific cod, and the Arctic spider crab (ACIA 2005).

A variety of ocean depth levels adds regional diversification. In the Arctic region, extended continental shelves, particularly along the Russian coastline (see also Chap. 1 for the ongoing claims by circumpolar states) result in rather shallow coastal waters. Across the North Pole, on the other hand, deep sea plateaus are cut by oceanic rims, providing for an entirely different environment. Biodiversity is clustered in areas of higher productivity with warmer waters, especially in the Barents and Chukchi Seas and the Bering Shelf, which host migratory seabirds, marine mammals, and some of the most important fisheries in the world (ACIA 2005).

Within coastal seas, pelagic and benthic organisms together provide for a wealth of ecosystems. Coastal regions also provide shelter, food, and breeding grounds for birds and mammals alike. Some species, such as the common eider, are dependent on benthic organisms in shallow waters. As this seabird also breeds along a vast range of Arctic coastlines, including among others the Barents region, Iceland, both of Greenland's coasts, wide stretches of the Hudson Bay, the Canadian and US Beaufort sea's coasts, and the Bering strait region with the Aleutian islands—it has been used as an indicator of the health of marine of marine environments (CAFF 2010).

2.2.2 Land-Based Impacts on the Marine Environment

Arctic landscapes cover a wide range of topography from bare rock, mountains and glaciers to swamps, meadows, and lowland plains (CAFF 2010). Wetlands cover almost 70 % of land masses in the Arctic region, contribute significantly to freshwater cycles and the exchange of atmospheric gases, including climate forcers, and provide a habitat for many bird species on their migratory routes (CAFF 2010).

Freshwater ecosystems form an important part of Arctic geography and are directly linked with the marine, saltwater ecosystems of the Arctic Ocean. They also span a range of diverse environments. Even in areas with rather low precipitation, freshwater ecosystems can be found, for instance in Arctic lakes, and include

[2] See also indicator Number 10 of CAFF 2010 for more details on each of the mentioned species of "Arctic sea-ice ecosystems".

one of the largest freshwater reserves in the world, the Greenlandic ice shield (see below, Sect. 2.3.1.2, for climate impacts on glaciers).

The Arctic region includes some of the largest rivers on the planet, leading an estimated 4,200 km^3 of freshwater together with about 221 million tons of sediment into the Arctic Ocean (AMAP 1998). Inflow from rivers into the Arctic Ocean represents about 2 % of the overall inflow, which is comparatively more than in other oceans (AMAP 1998).

Inputs by several large river systems, such as the Lena, Ob, and Yenisei in the Russian Arctic also provide pathways for pollutants to enter the Arctic Ocean (see Sect. 2.3.2).

Box 2.2 Biodiversity links between Arctic marine and land environments

Biodiversity thrives in the delta regions of these rivers, as they offer a range of various specific habitats. It is important to note that the marine environment is not entirely distinct from the terrestrial environment, as marine mammals, seabirds, and humans are dependent on both for their survival (ACIA 2005). Several species, including Atlantic salmon and some populations of polar bears (CAFF 2001), are specialized in migration between land/freshwater and oceanic habitats, for breeding and hunting grounds respectively, thereby linking the Arctic marine environment to land-based impacts, such as pollution.

2.3 Specific Threats

The Arctic environment suffers from a range of developments and human activities with increasingly adverse impacts. From a perspective directed towards managing these activities, it is beneficial to identify their respective origins and to classify them accordingly.

- From a global perspective (external), there are two broad areas in particular that exhibit Arctic-specific outcomes: climate change and chemical pollution.
- Focusing on the Arctic region itself (internal), there are five areas and activities that show prospects of rising impacts on the Arctic (marine) environment: natural resources, shipping, fisheries, tourism, and military activities.

2.3.1 Climate Change

The most overarching development with severe impacts on the Arctic environment is global climate change. The direct effects of climate change, as well as secondary effects from the increased use of Arctic marine resources, will significantly impact marine systems. Of the numerous risks to the Arctic environment that climate change presents, four particular environmental changes are highlighted here:

- Reduction of sea ice extent, thickness, and distribution,
- Melting of glaciers,
- Thawing of permafrost soil, and
- Ocean acidification

Warming in the Arctic is linked to global increases of greenhouse gas emissions from anthropogenic sources, which increased by 70 % between 1970 and 2004. Atmospheric carbon dioxide (CO_2) concentrations showed an increase of 35 % since the industrial revolution, while atmospheric methane concentrations more than doubled over that time (IPCC 2007a). Hence, anthropogenic sources account for greenhouse gas levels that by 2005 already exceeded the natural range of the past 650,000 years (IPCC 2007b).

As stated in the 2007 IPCC (Intergovernmental Panel on Climate Change) Fourth Assessment Report (AR4), climate change impacts in the polar regions over the next 100 years "will exceed the impacts for many other regions and will produce feedbacks that will have globally significant consequences". However, precise estimates and detailed understanding of the nature and extent of these impacts are still difficult to predict (IPCC 2007b). Models predict general warming in the Arctic with temperature increases ranging from about 3 °C to 6 °C by 2080, even using scenarios in which greenhouse gas emissions are projected to be lower than they have been for the past ten years (AMAP 2011a).

Box 2.3 Climate change and biodiversity

Species which are specially adapted to the harsh conditions of the Arctic region may suddenly find themselves competing with invasive species, where newly warmed Arctic waters create more widely habitable ecosystems. Changes in migration times and routes of birds and ocean mammals may occur, due both to warming and anthropogenic interference. Invasive parasites and pests can threaten both plant and animal populations (ACIA 2005). Such secondary effects of climate change will add stress on Arctic marine biodiversity.

Sea Ice Reduction

September sea ice extent has been declining during the period of 1979–2010 by an average of 11.5 % per decade (NSIDC 2010). Reduced Arctic sea ice extent, especially during the summer months, will rapidly alter the quality of the entire sea ice ecosystem and is expected to impact the entire Arctic marine food web. Sea ice is an important habitat for many Arctic species, including marine mammals such as polar bears, ringed seals, bowhead whales, belugas, and narwhals (AMAP 2011a; see also Sect. 2.2.1).

Ice-dependent species—both land and sea species—are expected to follow the ice edge as it melts and moves further north; however, the abundance of these species is expected to decline due to the rapid shifts in marine conditions (AMAP 2011a). Walruses are also directly threatened by sea ice loss, as the ice provides additional breeding grounds which are reduced and crowded spaces on the coast provide neither a place to raise young, nor sufficient food sources (Reimnitz et al. 1994).

However, it is important to note that some species, especially commercial fish (e.g., cod and herring in the North Atlantic and walleye pollock in the Bering Sea), are expected to benefit from increases of open water leading to increased productivity (Molenaar et al. 2008).

Seasonally occurring changes in the ice coverage of the Arctic region have becoming increasingly extreme in recent years as compared to available data in the earlier 20th century. The past decade is the warmest on record for global surface air temperature with some Arctic regions growing warmer at an even faster pace (IASC et al. 2011).

Shortly after the IPCC stated in its 2007 report that the Arctic could become ice-free in a business-as-usual scenario around the year 2100, Arctic sea ice extent fell far below these modelled estimates. The summer of 2012 brought the latest record of minimal sea ice extent (and thickness) in the Arctic after an already all-time low in recorded history in 2007: On 26 August 2012, the 5-day running average for Arctic sea ice extent was measured with 4.1 million km^2, almost 1.7 % (or 70,000 km^2) less than in September 2007, with the monthly average of 4.72 million km^2 about 38 % (or 2.94 million km^2) less compared to the 1979–2000 average (NSIDC 2012). The extent remained below the 2007 minimum for a total of 40 days (Fig. 2.1).

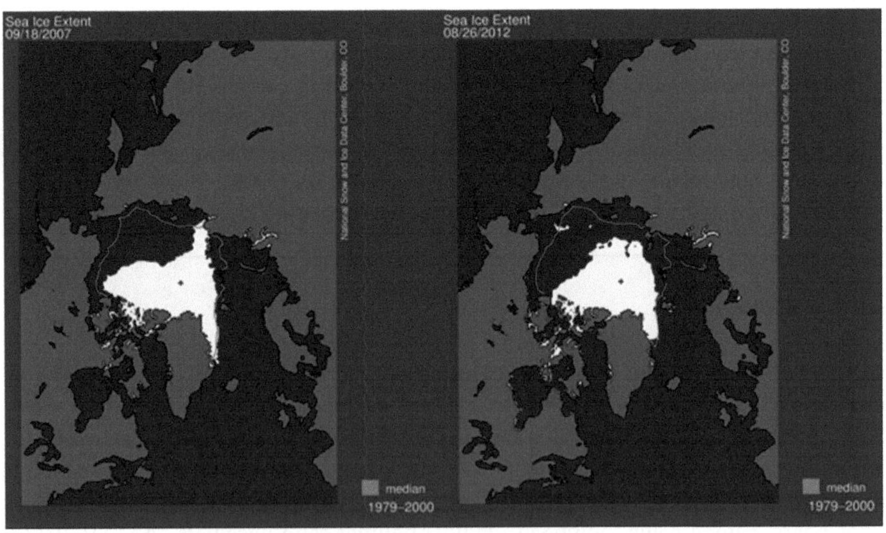

Fig. 2.1 Arctic sea ice extent minima 2007 and 2012 (NSIDC 2012)

Box 2.3 Local warming—the albedo effect

One of the feedback loops identified in relation to Arctic sea ice is the so called albedo effect. It is rooted in the principle that darker surfaces absorb more energy than brighter ones. As sea ice melts, reduction in albedo will likely create a positive feedback effect leading to further global warming (IPCC 2007b): With the opening up of areas in the Arctic Ocean through loss of sea ice cover, the amount of solar radiation absorbed rises, heating up the ocean even more. The higher surface temperatures contribute to further ice melt and slow the growth of new sea ice.

Another albedo-related threat is black carbon. Formed by the incomplete combustion of fossil fuels, biofuels, and biomass—mostly originating outside of the Arctic region—black carbon is a particulate matter with an extent less than 2.5 μm (micrometer) that is emitted into the atmosphere. Its consistency allows it to absorb light, darken surfaces, and thereby increase radiative forcing up to 0.9 W/m^2, second only to CO_2 with an estimated 1.66 W/m^2 (IPCC 2007b; SLCF Task Force 2011). Together with other pollutants that similarly have a powerful, but short-lived climate warming influence, such as tropospheric ozone and the powerful greenhouse gas methane, black carbon is defined as a 'short-lived climate pollutant' (SLCP).

In addition to effects from ice melt and increasing heat absorption, the above-average warming of the atmosphere over the Arctic region can be traced back to other particularities. In contrast to the tropics, for instance, a greater proportion of radiant energy warms the atmosphere above the Arctic, and the Arctic atmospheric layer is shallower (ACIA 2005).

Shrinking sea ice extent also coincides with the loss of ice volume and the loss of multi-year ice (NSIDC 2012). Thin ice melts more rapidly, indicating that the rate of sea ice melt is likely to continue to increase as sea ice continues to become thinner and thicker multi-year ice vanishes (AMAP 2011a). In addition to altering sea ice ecosystems, this ongoing ice loss has the potential to bring dramatic changes to coastal areas. As closed ice cover leads to a degree of protection of coastal lines, these areas will become vulnerable to increasing erosion by waves and storms from open water. A recent study found an average erosion rate of 0.5 m per year for over 60,000 km of sampled coast lines along the Arctic Ocean (Lantuit et al. 2012), with most segments from the Laptev Sea and the Eastern Siberian Sea, followed by the US and Canadian Beaufort Sea. While 89.2 % of the sampled segments (regardless of length) showed erosion rates between 0 and 2 m per year, about 3 % of the length coast lines were found to have a coastal erosion of over 3 m per year.(Lantuit et al. 2012).

Melting Glaciers and Rising Sea Levels

Another impact of climate change observed in recent years is the increasing melt of glaciers, particularly in Greenland. In contrast to sea ice melt, the melt of glacier-bound ice masses contributes to rising sea levels and impacts marine environments by altering the salinity of Arctic waters.

Over 80 % of Greenland's land mass is covered by enormous glaciers with ice sheets as thick as 2–3 km (AMAP 2011a). Average annual ice net mass loss was recently estimated for the years 2005 to 2006 to amount to around 200 (±50) billion tons (AMAP 2011a). In the summer of 2012, satellite imagery showed that more than 97 % of Greenlandic glacier surfaces had begun melting, the largest extent of surface melting observed in three decades of satellite observations.[3] Continued warming will lead to even further melt, although it is as yet unclear to what extent and how fast this will occur. Estimates suggests that total melt of the Greenlandic ice sheet would lead to a potential global sea level rise of 7.5 m (AMAP 2011a).

Rising sea levels can have severe impacts on coastlines and their inhabitants. For land-based species, such as sea birds and marine mammals that rely on bordering marine environments, coastal abrasion by increased erosive forces can lead to the loss of unique breeding spots and feeding grounds (IASC et al. 2011).

There is also concern that significant freshening could impact the thermohaline circulation of the world's oceans, which is a major driver of global weather patterns (IPCC 2007c). With regard to more immediate threats, the freshening of surface layers of the Arctic Ocean could lead to significant changes in the delivery and cycling of nutrients of surface waters, thereby influencing the amount and type of primary production (AMAP 2011a). Sensitive species, such as Greenland Halibut, react to changing salinity conditions in the water by moving their habitat to shelf areas (AMAP 2011a).

Greenhouse Gas Release by Melting Permafrost

In addition to the effects of global warming on oceanic conditions and marine environments, a major feedback loop could be triggered by climate change in land and marine environments. Apart from large areas of land being covered by glaciers (primarily in Greenland and Canada), the upper layer of most land masses of the circumpolar Arctic consists of permafrost—in existing climate conditions, this permanently frozen ground[4] can go as deep as 1,000 m. Permafrost soils provide a generally stable surface that is being used by land animals and birds (as breeding grounds) alike, and have enabled local communities to build necessary infrastructure and housing. It is not limited to upper layers of land but also exists in seabed formations, with the largest hydrate formation under the surface deemed to be on the East Siberian Shelf (AMAP 2011a).

The total coastline affected by permafrost in the northern hemisphere amounts to 407,680 km, thereby representing about 34 % of the world's coastline (Lantuit

[3] For images, see NASA Earth Observatory 2012a. Also, in July 2012, a piece of an estimated 120 ± 5 km^2 broke off the Petermann glacier in North-West Greenland that connects the Greenland ice shield with the Arctic Ocean (NASA Earth Observatory 2012b).

[4] The Permafrost Subcommittee of the National Research Council of Canada has established a definition that includes "ground (soil or rock and included ice and organic material) that remains at or below 0°C for at least two consecutive years" (National Research Council of Canada 1988).

et al. 2012). About two thirds of Arctic coastlines consist of permafrost grounds (IASC et al. 2011). These are particularly vulnerable to coastal erosion which is of growing concern due to the decreasing ice coverage of Arctic waters, as highlighted above. In addition, climate change has already brought temperatures at the top of the permafrost layer up by approximately 3 °C since 1980, decreasing the maximum area of frozen ground in the Northern Hemisphere by 7 % since 1900 (IPCC 2007a). The remaining total area in the Northern hemisphere covered by permafrost is estimated to amount to 18.8 million km^2 (Schuur and Abbott 2011).

These increasing temperatures on Arctic land masses have already had severe impacts on living conditions for fauna and flora: Warmer temperatures generally resulted in more greening of land and a shift of habitats of flora towards northern environments, including shrubs and other plants. Reduced permafrost also results in more Arctic wetlands, which release carbon and methane previously contained in the frozen soil into the atmosphere. As regional studies found, the effect of greening Arctic landscapes with greater photosynthetic activity and carbon uptake can offset these releases temporarily, but not completely (AMAP 2011a). Thawing permafrost can also cause an initial expansion of surrounding lakes and groundwater, followed by drainage and disappearance of lakes, which has been detected in Alaska and Siberia (IPCC 2007a).

With regard to the human dimension, thawing permafrost with its consequences for the consistency of the grounds can ultimately lead to the destabilization of building sites and have significant impacts on the coastal lines along the circumpolar shores as well as for indigenous and other local communities (see Chap. 4 for further detail). Permafrost has also sometimes been used as a natural insulation for landfills and containment holding facilities; thawing could now lead to a contamination of ground water and large clean-up costs (IPCC 2007c).

However, the impact of global warming on these frozen grounds goes even beyond the necessity of climate adaptation for communities, as it also releases previously bound carbon compounds, such as carbon dioxide and methane into the air. To stress the importance and possible impact of this reaction: Due to the vast amount of permafrost coverage (see numbers above), it is estimated that worldwide about 1,700 gigatons of carbon are bound in permafrost deposits (Schuur and Abbott 2011)—a number put into perspective by the world's largest emitter's (China's) numbers from 2010, about 7.285 gigatons of CO_2 emitted from fuel combustion (IEA 2012). While current annual emissions from the ice complex along the Siberian coastline, for instance, are estimated at 44 ($\pm$10) megatons of carbon (Vonk et al. 2012), the ever faster cycle of permafrost thawing and subsequent release of stored greenhouse gases is feared to potentially lead towards a climate change 'tipping point', greatly accelerating planetary warming.

Ocean Acidification

The most direct impact of climate change on the Arctic marine environment is caused by acidification. Ocean acidification results from a gas exchange between the oceans and atmosphere, whereby CO_2 dissolves in the water and decreases its pH level. Generally, oceans serve as a sink for CO_2, but increased anthropogenic atmospheric CO_2 concentrations have already led to a higher rate and scope of ocean acidification. This is of particular importance to the Arctic Ocean, since CO_2 is more soluble in cold water. Thus, due to its low water temperature, the Arctic Ocean faces more rapid rates of increasing acidification by absorbing more carbon dioxide than other oceans do (Robbins 2012).

Ocean acidification can lead to a reduction in the diversity and abundance of calcareous organisms, an important marine food source, and thereby affect the rest of the Arctic food chain (CAFF 2010). A combination of the aforementioned effects of climate change further increases the acidification process: The melting of sea ice exposes greater areas of the Arctic Ocean to the atmosphere, and freshwater entering the Arctic Ocean from melting glaciers increases the ocean's potential for CO_2 dissolution, while decreasing its buffering ability.

2.3.2 Chemicals and Air Pollution

Another global threat to the Arctic environment stems from a complex system of interdependencies within air and ocean currents as well as meteorological particularities. The Arctic region has proven to be a sink for pollutants from around the globe, due to its atmospheric conditions: Low air, ground, and water temperatures have a severe impact on the reactivity of chemicals. Once transported into the region—be it via pathways in the air or the ocean, or by riverine discharge—chemicals remain largely in place. The breakdown of chemicals in the Arctic region is slowed down by low temperatures and limited solar radiant. This poses a threat for animals and human beings alike. Accumulation of chemicals over time and rising through the food chain can increase to toxic levels, threatening large predators such as polar bears as well as local communities living on a subsistence lifestyle.

Several kinds of contaminants arrive in the Arctic environment from around the globe. For instance, persistent organic pollutants (POPs) and heavy metals (e.g., mercury) are mainly produced in warmer climates, volatilise, and then spread to the Arctic region through wind, water, and migratory species.

Box 2.4 Transport of contaminants into the Arctic

Air transport is a fast pathway for volatile contaminants and for contaminants that attach to particles. Patterns tend to favour transport of air masses from polluted regions in Europe and Asia during winter months.

Ocean currents continuously exchange water masses from the Arctic Ocean with Atlantic and Pacific waters. Declining ice coverage is likely to cause the Arctic Ocean to emit trapped contaminants back to the atmosphere and increases chemical exchange with the air.

Riverine inputs can also contribute significantly to the flow of pollutants into the Arctic Ocean. Russian rivers lead into Arctic regional seas, including 500,000 t of oil/oily substances every year. The Russian rivers Lena and Ob have carried higher amounts of mercury into the marine environment than atmospheric fluxes (Fisher et al. 2012).

Some non-volatile POPs, including brominated flame retardants (BFRs) for instance, are transported on other particles and thus rely on their transport processes to reach the Arctic. Once in the region, POPs then bio-accumulate in the Arctic marine food web, including humans, and can be stored in layers of ice and permafrost. The latter can lead to so-called secondary emissions, as further melting or thawing could release the POPs now locked in sea ice directly into the food chain (ACIA 2005). For some compounds, such as PCBs and mercury, the levels in some groups of people and wildlife populations are high enough to cause concern about health effects (AMAP 2011b).

Impacts on the environment from contaminants can be severe, as POPs and heavy metals include a number of anthropogenic and natural substances that are toxic to humans and animals under certain circumstances. Populations and ecosystems often experience the impact of several stressors at the same time, which can increase their vulnerability towards them (AMAP 2009).

Climate change is likely to affect both sources and pathways of POPs and mercury through changes in wind patterns or ocean currents and precipitation. Permafrost and glacier melt may also result in higher re-emissions of mercury and other contaminants. However, it is difficult to predict whether long-term climate change will lead to generally increased or decreased loads, as there are processes working in both directions. In terms of affecting long-term levels of and impacts from contaminants in the Arctic, anthropogenic emissions of greenhouse gases may become as important as emissions of the contaminants themselves (AMAP 2004).

Other forms of transboundary air pollution contribute to Arctic haze, a reddish-brown fog in the lower atmosphere at high northern latitudes. It is caused by a mixture of sulphate, black carbon, nitrogen oxides (NOx), sulphur dioxide (SO_2) and other contaminants. These aerosol particles provide a transport pathway for pollution into the Arctic and can also contribute to climate change (ACIA 2005). Black carbon, for example, reduces the albedo of Arctic snow and ice and accelerates warming (see also Sect. 2.3.1 on black carbon).

Industry in and around the Arctic also contributes to acidification and contamination, especially locally. This includes severely contaminated areas with major forest damage around the copper-nickel smelters on the Kola Peninsula and at Norilsk in Siberia (EEA 2007). Highly acidified soils are not able to support plant life.

Box 2.5 Origins of contaminants

Many POPs have been produced for technical applications (e.g., PCBs, BFRs, organic pesticides, PFOs) or are created when the technical products break down (e.g., DDE from DDT). Others are by-products in production of technical products or in various combustion processes (e.g., dioxins and furans).

Mercury and other heavy metals are released via mining, metal processing, or through the respective products. Mercury is also mobilized through coal combustion and also occurs naturally. Re-emissions account for about one third of emissions to the atmosphere and are hard to distinguish from natural sources (e.g., mercury released in forest fires).

2.3.3 Fisheries

The increasing relevance of fisheries activities for the Arctic environment is an indirect impact that can be attributed to climate change. It can be broken down into two aspects:

- Following declining sea ice coverage, Arctic waters provide larger areas for fishing vessels that can also be accessed during longer periods of the season.
- Increasing water temperatures shift inherent temperature gradients for certain species, allowing some to move their habitats further north,[5] but forcing other species to migrate into colder regions or threaten them to become extinct.

These shifts over time add urgency to existing problems in the Arctic marine environment, such as overfishing. Albeit providing only a mere 2.6 % of global fish catches (Rudloff 2012), Arctic fish stocks provide for a substantial part of the European Union (EU)'s supply: For instance, for Norwegian and Icelandic fisheries, the EU was by far the major export destination with 80 and 60 % of fish catches respectively (Rudloff 2012).

The negative impacts of overfishing are numerous. Overfishing can reduce the size of the stock not only temporarily, but can also distort its age structure, for instance by reducing the number of adult fish to an extent which threatens the longer term viability of a stock. As a result of continuously unsustainable quotas or non-compliance with existing quotas by illegal, unregulated, and unreported (IUU) fisheries, stocks of North American cod and Alaska pollock in the Central Bering Sea faced depletion already in the early 1990s (Burnett et al. 2008). More recent data shows that more than half of the Northeast Atlantic regional stocks of cod, haddock, whiting, and saithe are threatened with collapse (UNEP 2005).

[5] See, for instance, how the increase in mackerel stocks' abundance rapidly increased Icelandic catches of this species from 2005 on (European Commission 2012). Following the increase of Icelandic quotas, the EU is considering sanctions on Icelandic fishing boats (so called "Mackerel Wars"). See recently Davies 2013.

Apart from impacts to respective fisheries industries, collapsing fish stocks may also have enormous consequences for the Arctic marine environment, the species and ecosystems of which are delicately balanced. Food chains in Arctic ecosystems are usually very simple. Hence, a disruption of a single link in the food chain—for instance, by over-exploitation of stocks—could severely affect the rest of the system. The aforementioned impacts on water temperatures could, however, also lead to beneficial results: Moderate temperature increases are likely to benefit some commercial fish stocks that are currently threatened as well as increase habitat for some species such as cod and herring (IPCC 2007c).

The overall net effect on Arctic fish stocks and commercial fisheries is still uncertain. For one thing, practical approaches for the integrated management of fisheries and the adaptation of management structures will play a role as the effects of climate change continue to emerge (ACIA 2005). Possible conflicts with regional fisheries in the Arctic could arise from the mere physical interference with or from environmental contaminants caused by economic activities, such as transportation and the exploration and exploitation of natural (offshore) resources.

2.3.4 Shipping

The risks from increased shipping in the Arctic region may be considered as indirect effects of climate change. Traditionally, shipping in harsh Arctic conditions and weather (including limited visibility and navigation, sea ice coverage, and a lack of natural light), has been limited to supply vessels for regional settlements, as well as research and fishing vessels in some areas.

Recent developments, however, show several factors that already lead to increased shipping activities in the circumpolar region:

- Sea ice loss (making routes open longer/shipping safer) makes transports economically more viable
- Exploration and development of (offshore) natural resources entails an increase in traffic to support build up and continuing supply
- Fish stocks migrating north (see above) requires more fishing vessels in the Arctic

The economic implications of these developments are of interest to all Arctic coastal states (AMSA 2009). Reductions in sea ice already allow increased shipping within the Arctic marine area, opening up potential routes and widening the time frame available for shipping along them. The historically inaccessible Northwest Passage, for instance, was for the first time in history navigable in 2007 (ESA 2007), and the seasonally accessible Northern Sea Route, was open for five months in 2011 (Corell et al. 2012). After 34 vessels along the Northern Sea Route in 2011, 2012 saw the highest number of 46 vessels—with over 1.25 million tons of cargo transported a more than 50 % increase compared to the previous year (Barents Observer 2012).

With ongoing exploration and exploitation activities in the sector of offshore hydrocarbon resources (see also Sect. 2.3.5), the supply of drilling wells and their maintenance goes hand-in-hand with increased shipping activities by drilling vessels and support barges.

Finally, increasing water temperatures lead to a shift in natural barriers for migratory fish stocks (see Sect. 2.3.3). Following these stocks, fishing vessels will also need to use Arctic waters more frequently and/or for extended periods of time—due to increasing accessibility.

The expected future impacts of increased shipping activities on the Arctic marine environment can be narrowed down to three major influences:

- Operational spills and discharges,
- Accidental spills and discharges, and
- Impacts on marine mammals (including noise)

Increased use of Arctic sea routes is most likely to have positive impacts on shipping, but is yet limited by high operational costs (e.g., due to the need for icebreakers) and other hurdles mentioned above. However, the shipbuilding industry in participating countries could benefit from an increase in demand. On the other hand, the overall increase in traffic holds significant risks for the environment, not only during general operations, but also in the case of accidents or other emergencies (Brigham 2011).

Emergencies can quickly occur in the extreme environments of most Arctic regions throughout the year. The weather and oceanic conditions in the Arctic are generally harsh and difficult to account for. Depending on the season, few hours of daylight may be available—if any—and in areas of broken sea ice coverage larger pieces of floating ice can pose a threat to ships with low or no polar class level constructions. In addition to darkness and climatic conditions, the distances from coastal installations not only make search and rescue operations much more difficult to conduct, but also make clean-up challenging in the case of a spill (AMSA 2009).

Similar conditions, however, can result in different viabilities for shipping routes in Arctic waters: For instance, the scattered Canadian Arctic Archipelago covers a distance of about 2,400 km and leads to several possible "Northwest Passages", the use of which is heavily dependent not only on weather and ice conditions, but also on the respective ship's draft.[6] The Russian regional shelf seas—from the Russian border with the US to Norwegian territory—the Chukchi Sea, the East Siberian Sea, the Laptev Sea, the Kara Sea, and the Barents Sea are characterized in their respective coastal areas by rather shallow water depths, being in between 58 m (East Siberian Sea) and 578 m (Laptev Sea at its northern limit)

---

[6] PAME (2009).

only (AMSA 2009), which influences biodiversity as well as use for shipping activities along the Northern Sea Route.

Recent findings suggest that in the near future (up to 2020), only the Northern Sea Route is expected to become a viable trans-Arctic route, influencing the Bering Strait Region as well. However, an increase in supply traffic is also expected in Canadian Northern Communities (AMSA 2009).

Shipping along these routes has impacts on Arctic marine, air, and coastal environments. It is expected to negatively impact migratory marine mammals that use these routes, as well as increase the risk of oil spills (ACIA 2005). Shipping also contributes to the degradation of air quality from the release of carbon monoxide, nitric oxide, and other chemical substances from ships' combustion engines (Granier et al. 2006). During summer months, surface ozone concentrations in the Arctic could be enhanced by two or three times in the next decades as a result of ship operations through the northern passages (Granier et al. 2006).

2.3.5 Oil and Gas Extraction

Parts of the Arctic environment also face threats from future exploratory drilling for and exploitation of oil and gas resources. According to a much cited US Geological Survey study, approximately 13 % of the world's undiscovered oil resources, 30 % of the world's undiscovered gas resources, and 20 % of undiscovered natural gas liquids are estimated to be located in Arctic region, about 84 % of which are located offshore (Gautier et al. 2008). Direct environmental impacts in the Arctic from energy extraction, including drilling, infrastructure development, and possible accidents, pose a number of threats to ecosystems and communities. Oil pollution poses a particular threat to the fragile Arctic marine environment, which recovers slowly due to low temperatures. Natural recovery from spills is slower due to shorter growing seasons and slower growth rates.

Generally, risks can be grouped into:

- Operational risks, such as discharges and emissions from drilling platforms and transport vessels alike, and
- Accidental risks, such as oil spills

Although many remote onshore sites in Arctic regions are—just like offshore sites—dependent on sea routes for supply and transportation, (exploratory and production) offshore drilling itself can have a much bigger impact on marine habitats. The biggest threats from offshore hydrocarbon exploitation to the Arctic marine area are related primarily to accidental risks such as oil spills.

Oil spills can occur during oil extraction, storage, or transportation from subsea exploration or production and poorly maintained infrastructure in sub-sea pipelines. So far, there have been no major oil spills in the Arctic. However, should this happen—especially during winter months—rescue and clean-up actions in case of accidents are difficult to impossible, due to harsh climate conditions, usually isolated drilling locations, and a lack of effective removal methods in remote icy areas.

Although some climatic conditions might assist with clean up—for instance, ice contains oil, making it easier to prevent further spreading and make the removal more effective –limited experience to date with clean-up measures under Arctic conditions has lead to a shortage of best practices.[7] There is also concern that if a spill from an uncontrolled well in an ice-free area occurs late in the Arctic summer, ice conditions could change so quickly as to prevent drilling a relief well until the following year (Schmidt 2012).

Operational risks are also of concern. Possible discharges of produced waters, drilling liquids, or chemicals from drilling and extracting facilities and building new infrastructure to support operations would have all environmental impacts. Oil and gas flaring releases black carbon emissions, which can reduce albedo and thus increase the rate of regional warming within the Arctic (see Sect. 2.3.1). Air pollutants, such as NOx, SO_2, VOCs, CO_2, methane, and particulate matter (PM), are released into the atmosphere by fuel combustion for onsite power generation, well testing, gas flaring, and other operational leaks. These substances contribute to Arctic haze and have the potential to fasten ice melt (see above Sect. 2.3.2).

In addition, noise from oil and gas activities can interfere with marine animals and temporarily displace them from their habitats. Seismic exploration has affected the migration patterns of bowhead whales and reduces the accessibility of indigenous hunters to their game (CAFF 2001). It may also cause polar bears to abandon their dens and thereby increase cub mortality (CAFF 2001). The effects of drilling activity, pipelines, and subsurface installations on marine communities and seafloors vary and depend on the communities present and their level of sensitivity to disturbances. The geological composition of the sea floor appears to recover from exploratory drilling within a year in some cases (Corrêa et al. 2009).

2.3.6 Tourism

Despite all aforementioned changes in and impacts on the Arctic environment, the circumpolar region is still overall regarded as pristine. As a consequence, and following increased interest in 'adventure tourism', the popularity of Arctic tourism has increased greatly over the past two decades. Marine tourism in the Arctic is highly diversified and *inter alia* driven by tourists looking for sightseeing and observing wildlife species in their natural habitats (AMSA 2009). Concerns over the impacts of climate change and the perception that the Arctic environment's

[7] See, for instance, the EPPR working group's presentation at the SAO meeting in Haparanda (Bjerkemo 2012).

landscapes and wildlife are endangered are further driving demand for tourism services in these areas.

Tourism activities have increased in land-based hotels as well as on ship cruises. The number of nights spent at hotels in Greenland increased from about 180,000 in 2002 to more than 235,000 in 2008. Svalbard—an Arctic archipelago off Norway's coast—saw numbers rising from around 30,000 in 1995 to 89,000 in 2008 (Emmerson and Lahn 2012). Cruise passengers landing onshore increased in Svalbard from about 37,000 in 1996 to nearly 70,000 in 2003, while the number of cruise ship landing sites nearly tripled between 1996 and 2010 (Evenset and Christensen 2011).

Tourism activities have the potential to impact both land and marine environments in the Arctic region. For example, tourists can cause significant disturbances for nesting or breeding birds and haul-out sites of marine mammals such as ringed seals or walruses. Pathways of tourist groups to 'points of interest' (viewing points on colonies of birds, for instance) can leave Arctic flora trampled down; in addition, litter on the visited sites remains long after the temporary disturbances are gone.

Marine-based tourism accounts for the largest segment of the Arctic tourism industry in terms of numbers of persons, geographic range, and types of recreational activities (AMSA 2009). Due to limited housing capabilities in the high North, as well as the focus on ships being the main means of transport in the Arctic area, the vast majority of tourists visit Arctic regions on cruise vessels. Cruise ship tourism mostly takes place in areas around Greenland, Iceland, Norway (including Svalbard), and Alaska. The impacts on the marine environment mostly pertain to the aforementioned section on shipping. Particular risks arise where vessels used for tourist operations in Arctic waters do not meet needs for protection against floating ice, equipment for confinement of oil, or waste storage capacities.

2.3.7 Nuclear and Radioactive Waste (Including Military Use)

In general, the Arctic is considered a region of "particular vulnerability" to radioactive contamination (AMAP 2010). Despite radioactive particles from nuclear explosions having decreased since the end of atmospheric nuclear testing in 1963, there has been concern that without a nuclear-weapons-free zone agreement, the Arctic could be threatened by nuclear dumping and the expansion of nuclear activities in the Barents Sea region (ADHR 2004).

A large portion of the dumping, from waste and reactors, can be attributed to the Russian Federation—partially inherited from the former Soviet Union—while both the former Soviet Union and the US are largely responsible for pollution from nuclear testing, with France, China, and the United Kingdom also contributing (Bøhmer et al. 2001).

The specific environmental impacts and risks of military and nuclear waste in the Arctic marine environment depend upon the type of waste and containment.

Contamination from radioactive materials can persist for long periods in soils and plants, and may be revealed in higher concentrations further up the food chain. Radioactive contamination poses particular threats to marine ecology and fisheries with risks increasing where waste settles, rather than diluting and dispersing (Nuclear Threat Initiative 2010). Arctic indigenous peoples are also at risk from exposure from radioactive contaminants. Climate change and its impacts on temperature, permafrost, erosion, precipitation, weather events, sea ice, and oceanic circulation could alter radioactive uptake and distribution in the Arctic (AMAP 2010).

Still, studies have demonstrated that no significant amounts of radioactive materials have migrated from dumping sites and that releases from solid radioactive waste have been small and localized (AMAP 2010). Long-term monitoring has demonstrated that radioactivity is declining, but also stressed the need for significant hazardous operations in relation to the management of spent nuclear fuel (AMAP 2010).

In addition, Russian plans for expanding the use of mobile floating nuclear power plants in the Barents region are ongoing—despite safety concerns from environmental groups (Nikitin and Andreyev 2011). This will ensure ongoing discussions about the use of radioactive materials in, and their impacts on the Arctic environment. For instance, an assessment of whether the current international legal framework and safety standards are applicable and appropriate for transportable nuclear power plants (with particular attention given to floating reactors) was recently conducted by the IAEA (IAEA 2012).

2.4 Conclusion

The Arctic environment is indeed unique and at a point in time that requires international attention and continued efforts to address the many challenges it faces.

The continuing loss of sea ice coverage proves to be not only an imminent threat to species' habitats, but may also trigger climate feedback loops that will hasten the change in the region even further. Climatic changes in the Arctic are linked to global changes such as sea level rise and oceanic circulation, with potentially severely adverse affects. Warming water temperatures could shift species' distribution and confront highly specialized Arctic species with competition from invasive species. Pollutants that enter the region via air, ocean, or river pathways have a disproportionate impact on the Arctic and are now being released from decades of deposits in ice and permafrost. Increasing human activities in fisheries, shipping, and the exploration and exploitation of hydrocarbons add even more pressure to Arctic marine ecosystems. Growing numbers of tourists visit the Arctic to see one of the last pristine environments.

This chapter has sought to provide an overview of answers to the questions: What exactly is at risk in the Arctic environment, why is it at risk, and why does it matter? Once familiar with the urgency that lies in protecting the Arctic marine environment, a logical follow up question comes to mind: What do we do about it?

In the following chapters, specific threats will not only be further elaborated upon, but opportunities, instruments, and approaches will be identified and analyzed to begin answering this question. Particular attention is needed to describe both the influence of global and regional environmental developments and possible gaps in the legal and policy framework for the marine Arctic.

While complex questions are often used as excuses to postpone solutions, the Arctic region's complexity and uniqueness should not be invoked at its expense: There is an urgency and need to act. The diversity of habitats and need to adapt to ongoing changes demand quick responses in governance, however, the existing Arctic patchwork of differing national interests, environmental threats, regulatory approaches, and international fora, implies that there are no simple answers.

References

ACIA (2005) Arctic Climate Impact Assessment. Cambridge University Press, New York

ADHR (2004) Arctic Human Development Report. Stefansson Arctic Institute, Akureyri

AMAP (1998) AMAP Assessment Report: Arctic Pollution Issues. Arctic Monitoring and Assessment Programme, Oslo

AMAP (2004) Assessment 2002: Persistent Organic Pollutants in the Arctic. Arctic Monitoring and Assessment Programme, Oslo

AMAP (2009) AMAP Assessment 2009: Persistent Organic Pollutants in the Arctic. Arctic Monitoring and Assessment Programme, Oslo

AMAP (2010) Assessment 2009: Radioactivity in the Arctic. Arctic Monitoring and Assessment Programme, Oslo

AMAP (2011a) Snow, Water, Ice and Permafrost in the Arctic (SWIPA): Climate Change and the Cryosphere. Arctic Monitoring and Assessment Programme, Oslo

AMAP (2011b) AMAP Assessment 2011: Mercury in the Arctic. Arctic Monitoring and Assessment Programme, Oslo

AMSA (2009) Arctic Marine Shipping Assessment 2009 Report. Arctic Council, Apr 2009

AOR (2011) Phase I report (2009–2011) of the Arctic Ocean review (AOR) project. <www.pame.is>. Accessed 1 Feb 2013

Barents Observer (2012, 23 Nov) "46 Vessels Through Northern Sea Route." Barents Observer. 23 Nov. <http://barentsobserver.com/en/arctic/2012/11/46-vessels-through-northern-sea-route-23-11>. Accessed 20 Feb 2013

Bjerkemo O (2012) Recommended Practices in the Prevention of Marine Oil Pollution in the Arctic (RP3)". Presented at the Arctic Council SAO Meeting, 13 Nov, Haparanda. <http://www.arctic-council.org/index.php/en/about/documents/category/404-presentations?download=1494:4-1-recommended-practices-in-the-prevention-of-marine-oil-pollution-rp3-eppr>

Bøhmer N, Nikitin A, Kudrik I, Nilsen T, McGovern MH, Zolotkov A (2001) The Arctic Nuclear Challenge, vol. 3—2001. Bellona Report. The Bellona Foundation.

Brigham L (2011) Russia Opens Its Maritime Arctic. In: Proceedings of U.S. Naval Institute (May), pp 50–54

Burnett M, Dronova N, Esmark M, Nelson S, Rønning A, Spiridonov V (2008) Illegal Fishing in Arctic Waters. WWF International Arctic Programme

CAFF (2001) Arctic Flora and Fauna—Status and Conservation. Conservation of Arctic Flora and Fauna Working Group, Helsinki. <http://library.arcticportal.org/1295/1/AFF%2DStatus%2Dand%2DTrends.pdf>

CAFF (2010) Arctic Biodiversity Trends 2010—Selected indicators of change. CAFF International Secretariat, Akureyri. <http://www.arcticbiodiversity.is/images/stories/report/pdf/Arctic_Biodiversity_Trends_Report_2010.pdf>

Cavalieri S, McGlynn E, Stoessel S, Stuke F, Bruckner M, Polzin C, Koivurova T, Sellheim N, Stepien A, Hossain K, Duyck S, Nilsson A (2010) Arctic Footprint Policy Assessment Final Report. Ecologic Institute, Berlin. 21 Dec 2008

Corell R, Barry T, Earner J, Hislop L, Kullerud L, Melillo J, Nellemann C, Reiersen L, Samsetz J, Pearce F (2012) UNEP Yearbook Arctic 2012. The View from the Top [Draft], UNEP

Corrêa ICS, Toldo EE, Toledo FAL (2009) Impacts on Seafloor Geology of Drilling Disturbance in Shallow Waters. Deep Sea Res Part II: Top Studies Oceanogr 56(1-2):4–11

Davies C (2013) Fishermen Back Sanctions Against Iceland over Mackerel Catch. The Guardian. 6 Jan <http://www.guardian.co.uk/environment/2013/jan/06/fishermen-sanctions-iceland-mackerel-catch>. Accessed 20 Feb 2013

EEA (2007) Air Pollution in Europe 1990–20042/2007. European Environment Agency, Copenhagen

Emmerson C, Lahn G (2012) Arctic Opening: Opportunity and Risk in the High North. Chatham House, Lloyd's. <http://www.lloyds.com/news-and-insight/risk-insight/reports/arctic-report-2012>. Accessed 6 Feb 2013

ESA (2007) Satellites witness lowest Arctic ice coverage in history. European Space Agency. 14 Sept. <http://www.esa.int/Our_Activities/Observing_the_Earth/Envisat/Satellites_witness_lowest_Arctic_ice_coverage_in_history>

European Commission (2012) Icelandic Fisheries: A Review. European Commission, Copenhagen. <http://www.europarl.europa.eu/committees/en/pech/studiesdownload.html?languageDocument=EN&file=73431>

Evenset A, Christensen GN (2011) Environmental impacts of expedition cruise traffic around Svalbard. AECO, Oslo. <http://www.aeco.no/documents/Finalreport.pdf>

Fisher JA, Daniel JJ, Anne LS, Helen MA, Alexandra S, Elsie MS (2012) Riverine Source of Arctic Ocean Mercury Inferred from Atmospheric Observations. Nature Geoscience 5(7):499–504. doi:10.1038/ngeo1478

Gautier DL, Bird K, Charpentier R, Houseknecht D, Klett T, Pitman J, Moore T, Schenk C, Tennyso M, Wandrey C (2008) Circum-Arctic Resource Appraisal: Estimates of Undiscovered Oil and Gas North of the Arctic Circle. U.S. Geological Survey (USGS). <http://pubs.usgs.gov/fs/2008/3049/>

Granier C, Niemeier U, Jungclaus J, Emmons L, Hess P, Lamarque J, Walters, S, Brasseur G (2006) Ozone Pollution from Future Ship Traffic in the Arctic Northern Passages. Geophysical Research Letters 33(13)

IAEA (2012) Nuclear Safety Review for the Year 2012, GC (56)/INF/2. IAEA. <http://www.iaea.org/About/Policy/GC/GC56/GC56InfDocuments/English/gc56inf-2_en.pdf>

IASC, LOICZ, AMAP, IPA (2011) State of the Arctic Coast 2010—Scientific Review and Outlook. International Arctic Science Committee, Land-Ocean Interactions in the Coastal Zone Project, Arctic Monitoring and Assessment Programme, and International Permafrost Association. Helmholtz-Zentrum, Geesthacht, Germany. < http://arcticcoasts.org/>

IEA (2012) CO_2 Emissions from Fuel Combustion—Highlights, 2012 edn. <http://www.iea.org/publications/freepublications/publication/CO2emissionfromfuelcombustionHIGHLIGHTS.pdf>

IPCC (2007a) Climate Change 2007: Working Group I: The Physical Science Basis. Contribution of Working Group I to the Fourth Assessment Report of the Intergovernmental Panel on Climate Change. In: Solomon S, Qin M, Manning M, Chen Z, Marquis M, Averyt KB, Tignor M, Miller HL (eds) Cambridge University Press, Cambridge

IPCC (2007b) Climate Change 2007: Synthesis Report. Contribution of Working Groups I, II and III to the Fourth Assessment Report of the Intergovernmental Panel on Climate Change. Core Writing Team. In: Pachauri RK, Reisinger A (eds) IPCC, Geneva

IPCC (2007c) Climate change 2007: Impacts, Adaptation and Vulnerability. Contribution of Working Group II to the Fourth Assessment Report of the Intergovernmental Panel on Climate Change. In: Parry ML, Canziani OF, Palutikof JP, van der Linden PJ, Hanson CE (eds) Cambridge University Press, Cambridge

Lantuit H, Overduin PP, Couture N, Wetterich S, Aré F, Atkinson D, Brown J (2012) The Arctic Coastal Dynamics Database: A New Classification Scheme and Statistics on Arctic Permafrost Coastlines. Estuaries Coasts 35:383–400

Molenaar EJ, Corell R, Koivurova T, Cavalieri S (2008) Arctic Transform—Introduction to the background papers. Arctic TRANSFORM

NASA Earth Observatory (2012a) Satellites Observe Widespread Melting Event on Greenland. NASA Earth Observatory. <http://earthobservatory.nasa.gov/NaturalHazards/view.php?id=78607>

NASA Earth Observatory (2012b, 18 July) "More Ice Breaks off of Petermann Glacier." NASA Earth Observatory. July 18. <http://earthobservatory.nasa.gov/IOTD/view.php?id=78556>

National Research Council of Canada (1988) Technical Memoranmdum No. 142

Nikitin AK, Andreyev L (2011) Floating Nuclear Power Plants. The Bellona Foundation. <http://www.bellona.org/filearchive/fil_fnpp-en.pdf>

NOAA (2012) NOAA Arctic Theme Page. <http://www.arctic.noaa.gov/index.shtml>

NSIDC (2010, 4 Oct) Arctic Sea Ice News & Analysis—Weather and feedbacks lead to third-lowest extent. National Snow and Ice Data Center. <http://nsidc.org/arcticseaicenews/2010/10/weather-and-feedbacks-lead-to-third-lowest-extent/>

NSIDC (2012, 27 Aug) Arctic sea ice extent breaks 2007 record low. National Snow and Ice Data Center (NSIDC). <http://nsidc.org/arcticseaicenews/2012/08/arctic-sea-ice-breaks-2007-record-extent/>

Nuclear Threat Initiative (2010) Russia: Spent Fuel and Radioactive Waste. <http://www.nti.org/db/nisprofs/russia/naval/waste/wasteovr.htm>

PAME (2009) Names 5–6 different routes along the Canadian Arctic Archipelago

Reimnitz E, Dethleff D, Nürnberg D (1994) Contrasts in Arctic Shelf Sea-ice Regimes and Some Implications: Beaufort Sea and Laptev Sea. Marine Geol 119:215–225

Robbins L (2012) Studying Ocean Acidification in the Arctic Ocean. United States Geological Survey. <http://pubs.usgs.gov/fs/2012/3058/pdf/fs20123058.pdf>

Rudloff B (2012) The EU as fishing actor in the Arctic. Stocktaking of institutional involvement and existing conflicts. 2010/02. Working Paper, FG 2. SWP, Berlin. <http://www.swp-berlin.org/fileadmin/contents/products/arbeitspapiere/Rff_WP_2010_02_ks.pdf>

Schmidt CW (2012) Offshore Exploration in the Arctic: Can Shell's Oil-Spill Response Plans Keep Up? Environ Health Perspect 120(5):194–199

Schuur EAG, Abbott B (2011) Climate Change: High Risk of Permafrost Thaw. Nature 480:32–33

SLCF Task Force (2011) An Assessment of Emissions and Mitigation Options for Black Carbon for the Arctic Council. Arctic Council, May 2011. <http://www.arctic-council.org/index.php/en/about/documents/category/7-working-groups-scientific-reportsassessments?download=135:technical-report-of-the-arctic-council-task-force-on-short-lived-climate-forcers>

UNEP (2005) Ecosystems and Human Well-being: Current State and Trends, Vol. 1. The Millenium Ecosystem Assessment Series. UNEP, Washington, DC. <http://www.unep.org/maweb/en/Condition.aspx>

Vonk JE, Sánchez-García L, van Dongen BE, Alling V, Kosmach D, Charkin A, Semiletov IP et al (2012) Activation of Old Carbon by Erosion of Coastal and Subsea Permafrost in Arctic Siberia. Nature 489:137–140

Chapter 3
Environmental Governance in the Marine Arctic

Susanah Stoessel, Elizabeth Tedsen, Sandra Cavalieri, and Arne Riedel

Abstract This chapter presents an overview of the existing institutional and legal framework at the international and regional levels relevant to environmental governance in the marine Arctic. Examples of both formal and informal environmental governance in the Arctic marine area are presented to demonstrate how different approaches can be used to address particular issues. The chapter highlights the complex interplay of approaches applicable at local, regional, and international scales rather than identifying gaps at the sectoral level. It describes the landscape of governance approaches in place in the Arctic to stimulate discussion regarding the future of environmental governance in the marine Arctic. Many existing institutions and governance mechanisms were developed under political and environmental circumstances that were vastly different from today's reality. As the Arctic is undergoing drastic changes that will spark unprecedented activity in the region, a rethinking of existing structures is required to appropriately address emerging issues.

Based on De Roo C, Cavalieri S, Wasserman M, Knoblauch D, Baush C, Best A (2008) Background Paper: Environmental Governance in the Marine Arctic. Arctic TRANSFORM

S. Stoessel · S. Cavalieri · A. Riedel
Ecologic Institute, Berlin, Germany
e-mail: susanah.stoessel@ecologic.eu

S. Cavalieri
e-mail: sandra.cavalieri@ecologic.eu

A. Riedel
e-mail: arne.riedel@ecologic.eu

E. Tedsen (✉)
Ecologic Institute, 1630 Connecticut Avenue NW, Suite 300, Washington, DC 20009, USA
e-mail: elizabeth.tedsen@ecologic.eu

E. Tedsen et al. (eds.), *Arctic Marine Governance*,
DOI: 10.1007/978-3-642-38595-7_3,

3.1 Introduction

Beyond the biophysical environment, the Arctic marine area is also geopolitically unique. The Arctic Ocean is an almost enclosed ocean basin surrounded by land, including both nation-state territories as well as international waters. Most of the ocean is within the sovereign territory of the five Arctic coastal states—Canada, Denmark/Greenland, Norway, Russia, and the United States (US)—or is part of their Exclusive Economic Zones (EEZs). Under the United Nations (UN) Law of the Sea Convention (LOS Convention 1982), the centre of the Arctic Ocean (in addition to a few other pockets around its perimeter) is defined as the 'high seas'—*de facto* international water that attracts growing interest from a wide range of nations beyond the Arctic borders. European countries, especially the United Kingdom, France, and Germany, as well as China, Japan, and South Korea are interested in science, energy, and transportation in the Arctic. This growing external interest increases the complexity of policy making in the region (Nordregio 2007).

Melting sea ice will not only result in changes to flora and fauna, but will allow unprecedented access for activities such as shipping, fisheries, and exploitation of offshore hydrocarbon resources, bringing additional threats to species and ecosystems, causing significant impacts for local people and their ways of life, and creating new governance challenges. The way in which actors approach these new challenges is an opportunity to learn from past experience and set precedents for effective regional and international governance regimes (Chapin and Hamilton 2009; Arctic Governance Project 2010). Coherent governance structures at the local, regional, and international levels are especially important in the Arctic as the region both strongly impacts and is impacted by global systems.

This chapter presents an overview of the existing institutional and legal framework at the international and regional levels relevant to environmental governance in the marine Arctic. Examples of both formal and informal environmental governance in the Arctic marine area are presented to demonstrate how different approaches can be used to address particular issues. The chapter highlights the complex interplay of approaches applicable at local, regional, and international scales rather than identifying gaps at the sectoral level.

Sector-based regulation is a necessary component of environmental protection and can provide effective management solutions, but does not always adequately address integrated issues and complex environments. Governance frameworks for marine issues have historically been sectoral in nature, resulting in fragmentation, gaps, and inefficiencies (Cavalieri et al. 2011). In many cases, existing institutional structures are ill-suited for the management of shared and connected resources (Millennium Ecosystem Assessment 2005). Certain sectoral issues such as fisheries, shipping, and offshore hydrocarbon extraction are covered in detail in other chapters of this book.

As seen in the literature and recent policy developments, environmental governance increasingly aims to follow a place-based, ecosystem-based approach (Young et al. 2007; see also Box 3.1 below). However, the practical steps needed to implement the principles of ecosystem-based management will undoubtedly vary based on the specific activities and ecosystems to be addressed. As such, it is important to recognise the value of tailoring and of combining multiple approaches to achieve effective governance for the Arctic marine area.

Effective policies for Arctic marine governance are but one part of a broader Arctic policy framework for environmental governance. Although this book focuses on the marine environment specifically, it should be stressed that terrestrial and marine ecosystems are highly interconnected and that governance approaches should take these interactions and interdependencies into account, especially in the coastal zone.

3.2 Environmental Governance

Governance is the continuous process through which society's goals are defined and met (Jachtenfuchs and Kohler-Koch 2004; Juda and Hennessey 2001). It incorporates a wide range of actors, institutions, and mechanisms that "serve to alter and influence human behaviour in particular directions" to meet societal goals and interests (Juda 1999). Governance encompasses laws and regulations, government policies and institutions, market forces, private sector activities, community actions, and more (Rogers and Hall 2003).

In the environmental context, governance refers to the processes, decision making, and mechanisms by which actors and institutions influence environmental outcomes (Lemos and Agrawal 2006). It encompasses a range of political, social, economic, and administrative systems that are in place to develop and manage natural resources and the delivery of ecosystem services at different levels of society (Rogers and Hall 2003). Policy choices made within these systems define and evaluate opportunities, rules, and acceptable behaviour that result in the management and use of natural resources (Juda and Hennessey 2001).

Governance in the Arctic region was summarized by the 2010 Arctic Governance Project:

> Governance systems emerge to address a variety of societal needs, ranging from the production of public goods (e.g., maintaining healthy populations of living resources subject to human harvesting), to avoidance of public bads (e.g., preventing dangerous climate change or the degradation of large marine ecosystems), internalization of externalities (e.g., curbing the spread of contaminants across borders, avoiding the environmental impacts of oil spills), and protection of human rights (e.g., strengthening the right to self-determination of indigenous peoples) (Arctic Governance Project 2010).

Due to the inherent complexity of natural resource use and the particular difficulty of natural resource management where visible boundaries are lacking between marine ecosystems and vast areas of international waters, governance of the Arctic marine environment involves a myriad of approaches for managing human impacts on the environment. Approaches range from targeting a single species, sector, or issue (e.g., oil pollution) to broader cross-cutting strategies. Depending on the context, various actors are involved from the local to the international level, with participation from stakeholders holding diverse perspectives. Also depending on the context, management occurs through legally binding hard law or non-legally binding soft law measures, with varying levels of enforcement. It is important to note at the outset that legally binding approaches are not necessarily preferable to non-binding approaches.

As described later in this chapter, these various approaches to management in the marine Arctic are often interrelated, overlapping, and at times conflicting. National regulations often differ among countries, causing confusion within sectors. There are also differences in the primary objectives of government agencies and international conventions, resulting in conflicting mandates. Management in the marine Arctic will be further challenged by the impacts of global climate change, as access to and distribution of Arctic resources changes. It is impossible to create an exhaustive list of existing approaches, and often multiple approaches are combined in a single management example. However, it is important to recognise that a spectrum of approaches exists, and that a combination of them could provide the foundation for improved, flexible frameworks for good governance in the marine Arctic.

Box 3.1 Ecosystem-based management (EBM)

Martha McConnell and Dorothée Herr (*IUCN Polar Programme*)

Maintaining the sustainability of the Arctic's natural and economic resources with the region's cultural integrity requires a flexible, adaptive, and interdisciplinary management approach such as ecosystem-based management.

Ecosystem-based management (EBM) has the potential to provide an organizing framework for decision making in the Arctic by balancing potentially competing priorities and interests while supporting ecosystem resilience to maintain ecological functions and services. Such an approach considers the resilience of an ecosystem primary in order to ensure sustainable economic and social systems. EBM recognizes that humans and their activities are an integral part of ecosystems as a whole. More specifically, EBM assesses and monitors the effects of multiple stressors affecting the same ecosystem in order to facilitate efficient and adaptive science-based decisions. Through multi-stakeholder engagement, the process also helps Arctic inhabitants adapt to a changing ecological and socioeconomic landscape. The following definition of EBM is modified from Laughlin and Speer's (2011) use of the term:

Ecosystem-based management is the comprehensive, integrated management of human activities based on best available scientific knowledge about ecosystems and their dynamics, in order to identify and take action on influences which are critical to the health of ecosystems, thereby achieving sustainable use of ecosystem goods and services and maintenance of ecosystem integrity.

Ecosystem-based management:

- Manages geographic areas that are defined by ecological criteria;
- Balances and integrates the conservation and sustainable use of ecosystems, considering the inter-relationship among living and non-living components;

- Addresses the cumulative impacts of multiple human activities rather than individual sectors;
- Incorporates scientific, traditional, and local knowledge, and is an inclusive process that encourages participation from public and private sectors, indigenous peoples, and other stakeholders;
- Recognizes that a dynamic environment requires a flexible and adaptive management approach.

Arctic states are supporting marine EBM as a goal at the global, regional, and national levels. All eight Arctic states have agreed to one or more of the following global agreements and resolutions that call for EBM or ecosystem approaches: the 2002 World Summit on Sustainable Development Plan of Implementation; the UN Fish Stocks Agreement (1995), the Convention on Biological Diversity (1992); as well as numerous UN resolutions on sustainable fisheries and oceans issues. A report by the Norwegian Polar Institute describes the important progress made by Canada, Denmark/Greenland, Finland, Iceland, Norway, Russia, and the US in implementing ecosystem approaches to oceans management (Hoel 2009). A recent challenge has been to identify how this progress can be enhanced within the Arctic Council.

Through the Arctic Council network, EBM is a guiding principle informing the work of the CAFF (Conservation of Arctic Flora and Fauna) working group, and is reflected in both the Arctic Marine Strategic Plan (Arctic Council 2004) and the approach taken by the Circumpolar Biodiversity Monitoring Programme (CBMP). Other relevant Arctic Council projects include PAME's (Protection of the Arctic Marine Environment working group) work on Large Marine Ecosystems (LMEs), as well as the Best Practices in Ecosystem-based Oceans Management in the Arctic (BePOMAr) document that was endorsed at the 2009 Arctic Council Ministerial Meeting (Hoel 2009).

In May 2011, the Arctic Council Ministers called for the establishment of an expert group on EBM for the Arctic environment. Composed of government experts from Arctic states and representatives from the Arctic Council's permanent participants and working groups, the Arctic EBM expert group was tasked with fostering a common understanding of EBM and its principles across the Arctic Council and providing guidelines or recommendations for advancing EBM in the coastal, marine, and terrestrial ecosystems of the Arctic. The findings and recommendations of the expert group will be considered by the Senior Arctic Officials before the end of the Swedish Chairmanship in May 2013.

Similar to the Arctic states, the EU has also promoted an EBM approach through, for example, its Integrated Maritime Policy (European Commission 2007) and its Marine Strategy Framework Directive (MSFD 2008), and the EU's 2012 Arctic Joint Communication (European Commission and High Representative 2012) notes that the EU will engage on EBM in the Arctic.

Successful environmental governance entails recognition of the interconnected nature of resources. Flexibility in approaches to governance is needed for both complex resource management issues as well as for changing human and natural conditions, such as those brought about by climate change. The broad range of mechanisms encompassed by governance principles may allow for solutions for governance where 'one size doesn't fit all' (Cleaver et al. 2005).

3.3 Legal and Policy Framework

Management of human activities and impacts in marine Arctic ecosystems is undertaken through a complex array of international treaties and programmes, bilateral agreements, national and sub-national laws, and non-governmental and governmental initiatives. Existing Arctic marine governance involves both hard and soft law mechanisms, state and non-state actors, as well as innovative initiatives[1] that incorporate a variety of stakeholders, including indigenous peoples (Young 2002). The following presents a summary overview of the global and regional instruments and bodies involved in governance of the marine Arctic environment.[2]

3.3.1 Global Agreements and Institutions

Global treaties relevant to marine Arctic ecosystems are numerous and address issues ranging from the establishment of protected areas and species protection to reducing greenhouse gas emissions and pollution prevention.

The **UN Convention on the Law of the Sea** (LOS Convention 1982) provides the basic framework for jurisdiction of and resource control in marine areas. The LOS Convention addresses issues related to the protection of the marine environment, with respect to marine pollution, land-based pollution, dumping, and fisheries. The Convention confirms and designates coastal state authority to create and enforce laws to control marine pollution within their national territories and EEZs, designating minimum standards for dumping regulations. The only direct reference to the Arctic is in Article 234, which establishes the right of coastal states to legislate for the "prevention, reduction and control of marine pollution from vessels in ice-covered areas" in their EEZ.

The LOS Convention, despite providing the basic legal framework for the law of the sea, does not claim to cover all aspects of ocean governance, and refers to

[1] For example, co-management schemes have been cited in particular as being innovative governance mechanisms.

[2] See Annex for a comprehensive overview of global treaties, conventions, and agreements relevant to Arctic marine ecosystems.

other international instruments and bodies that have competence in this area. It is important to note that the challenge of managing pollution in a transboundary context involves states (within their national territories and EEZ) as well as international space (outside national jurisdiction) in the Arctic Ocean.

Other global conventions dealing specifically with marine pollution include the **International Convention for the Prevention of Pollution from Ships (MARPOL 73/78)** which is the primary international convention for preventing pollution of the marine environment by ships and addresses oil, chemicals, harmful substances in packaged form, sewage, garbage, and air pollution. MARPOL 73/78 (1973/1978) also designates 'special areas' and 'particularly sensitive sea areas' (PSSAs) that are potentially especially vulnerable to vessel-based pollution and therefore require more stringent protection measures through, e.g., routing measures and discharge and equipment requirements. Antarctica was designated a special area in a 1990 amendment, but the Arctic has no areas with this designation. Inadequate compliance with MARPOL 73/78 standards has been cited as a shortcoming due to treaty enforcement through a vessel's flag state, and whereby ships choose to register with states having low enforcement (Rothwell 2000, 2012).

Other agreements relevant to marine pollution include the **Convention on the Prevention of Marine Pollution by Dumping of Wastes and Other Matter** (London Convention 1972) and London Protocol (1996) which regulate dumping in marine areas. The London Convention and London Protocol are particularly relevant to protection of the Arctic marine environment, as there have been problems with dumping of wastes (including radioactive waste) in the Arctic Ocean.[3]

Land-, sea-, and air-based pollution remain major concerns for the fragile ecosystems of the marine Arctic since pollution and contaminants, particularly from sources in lower latitudes, accumulate in the Arctic and adversely affect its inhabitants and marine life. It was therefore a major success when the Stockholm Convention on Persistent Organic Pollutants (2001) was adopted after considerable advocacy efforts by Arctic indigenous peoples' organizations and the Arctic Council. The Convention recognised the negative effect that persistent organic pollutants (POPs) have on humans and the environment, initially banning 12 toxic pollutants. In addition, each party is required to develop a national implementation plan for the reduction of POPs (UNEP/GRID-Arendal 2006). Although all Arctic states have signed the Convention, the US has not yet ratified it and Denmark has entered a territorial exclusion with respect to Greenland.

The **Convention on Long-range Transboundary Air Pollution** (LRTAP Convention 1979) also focuses on the reduction and prevention of air pollution, with specific requirements laid out in several protocols, each concerning different pollutants. Arctic-relevant pollutants are, for instance, covered by the POP

[3] Russia has been cited in particular as utilising the Arctic for dumping radioactive waste (see Chap. 2).

Protocol (1998) and the Gothenburg Protocol on Acidification, Eutrophication, and Ground-level Ozone (1999) (amended in 2012 to include limits for black carbon). Protocols for other pollutants which are covered by the Gothenburg Protocol will be considered as terminated once all of their respective parties have ratified, accepted, approved, or acceded to the Gothenburg Protocol (Gothenburg Protocol 1999, art. 18). This includes two protocols on sulphates (Helsinki Protocol 1985; Oslo Protocol 1994), on nitrogen emissions (Sofia Protocol 1988) and on volatile organic compounds (VOCs; Geneva Protocol 1991).

The basic instrument with regard to international action on the issue of climate change is the **UN Framework Convention on Climate Change** (UNFCCC 1992) and its Kyoto Protocol (1997). Signatories to the latter include the EU and its Member States who, along with other so-called Annex I Parties, signed up to binding, specific, quantified emissions reductions listed in Annex B of the Protocol for the timeframe of 2008–2012. During the Conference of the Parties (COP) in December 2012, parties agreed to a second commitment phase for the Kyoto Protocol from 2013 to 2020.[4] Canada has withdrawn from the Protocol while others seem to lack interest in binding reduction targets. As of January 2013, the participation of the Russian Federation—along with Belarus, Ukraine, and Kazakhstan –in the second commitment period is unclear (Murray 2013). The US is a party to the UNFCCC, but has not ratified the Kyoto Protocol. A new "outcome with legal force" is to be concluded under the UNFCCC by 2015 and is supposed to take effect by 2020.[5]

While neither of these instruments contain Arctic-specific rules or requirements (due to the global distribution of greenhouse gases and global participation within the convention), the Arctic is one of the regions affected the most by climate change: the decrease of annual sea ice, the melting of glaciers, and the thawing of permafrost soils already shows severe impacts on the Arctic environment (see also Chap. 1).

As a more recent initiative in this context, the **Climate and Clean Air Coalition** (CCAC) targets short-lived climate pollutants (SLCPs)—covering greenhouse gases under the Kyoto Protocol such as methane and HFCs (hydrofluorocarbons), but also black carbon and other SLCPs. The actions of the CCAC complement efforts underway within the UNFCCC, aiming to leverage high-level political will among its partners to effectively scale-up existing activities to reduce these emissions (e.g., reduce methane and black carbon emissions from oil and natural gas production). The CCAC was launched by six countries and the UN

[4] Outcome of the work of the Ad Hoc Working Group on Further Commitments for Annex I Parties under the Kyoto Protocol at its sixteenth session, UNFCCC Decision 1/CMP.7, 11 Dec 2011, UN Doc. FCCC/KP/CMP/2011/10/Add.1.

[5] Establishment of an Ad Hoc Working Group on the Durban Platform for Enhanced Action, UNFCCC Decision 1/CP.17, 11 Dec 2011, UN Doc. FCCC/CP/2011/9/Add.1.

Environment Programme in February 2012 and is a voluntary partnership with an international scope. As of its first anniversary, the Coalition is comprised of more than 50 partners, including 27 countries, the European Commission, 22 non-governmental organizations, and six intergovernmental organizations. While the initiative is still in its early stages, a key focus of the CCAC's outreach activities has been the Arctic region, as black carbon is a contributing factor to the rapidly melting polar ice cap. Three Arctic states—Canada, Sweden, and the US—were among the founding partners of the CCAC, with Denmark, Finland, and Norway joining soon afterwards. The CCAC and Arctic Council have not yet formally engaged in cooperative activities, however, through efforts of Arctic Council Member States who are also CCAC Partners, there is informal cooperation to advance the shared objectives.

Land-based sources of marine pollution still lack regulation both in the Arctic and around the world. Efforts of the UN with the **Global Programme for Action for the Protection of the Marine Environment from Land-based Activities,**[6] adopted in 1995, and the Arctic Council's **Regional Programme for Action for the Protection of the Arctic Marine Environment from Land-based Activities,**[7] a regional effort adopted in 1998, are possible first steps toward legally binding measures, but remain ultimately non-binding in nature.

Fisheries management on a global and non-species-specific level is regulated through the UN Fish Stocks Agreement (1995) and complemented by the soft law **FAO Code of Conduct for Responsible Fisheries** (FAO 1995). All Arctic states are parties to the Fish Stocks Agreement (for more, see Chap. 5).

Although wildlife management and protection take place largely at the national level, there are a number of relevant international instruments geared toward this purpose. Species-specific initiatives include the **International Convention on the Regulation of Whaling** (ICRW 1946) and the **International Agreement for the Conservation of Polar Bears** (Polar Bear Agreement 1973). Frameworks like the **Convention on the Conservation of Migratory Species of Wild Animals** (CMS 1979) aim at broader protection of wildlife. Although the CMS Convention has no particular focus on the Arctic, there are numerous migratory species that inhabit the Arctic for parts of the year (UNEP/GRID-Arendal 2006). Only four of the eight Arctic countries[8] are parties, however, limiting its efficacy. The CITES Convention (1973) (see Box 3.2) regulates the trade of currently endangered species and those that could be threatened if no action is taken. It requires a permit for the export and import of listed species and certain species, if designated, may not be traded at all.

---

[6] See <http://www.gpa.unep.org./>

[7] See <http://www.pame.is/index.php/regional-program-of-action>

[8] Denmark, Finland, Norway, and Sweden.

Box 3.2 Polar bear management and climate change challenges

After mounting evidence in the 1960s that the harvesting of polar bears was endangering populations, the five polar bear range states (Canada, Denmark/Greenland, Norway, Russia, and the US) entered into the Polar Bear Agreement in 1973. The agreement established research coordination and conservation of polar bears through the preservation of polar bear habitat. Furthermore, it prohibited the 'taking'[9] of polar bears except for scientific and indigenous subsistence purposes. The agreement represented a historic international cooperation during the Cold War era. The Polar Bear Agreement outlines framework goals and the contracting parties have an obligation to advance these in accordance with national law. Implementation of management frameworks for polar bear stocks uses varying approaches (e.g., prohibitions on takings, quota systems, action plans).

Until recently, polar bear stocks were generally considered to be stable in the Arctic, and the Polar Bear Agreement and national frameworks—in conjunction with bilateral cooperation[10]—deemed effective. However, recent climate change-induced environmental effects have altered this outlook. In 2005, the International Union for Conservation of Nature (IUCN) Polar Bear Specialist Group (PBSG) unanimously recommended that the species be classified as 'vulnerable' according to IUCN Red List criteria (see IUCN 2013) and in 2009, the five range states began development of a coordinated, range-wide action plan to guide polar bear conservation efforts.[11] This joint circumpolar action plan covers the entire circumpolar range of polar bears.

In 2009, the US recommended that the polar bear be transferred from listing under Appendix II of the CITES Convention, which limits international trade of polar bears and their parts, to Appendix I, shifting trade restrictions from the requirement of a CITES permit for import or export to a prohibition of international commercial trade in polar bear specimens. The proposal was rejected at the CITES COP in 2010, with the opposition led by Canada, the only country that allows polar bear harvesting for export purposes (CITES 2010; Cummings and Siegel 2012; Telecky and Smith 2012). In October 2012, to the dismay of several indigenous groups and

[9] 'Taking' is defined by the Agreement as hunting, killing, and capturing [art. 1(2)].

[10] See Polar Bear Management Agreement for the Southern Beaufort Sea; Memorandum of Understanding between Environment Canada and the United States Department of the Interior for the Conservation and Management of Shared Polar Bear Populations (2008); Agreement on the Conservation and Management of the Alaska-Chukotka Polar Bear Population; Memorandum of Understanding for the conservation of the Baffin Bay and Kane Basin sub-populations (Meeting of the parties to the 1973 Agreement 2011).

[11] Meeting of the parties to the 1973 Agreement 2011.

some environmental organizations (e.g., WWF 2008), the US again submitted a proposal to uplist polar bears, however, this was rejected by the Parties in March 2013. Those in opposition emphasized that the current Appendix II listing is appropriate given that the main driver threatening polar bears is climate change and not international trade. During the March 2013 vote, the EU offered a compromise position to regulate trade with export quotas and a tagging system. The compromise was rejected and all 27 EU Member States abstained from voting because they failed to agree on a common position.

The evolving management of the polar bear demonstrates the complexity of options surrounding a single species and its habitat. It highlights that Arctic nations are influenced by political decisions taken outside the Arctic and at the same time, suggests decisions taken outside the Arctic could impact governing decisions for the region as a whole.

The **Convention on Biological Diversity** (CBD 1993) marks a departure from issue-specific agreements, concentrating on the conservation and sustainable use of biological diversity as well as the fair use of its resources. For the *components* of biological diversity, the CBD applies only in areas within the national jurisdiction of each party. However, with respect to *processes and activities* impacting biological diversity, the CBD applies regardless of where their effects occur, provided that they are carried out under a party's jurisdiction or control. This would include processes and activities carried out in Arctic waters including the high seas. The CBD includes a strong emphasis on the establishment of protected areas, which are utilised to conserve certain species or areas that have unique biodiversity or hold special importance.

Although no specific programme dealing with the Arctic environment exists within the CBD and the treaty text does not address marine and coastal biodiversity, the 'Jakarta Mandate on Marine and Coastal Biological Diversity' (1995) is relevant for the protection of Arctic marine species as it focuses on priority areas for action regarding marine and coastal biodiversity. Additionally, the CBD Secretariat signed a Memorandum of Cooperation with the Secretariat of the Arctic Council's working group on **Conservation of Arctic Flora and Fauna** (CAFF) in April 2009. This cooperation is intended to build and share knowledge, create awareness, and enhance capacity for implementation of the CBD in the Arctic region. The CBD helps place Arctic biodiversity within a global framework, while CAFF helps inform the CBD on the status and trends of biodiversity in the region.

Other global instruments that provide for the designation of protected areas are the **Convention for the Protection of the World Cultural and Natural Heritage** (UNESCO 1972) and the Ramsar Convention on Wetlands (1971), along with many others, including the above mentioned ICRW. One of the largest protected Ramsar wetlands in the world is Queen Maud Gulf in Nunavut, Canada, and numerous other Ramsar, as well as Natural Heritage, sites exist throughout Greenland, Scandinavia, Siberia, and on Svalbard and Iceland.

In 2013, a new global treaty for reducing mercury emissions was negotiated. The **Minamata Convention on Mercury** provides controls and reductions for products, processes, and industries where mercury is used or released, using a combination of legally binding and voluntary approaches, including trade measures, national strategies, and technologies (UNEP 2013). Over 140 states took place in the negotiations, which were launched at the 2009 session of the UNEP Governing Council. The Minamata Convention on Mercury will be open for signature at a Ministerial meeting in Japan in October 2013. The Inuit Circumpolar Council (ICC) has noted that it is pleased with the efforts to reduce mercury levels, an issue of particular concern for Arctic peoples (Alaska Dispatch 2013) (Table 3.1).

3.3.2 Regional and Sub-Regional Regimes

The marine Arctic is governed by the global instruments described above, as well as regional and sub-regional initiatives, numerous bilateral agreements, and national legislation. Of particular importance is the **Arctic Environmental Protection Strategy** (AEPS 1991) by the eight Arctic countries, a regional initiative stemming from broad recognition of the need for more international cooperation in the Arctic. The **Arctic Council** was created in 1996 to strengthen the AEPS as an intergovernmental forum for discussion and policy making for the Arctic environment, as well as monitoring trends in the Arctic environment. The Arctic Council and its six working groups have effectively brought together actors and stakeholders in the Arctic to address environmental issues. The Arctic Council's key strengths lie in its engagement with indigenous peoples and its ability to produce pan-Arctic scientific assessments that raise the visibility of Arctic environmental issues. The Council has produced many important scientific assessments, such as the Arctic Climate Impact Assessment (ACIA 2005), Arctic Biodiversity Assessment, and Arctic Marine Shipping Assessment (AMSA 2009), which have brought attention to important regional issues and promoted policymaking. PAME is currently leading the Arctic Ocean Review (AOR) project and assessing the status, trends, and recommendations for Arctic marine governance. The Council also serves as the most important forum for discussion between and within countries on many overarching issues (both sectoral and cross-sectoral).

The Arctic Council is a non-regulatory body. However, it is beginning to show tendencies towards becoming a decision making body: While political and academic spheres debated about whether it should be more policy-oriented or eventually have regulatory capacities, in 2011 the Member States of the Arctic Council concluded a search and rescue agreement (Arctic SAR Agreement) under its auspices; however, although the agreement will be binding once the states have implemented it into their respective national legal regimes, the Arctic Council itself remains without regulatory powers. Taking this model approach forward, its member states put the development of a binding agreement on the agenda concerning preparedness and response to oil pollution, to be concluded at the Council's May 2013 Ministerial Meeting. The Arctic Council's role in the governance of the area could be described as a decision-shaping body evolving into a negotiating forum (Munk-Gordon 2012).

Table 3.1 Participation of Arctic countries in global treaties

	Marine	Pollution			Environment				Biodiversity
	UNCLOS	MARPOL[1]	London convention	POPs	CMS	Heritage	Ramsar	Fish stocks	CBD
	1982	1973/1978	1972	2001	1979	1972	1971	1995	1992
Canada	Yes	Yes[b]	Yes[a]	Yes	*No*	Yes	Yes	Yes	Yes
Denmark	Yes	Yes[b]	Yes[a]	Yes[2]	Yes	Yes	Yes	Yes	Yes
Finland	Yes	Yes[b]	*No*	Yes	Yes	Yes	Yes	Yes	Yes
Iceland	Yes	Yes	Yes[a]	Yes	*No*	Yes	Yes	Yes	Yes
Norway	Yes	Yes[b]	Yes[a]	Yes	Yes	Yes	Yes	Yes	Yes
Russia	Yes	Yes[b]	Yes	Yes	*No*	Yes	Yes	Yes	Yes
Sweden	Yes	Yes[b]	Yes[a]	Yes	Yes	Yes	Yes	Yes	Yes
USA	*No*	Yes	Yes	*No*	*No*	Yes	Yes	Yes	*No*
Total Arctic countries	7	8	7	7	4	8	8	8	7
(EU)	(Yes)	(No)	(No)	(Yes)	(Yes)	(No)	(No)	(Yes)	(Yes)
Total parties globally	164	152	86	178	117	190	163	80	193

Source Adapted from UNEP/GRID-Arendal (2006)
As of 3 December 2012
[1]Annexes I and II
[2]This ratification was with territorial exclusion of Greenland
[a]Indicates that this country has also ratified the 1996 Protocol of the London convention
[b]Indicates that this country has ratified all of the Annexes to MARPOL

Other state-linked, regional groups with more focused interests or limited mandates include the Nordic Council of Ministers, the Northern Dimension, the Conference of Arctic Parliamentarians, International Arctic Science Committee, the Barents Euro-Arctic Council, and many other regionally based organizations.

Also at the regional level, a Canadian initiative resulted in the drafting of the **Guidelines for Ships Operating in Arctic Ice-covered Waters** (Arctic Shipping Guidelines 2002) in the International Maritime Organization (IMO), that were extended in their applicability to Antarctic waters in 2009, now called the **Guidelines for Ships Operating in Polar Waters** (Polar Shipping Guidelines 2009). These guidelines outline non-binding principles and standards for construction, equipment, and safety procedures for ships in polar regions. Albeit their non-binding status, these guidelines have advanced several national initiatives in Arctic countries with regard to shipping safety (Vidas 2000). Since 2010, a mandatory 'Polar Code' has been under development through the IMO and is planned to contain aspects of shipping safety as well as environmental requirements, for instance with regard to discharges and emissions, but also covering liabilities and insurance requirements. A final decision on the Polar Code is not to be expected before 2014 implementation via existing IMO agreements could defer its entry into force years later.

Box 3.3: Comparing environmental governance in the Arctic and Antarctic

A number of studies have broached the topic of using the Antarctic governance framework as a model for a future regime in the Arctic (see Lennon 2008). The Arctic and Antarctic environments are both high-latitude, circumpolar regions with extreme environmental conditions. Both regions have large deposits of natural resources such as coal, natural gas, and offshore oil reserves. Furthermore, because most of the Central Arctic Ocean is outside of national jurisdiction, arguably, the threat of claiming territories in the Arctic is similar to the situation in the Antarctic before the Antarctic Treaty (1959) was established.

Major differences between the two regions exist, however. The Arctic is primarily oceanic, whereas the Antarctic primarily consists of an ice-covered land mass. The Arctic is characterized by the presence of indigenous populations, whereas the Antarctic has virtually no permanent residents (Millennium Ecosystem Assessment 2005). Arctic nations have active territorial and marine claims in the Arctic, whereas territorial claims are on hold in the Antarctic, as stated in the Antarctic Treaty, signed 50 years ago.

The Antarctic environment is governed primarily through the legally binding **Antarctic Treaty** and its accompanying Protocol on Environmental Protection to the Antarctic Treaty (1991), together with two separate conventions on the Conservation of Antarctic Seals (1972) and the Conservation of Antarctic Marine Living Resources (1980) and some 200 recommendations adopted at the Antarctic Treaty Consultative Meetings, altogether referred to as the Antarctic Treaty System (ATS). These constitute a regional cooperative effort in the sense that the treaty system exclusively addresses one region (Vidas 2000). Emphasising that Antarctica should be used exclusively for peaceful purposes, the Antarctic Treaty promotes scientific research and international cooperation and the 1991 Protocol essentially designates the region as a nature reserve, prohibiting claims to mineral deposits, and regulating waste management and marine pollution (Antarctic Treaty 1959, arts. 1–3). The **Convention on the Conservation of Antarctic Marine Living Resources**, one of the agreements belonging to the ATS, utilises a precautionary and ecosystem-based approach to regulate in particular krill fishing and all marine resources excluding seals and whales, which are governed by other instruments (Rayfuse 2008). Despite the differences between the Antarctic and the Arctic, some aspects of the ATS, particularly its focus on security and peace, could serve as a model for a future Arctic regime.

Aside from comprehensive Arctic regimes, sub-regional, multilateral cooperation has also been cited as an important component of Arctic marine governance (Vidas 2000). Intergovernmental initiatives addressing a portion of the Arctic include the **Barents Euro-Arctic Region (BEAR)** with its associated council and the **Norwegian/Russian Commission on Environmental Protection**.

In the Barents region, bilateral agreements have often preceded multilateral agreements. In 1975, the Soviet-Norwegian Fishing Commission was established and fisheries agreements in the Barents Sea were agreed between Norway and the former Soviet Union. Since that time, Norway and Russia have worked together to manage fish stocks through the Joint **Norwegian-Russian Fisheries Commission** (Stokke 2001; UNEP 2004). In 1988, the **Joint Norwegian-Russian Commission on Environmental Protection** was established as a bilateral, intergovernmental commission focused on environmental protection through control of economic activities (e.g., petroleum-related operations and oil refuelling from ship to ship). Scientific cooperation supports Norwegian-Russian cooperation on fisheries and environmental protection, such as through the Knipovich Polar Research Institute of Marine Fisheries and Oceanography and Bergen-based Institute of Marine Research, which work together to provide scientific data used to help sustainably harvest fish stocks in the Barents Sea, and through the Research Council of Norway and the Russian Foundation for Basic Research, fostering collaboration to strengthen knowledge-based, integrated management of the Barents Sea (Mila 2012). With stocks in the Barents Sea at an all-time high, Norwegian-Russian cooperation has shown that concerted action against illegal, unreported, and unregulated (IUU) fishing can indeed be successful (Fisheries.no 2011).

Other sub-regional legally binding approaches include the six-nation **Convention on the Conservation and Management of Pollock Resources in**

the Central Bering Sea (CBS 1994) and the **Convention on the Protection of the Marine Environment of the North-East Atlantic**[12] (OSPAR Convention 1992). A specific aim of the OSPAR Convention (Annex V) is to apply an "integrated ecosystem approach', although fisheries management and maritime transport are outside of its mandate.[13] OSPAR has also promoted the establishment of a network of marine protected areas, four of which, as of 2007, are in the Arctic[14] (OSPAR Commission 2007). Together with the North East Atlantic Fisheries Commission (NEAFC), OSPAR is currently working to describe 10 areas in the northeast Atlantic that have been jointly identified as meeting the CBD scientific criteria for ecologically or biologically significant areas (EBSAs), in preparation for their possible inclusion in the CBD Repository of EBSAs.

The OSPAR Commission has signed five MOUs (memorandum of understanding) to enhance collaboration and coordination with the International Council for the Exploration of the Sea (ICES)—for scientific information; the European Environment Agency (EEA)—for compatibility in data collection and assistance with information dissemination; the UN Economic Commission for Europe (UNECE)—to provide data and analysis of airborne pollutants from regional monitoring centres across the OSPAR area; the International Seabed Authority (ISA)—to help ensure appropriate coordination of measures in order to reconcile the development of mineral resources with protection of the marine environment; and NEAFC—to promote mutual cooperation towards the conservation and sustainable use of marine biological diversity. In addition, the OSPAR Commission and the International Atomic Energy Agency (IAEA) have established Practical Arrangements on sharing data on concentrations of radioactive substances in the OSPAR Maritime Area.

The **North Atlantic Marine Mammal Commission (NAMMCO)** has been recognized for its ecosystem-based approach to marine mammal protection, involving state and non-state actors and including indigenous populations (Young 2002). NAMMCO is a sub-regional cooperation and co-management framework for whales, seals, and walruses among Norway, Iceland, Greenland, and the Faroe Islands (AHDR 2004).

Norway, Iceland, Greenland, and the Faroe Islands manage beluga whales through the sub-regional NAMMCO, while the US and Canada rely on **co-management agreements** between indigenous communities and federal agencies (e.g., Alaska Beluga Whale Committee and the Indigenous People's Council

[12] The OSPAR Maritime Area covers the northeast Atlantic and therefore includes but is not limited to part of the marine Arctic area referred to in this paper; OSPAR Region 1 covers part of the marine Arctic.

[13] Fisheries management lies within the mandate of the relevant Regional Fisheries Management Organisations (RFMOs) and EU and domestic authorities. For initiatives related to shipping, OSPAR relies primarily on the IMO, an observer to the Convention with which it has an Agreement of Cooperation.

[14] The four marine protected areas in the Arctic were nominated by Norway.

Table 3.2 Examples of informal approaches focused on environmental governance

Type of informal approach	Description
Government initiatives involving non-governmental groups	Co-management of resources and wildlife: empowering local communities by allowing independent choice of practices and ways to achieve government goals
Cooperation between researchers and local communities	Involving researchers and local communities in providing vital information for projects and bridging the gap between science and local knowledge
Cooperation between NGOs and local communities	Involving NGOs as donors and project managers and local communities in implementing project objectives
Initiatives governed by local communities	Involving local communities who have sought alliances for a common cause within their community and inside or outside their region
Cooperation initiatives between researchers or research institutes	Involving researchers and institutes with a common research interest in the Arctic

for Marine Mammals in the US and Nunavut Wildlife Management Board in Canada). These signed agreements are non-binding, but outline agreed principles of management and methods of communication and collaboration. Although some argue that these agreements do not sufficiently transfer power to indigenous peoples, co-management has been widely applauded as an effective tool to increase user participation (ACIA 2005) and has resulted in increased knowledge about species' health and distribution for hunters and scientists (Fernandez-Gimenez et al. 2006).

3.3.3 Informal Approaches and Initiatives

In addition to these legally binding and non-legally binding approaches to environmental governance at the international and regional levels, there are a large number of informal approaches and initiatives that play important roles. Common characteristics of such informal approaches are: a lesser degree of institutionalisation, cooperation emerging on an ad-hoc basis, less complex decision making processes, and less formal cooperation structures, such as verbal agreements.

Roughly five different types of informal approaches to environmental governance in the Arctic region can be distinguished, as shown in Table 3.2.

3.4 Analysis of Governance Shortcomings

Identifying general shortcomings in Arctic environmental governance can help to identify areas for (transatlantic) policy action. In order to address future challenges and governance needs in the Arctic marine area, the literature suggests different ways forward, as well as areas requiring further examination. The following section discusses certain weaknesses—and strengths—of the existing legal and policy framework in light of current conditions and foreseeable changes.

Environmental governance in the marine Arctic is characterised by a patchwork of rules and institutions that reflects the mix of national jurisdictions and international space in the region, as well as its historical realities. As shown, there is currently no overarching governance body specifically mandated to adopt and enforce legally binding rules for the marine Arctic; rather, governance mechanisms come from a variety of sources. The primary Arctic institution—the Arctic Council—does not have the mandate to develop or impose legally binding obligations on its participants, although, as mentioned, the Council has recently begun taking a stronger role in policy formulation by helping to negotiate binding treaties. This means that capacity for legally binding governance measures currently lies primarily in the hands of the Arctic states themselves and through any bilateral and multilateral initiatives they might undertake.

According to the Arctic Governance Project, a strengthening of the complex of institutions and arrangements is needed "to form an interlocking suite of governance systems for the Arctic in which the idea of stewardship is central and the whole is greater than the sum of the parts" (Arctic Governance Project 2010). Absent a more integrated approach, current governance systems are not set up to manage for resilience in the face of the uncertain but inevitable and rapid changes that are occurring in the region. There is a high risk that a 'business as usual' approach will be the default position regarding both national and international waters. Encouragingly, although there is more urgency related to climate change now and less time to implement responses, the ability to establish adaptive management systems is greater now than it was 20 years ago.

Integrated, cross-sectoral governance is still evolving, for instance, through the work of the Arctic Council. The Arctic Council is promoting cross-sectoral approaches for integrated management of Arctic resources through several programmes such as the Arctic Marine Strategic Plan and the Integrated Oceans Management project and several working groups of the Arctic Council, notably PAME and AMAP, have made specific recommendations on cross-sectoral Arctic governance issues. Given the lack of another pan-Arctic forum, and its leading work to date, the Arctic Council may be the best means to implement an ecosystem-based management approach requiring coordination across sectors and countries. The Council's scientific activities and forum for dialogue have the potential to facilitate greater state action, but modifications to its current mandate may be necessary.

While sector-specific policies are critical for managing fishing, hydrocarbon, and shipping activities, a holistic, EBM approach is necessary to ensure that adequate environmental safeguards are established in the marine Arctic. The development of such an approach is fundamentally international in nature and should be based on a system of international principles, standards, and rules that address the interactions and interdependencies among countries, stakeholders, and institutions in the context of climate change.

3.5 Perspectives on the Way Forward: Policy Pathways[15]

As useful as it is to consider a set of policy options, it is perhaps even more fitting to approach environmental governance in terms of policy pathways. The idea of pathways particularly allows for an adaptive management approach and an evolution of policy over time. For example, a pathway approach could enable a precautionary beginning to environmental protection—e.g., moratoriums on certain activities in specific regions or protected areas and then, where warranted by the evidence, a gradual easing of environmental restrictions. Especially in the face of so much uncertainty, there is a need to adopt a precautionary approach regarding the Arctic environment. Chapin and Hamilton (2009) note:

> A key choice in approaching environmental governance is the extent to which a precautionary approach should be undertaken. Typically, environmental rules are developed only after problems of overexploitation or destruction of the natural environment and its resources have already emerged. Such an approach is also possible in the modern Arctic context, but the relatively pristine nature of the Arctic, its heightened fragility and past failures elsewhere point to a need to take a higher level of precaution.

Thinking in terms of pathways also encourages a plurality of approaches within and among the governing institutions involved, perhaps informed by shared principles and improved through dialogue on emerging best practices. Pathways can emerge, diverge, and merge as needed.

Figure 3.1 illustrates possible scales of various environmental governance activities. Consistent with the idea of policy pathways, it is possible to envision multiple starting points on any particular issue. It is also possible to identify approaches that can be implemented relatively quickly as a means of developing best practices and measures in the near-term.

There are several components of a multi-pronged approach that need to be further developed to improve environmental governance at local, national, bilateral, and international scales. These approaches include those that are currently underway through sector-based approaches, national approaches, and multilateral approaches as well as the potential development of a new set of instruments to improve multilateral cooperation. Chapin et al. (2006) argues that institutions should be identified that are poised to implement policies at appropriate scales (or governance levels) and that different issues should be dealt with at different levels.

[15] Many of the ideas in this section came from the paper prepared by the Arctic TRANSFORM Project's Environmental Governance Working Group, co-chaired by Dr. Stuart Chapin and Dr. Neil Hamilton (2009). <http://arctic-transform.org/download/EnvEX.pdf>. Accessed 11 Feb 2013.

Activity	Possible scales (easier / less ambitious) ← → (more difficult / ambitious)
Scientific research	species-specific – LME research – full Arctic Ocean assessment
Management level	local/regional – national – bilateral – multilateral
Marine protection	species/stocks – LMEs and MPAs – reserve networks – ocean
Legal structures	soft law –national regulation – bilateral agreement – multilateral treaty

Fig. 3.1 Governance scales

3.5.1 Principles of Environmental Governance[16]

The following set of core governance principles can provide a foundation for effective governance of Arctic marine ecosystems:

- *The principle of fit*: Create arrangements that avoid or minimize spatial and temporal mismatches among biophysical systems, socioeconomic activities, and governance practices. Multi-level governance is an example of this principle. Different system components operate at different scales, and effective regime design implies attention to relevant scales.
- *The principle of multiple use*: Develop integrated approaches that can mediate among different uses of marine resources and establish priorities when such uses are incompatible.
- *The principle of cooperation*: Ensure that all interested stakeholders have a voice in decision making and decisions are made in a transparent fashion at the appropriate level of governance.
- *The principle of adaptive management*: Design and operate governance systems to promote adaptation and social learning as knowledge improves regarding the relevant biophysical systems, human activities, and their interactions.
- *The principle of policy flexibility*: Marine ecosystems in the Arctic are changing rapidly, and ecosystem functions requiring protection will be different in the future than the present. Attention solely to issue-based threats is thus highly unlikely to be effective unless framed within an overarching context. Resilience and learning are significant elements of this principle, and EBM is an example approach.
- *The principle of precaution*: Frameworks for environmental protection should recognise that preserving healthy ecosystems and functioning ecosystem

[16] This list of principles was developed by the Arctic TRANSFORM Project's Environmental Governance Working Group, co-chaired by Dr. Stuart Chapin and Dr. Neil Hamilton.

services requires a precautionary approach, especially in conditions as vulnerable and relatively pristine as those found in the Arctic. This would ideally entail putting regulations in place *before* human activities increase.

3.5.2 Conclusion and Questions for Discussion

To conclude, four key questions are posed to serve as a starting point for identifying transatlantic policy options for governance in the Arctic marine area. Alongside consideration of the pathways and principles laid out above, the following questions provide a possible starting point for further reflection and discussion on environmental governance needs and opportunities:

- *Uniqueness*: What are the unique opportunities and threats in the Arctic marine area that could guide the adaptation of governance regimes in light of future changes?
- *Content*: Where are there gaps and overlaps in the current governance structure?
- *Approaches*: What are the advantages and trade-offs of the various possible approaches? (e.g., flexibility versus enforceability)
- *Transatlantic contribution*: How can transatlantic policies contribute to effective environmental governance in the marine Arctic and to adaptation to climate change?

This chapter has highlighted the complexity of Arctic environmental governance from multiple perspectives. It is an attempt to describe the landscape of governance approaches in place in the Arctic and to stimulate discussion regarding the future of environmental governance in the marine Arctic. Many existing institutions and governance mechanisms were developed under political and environmental circumstances that were vastly different from today's reality. As the Arctic is undergoing drastic changes that will spark unprecedented activity in the region, a rethinking of existing structures is required to appropriately address emerging issues.

References

ACIA (2005) Arctic climate impact assessment. Cambridge University Press, New York

AEPS (1991) Arctic Environmental Protection Strategy, 14 Jan 1991, 30 I.L.M. 1624

AHDR (2004) Arctic Human Development Report. Stefansson Arctic Institute, Akureyri

Alaska Dispatch (2013) Mercury treaty adopts legal framework welcomed by Arctic indigenous peoples.<http://www.alaskadispatch.com/article/mercury-treaty-adopts-legal-framework-welcomed-arctic-indigenous-peoples>. Accessed 20 Feb 2013

AMSA (2009) Arctic Marine Shipping Assessment 2009 Report. Arctic Council, Apr 2009

Antarctic Treaty (1959) The Antarctic Treaty, 1 Dec 1959, 402 U.N.T.S. 71. Entered into force 23 June 1961

Arctic Council (2004) Arctic Marine Strategic Plan, 24 Nov 2004. Available at <http://www.pame.is/images/stories/AMSP_files/AMSP-Nov-2004.pdf>. Accessed 5 Feb 2013

Arctic Governance Project (2010) Arctic Governance in an Era of Transformative Change: Critical Questions, Governance Principles, Ways Forward. <http://arcticgovernance.custompublish.com/the-arctic-governance-project.142454.en.html>. Accessed 8 Feb 2013

Arctic SAR Agreement (2011) Agreement on Cooperation on Aeronautical and Maritime Search and Rescue in the Arctic, 12 May 2011, 50 I.L.M. 1119 (2011). Entered into force on 19 Jan 2013

Arctic Shipping Guidelines (2002) Guidelines for Ships Operating in Arctic Ice-Covered Waters, IMO MSC/Circ. 1056, MEPC/Circ. 399, 23 Dec 2002

Cavalieri S, Reid A, Lang S, McGlynn E, Cicin-Sain B, Balgos M, Orbach M et al (2011) Policy recommendations for improved EU and US cooperation in maritime governance. <http://www.calamar-dialogue.org/sites/default/files/Policy_recommendations_full.pdf>. Accessed 8 Feb 2013

CBS Convention (1994) Convention on the Conservation and Management of Pollock Resources in the Central Bering Sea, 16 June 1994, 34 I.L.M. (1995). Entered into force 8 Dec 1995

Chapin FS et al (2006) Building Resilience and Adaptation to Manage Arctic Change. AMBIO J Human Environ 35(4):198–202

Chapin S, Hamilton N, Arctic TRANSFORM Environmental Governance Working Group (2009) Policy options for Arctic environmental governance. Prepared by the Environmental Governance Working Group. Arctic TRANSFORM. <http://arctic-transform.org/download/EnvEX.pdf>. Accessed 8 Feb 2013

CITES (2010) Consideration of Proposals for Amendment of Appendices I and II. In: 15th Meeting of the Conference of the Parties, Doha, 13–25 Mar 2010

CITES Convention (1973) Convention on International Trade in Endangered Species of Wild Fauna and Flora, 3 Mar 1973, 993 U.N.T.S. 243. Entered into force 1 July 1975

Cleaver F, Franks T, Boesten J, Kiire A (2005) Water governance and poverty: what works for the poor?. Bradford Centre for International Development, University of Bradford, Bradford

CMS Convention (1979) Convention on the Conservation of Migratory Species of Wild Animals, 23 June 1979, 1651 U.N.T.S. 33. Entered into force 1 Nov 1983

Convention on Biological Diversity (1992) Convention on Biological Diversity, 5 June 1992, 1760 U.N.T.S. 79. Entered into force 29 Dec 1991

Convention on the Conservation of Antarctic Seals (1972) Convention on the Conservation of Antarctic Seals, 1 June 1972, 11 I.L.M. 251 (1972). Entered into force 11 Mar 1978

Convention on the Conservation of Antarctic Marine Living Resources (1980) Convention on the Conservation of Antarctic Marine Living Resources, 20 May 1980, 19 I.L.M. 841 (1980). Entered into force 7 Apr 1982

Cummings B, Siegel K (2012) A critical examination of the inconsistent status designations of the polar bear under United States, Canadian, and international legal regimes. In: IPY 2012 Conference. Montreal. <http://132.246.11.198/2012-ipy/Abstracts_On_the_Web/by_theme.html>. Accessed 8 Feb 2013

De Roo C, Cavalieri S, Wasserman M, Knoblauch D, Baush C, Best A (2008) Background Paper: Environmental Governance in the Marine Arctic. Arctic TRANSFORM

European Commission (2007) Communication from the Commission to the European Parliament, The Council, the European Economic and Social Committee and the Committee of the Regions: An Integrated Maritime Policy for the European Union, 10 Oct 2007, COM (2007) 575 final

European Commission and High Representative (2012) Joint communication to the European Parliament and the Council. Developing a European Union Policy towards the Arctic region: progress since 2008 and next steps. 26 June 2012, JOIN (2012) 19 final

FAO (1995) Code of Conduct for Responsible Fisheries. Adopted by the Twenty-eight Session of the FAO conference, 31 Oct 1995

Fernandez-Gimenez M, Huntington H, Frost K (2006) Integration or co-optation? Traditional knowledge and science in the Alaska Beluga Whale Committee. Environ Conserv 33(4):306–315

Fish Stocks Agreement (1995) Agreement for the implementation of the provisions of the United Nations convention on the law of the sea of 10 December 1982 relating to the conservation and management of straddling fish stocks and highly migratory fish stocks, 4 Aug 1995, 2167 U.N.T.S. 3. Entered into force 11 Dec 2001

Fisheries.no (2011) Fisheries no. Norway. <http://www.fisheries.no/resource_management/International_cooperation/Fisheries_collaboration_with_Russia/>. Accessed 16 Nov 2012

Geneva Protocol (1991) Protocol to the 1979 Convention on LRTAP concerning the Control of Emissions from Volatile Organic Compounds or their Transboundary Fluxes, 18 Nov 1991, 2001 U.N.T.S. 187. Entered into force 29 Sept 1997

Gothenburg Protocol (1999) Protocol to the 1979 Convention on Long-range Transboundary Air Pollution to Abate Acidification, eutrophication and Ground-level Ozone, 30 Nov 1999, EB.AIR/1999/1. Entered into force 17 May 2005

Helsinki Procotol (1985) Protocol to the 1979 Convention on Long-Range Transboundary Air pollution on the Reduction of Sulphur Emissions or their Transboundary Fluxes by at least 30 percent, 14 June 1985, 1480 U.N.T.S. 215. Entered into force 2 Sept 1987

Hoel AH (2009) Best practices in ecosystem-based oceans management in the Arctic. Report no. 129 of the Norwegian Polar Institute, Polar Environmental Centre

ICRW (1946) International Convention for the Regulation of Whaling, 2 Dec 1946, 161 U.N.T.S. 72. Entered into force 10 Nov 1948

IUCN (2013) The IUCN Red List of Threatened Species. International Union for the Conservation of Nature. <http://www.iucnredlist.org/details/6335/0>. Accessed 11 Feb 2013

Jachtenfuchs M, Kohler-Koch B (2004) The dynamics of European integration: Why and when EU institutions matter. Palgrave, Basingtoke

Jakarta Mandate (1995) Jakarta Mandate on the Conservation and Sustainable Use of Marine and Coastal Biological Diversity, Nov 1995, COP 2 Decision II/10

Juda L (1999) Considerations in developing a functional approach to the governance of large marine ecosystems. Ocean Dev Int Law 30:89–125

Juda L, Hennessey T (2001) Governance profiles and the management of the uses of large marine ecosystems. Ocean Dev Int Law 32:43–69

Kyoto Protocol to the United Nations Framework Convention on Climate Change (1997) 11 Dec 1997, 2303 U.N.T.S. 148. Entered into force 16 Feb 2005

Laughlin T, Speer L (2011) IUCN/NRDC workshop on ecosystem-based management in the Arctic marine environment. Workshop report. International Union for the Conservation of Nature and Natural Resources Defense Council, Rekjavik

Lemos M, Agrawal A (2006) Environmental governance. Ann Rev Environ 31:297–325

Lennon E (2008) A tale of two poles: A comparative look at the legal regimes in the Arctic and the Antarctic. Sustain Dev Law and Policy 8(3)

London Convention (1972) Convention on the Prevention of Marine Pollution by Dumping of Wastes and Other Matter. 29 Dec 1972, 1046 U.N.T.S. 120. Entered into force 30 Aug 1957

London Protocol (1996) Protocol to the Convention on the Prevention of Marine Pollution by Dumping of Wastes and Other Matter, 7 Nov 1996, 36 I.L.M. 7 (1997). Entered into force 24 Mar 2006

LRTAP Convention (1979) Convention on Long-range Transboundary Air Pollution, 13 Nov 1979, 1302 U.N.T.S. 217. Entered into force 16 Mar 1983

LOS Convention (1982) United Nations Convention on the Law of the Sea, 10 Dec 1982, 1833 U.N.T.S. 396. Entered into force 16 Nov 1994

Marine Strategy Framework Directive (2008) Directive 2008/56/EC of the European Parliament and the Council of 17 June 2009 establishing a framework for community action in the field of marine environmental policy, June 17 2009, 2008 O.J. (L 164)

MARPOL 73/78 (1973/1978) International Convention for the Prevention of Pollution from Ships, 2 Nov 1973, 2 I.L.M. 1319 (1973) as modified by the 1978 Protocol Relating to the International Convention for the Prevention of Pollution from Ships [17 Feb 1978, 17 I.L.M. 546 (1978)] and the 1997 Protocol to Amend the International Convention for the Prevention of Pollution from Ships (26 Sept 1997) and as regularly amended. Entry into force varies for each Annex. At the time of writing Annexes I-VI were all in force

Meeting of the parties to the 1973 Agreement (2011) Meeting of the parties to the 1973 Agreement on the Conservation of Polar Bears. Outcome of Meeting. Iqaluit, Nunavut, Canada. 24–26 Oct 2011 <www.polarbearmeeting.org/content.ap?thisId=500040871>. Accessed 8 Feb 2013

Mila M (2012) The research council strengthens research on the environment and resources of the Barents Sea. The Research Council of Norway. <http://www.forskningsradet.no/prognett-geopolitikknord/Nyheter/The_Research_Council_strengthens_research_on_the_environment_and_resources_of_the_Barents_Sea/1253978601513/p1253961979600>. Accessed 16 Nov 2012

Millenium Ecosystem Assessment (2005) Ecosystems and human well-being: current state and trends. Island Press, Washington

Munk-Gordon (2012) Canada as an Arctic power: Preparing for the Canadian Chairmanship of the Arctic Council (2013–2015). The Munk-Gordon Arctic Security Program. <http://www.gordonfoundation.ca/sites/default/files/publications/CanadaasanArticPower_1.pdf>. Accessed 8 Feb 2013

Murray J (2013) Russian row over Kyoto extension rumbles on. Business Green. <http://www.businessgreen.com/bg/news/2239407/russian-row-over-kyoto-extension-rumbles-on>. Accessed 8 Feb 2013

Nordregio (ed) (2007) People and politics of the Arctic. J Nordregio 4(7)

Oslo Protocol (1994) Protocol to the 1979 Convention on Long-Range Transboundary Air Pollution on Further Reduction of Sulphur Emissions, 14 June 1994, 122 U.N.T.S. 2030. Entered into force 5 Aug 1998

OSPAR Commission (2007) 2006 Report on the status of the OSPAR network of marine protected areas

OSPAR Convention (1992) Convention for the Protection of the Marine Environment of the North-East Atlantic, 22 Sept 1992, 32 I.L.M. 1072 (1993). Entered into force 25 Mar 1998

Polar Bear Agreement (1973) International Agreement on the Conservation of Polar Bears, 15 Nov 1973, 13 I.L.M. 3 (1974). Entered into force 26 May 1976

Polar Shipping Guidelines (2009) Guidelines for ships operating in polar waters. IMO Assembly Resolution A, 1024(26), 2 Dec 2009

POPS Protocol (1998) Protocol to the 1979 Convention on LRTAP on Persistent Organic Pollutants (POPs), 24 June 1998, 2230 U.N.T.S. 79. Entered into force 23 Oct 2003

Protocol on Environmental Protection to the Antarctic Treaty (1991), 4 Oct 1991, 30 I.L.M. 1461 (1991). Entered into force 14 Jan 1998

Ramsar Convention (1971) Convention on Wetlands of International Importance Especially as Waterfowl Habitat, 2 Feb 1971, 996 U.N.T.S. 245. Entered into force 21 Dec 1975

Rayfuse R (2008) Protecting marine biodiversity in polar areas beyond national jurisidiction. Rev Eur Community Int Environ Law 17(1):3–13

Rogers P, Hall AW (2003) Effective water governance. Global Water Partnership Technical Committee (TEC) background papers no. 7

Rothwell D (2000) Global environmental protection instruments and the polar marine environment. In: Vidas D (ed) Protecting the Polar Marine Environment: Law and Policy for Pollution Prevention. Cambridge University Press, Cambridge

Rothwell D (2012) Legal Challenges for Maritime Operations in the Southern Ocean. In: 2012 Comité Maritime International Beijing Conference, 18 Oct 2012

Sofia Protocol (1988) Protocol to the 1979 Convention on Long-Range Transboundary Air Pollution Concerning the Control of Emissions of Nitrogen Oxides or their Transboundary Fluxes, 31 Oct 1988, 28 I.L.M. 212 (1989). Entered into force 14 Feb 1991

Stockholm Convention on Persistent Organic Pollutants (2001) 2256 U.N.T.S. 119. Entered into force 17 May 2004

Stokke OS (2001) Managing fisheries in the Barents Sea loophole: Interplay with the UN fish stocks agreement. Ocean Dev Int Law 32:241–262

Telecky T, Smith Z (2012) The science and policy interface in a United Nations Treaty: Polar bears and CITES. In: IPY 2012 Conference. Montreal. <http://132.246.11.198/2012-ipy/Abstracts_On_the_Web/by_theme.html>. Accessed 16 Nov 2012

UNEP (2004) Matishov G, Golubeva N, Titova G, Sydnes A, Voegele B. (eds) Barents Sea, GIWA Regional Assessment 11. Kalmar. University of Kalmar, Sweden

UNEP (2013) Minamata Convention Agreed by Nations. UNEP News Centre. <http://www.unep.org/newscentre/default.aspx?DocumentID=2702&ArticleID=9373>. Accessed 20 Feb 2013

UNEP/GRID-Arendal (2006) Background report for the seminar on multilateral environmental agreements and their relevance to the Arctic, Arendal, Norway, 21–22 Sept 2006

UNESCO (1972) Convention Concerning the Protection of the World Cultural and Natural Heritage, 16 Nov 1972, 1037 U.N.T.S. 151. Entered into force 17 Dec 1975

United Nations Framework Convention on Climate Change (1992) 1771 U.N.T.S. 107. Entered into force 21 Mar 1994

Vidas D (2000) The polar marine environment in regional cooperation. In: Vidas D (ed) Protecting the Polar Marine Environment: Law and Policy for Pollution Prevention. Cambridge University Press, Cambridge

WWF (2008) A new sea: The need for a regional agreement on management and conservation of the Arctic marine environment. World Wildlife Fund. <http://wwf.panda.org/what_we_do/where_we_work/arctic/publications/?122260/A-New-Sea-The-Need-for-a-Regional-Agreement-on-Management-and-Conservation-of-the-Arctic-Marine-Environment>. Accessed 15 Mar 2008

Young O (2002) Arctic governance: Preparing for the next phase. Presented at the Arctic Parliamentary Conference Tromsø 11-13 Aug 2002. <http://www.arcticparl.org/resource/images/conf5_scpar20021.pdf>. Accessed 29 Apr 2008

Young O, Osherenko G, Ekstrom J, Crowder LB, Ogden J, Wilson JA, Day JC, Douvere F, Ehler CN, McLeod KL, Halpern BS, Peach R (2007) Solving the crisis in ocean governance: Place-based management of marine ecosystems. Environment 49(4):20–32

Chapter 4
Arctic Indigenous Peoples and the Challenge of Climate Change

Adam Stepien, Timo Koivurova, Anna Gremsperger, and Henna Niemi

Abstract This chapter presents climate change impacts on indigenous traditional harvesting, cultures, identities, traditional knowledge, economies, societies, health, and infrastructure in light of overall socioeconomic and political changes in the Arctic. Responses to these stressors can be autonomous (e.g., ad-hoc responses within communities) or planned (i.e., governmental strategies). Responses are evaluated here in light of the predominant scientific and political discourse on vulnerability and adaptive capacity. This dominant vulnerability-adaptation approach has had a major influence on policy developments and research, though requires greater problematization and critical overview. Therefore, notions of intervention, trusteeship, power, and the use of the language of crisis are discussed. As an outcome of these deliberations, further and genuine empowerment is presented as a primary response to climate change impacts and adaptation challenges.

Based on Koivurova T, Tervo H, Stepien A (2008) Background Paper: Indigenous Peoples in the Arctic. Arctic TRANSFORM. This chapter has been restructured and expanded with recent scholarship and in-depth discussion on the concepts of vulnerability and adaptation.

A. Stepien · T. Koivurova · A. Gremsperger · H. Niemi (✉)
Arctic Centre, University of Lapland, Yliopistonkatu 8, 96300 Rovaniemi, Finland
e-mail: htervo@ulapland.fi

A. Stepien
e-mail: adam.stepien@ulapland.fi

T. Koivurova
e-mail: timo.koivurova@ulapland.fi

A. Gremsperger
e-mail: annagremsperger@gmail.com

E. Tedsen et al. (eds.), *Arctic Marine Governance*,
DOI: 10.1007/978-3-642-38595-7_4,

4.1 Introduction

The aim of this chapter is to present Arctic indigenous peoples in relation to the changing climate and marine environment. This overview is based largely on scientific and policy discussions on vulnerability and adaptation. The objective is not to present climate change impacts to indigenous peoples and adaptation options in detail, but rather to draw a general picture of climate change challenges alongside currently advocated approaches and ideas for adaptation. This allows for recommendations with an emphasis on empowerment as a primary response to holistically-viewed Arctic change. As this book focuses on the Arctic marine environment, issues typical for coastal, mainly Inuit, communities are highlighted.

The Arctic region is home to many culturally diverse indigenous peoples who have been exposed to a wide range of changes and pressures that include economic and cultural globalization, colonial policies, modernization, industrialization, major development projects, modern transport, and communication. These processes have brought about new industries, the emergence of mixed economies, acculturation, and social problems—producing both positive and negative outcomes for indigenous peoples. Climate change poses new and additional challenges for indigenous communities and facilitates creation of a 'total environment of change' (Moerlein and Carothers 2012).

Indigenous peoples have traditionally been adaptive and resilient to change. However, the scope and pace of climate change are seen as being beyond indigenous adaptive capacity—capacity that is already undermined by existing social pressures. The concepts of vulnerability, adaptive capacity, and adaptation must be viewed within the context of existing political and legal systems, which influence the ability of communities to cope with changing environmental conditions. In particular, traditional harvesting in marine areas is under threat, as these practices usually depend on sea ice and the availability of species. Culture, traditional ecological knowledge (TEK), economy, and health are bound with harvesting activities and are expected to undergo serious disturbances under climate change. Other major affected areas include community infrastructure and housing.

This chapter opens with an introduction to Arctic indigenous peoples and their situation in the face of climate change. Next, climate change impacts are presented, followed by a discussion on adaptive capacity and on autonomous and planned adaptation options, including an example from Alaska. This is juxtaposed with a critical discussion on the dominant adaptation discourse and concluded with certain options for moving forward.

The term 'indigenous' is used throughout the paper, although other terms—such as 'natives', 'numerously small peoples', or 'original peoples'—are also applied around the circumpolar North. As the United Nations (UN) Declaration on the Rights of Indigenous Peoples (UNDRIP 2007) uses 'peoples' as a leading term, this is preferred to 'people', 'populations', or 'groups'—although these terms can be used as synonyms without prejudice.

4.2 Arctic Indigenous Peoples

The Arctic region is home to many indigenous peoples with diverse cultural, social, economic, and historical backgrounds, including the Inuit of Russia, Alaska [United States (US)], Canada, and Greenland, Aleut, North American Natives (Athabascans, Gwitch'n, Métis), Sámi people of Fennoscandia, and numerous groups in Russia (*inter alia* Chukchi, Nivkhi, Sámi, Eveny, Evenky, and Nenets). Indigenous peoples constitute 10 % of the total Arctic population of 4 million (Nuttall 2000; IPCC 2007).

The definition of 'indigenous peoples' can be problematic. Indigenous peoples themselves are uncomfortable with definitions and emphasize (for instance in the UN Permanent Forum on Indigenous Issues) self-identification and commonsensical approaches (UNPFII n.d.). Nevertheless, international documents contain certain defining characteristics, as is the case with Special Rapporteur Martinez Cobo's report or the International Labour Organization (ILO) Convention No. 169 Concerning Indigenous and Tribal Peoples (1989) (Thronberry 1994; UNPFII n.d.). Indigenous peoples are defined as non-dominant groups descending from populations inhabiting certain areas before the time of conquest, colonization, or establishment of present state borders. These groups identify themselves as indigenous, have retained some of their own social, economic, cultural, and political institutions, and have maintained special relationships with their land and environment (ILO 1989, art. 1). The historic legacies and current positioning of indigeneity have bearing on climate change vulnerability, adaptation options, and empowerment (AHDR 2004).

Indigenous communities are often viewed from the perspective of *inter alia*, traditional livelihoods, cultures with close connections to the environment, and disadvantaged minority positions. While these characteristics have salience for vulnerability, other equally important factors include the development of mixed economies, technological progress, and colonial legacies. Policies for resettlement or assimilation and large-scale industrial and infrastructure projects have in some cases been imposed without consideration of indigenous social structures and within frameworks of colonial trusteeship and paternalism (Young 1995; Moerlein and Carothers 2012). Discriminatory practices, acculturation, usurpation of ownership and management of natural resources, and a rise in poverty are undeniable and persistent parts of the Arctic's historical and political landscape. Recent developments have aimed at enabling greater empowerment, participation, and the resolution of past injustices towards indigenous groups, though have in some cases resulted in producing new, complex co-management[1] and decision making frameworks that can be overwhelming for indigenous communities (Ford et al. 2010b).

Moreover, the Arctic no longer serves primarily as a harvesting ground for indigenous hunters, but has become a region that is increasingly crowded with other activities such as transport, shipping, resource extraction, commercial

[1] Co-management regimes are governance systems where resources, the environment, conservation, or land use are managed via joint (national, local, and indigenous) institutions and with strong participation from regional, local, community, and indigenous actors in policy making.

fishing, tourism, and nature conservation (AHDR 2004). Climate change is an additional factor affecting livelihoods that are already under pressure. Therefore, to fully understand climate impacts on indigenous peoples, these must be viewed within the context of modern indigenous societies and pertinent economic and political factors (Smit and Wandel 2006; Ford et al. 2006; Njåstad et al. 2009).

4.2.1 Traditional Harvesting and Mixed Economies

Understanding Northern mixed economies is a key component of assessing vulnerability and adaptive capacity (ACIA 2005; Pearce et al. 2012). Mixed economies are a combination of formal economies based on cash flows and traditional methods of acquiring food, clothing, and commodities. Harvesting and pastoralism remain important parts of Arctic livelihoods, with significance not only for health and food security, but also for culture and social ties. Today, however, complete reliance on harvesting is no longer possible due to changes in lifestyles, new standards of living, and declines in natural resources (Nuttall 2002; AHDR 2004; Ford et al. 2006; Poppel et al. 2007). Further, traditional subsistence is now subject to a range of regulations and resource management regimes, harvesting quotas, and access limitations (Nuttall 2000; ACIA 2005; Ford et al. 2006).

Financial resources support contemporary lifestyles, as well as modern harvesting in the Arctic, where costly technology such as snowmobiles, rifles, and GPS are now used (ACIA 2005; ICC 2005; IPCC 2007). In mixed economies, money is acquired through wage income (often through government jobs or resource extraction), governmental transfers, leisure hunting tourism, and selling of harvested goods and handicrafts (Nuttall 2000; AHDR 2004; Ford et al. 2010b).

Indigenous peoples' involvement in the formal, global economy creates both opportunities and challenges for adaptation governance. Extraction of hydrocarbons and minerals, tourism, military facilities, commercial fishing, and transport can offer the revenues required to support adaptation actions, but at the same time, make communities more dependent on world markets and undermine the long-term viability of traditional livelihoods. While many indigenous communities endorse industrial development, particularly when having legal control over land and resources, there is at the same time also strong opposition due to anticipated negative effects on environment and indigenous cultures (ACIA 2005; IPCC 2007; Bone et al. 2011).

An example of an external policy claimed to have bearing on indigenous mixed economies is the European Union's (EU) ban on the import and marketing of seal products (European Community 2009; Hossain 2012). The ban is seen to have resulted in adverse impacts to Arctic indigenous livelihoods and, consequently, the future adaptive capacity of Arctic communities. Although the EU's ban included an exemption for products coming from traditional indigenous seal hunting, it nonetheless caused a collapse of the market for seal products and thus limited potential Inuit income from traditional activities (European Parliament 2012). From the Inuit perspective, the EU trade regulation was an additional stressor in the Arctic's 'total environment of change'.

4.2.2 Challenges for Indigenous Societies and Culture

For centuries, newcomers to the Arctic have introduced new lifestyles, technologies, culture, and food, along with new policies and industries, which have increasingly affected all aspects of life in the Arctic (AHDR 2004; ACIA 2005; Pearce et al. 2009b).

Following such modernization and changes, unemployment and poverty have emerged as increasing phenomena in indigenous communities. Social and health problems such as alcohol and drug abuse, domestic violence, depression, and high suicide rates need to be taken into consideration when studying climate change impacts and vulnerability (AHDR 2004; ACIA 2005; Poppel et al. 2007).

For most Arctic peoples, family and communal ties are of great importance (Poppel et al. 2007), although traditional family structures are changing, with the nuclear family replacing the multi-generational model (AHDR 2004; Ford et al. 2006). Important parts of community identity and integrity, such as sharing and reciprocity systems and the position of elders, are in decline.

Problems with transmission of knowledge, culture, and language to younger generations are visible throughout the Arctic. For instance, changes to lifestyles, boarding schools, and educational systems have induced acculturation. They have, however, also created indigenous elites and economic opportunities (AHDR 2004; ACIA 2005; Furgal and Seguin 2006). Differentiated access to modern technologies has introduced additional inequalities (ACIA 2005; Ford et al. 2006; Pearce et al. 2010).

4.2.3 Political and Legal Framework

Political systems and indigenous governance vary throughout the Arctic. Past colonial policies aimed to assimilate indigenous peoples, but modern states' approaches have gradually shifted towards endorsing indigenous knowledge and values, protecting cultures, and introducing more participatory structures of governance (AHDR 2004). Today, trends of globalization, democratization, devolution, and human rights promotion are common in the Arctic region, although to varying degrees (for instance, in the Russian North, some opposing trends occur, especially regarding devolution). In areas where indigenous peoples constitute a majority, models of public government have been used (e.g., in Greenland and Nunavut), while in other areas, dual systems of governance prevail and indigenous and public governments coexist (e.g., Nordic Sámi Parliaments, the representative bodies of the Sámi). Powers have been redistributed to indigenous peoples in arrangements varying from granting legislative (e.g., agreements with Yukon First Nations) or decision making responsibilities, to providing consultation rights. In North America, land claims agreements have been a significant means of redistributing power and securing rights to territories and resources (AHDR 2004; Tennberg 2012).

Increasing recognition of human rights, including collective rights, such as self-determination, group rights, and land rights is visible both internationally and in the Arctic. Universal, European, and American human rights instruments [including the International Covenant on Civil and Political Rights (ICCPR 1966), International Covenant on Economic, Social, and Cultural Rights (1966), the European Convention on Human Rights (1950), and the Convention on Elimination of All Forms of Racial Discrimination (1966)] are generally applicable to indigenous peoples, although the acceptance of instruments differs among Arctic nations. The ICCPR has been ratified by all of the Arctic states and both Article 27 concerning minority rights and Article 1 concerning self-determination are seen by the Human Rights Committee as applicable to indigenous peoples, including their right to culture, traditional ways of life, land, and resources (HRC 1994, 1999/200/2002/2004). In addition, indigenous-specific instruments have been developed: in particular, ILO Convention No. 169 on Indigenous and Tribal Peoples in Independent Countries (ILO 1989), which has been ratified by Denmark and Norway (with Finland and Sweden considering ratification) and the UN Declaration on the Rights of Indigenous Peoples (UNDRIP 2007), which has been endorsed by all of the Arctic states except the Russian Federation. The goal of the currently negotiated draft Nordic Sámi Convention (see Koivurova 2008) is to harmonize and safeguard Sámi rights, including rights to marine resources.

Land and resource rights are critical for indigenous groups facing environmental change and resource extraction pressures (AHDR 2004). ILO Convention 169 (ILO 1989 arts. 14, 15) and UNDRIP give land and resources high priority, addressing recognition of ownership and possession of lands traditionally occupied, the requirement of indigenous free, prior, and informed consent, and states' positive obligation to identify and protect indigenous lands.

National developments in the Arctic have followed international trends. For example, the 2005 Finnmark Act in Norway gives the Sámi Parliament certain influence over management and titling of public lands in Norway's northernmost county (Bull 2008). For the most part, Arctic land and resources are publicly owned, although this picture is diverse and complex. In Alaska, for instance, 14 % of land is owned by Native Corporations, while in Russia, post-Soviet land ownership is often unclear due to complex legislation and legacies of Soviet collectivisation. Perhaps the most multifarious situation is in Canada, where a variety of treaties, land claim agreements (including the Nunavut Land Claim Agreement and Inuvialuit Final Agreement) create a patchwork of co-management and ownership, with situations differing greatly between provinces and territories (AHDR 2004).

4.2.4 Arctic Cooperation

Both the Arctic Council and Barents Euro-Arctic Region work to engage indigenous peoples and perform substantial work on the issues of the climate change and adaptation (Koivurova and Hasanat 2009; Hasanat 2010). Since 1996, indigenous organizations are to be fully consulted in the Arctic Council as permanent participants, a unique status within the international context. By participating in all of the activities of the Arctic Council, including working groups, indigenous organizations have been able to raise their concerns and emphasize the importance of their traditional knowledge, as with the Arctic Climate Impact Assessment (ACIA) and the Arctic Marine Shipping Assessment (ACIA 2005; AMSA 2009). The salience of the permanent participant model is visible in the demands to adopt a similar solution within the Barents Euro-Arctic Region (Koivurova and Heinämäki 2006; Koivurova 2011b).

4.3 Climate Change Impacts, Stressors, and Indigenous Vulnerability

Rapid non-linear environmental changes are expected to push the social-ecological systems of the Arctic beyond tipping points (Pearce et al. 2010; Carmack et al. 2012). Following the ACIA and the International Polar Year 2007–2009, a surge of research on Arctic community vulnerability and adaptation was seen (Ford 2009). The UN Framework Convention on Climate Change (UNFCCC 1992) has introduced vulnerability language by calling for attention to particularly vulnerable regions such as the Arctic (UNFCCC 1992, art. 3.2; Smit and Wandel 2006). Vulnerability approaches have dominated the scientific and political discourse on climate change impacts (Pearce et al. 2009b; Tennberg 2012; Cameron 2012).

Vulnerability can be understood as a function of the "exposure of the community to climatic conditions and adaptive capacity to deal with such conditions" (Smit and Pilifosova 2003) with both human and bio-physical determinants playing a role at multiple spatial and temporal scales (see also Ford 2009). When considering vulnerability, researchers examine who is vulnerable and how, to what stresses, in what way, and what the local capacity to adapt is. Social, economic, political, and ecological features (experienced diversely at a community level) are assessed. Understanding of climate change impacts has been influenced by indigenous observations and TEK, supplementing scientific observations (Smit and Wandel 2006; Pearce et al. 2010).

Climate change directly impacts traditional livelihoods and has indirect effects on indigenous societies. For Arctic indigenous peoples, climate change affects hunting, whaling, and fishing activities as sea ice, weather conditions, and the availability and health of harvested species change – all crucial aspects of successful and relatively safe harvesting. Furthermore, changes or decreases in harvesting activities have implications for indigenous economies, societies, cultures, and health. Not all observed and predicted climatic changes are necessarily negative, but overall impacts are seen as having increasingly adverse effects on indigenous livelihoods (ACIA 2005; Moerlein and Carothers 2012; see Fig. 4.1 for a summary of impacts).

4.3.1 Primary Impacts on Livelihoods, Harvesting, Health, and Infrastructure

Traditional harvesting is still seen as a necessary and important element of Arctic indigenous lifestyles and cultures (Ford et al. 2006, 2010b; Pearce et al. 2010).

Sea ice, on which many traditional harvesting activities depend, is an integral part of life for many Arctic peoples and is tied to some of climate changes' major impacts. The health, location, and abundance of harvested animals are adversely affected by disappearing sea ice, on which species depend, as well as changing

Elements of climate change	*Harvesting/ availability of resources*	*Infrastructure and transport*	*Economy*	*Society and culture*	*Health*
Changing ice and snow conditions	-Less access to certain, particularly ice-dependent, species -More polynyas make fishing and duck hunting easier -Changes in harvesting cycles -Difficulties building igloos	- Thinning ice reduces available travel time and increases risks - Challenges for ice-dependent communities to maintain connections with population centres - Better sea access for coastal settlements	- Travelling and harvesting become more expensive - Replacing country food with imported food creates additional expenses	- Ice, an integral part of indigenous cultures, is lost - Risk of loss of hunting cultures - Sharing systems may decline	- Possible rise in accidents during harvesting and travel on ice - Loss of country food, which is often healthier than imported products - Increased risk of health problems such as diabetes, obesity, or cardiovascular diseases
Warming waters	-Sea mammals and fish species shift locations and change in numbers	- In some cases, necessary travelling of longer distances to reach resources increases costs and hazards	- Loss of important food sources (as above)	- Decrease in hunting and fishing success and/or frequency of harvesting activities	- Less country food (as above)
Thawing permafrost, coastal erosion, rising sea level, and decreasing river run-off	-Decreased fish and migratory duck populations	- Damage to houses, roads, power lines, airstrips, water supplies and sewage systems - Community relocation - Damage to permafrost or ice-based food storage - Changes to water supply	- Rising costs of infrastructure maintenance - High costs of settlements relocation - Higher costs of food preservation	- Damage to sacred sites - Estrangement to the environment and changing landscapes - Traditional knowledge no longer applicable	- Possible rise in accidents

Fig. 4.1 Selected impacts of climate change on coastal indigenous communities. Only the most vital of many concurrent factors have been mentioned and taken out of the broader social-economic-political context. Table is based partly on tables available in ACIA 2005

General warming, intensification of extreme weather events, and other impacts	- Mitigation options limited due to impacts to terrestrial species (e.g., reindeer herding affected by snow and river conditions) - Worsened health of harvested animals - Shorter and/or less frequent harvesting trips	- Uncertainty of the weather, more storms - Houses not prepared for warmer average temperatures - Damage to infrastructure and housing	- Rising costs of infrastructure maintenance - Savings on heating - More jobs due to development of agriculture, tourism, shipping, and resource extraction - Increased need for expensive technologies to ensure travel safety	- Traditional knowledge no longer useful - Migration of settlers from the south may cause acceleration of acculturation - Effects on transfer of traditional knowledge and skills to younger generations	- Invasive insect and animal species bringing new diseases - Inadequate housing - UV radiation - Rise in accidents - Less country food (as above)
Autonomous adaptations	- Changing timing and frequency of harvesting - Harvesting new species - Travelling longer distances - Use of modern technologies - Carrying more equipment	- Acquiring information on changing snow/ice conditions - Ad-hoc construction strengthening - Changing modes of transportation	- Supplementing country food with store food - Engaging to a greater extent in wage economy		
Planned adaptations	- Harvester support programmes - More flexible regulatory frameworks responsive to changes in environment	- Need for better information - Spatial planning adjusted to changing environments - Funding for infrastructure renewal - Relocation - Improving search and rescue	- Job creation in the North - Community freezer programmes - Facilitating cash acquisition via traditional harvesting	- Supporting the transfer of traditional knowledge - Cultural programmes	- Better information on health impacts - Health programmes responsive to additional stressors - Better surveillance
General actions	- Need for general empowerment - Better designed decision making processes - Research on adaptation planning and implementation, sensitive to indigenous perceptions - Human rights-based approach - Integration of various policy fields				

Fig. 4.1 Continued

ocean productivity, altered water salinity due to melting ice and river run-off, and the emergence of new diseases and insects (ACIA 2005; ICC 2005; IPCC 2007). These changes additionally result in worsened quality of meat and hides and increased reliance on store food, which can be more costly and have negative dietary implications. Other climate change impacts to traditional harvesting activities range from changes in hunting logistics (e.g., impossibility of igloo construction) to communities being cut off from hunting areas owing to later autumn freeze-up and earlier spring melt (Nuttall 2002; Abate 2007; Poppel et al. 2007).

Hunting is seen as an increasingly dangerous activity as ice undergoes various transformations, such as the occurrence of more polynyas and areas of unusually thin ice. Under climate change, travelling becomes more dangerous, less frequent, and often requires either shorter or much longer distances, depending on local sea ice conditions and availability of harvested species. Hunters increasingly rely on modern technologies. Climate change and changing ice conditions also influence other livelihoods practiced by indigenous coastal communities such as reindeer herding and hunting and fishing (ACIA 2005; Abate 2007). Less commonly, communities' situations may improve; for example, due to newly gained access to large fish stocks (ACIA 2005; Ford et al. 2006).

Impacts on transportation in the Arctic, as noted in relation to harvesting activities, affect connections between communities and population centres (Pearce et al. 2010). Thawing permafrost and coastal erosion cause roads to disintegrate. In the spring and autumn, routes based on ice and snow cover are affected and in the summer, rivers may be un-navigable due to decreases in run-off. On the other hand, summer sea access for many coastal communities may improve (ACIA 2005; ICC 2005; IPCC 2007).

Construction of buildings and infrastructure such as sewage systems, airstrips, power lines, and roads built on permafrost and eroding coastlines are endangered or have already experienced climate change impacts, due not only to thawing permafrost but also to enhanced storm strength, river run-off, and flooding (ACIA 2005; IPCC 2007). Some communities may be confronted with the prospect of relocation (ICC 2005; Ford et al. 2006; Pearce et al. 2010).

For Arctic indigenous peoples, the health impacts from climate change are multiple and complex. Experts predict the introduction and expansion of new diseases and allergies due to the arrival of southern plant, insect, and animal species as well as changes in diet that may contribute to obesity, diabetes, and cardiovascular disease. Increases in sunburns due to UV radiation are expected (Furgal and Seguin 2006; ACIA 2005; ICC 2005; IPCC 2007). Freshwater supply and traditional methods of food preservation are also impacted by climate change, and the probability of contamination due to damage to pipelines and permafrost-based waste containers increases (ACIA 2005; ICC 2005; IPCC 2007). Conversely, health benefits are also possible, such as a reduction in winter mortality and cold-related injuries (IPCC 2007).

4.3.2 Impacts on Northern Economies, Societies, Cultures and Health

The formal, cash economy is an alternative to traditional harvesting and is an essential part of mixed economic systems, though many Arctic regions today experience high levels of unemployment and poverty (ACIA 2005; ICC 2005; IPCC 2007). Climate change may increase economic hardship due to rising living costs (infrastructure maintenance, harvesting costs, transport, increased demand for store food), despite expected savings in heating and insulation (ACIA 2005; IPCC 2007). Infrastructure damage by climate change will adversely impact access to communities, transport costs, trade, development of tourism, and job creation, endangering aspects of the indigenous formal economy (ACIA 2005; IPCC 2007; Pearce et al. 2010).

At the same time, increased agricultural and fisheries productivity and opportunities for resource extraction may create more jobs and promote investment in the region, although many of the jobs offered to indigenous communities are likely to be of low quality due to both disparities in education and training and the character of positions (ACIA 2005; IPCC 2007). Moreover, an influx of southern employees to fill new positions might increase acculturation pressures (AHDR 2004).

Changes to the Arctic landscape and harvesting activities are expected to influence social structures, affect communities' distinct ways of life, and even endanger survival of some groups as distinctive peoples (ACIA 2005; Bravo 2009). Although, as discussed in the previous section, social and economic impacts currently pose a greater threat to indigenous societies than climate change itself, climate change impacts are expected to increase the pressures and speed of certain negative processes (ACIA 2005; ICC 2005).

Social and cultural features perceived as vulnerable in the context of the 'total environment of change' are numerous, and many are especially connected to climate-induced impacts to traditional harvesting, impoverishment, out-migration, and changing landscapes (IPCC 2007; Pearce et al. 2010). Such social and cultural features include social and family ties, traditional sharing systems, and the authority of elders. Historical, sacral, and cultural sites, such as cemeteries, are threatened by the environmental impacts of climate change (AHDR 2004; ACIA 2005; Ford et al. 2006). Changes in landscape and declines in sea ice—a cultural, social, and economic component of indigenous coastal communities—affect the sense of place and belonging (ICC 2005). Limitations to traditional harvesting also have adverse impacts on self-esteem, tied to notions of lack of opportunities. Climatic pressures on culture, in connection with other social pressures, are expected to amount to a level of social crisis and to have effects on mental health. Critical impacts include depression, alcoholism, drug addiction, permanent unemployment syndromes, and rising suicide rates (ACIA 2005; Furgal and Seguin 2006).

Traditional knowledge provides community members with an understanding of their environment and binds them with nature and each other, but is claimed to be less applicable to an environment undergoing new and unprecedented changes (AHDR 2004; Abate 2007). Coupled with a decrease in harvesting activities, this causes alienation to land and disruptions to community relationships and intergenerational transfers of knowledge, values, and culture (ACIA 2005; IPCC 2007; Moerlein and Carothers 2012).

4.4 Adaptive Capacity and Proposed Responses to Climate Change

The hazardous impacts of climate change are seen as already exceeding national and community capacities to cope through use of traditional strategies and behaviours (Smit and Pilifosova 2003). There is a need for better and more feasible adaptation responses and options, both of an autonomous and planned character (Carmack et al. 2012; Smit and Wandel 2006; Hovelsrud and Smit 2010).

4.4.1 The Concepts of Adaptation and Adaptive Capacity

Adaptation can be described as a "process, action, or outcome in a system in order for the system to better cope with, manage or adjust to some changing condition, stress, hazard, risk, or opportunity" (Pearce et al. 2010). Adaptation is not a term specific to climate change, and generally accompanies all processes of human learning and development. It is a multi-stage, problem-solving process in response to situations where normal responses cease to be sufficient, but it is also a conflict-inducing challenge for social organization. Climate adaptation policy can be viewed as transition management of changes and impacts that are inherently uncertain (Smit and Wandel 2006; Tennberg 2012).

Adaptation may be either autonomous or planned (Smit and Wandel 2006). In the former, communities and societies adapt to changing environmental conditions on a continuous basis. Conversely, planned climate change adaptation measures, such as spatial planning or building code adjustments, are still in the early phases of development and implementation. Adaptation is neither a linear nor a homogenous process, with some communities or sectors adapting better or faster than others (Keskitalo 2010; Tennberg 2012).

Discussion on adaptive capacity has been instigated by the International Panel on Climate Change (IPCC), where adaptive capacity is defined as the "ability or potential of a system to respond successfully to climate variability and change, including adjustments in behaviour and in resources and technologies" (IPCC 2007). Concepts closely associated with adaptive capacity include coping ability, management capacity, stability, and flexibility, as well as resilience, conceptualized as the capacity of ecosystems and populations (or socio-ecological systems) to recover from stress or change (Tennberg 2012).

All elements of the vulnerability-adaptation-resilience scientific model are seen as place-specific and dynamic, and are subject to on-going alterations that are dependent on policies, environmental variations, individual and community values, and changing conditions within communities (Smit and Wandel 2006; Tennberg 2012).

Assessments of both vulnerability and adaptive capacity depend on knowledge of past experience with climatic conditions, traditional responses to climate variations, future climate change projections, and understanding of non-climatic factors (Pearce et al. 2009b). Adaptive capacity is dependent on factors such as managerial abilities, infrastructure, institutional and political environments, and access to financial, technological, and information resources. Kinship and social networks, social capital, strength of local institutions, and flexibility are also of importance. Broad community support for adaptation policies, willingness to change, and good linkages to external governance institutions are often necessary for successful adaptation. Socio-environmental systems need to be analysed holistically, on not only on a community level but on all interdependent scales (e.g., regional, national, international) (Smit and Wandel 2006; Carmack et al. 2012). Enhancing adaptive capacity requires addressing pertinent locally identified vulnerabilities and involving local stakeholders (Smit and Pilifosova 2003; Carmack et al. 2012).

4.4.2 Autonomous Adaptations

Research has shown Arctic communities to be highly adaptive and resilient, but also increasingly overwhelmed by environmental changes (AHDR 2004; Ford et al. 2010b; Pearce et al. 2010). Arctic indigenous peoples are accustomed to natural variability and unpredictability. For example, fish stocks and wildlife location and availability or sea ice extent can change significantly from year to year. Communities have developed flexible social structures (i.e., group size, multiple means of livelihood), which allow them to adapt without significant cultural loss (ACIA 2005). Nonetheless, the current and projected rate and scope of climate change in the Arctic presents a significant challenge in comparison to earlier experienced phenomena (ACIA 2005; Pearce et al. 2009b). Moreover, climate change will impact communities that are already affected by social, economic, cultural, and political pressures (ACIA 2005; AHDR 2004; Pearce et al. 2010). For example, inequality of access to traditional and monetary resources weakens social networks, and weaker social network in turn affect transmission of traditional skills and knowledge to younger generations (Moerlein and Carothers 2012; Ford et al. 2006; Bone et al. 2011).

A number of autonomous climate adaptation strategies are already being applied in the Arctic, although most are reactive and individual in nature. Traditional adaptation mechanisms relevant for climate change adaptive capacity include changing the timing and frequency of hunting activities, harvesting new species, and lengthening travelled distances to follow moving animals (Ford et al. 2006; Pearce et al. 2009a, 2010). While these traditional mechanisms already exist, time is still needed for learning new harvesting and livelihood practices in an altered environment. For example, hunters must prepare themselves for lengthier ventures, impossibility of igloo construction, and more frequent storms by taking additional equipment and carefully planning journeys. Moreover, hunters and herders must acquire current information on weather and ice conditions, utilize different modes of transportation, travel routes, hunting areas, and species harvested. Modern technologies (e.g., snowmobiles, GPS, satellite phones, all-terrain vehicles) play a significant role in such autonomous adaptations. Technological adaptations, however, make hunting also more costly (ACIA 2005; ICC 2005; IPCC 2007) (See Fig. 4.1 for the summary of impacts and adaptations).

4.4.3 Adaptation Planning and Governance

Autonomous adaptation measures at the community level need to be considered in a broader governance context and coupled with adaptation planning. Governmental programmes are seen as crucial for adaptive capacity-building and preservation of indigenous hunting cultures. Adaptation planning is hoped to further social and economic well-being by insulating communities from threats and enhancing realisation of opportunities connected with climatic and economic changes (Pearce et al. 2010; Lukovich and McBean 2009; Tennberg 2012). Although the field of adaptation planning appears to be underdeveloped, many Arctic municipalities and regions are already designing climate change adaptation plans (Tennberg 2012; Hovelsrud and Smit 2010).

Adaptation governance does not occur in a political, economic, or administrative vacuum and is closely tied to existing political and administrative structures and practices (Tennberg 2012). In addition to climate change impacts, many adaptation policy options simultaneously address pressing economic, social, cultural, and health concerns pertaining to northern communities and may bring near-term positive benefits even before the occurrence of major climate change impacts (Ford et al. 2010b; Pearce et al. 2010). Addressing climate change may be impossible in communities facing serious social problems (ACIA 2005; IPCC 2007; Smit and Wandel 2006). Due to the perceived urgency of climate change impacts, locating adaptation initiatives within existing decision making procedures is often advised; for instance, within established resource management systems, disaster preparedness planning, and sustainable development programmes. Adaptation initiatives should therefore be incremental, modifying rather than replacing current structures (Smit and Pilifosova 2003; Smit and Wandel 2006).

Proposals for culturally appropriate and acceptable adaptation strategies include: creation of forums between government decisionmakers and indigenous leaders; promotion of assessment, data gathering, and information, and integration or mainstreaming[2] of climate change adaptation into policies (e.g., land-use plan-

[2] Mainstreaming is a policy concept whereby a certain issue is considered across different policy areas so that each addresses aspects of the issue concerned.

ning, fish and wildlife management, water management). In order to successfully implement adaptation actions, ownership, use, and access to land and resources must be settled—an issue that continues to be a major challenge in many parts of the Arctic.

Adaptation strategies must meaningfully involve indigenous peoples at the outset and secure their participatory rights. Policies should be designed in a cooperative manner and founded on cooperative research to be both legitimate and potentially successful. Without local engagement, policies would be either based on inaccurate goals or meet with resistance from within communities (ACIA 2005; Ford et al. 2010b). This may require authorities to enhance the financial, human, and technical capacity of indigenous organizations and governments in order to enable them to draw up their own visions of adaptation (ACIA 2005).

Options for adaptation actions are numerous (Wenzel 2001; Ford et al. 2010b). For instance, programmes may support TEK transmission and training in use of modern technologies. Financial support for adaptation (e.g., harvester support or community freezer programmes) should be provided alongside enabling more cash acquisition through traditional harvesting via sport hunting, tourism, and commercial sale of hunting outcomes (Ford et al. 2006). Safety threats require: enhanced search and rescue capabilities, forecasting of weather, ice conditions, and hazards, and protection of major infrastructure. Integrated land use planning, moving new construction away from endangered areas, and promoting awareness of climate change are seen by researchers as vital measures to take (ACIA 2005; IPCC 2007; Tennberg 2012). Health issues are identified as particularly urgent. Researchers have called for more surveillance, establishing early warning systems, and culturally specific health assessments (Ford et al. 2010b).

Adaptation planning must be a dynamic process based on the on-going production of knowledge. Governments, institutions, researchers, and communities must perpetually monitor, evaluate, and redesign adaptation strategies as underlying conditions change. Consequently, legal and regulatory arrangements should also be incremental and flexible in nature (ACIA 2005). For instance, reassessment of hunting and fishing regulations (e.g., spatial limitations or quotas) as well as conservation laws may prove necessary as the conditions upon which they are premised are radically altered (Dahl et al. 2001; ACIA 2005). In more heavily bureaucratized Arctic governance environments, adaptive capacity depends on institutional flexibility. Despite the progress of co-management regimes, there is still a need for more flexible, multi-level governance schemes with the ability to serve as arenas of joint decision making, conflict resolution, and information sharing. Evolution of such flexible systems requires co-production of knowledge, exchange of best practices, local and decentralized decision making, and acknowledgment of traditional rights. However, governance changes are not expected to be smooth, as so far the prior reconstruction of governance in the North has been a difficult undertaking underpinned with inter-group conflicts (Dahl et al. 2001; Ford et al. 2010b).

In order to enhance adaptive capacity, education, and exchange of information on vulnerability, behavioural adaptation options are advocated with rising focus on indigenous youth and trans-generational transmission of culture and knowledge (Magga et al. 2011; Carmack et al. 2012). Diverse, audience-specific approaches and materials integrating indigenous history and culture are needed for education, communication, capacity-building, outreach, and training (Njåstad et al. 2009; Pearce et al. 2010; ACIA 2005).

Both the viability and transmission of TEK is increasingly seen by indigenous politicians and researchers as central to adaptation and resilience (Magga et al. 2011; Ford et al. 2010b). Integration of TEK and community observations with scientific knowledge and policy making is seen as crucial for production of applicable research and effective policies. Involving local people at all stages of research projects and taking into account their observations is intended to make research more holistic, participatory, and relevant for communities (Pearce et al. 2009a; Berkes and Jolly 2001; Bravo 2009).

Research has played and is to play a major, if not primary, role in the debate on Arctic climate change (Cameron 2012; Bravo 2009). Assessment of impacts in regard to long-term scenarios, locally-focused analysis of specific adaptation options, and comparative studies are encouraged (ACIA 2005; Smit and Wandel 2006; IPCC 2007) as well as studies on already implemented measures and adaptations processes (Tennberg 2012; Ford et al. 2010b).

Calls have been voiced for a comprehensive Arctic strategy dedicated to human security and human rights (Ford et al. 2010b; Lukovich and McBean 2009). There is a movement towards inclusion of human well-being, adaptation, and vulnerability in Arctic security discourse, or, alternatively, framing these issues within the language of human rights. For example, food security, when expressed as a right to food, produces a duty on the side of public authorities. From an adaptation standpoint, governments, in light of their human rights obligations, would need to tackle inequalities and provide resources for adaptation (Adger 2010; Ford et al. 2010a).

4.4.4 Barriers to Adaptation

Both autonomous adaptation and governmental adaptation planning have limitations. Major constraints include sociopolitical values, institutional capacity, and lack of economic, technological, or information resources (Ford et al. 2010a, 2010b). Local governments often lack the resources necessary to conduct adaptation activities (e.g., in the Russian Federation; see Tennberg 2012), and such deficiencies are even more visible in indigenous communities. Adaptive capacity depends on financial resources (e.g., ability to purchase equipment and engineering responses to infrastructural damages) and the ability to use modern technologies. Social values, too, which are often neglected in planning, are crucial and include the use of social networks for sharing risks, readiness to adjust behaviour, and transmission of traditional knowledge (Ford et al. 2006; Pearce et al. 2009b, 2010). Developing adaptation responses can also result in conflicts between diverse political, economic, legal, and conservation interests. Engagement in adaptation governance and planning also depends on different perception of risks by public authorities and communities (Tennberg 2012; Keskitalo 2010; ACIA 2005).

Indigenous people have become more dependent on the outside world, which increases their adaptive capacity, but, paradoxically, also amplifies their vulnerability to the impacts of climate change. A settled way of life may improve educational opportunities and increase material standards of living, but at the same time, people may become dependent on mechanized transportation and fossil fuels.

Today, some past adaptation options, such as reconstruction or community migration are either unavailable, limited, or very costly. Modern, permanent settlements have elaborate infrastructure, raising the costs of both maintenance and resettlement. For example, engineering solutions to climate change impacts on infrastructure and constructions in Canada's Northwest Territories were estimated at 200-420 million CND and relocation of one community of Inupiaq was predicted to cost 50 million CND (Ford et al. 2010b; Pearce et al. 2010). However, even where financial resources are available, relocation is often seen as an unacceptable option, particularly due to past indigenous experiences of forced and semi-forced population movements such as in the 1950s–1960s in Canada and in the Stalinist Soviet Union. Relocation also results in social problems or deepening of cultural loss (ACIA 2005; IPCC 2007; AHDR 2004).

Education is a priority issue in the Arctic if autonomous and planned adaptations are to be successfully implemented (Lukovich and McBean 2009). Weakening of TEK and its transmission to younger generations has already been observed and not only increases vulnerability by affecting autonomous and planned adaptations, but also adversely impacts indigenous cultural development (ACIA 2005; Ford et al. 2006; Tennberg 2012). Young people are not always interested in traditional livelihoods and there is a rising knowledge gap between older and younger generations. New methods of knowledge transfer, such as cultural programmes and trainings, and documentation of knowledge are seen by some as a remedy for this (Bone et al. 2011; Moerlein and Carothers 2012). Young people have also been viewed as not fully prepared to assume the responsibilities of older generation in indigenous politics and resource co-management (Ford et al. 2010b; Pearce et al. 2010).

Box 4.1 Alaskan experiences with adaptation governance

Over the past 50 years, Alaska has been warming at more than twice the rate of the rest of the US (USGCRP 2009). The US Government Accountability Office has reported that most of Alaska's 200 native villages have been affected by flooding and erosion, likely in relation to climate change, and relocation options have been considered for several communities (GAO 2004).

In general, experiences with adaptation governance around the circumpolar North are limited, and that is also the case for Alaska. In 2010, the Alaskan state government prepared a report on climate change impacts and possible adaptive action (Adaptation Advisory Group 2010), although the document has not been fully endorsed by state authorities. The report analyzes climatic changes and expected impacts (with conclusions similar to those presented earlier in the chapter) focusing particularly on fisheries, infrastructure, and traditional livelihoods. It contains adaptation options designed to reduce the vulnerability of Alaska's natural and human systems to climatic changes and examines potential economic opportunities (e.g., due to a longer summer tourism season or increased navigation).

Having future adaptation options as a point of reference, the strategy operationalized the notion of ecosystem services, seeking to sustain ecosystem services in order to meet essential needs for food, water, renewable resources, community stability and safety, and cultural well-being. Five areas for adaptation covering broad spectrum of governance were identified: fish and wildlife, fisheries management, wildland fire, freshwater management, and invasive and eruptive species (Adaptation Advisory Group 2010).

Actual implementation of adaptation policies in Alaska remains somewhat limited, although action has been initiated at the state, national, and local levels. Some funding, primarily from the US federal government, has been provided for further research on future impacts and for community adaptation. The US Department of Interior established the Alaska Climate Science Center (part of the US Geological Survey), which finances research on, *inter alia*, impacts on communities and models for future climate change-related phenomena such as coastal erosion (AK-CSC 2012). In 2011, the Alaska Native Fund was introduced to support, among other areas, the protection of cultures and communities from future climate change impacts. Community capacity-building or trainings are among the supported actions (MFPP 2012). Other initiatives include the university-based Marine Advisory Programme (MAP) funded by the State of Alaska and US Department of Commerce, which seeks to provide Alaskan coastal communities with advice on adaptation to the changing environment. MAP proposes, for example, utilizing new species in hunting grounds, engaging elders to teach youth adaptability, avoiding construction on frozen subsurfaces, rerouting transportation routes to avoid vulnerable permafrost areas, and identifying more secure sources of water (MAP website). Some Alaskan municipalities have also taken action to implement adaptation planning projects (see Feifel 2010 on the case of Homer).

4.5 Criticism Towards Vulnerability and Adaptation Approaches

Vulnerability and adaptation discourses are predominant in research and policy making regarding climate change impacts on Arctic indigenous communities. Participatory research and governance, co-management, and adaptive capacity-building actions are highly promoted. There is, however, also criticism to such approaches—to notions and narratives they produce and to the resulting exclusion of certain issues from the scientific and political discourse.

4.5.1 Crisis Narrative and Resilience Language

Bravo (2009) sees the climate change debate in the Arctic as dominated by resilience/instability/crisis narratives. These narratives seek to make abstract climate science tangible and morally relevant by utilizing the plight of vulnerable indigenous peoples, and therefore are charged with moral overtones and emotional rhetoric. Indigenous peoples, however, often reject being presented as populations on the edge of extinction and cultural collapse. Vulnerability discourses are seen as solidifying the victimization of indigenous communities (Lindroth 2011; Niezen 2003). Conversely, living from the land means an on-going process of negotiating one's position in a changing environment (Forbes and Stammler 2009), and adaptation itself is a crucial part of life, not a catastrophe. Still, developing an alternative climate change narrative free of crisis dimension appears to be a difficult task (Bravo 2009).

Consequently, the language applied in adaptation discourse is viewed as disturbing by some researchers. Ideas coming from natural science are applied to complex human societies and relationships in a highly politicized Arctic milieu. For example, adaptation and resilience debate is criticized for the introduction of reactive and path-dependent thinking and the notion of permanent instability—where the aim is to govern something as indeterminate and uncertain as climate change (Tennberg 2012; Carmack et al. 2012). Moreover, there is a paradox in the fact that at the same time that adaptive capacity of indigenous communities has decreased due to modernization, dependence on public transfers, and so-called 'technology-induced environmental distancing'[3] (Bone et al. 2011), the same modern technology and dependence on financial resources, wage economies, and state support are among proposed adaptations (Cameron 2012).

These applied terms, by taking a singular focus, can also reduce the visibility of other related problems. Adaptation discourse often averts issues of colonial legacies and injustices, not only making adaptation strategies potentially ineffective, but also possibly strengthening processes leading to dispossession, inequality, and further political marginalization (Cameron 2012; Bravo 2009).

Cameron (2012) sees a danger of confining our understanding of indigeneity in the context of climate adaptation to what is *local* and *traditional* (see also Tennberg 2012). As a result, in climate change debates indigenous groups are often excluded from discussions on issues located outside the *traditional* and *local* (although these notions are of major importance), such as hydrocarbon and mineral extraction, shipping, sovereignty, or militarization of the region (e.g., ICC

---

[3] 'Technology-induced environmental distancing' means that the more technology is applied in daily lives, the less the individual and community understand and interact with surrounding environment.

2009). The role of permanent participants in the Arctic Council can be seen here as a positive example, a step outside of the local/traditional limit and endorsing indigenous peoples as equal and capable actors and not only victims of change (Koivurova and Heinämäki 2006). Such a participatory and empowering approach should be more pronounced in governance structures.

4.5.2 Adaptation Governance as Intervention

Many adaptation policies are presented as new forms of state intervention—usually of a neoliberal, i.e., market-based and technical character. These mirror past well-intentioned, but currently criticized, interventions into people's lives and relationships with the environment. Adaptation ideas are often limited to technical approaches and economic and market logic, including cost-benefit, cost-effectiveness, and efficiency calculations. Economic concerns, such as high relocation and infrastructure costs, harvesting expenses, unemployment, and low wages, have indeed dominated the debate (Moerlein and Carothers 2012).

To make management of climate change impacts possible, risks need to be construed as manageable and governable. Communities should be encouraged to take responsibility and urgently adjust to changes (Tennberg 2012; Slezkine 1996; Cameron 2012). Due to notions of crisis and urgency, the *will to improve* indigenous lives (i.e., secure their well-being) may take a form of trusteeship, resembling colonial way of thinking (Cameron 2012; Escobar 1995; Li 2007).

The consequence of presenting environmental change and adaptation as purely technical problems is that it depoliticizes climate change and makes it subject to expert direction, often confined to existing systems of governance and excluding related issues that lie outside of purely technical discussion (such as resource extraction, commercial fisheries, or colonial legacies) (Cameron 2012; Bravo 2009). Governmental routines, including participatory procedures, may be seen as ways in which indigenous resistance to interventions is de-politicized, managed, and limited to bureaucratic forms (Tennberg 2012). External experts claim stakeholder and stewardship status in regard to resources, governance, and the environment of the region (Bravo 2009; Nuttall 2002).

4.5.3 Using Traditional Knowledge

There are also critical voices regarding the use of TEK in predominant research and policy making. There are problems with translation, representation, and de-contextualisation of TEK as something produced in interactions between communities and researchers where information based on TEK is taken out of holistic spiritual systems and local practices and used as data for western science (Cameron 2012). Often, terms and ideas used in research are not in line with the actual perceptions in communities and are rather abstract, exogenous, managerial, or scientific concepts like *TEK*, *reindeer management*, or *climate change* (Forbes and Stammler 2009; Moerlein and Carothers 2012; Njåstad et al. 2009). Moreover, climate change perceptions are shaped not exclusively by community observations, but also by interpretations and ideas coming from outside. Views of natural hazards and their impacts and likelihood are socially constructed through research, politics, and media, and therefore, economic, social, and political conditions have bearing on perception of and responses to such impacts (Ford et al. 2006; Bravo 2009).

Criticism of vulnerability-adaptation approaches, at least within this chapter, is not intended to dismiss discourse on the need for adaptation to a changing environment. Even if narratives on cultural collapse often appear exaggerated, indigenous communities indeed depend on traditionally used resources, with broader implications for society and culture (Bravo 2009). Vulnerability-adaptation research and recommendations form the only relatively coherent and practically applicable proposition so far for responding to climate change. However, there is a need to identify and problematize the shortcomings of the vulnerability-adaptation-resilience approach—as it has been directing policy making—and to supplement it by a broader view of human communities and cultures than that offered by human ecology studies (Cameron 2012). Consequently, as the debate on community adaptation cannot be confined to technical terms, community participation and empowerment of indigenous peoples need to play central roles.

4.6 Empowerment as a Primary Response

An overview of climate change impacts, vulnerability, and adaptive capacity studies, autonomous and planned adaptations and barriers, as well as criticism to current climate change discourses, all lead to the conclusion that further and more genuine empowerment need to be a primary response to climate change and to interconnected socioeconomic-political stressors, even before planned adaptation actions are taken. This may offer Arctic governance a safe passage between the need for active adaptation policies and the danger of new state interventionism and paternalistic policies. Without genuine empowerment, truly inclusive decision making and meaningful participation, adaptation, and adaptive capacity-raising efforts may prove ill-conceived, ineffective, costly, or turn into a continuation of a colonial trusteeship project. The voices critical of the adaptation and vulnerability discussion also highlight that research, even if it incorporates traditional knowledge, and science-based policy making cannot offer an ultimate response to climate change challenges. In order to be effective and legitimate, these need to be combined with the participation of local actors. Furthermore, meaningful participation in decision making has to be based on capacity-building.

Any deliberations on policy options for adaptation for indigenous peoples need to begin with the needs, perspectives, and perceptions of indigenous peoples and communities. Advocated holistic, community-based, and participatory research focused on application of traditional knowledge (Cochran et al. 2009) cannot take place without empowerment of peoples and communities. Empowerment, to be meaningful, should refer also to challenging issue areas such as self-determination, sovereignty, lands, and resources, and not only to hunting practices and cultural activities (Cameron 2012).

Empowerment allows communities to participate to a greater extent in research activities and have greater control on how their traditional knowledge is produced and used. In decision making, empowerment also helps guarantee effective participation. Communities that are more autonomous in terms of financial and human resources can be more active in the design and implementation of strategies and actions. Resolved land rights issues establish clearer frameworks for discussion and decisions on how policies are implemented. They allow communities to plan better and provide those engaged in harvesting with more certainty of their access rights and flexibility when the adjustments in light of changing environment are needed. Efficient co-management systems may allow for faster response to changing conditions, for example, regarding limit-setting on the take of fish and wildlife (see e.g., MAP website).

Strong, capable community leadership can, in an informed manner, enter into negotiations with authorities, especially in cases where major actions such as infrastructure reconstruction and relocation are to be taken. That is of primary importance if the history of improving Arctic indigenous lives by force, with decisions made in southern capitals (as was the case with relocations occurring in the past, for instance in Canada and the former Soviet Union) is not to be repeated in the case of adaptation policies. The need for participation will rise with the increase in climate change impacts and development of adaptation policies.

Climate change could, in fact, perversely enhance indigenous empowerment and participation in governance. Arctic indigenous activists attempt to utilize the discussion on climate change to strengthen advocacy of their rights and to empower their peoples and organizations (Bravo 2009; ICC 2005, 2009). Adaptation governance is seen as an opportunity for facilitating social learning, public participation, and strengthening deliberative democracy (Tennberg 2012).

There is clearly a trend and a normative pressure towards indigenous empowerment and participation in the Arctic, which is expected to also have positive effects on communities' adaptive capacities. Available short-term options include strengthening co-management systems, adjusting participation mechanisms to indigenous capacities and traditions, providing communities with feedback on the outcomes of their participatory engagements, utilizing indigenous rights frameworks, and applying a human rights-based approach.

4.6.1 Co-management, Participatory Capacities, and Clear Outcomes of Participatory Engagement

The vital elements of this empowerment trend are self-government, self-determination, strengthening local governments, and co-management arrangements, including management and use of land and resources. Newly developed institutions can provide communities with the possibility to transmit their concerns and enhance learning and self-organization processes (Berkes and Jolly 2001; ACIA 2005).

Management of marine natural resources is a particularly sensitive issue. Indigenous territorial use and perception, where waters and ice are an extension of traditional territory, should be taken into consideration in the restructuring of governance systems. For example, owing to the Nunavut Land Claims Agreement, local communities have the possibility to influence offshore developments (NLCA 1993: para.11.1.2; Royal Commission on Aboriginal Peoples 1996).

Nevertheless, what is understood by participation and empowerment as well as the practical implementation of these principles remains vague (Pearce et al. 2010). First, it is often not clear what changes follow indigenous contributions, as is the case in the Canadian Arctic or with consultations between Sámi Parliaments and Nordic governments. Secondly, there can be too much consultation, with indigenous institutions overburdened with complex procedures and processing hundreds of documents, without a clear picture of the consultation outcome. Balancing between providing indigenous groups with the genuine possibility to be fully consulted in an informed manner and the capacities of these actors, needs to be taken into account when designing decision making procedures (Henriksen 2008, 2010; Huntington et al. 2012).

Devolution is not a straightforward or clearly empowering process. Neither does bringing power closer to the communities and putting it in the hands of indigenous politicians automatically result in the improvement of social, economic, and political conditions. Experiences of Nunavut with mismanagement, corruption, inefficient spending, and questionable policies show that empowerment is a complex and long-term process and not a clear-cut solution (Loukacheva and Garfield 2009).

4.6.2 Indigenous Rights

The application of indigenous rights, which has evolved within human rights mechanisms over recent decades (Niezen 2003), is seen as having empowerment potential. Indigenous rights extend beyond the boundaries of traditional and local livelihoods and provide for indigenous involvement in issues of resource extraction or shipping (Hansen and Bankes 2008; Lukovich and McBean 2009; Cameron 2012). In the 2009 Arctic TRANSFORM project, the Indigenous Peoples Working Group argued for treating indigenous peoples not as stakeholders, but rather as rights-holders and incorporating rights language into Arctic governance, for example by establishing an Indigenous Rights Review Working Group under the Arctic Council, which would analyze legal and institutional adaptation barriers from the human rights perspective (Cochran et al. 2009).

Indeed, climate change impacts are increasingly presented as an issue of human rights (Humphreys 2010) with an obligation of states, the international community, and industries (seen as duty-bearers) to assist those affected. The aforementioned Arctic TRANSFORM Working Group also proposed establishing an Arctic Trust Fund, through which those benefiting from exploitation of Arctic resources—particularly extractive industries—would counterbalance risks these activities create (Cochran et al. 2009).

Indigenous peoples around the circumpolar North have attempted to utilize human rights avenues, including the Human Rights Committee or the European Court of Human Rights (Koivurova 2011a). A prominent example is the unsuccessful, but influential within climate change debates, Inuit Circumpolar Council's (ICC) petition to the Inter-American Commission on Human Rights, where US "acts and omissions" in climate policy were argued to have violated Inuit's human rights (ICC 2005; Koivurova 2008; Abate 2007). In addition to developments in litigation, the Human Rights Council recognized climate change as "an immediate and far-reaching threat to people and communities around the world and [which] has implications for the full enjoyment of human rights" (HRC 2008).

4.7 Conclusion: A Holistic Response

Empowering Arctic communities in the current social, economic, environmental, and political landscape of the region is challenging. Available literature and governance practices continue to offer few practical propositions or concrete best practices. Ideally, empowerment, co-management, and participatory decision making should be tailor-made to communities' individual capacities and perpetually re-evaluated and attuned to changing conditions. Participatory efforts need to have clear relevance in decision making processes or risk turning into purely formal acts resulting in disappointment rather than empowerment.

Adaptation interconnected with empowerment needs to be a holistic endeavour. Increasing adaptive capacity is not only foreseen to address climate change impacts, but may potentially help respond to other present-day challenges. Addressing climate change impacts cannot become the only goal of implemented policies as, for instance, poverty alleviation or protection of culture could be equally important (Ford 2009). Rather, engagement should take place alongside planning and engineering solutions and at multiple scales, including the national, regional, and community levels and addressing diverse challenges, with values pertinent to each society playing a crucial role in choosing adaptation strategies (Hovelsrud and Smit 2010).

Financial constraints are one of the primary barriers to implementing adaptation measures. The high costs of adaptation will continue to be an ongoing problem in policy making, but if the desired goal is protection of invaluable indigenous cultures and preserving settlements in the Arctic, policymakers need to accept that certain costs need to be borne.

Adaptation to climate change and concern for particular vulnerability of indigenous peoples are nowadays increasingly presented as issues of human rights and human security (Government of Canada 2009; Ford et al. 2010b; Lukovich and McBean 2009). These approaches have advocacy potential to trigger greater action on the side of duty-bearers and security-providers such as states, public authorities, private companies, or NGOs. At the same time, however, the language of securitization and perception of indigenous peoples exclusively as victims can be seen as introducing old notions of paternalism and trusteeship, and therefore must be applied with caution (Bravo 2009; Cameron 2012; Pearce et al. 2012).

Adaptation policies are already being developed and implemented (Tennberg 2012) and they need to be monitored, evaluated, and reflected upon on an ongoing basis. Therefore, additional research on actual adaptation planning and implementation is needed.

The use of a common definition of adaptation in policy making needs to be critically questioned. The meaning of adaptation—a complex term in itself—within diverse cultural settings has to be better understood by decisionmakers (Cochran et al. 2009). Different concepts of climate change impacts and adaptation may exist even in neighbouring communities, especially as climate change in the Arctic is a heterogeneous phenomenon (Pearce et al. 2010).

Research and sharing experience on current adaptation planning is of major significance broadly, as well as between the US and the EU (including EU Member States). Partners from both sides of the Atlantic could learn much from one another regarding indigenous governance and adaptation. Alaskan policies could prove particularly useful for EU policymakers as they represent decades of experience with various indigenous governance arrangements (e.g., North Slope Borough set-up and land claims agreements). Greater cooperation and dialogue may contribute to sensitizing various actors—particularly on the side of the EU, which has been perceived as a regional outsider—to the specificity of economic, social, cultural, and political developments in the indigenous North. On the other hand, the US could utilize the EU's experiences with its progressive climate change policies, including adaptation actions at multiple levels.[4]

The Arctic is seen as a harbinger of change, but also a potential test-bed for developing creative responses, adaptive policies, resilient communities, capacity for management, responsible governance, and community-based sustainable (i.e., not hypothesis-testing) research. It is believed that in the Arctic, such tools for managing and coping with emerging global issues may best be developed (Carmack et al. 2012; Berkes and Jolly 2001). However, Arctic communities cannot be seen as locations of 'adaptive experiments' for the benefit of global governance and their own needs should always come to the fore (Carmack et al. 2012; Tennberg 2012).

References

Abate RS (2007) Climate change, The United States, and the impacts of Arctic melting: A case study in the need for enforceable international environmental human rights. Symposium: Climate Change Liability and the Allocation of Risk. Stanford J Int Law 43A:3–76

[4] See e.g, the EU Commission's 2009 White Paper on climate change adaptation (European Commission 2009), as well as Mettiäinen (2012) on an example of regional adaptation strategy supported by the EU funding.

ACIA (2005) Arctic Climate Impact Assessment. Cambridge University Press, New York

Adaptation Advisory Group (2010) Alaska's climate change strategy: Addressing climate change in Alaska. Final Report Submitted by the Adaptation Advisory Group to the Alaska Climate Change Sub-Cabinet (Final Draft Report). Alaska Department of Environmental Conservation. <http://www.climatechange.alaska.gov/aag/aag.htm>. Accessed 20 Aug 2012

Adger WN (2010) Climate change, human well-being and insecurity. New Polit Econ 15(2):275–292

AHDR (2004) Arctic Human Development Report. Stefansson Arctic Institute, Akureyri

AK-CSC (Alaska Climate Science Centre) (2012) Annual Action Plan FY2012. <http://csc.alaska.edu/AK%20CSC%20annual%20action%20plan%202012.pdf>. Accessed 7 Nov 2012

AMSA (2009) Arctic Marine Shipping Assessment 2009 Report. Arctic Council, US 2009

Berkes F, Jolly D (2001) Adapting to climate change: Social-ecological resilience in a Canadian western Arctic community. Conserv Ecol 5(2):18–32

Bone C, Alessa L, Altaweel M, Kliskey A, Lammers R (2011) Assessing the impacts of local knowledge and technology on climate change vulnerability in remote communities. Int J Environ Res Public Health 8(3):733–761

Bravo MT (2009) Voices from the sea ice: the reception of climate impact narratives. J Hist Geogr 35:256–278

Bull KS (2008) Historisk fremstilling av retten til fiske i havet utenfor Finnmark. Official Norwegian Reports, NOU 2008:5

Cameron ES (2012) Securing Indigenous politics: A critique of the vulnerability and adaptation approach to the human dimensions of climate change in the Canadian Arctic. Global Environ Change 22:103–114

Carmack E, McLaughlin F, Whiteman G, Homer-Dixon T (2012) Detecting and coping with disruptive shocks in Arctic marine systems: a resilience approach to place and people. Ambio 41:56–65

Cochran P, Nuttal M, Arctic TRANSFORM Indigenous Peoples Working Group (2009) Policy Options for Arctic Environmental Governance. Prepared by the Indigenous Peoples Working Group. Arctic TRANSFORM. <http://arctic-transform.org/download/IndPeEX.pdf>

Dahl J, Hicks J, Jull P (eds) (2001) Nunavut. Inuit regain control of their lands and their lives. IWGIA, Copenhagen

Escobar A (1995) Encountering development: The making and unmaking of the third world. Princeton University Press, Princeton

European Commission (2009) Adopting to climate change: Towards a European framework for action (White Paper), 1 Apr 2009, COM (2009) 147 final

European Community (2009) Regulation No. 1007/2009 of the European Parliament and of the Council of 16 Sept 2009 on trade in seal products, 31 Oct 2009, 2009 O.J. (L 286/36)

European Convention on Human Rights (1950) European Convention for the Protection of Human Rights and Fundamental Freedoms, 4 Nov 1950, 213 U.N.T.S. 221. Entered into force 3 Sept 1953

European Parliament (2012) The impact of the EU seal ban on the Inuit population in Greenland. Seminar chaired by Pat the Cope Gallagher, Member of the European Parliament, Brussels

Feifel K (2010) Homer, Alaska Climate Action Plan. Case study on a project of the City of Homer and ICLEI. Product of EcoAdapt's State of Adaptation Program. <http://www.cakex.org/case-studies/711>. Accessed 12 Nov 2012

Forbes BC, Stammler F (2009) Arctic climate change discourse: the contrasting politics of research agendas in the West and Russia. Polar Res 28:28–42

Ford JD (2009) Dangerous climate change and the importance of adaptation for the Arctic's Inuit population. Environ Res Lett 4

Ford JD, Smit B, Wandel J (2006) Vulnerability to climate change in the Arctic: A case study from Arctic Bay, Canada. Global Environ Change 16:145–160

Ford JD, Berrang-Ford L, King M, Furgal C (2010a) Vulnerability of Aboriginal health systems in Canada to climate change. Global Environ Change 20:668–680

Ford JD, Pearce T, Duerden F, Furgal C, Smit B (2010b) Climate change policy responses for Canada's Inuit population: The importance of and opportunities for adaptation. Global Environ Change 20:177–191

Furgal CH, Seguin J (2006) Climate change, health and vulnerability in Canadian northern aboriginal communities. Environ Health Perspect 114(12). doi:10.1289/ehp.8433

GAO (2004)Testimony before the committee on Appropriations, U.S. Senate: ALASKA NATIVE VILLAGES—Villages Affected by Flooding and Erosion Have Difficulty Qualifying for Federal Assistance. Government Accountability Office. <http://www.gpo.gov/fdsys/pkg/GAOREPORTS-GAO-04-895T/pdf/GAOREPORTS-GAO-04-895T.pdf>. Accessed 7 Nov 2012

Government of Canada (2009) Canada's Northern Strategy: Our north, our heritage, our future. Ottawa: Minister of Indian Affairs and Northern Development and Federal Interlocutor for Métis and Non-Status Indians

Hansen KF, Bankes N (2008) Human rights and Indigenous Peoples in the Arctic: What are the implications for oil and gas industry. In: Mikkelsen A, Langelle O (eds) Arctic Oil and Gas: Sustainability at Risk?. Routledge, New York, pp 291–316

Hasanat W (2010) Cooperation in the Barents Euro-Arctic region in the light of international law. Yearb Polar Law 2:279–309

Henriksen J (ed) (2008) Sámi self-determination: Scope and implementation. Galdu Čala. J Indigenous Peoples Rights 2

Henriksen J (ed) (2010) Sámi self-determination: Autonomy and economy—The authority and autonomi of Samediggi in the health and social services sector. Galdu Čala. J Indigenous Peoples Rights 2

Hossain K (2012) The EU ban on the import of seal products and the WTO regulations: neglected human rights of the Arctic indigenous peoples? Polar Record. doi:<http://dx.doi.org/10.1017/S0032247412000174>

Hovelsrud GK, Smit B (eds) (2010) Community adaptation and vulnerability in Arctic regions. Springer, Berlin

HRC (1994) General Comment No. 23 (50th Session, 1994) by the Human Rights Committee, UN Doc. HRI/GEN/1/Rev 3

HRC (1999/2000/2002/2004) Human Rights Committee. Concluding Observations (multiple): Canada (UN doc. CCPR/C/79/Add.105 (1999)); Mexico (UN Doc. CCPR/C/79/Add.109 (1999)); Norway (UN Doc. CCPR/c/79/Add.112 (1999)); Australia (UN Doc. CCPR/CO/69/AUS (2000)); Denmark (UN Doc. CCPR/CO/70/DNK (2000)); Sweden (UN Doc. CCPR/CO/74/SWE (2002)). Finland (UN Doc. CCPR/CO/82/FIN (2004))

HRC (2008) Human Rights Council. Resolution 7/23 on human rights and climate change. <http://ap.ohchr.org/documents/E/HRC/resolutions/A_HRC_RES_7_23.pdf>. Accessed 28 Aug 2012

Humphreys S (2010) Human rights and climate change. Cambridge University Press, Cambridge

Huntington HP, Lynge A, Stotts J, Hartsig A, Porta L, Debicki CH (2012) Less ice, more talk: The benefits and burdens for Arctic communities of consultations concerning development activities. Carbon Climate Law Rev 1:33–46

ICC (2005) Petition to the Inter American Commission on human rights, violations resulting from global warming caused by the United States, 7 Dec 2005 by Sheila Watt-Cloutier with support of Inuit Circumpolar Council. <http://www.inuitcircumpolar.com/files/uploads/icc-files/FINALPetitionICC.pdf>. Accessed 20 Aug 2012

ICC (2009) A Circumpolar Inuit Declaration on Sovereignty in the Arctic. Inuit Circumpolar Council. Apr 2009

ICCPR (1966) International Covenant on Civil and Political Rights, 16 Dec 1966, 999 U.N.T.S. 17. Entered into force 23 Mar 1976

ILO Convention No. 169 (1989) Convention concerning Indigenous and Tribal Peoples in Independent Countries, 27 June 1989, 28 I.L.M. 1382 (1989). Entered into force 5 Sept 1991

International Covenant on Economic, Social, and Cultural Rights (1966) 16 Dec 1966, 993 U.N.T.S. 3. Entered into force 3 Jan 1976

International Convention on the Elimination of All Forms of Racial Discrimination (1966) 7 Mar 1966, 660 U.N.T.S. 195. Entered into force 4 Jan 1969

IPCC (2007) Climate change 2007: Impacts, Adaptation and Vulnerability. Contribution of working group II to the fourth assessment report of the intergovernmental panel on climate change.In: Parry ML, Canziani OF, Palutikof JP, van der Linden PJ, Hanson CE (eds) Cambridge University Press, Cambridge

Keskitalo ECH (ed) (2010) Developing adaptation policy and practice in Europe: Multi-level governance of climate change. Springer, Berlin

Koivurova T (2008) The draft Nordic Sámi Convention: Nations working together. Int Commun Law Rev 10:279–293

Koivurova T (2011a) Jurisprudence of the European court of human rights regarding Indigenous Peoples: Retrospect and prospects. Int J Minori Group Rights 18:1–37

Koivurova T (2011b) The status and role of Indigenous Peoples in Arctic international governance. Polar Law Yearb 3:169–192

Koivurova T, Hasanat W (2009) Climate policy of the Arctic Council. In: Koivurova T, Heinämäki L (2006) The participation of Indigenous Peoples in international norm-making in the Arctic. Polar Record 42(221):101–109. In: Koivurova T, Keskitalo C, Bankes N (eds) Climate Governance in the Arctic. Springer, Berlin

Koivurova T, Heinämäki L (2006) The participation of Indigenous Peoples in international norm-making in the Arctic. Polar Rec 42(221):101–109

Li TM (2007) The will to improve: Governmentality, development, and the practice of politics. Duke University Press, Durham

Lindroth M (2011) Paradoxes of power: Indigenous Peoples in the Permanent Forum. Cooperation Conflict 46(4):542–561

Loukacheva N, Garfield MD (2009) Sustainable human rights and governance: The quest of an Arctic entity in transition. Polar Law Yearbook 1:283–305

Lukovich JV, McBean GA (2009) Addressing human security in the Arctic in the context of climate change through science and technology. Mitig Adapt Strat Glob Change 14:697–710

Magga OH, Mathiesen SD, Corell RW, Oskal A, Benestad R, Bongo MP, Burgess P, Degteva A, Etylen V (2011) Reindeer herding, traditional knowledge and adaptation to climate change and loss of grazing land. EALAT project executive summary. Arctic Council and International Centre for Reindeer Husbandry. Fagtrykk Ide AS, Alta

MAP (n.d.) Marine Advisory Programme. <http://seagrant.uaf.edu/map>. Accessed 8 Nov 2012

Mettiäinen I (2012) Knowledge to action question in regional climate change policy formation. In: Adam S, Mettiäinen I, Soppela P (eds) Science-Policy Interface – Societal Impacts of Arctic Research. Abstract book. The 10th Annual Seminar of the ARKTIS Arctic Doctoral Programme and the Final Seminar of the project “Knowledge in Environmental Planning and Decision-making” (2010-2012). 15-16 Mar 2012, Rovaniemi, Finland. Arctic Centre. University of Lapland. University of Lapland Printing Centre, Rovaniemi. pp 67

MFPP (2012) Funding opportunity—Alaska native fund could speed climate adaptation training. Model Forest Policy Programme. <http://www.mfpp.org/?p=2326>. Accessed 8 Nov 2012

Moerlein KJ, Carothers C (2012) Total environment of change: Impacts of climate change and social transitions on subsistence fisheries in Northwest Alaska. Ecol Soc 17(10)

Niezen R (2003) The Origins of Indigenism. University of California Press, Berkeley

Njåstad B, Kelman I, Rosenberg S (2009) Vulnerability and adaptation to climate change in the Arctic (VACCA). Brief Report Series No. 12. Norwegian Polar Institute and Arctic Council Sustainable Development Working Group. Tromso: Norwegian Polar Institute

NLCA (1993) Nunavut Land Claims Agreement, extract. <http://npc.nunavut.ca/eng/npc/article11.html>. Accessed 5 May 2008

Nuttall M (2002) Protecting the Arctic, Indigenous Peoples and cultural survival. Routledge, London

Nuttall M (2000) Indigenous Peoples, self-determination and the Arctic environment. In: Nuttall M, Callaghan TV (eds) The Arctic—Environment, people, policy. Harwood Academic, Amsterdam

Pearce TD, Ford JD, Laidler GJ, Smit B, Duerden F, Allarut M, Andrachuk M, Baryluk S, Dialla A, Elee P, Goose A, Ikummaq T, Joamie E, Kataoyak F, Loring E, Meakin S, Nickels S, Shappa K, Shirley J, Wandel J (2009a) Community collaboration and climate change research in the Canadian Arctic. Polar Res 28:10–27

Pearce T, Smit B, Duerden F, Ford JD, Goose A, Kataoyak F (2009b) Inuit vulnerability and adaptive capacity to climate change in Ulukhaktok, Northwest Territories, Canada. Polar Rec 46:157–177

Pearce T, Ford JD, Duerden F, Smit B, Andrachuk M, Berrang-Ford L, Smith T (2010) Advancing adaptation planning for climate change in the Inuvialuit Settlement Region (ISR): a review and critique. Reg Environ Change 11:1–17

Pearce T, Ford JD, Caron A, Kudlak BP (2012) Climate change adaptation planning in remote, resource-dependent communities: An Arctic example. Reg Environ Change. doi:10.1007/s10113-012-0297-2

Poppel B, Jack K, Gérard D, Larissa A (2007) Survey of living conditions in the Arctic (SLiCA): Results. Institute of Social and Economic Research, University of Alaska Anchorage, Anchorage

Royal Commission on Aboriginal Peoples (1996) Restructuring the relationship. Canada Communication Group Publishing, Part two. Report

Slezkine Y (1996) Arctic mirrors: Russia and small Peoples of the North. Cornell University Press, Ithaca

Smit B, Pilifosova O (2003) From adaptation to adaptive capacity and vulnerability reduction. In Smith JB, Klein RJT, Huq S (eds) Climate change, adaptive capacity and development. Imperial College Press, London

Smit B, Wandel J (2006) Adaptation, adaptive capacity and vulnerability. Global Environ Change 16:282–292

Tennberg M (ed) (2012) Governing the Uncertain. Adaptation and Climate in Russia and Finland. Springer, Dordrecht

Thronberry P (1994) On some implications of the UN Declaration on Minorities for Indigenous Peoples. In: Gayim E, Myntti K (eds) Indigenous and Tribal Peoples Rights—1993 and after. University of Lapland, Rovaniemi

UNDRIP (2007) United Nations Declaration on the Rights of Indigenous Peoples. General Assembly Resolution. New York, 13 Sept 2007. A/RES/61/295

United Nations Framework Convention on Climate Change (1992) 9 May 1992, 1771 U.N.T.S. 107. Entered into force 21 Mar 1994

UNPFII (United Nations Permanent Forum on Indigenous Issues) (n.d.) Who are indigenous peoples? Factsheet. United Nations

USGCRP (2009) U.S. Climate Impacts Report—Regional Climate Change Impacts. United States Global Change Research Program. <http://globalchange.gov/images/cir/pdf/alaska.pdf>. Accessed 7 Nov 2012

Wenzel G (2001) Inuit subsistence and hunter support in Nunavut. In: Dahl et al (eds) Nunavut. Inuit regain control of their lands and their lands. International Work Group for Indigenous Affairs (IWGIA), Copenhagen

Young E (1995) Third world in the first: Development and Indigenous Peoples. Routledge, London

Part II
Impacts and Activities in the Marine Arctic

Chapter 5
Status and Reform of International Arctic Fisheries Law

Erik J. Molenaar

Abstract Marine capture fisheries are among the maritime uses that are expected to expand and intensify in the marine Arctic. Fishing could intensify in existing fishing areas and expand into areas where marine capture fisheries have never taken place. This chapter assesses the adequacy of the current international legal and policy framework for Arctic fisheries conservation and management in light of the current and expected impacts of global climate change on the marine Arctic. It provides an overview of the international legal and policy framework as well as national regulation and policy, identifies the main gaps therein, and suggests options for addressing them. These options include increased efforts in the sphere of research and data gathering, national regulation aimed at avoiding unregulated fishing, fisheries arrangements between Arctic Ocean coastal states, and a new regional fisheries management organization (RFMO) or Arrangement for part of the (Central) Arctic Ocean. Separate attention is devoted to the potential for cooperation between the European Union and the United States in this regard.

Writing this chapter was facilitated by funding from the Netherlands Polar Programme. The author is grateful for comments by Ted McDorman and others on an earlier version of the chapter. This chapter builds on Molenaar EJ, Corell R (2009) Background Paper. Arctic Fisheries. Arctic TRANSFORM.

E. J. Molenaar (✉)
Netherlands Institute for the Law of the Sea (NILOS), Utrecht University, Utrecht, The Netherland
Faculty of Law, University of Tromsø, Tromsø, Norway
e-mail: E.J.Molenaar@uu.nl

E. Tedsen et al. (eds.), *Arctic Marine Governance*,
DOI: 10.1007/978-3-642-38595-7_5, © Springer-Verlag Berlin Heidelberg 2014

5.1 Introduction

Marine capture fisheries are among the maritime uses that are expected to expand and intensify in the marine Arctic. Fishing could intensify in existing fishing areas and expand into areas where marine capture fisheries have never taken place so far. This chapter assesses the adequacy of the current international legal framework for Arctic fisheries conservation and management in light of the current and expected impacts of global climate change on the marine Arctic. After providing some context and background information on Arctic fish stocks, fisheries, and climate change in Sect. 5.2, an overview of the international legal and policy framework for Arctic fisheries management is presented in Sect. 5.3. Section 5.4 then devotes attention to national regulation and policy. Gaps in the international legal and policy framework and national regulation and options for addressing them are covered succinctly in Sect. 5.5 and the potential for EU–US cooperation is examined in Sect. 5.6. Some conclusions are offered in Sect. 5.7.

This chapter relates to marine capture fisheries that target 'fishery resources', which are defined as fish, molluscs, crustaceans, and (other) sedentary species. Inland fisheries, aquaculture, and harvesting of marine mammals are thus excluded. For the purposes of this chapter, the term regional fisheries management organization (RFMO) includes a so-called 'Arrangement', which is understood to be a bilateral or (sub-)regional cooperative mechanism other than an intergovernmental organization, but otherwise has in principle the same characteristics as an RFMO.[1] Due to its predominantly sectoral perspective, the international component of this chapter will be restricted to instruments and bodies that relate to, or pursue, conservation as well as management of fishery resources. No attention will therefore be paid to those that focus exclusively on conservation of species and habitats by various means, including by the regulation of international trade.

As explained in Chap. 1, there are no generally accepted geographical definitions for the terms 'Arctic' and 'Arctic Ocean' and for the purposes of this book, the term 'Arctic' has an identical meaning as the term 'AMAP area' adopted by the Arctic Monitoring and Assessment Programme (AMAP) of the Arctic Council (AMAP 1997). The waters within the AMAP area are in this chapter referred to as the 'marine Arctic'. The 'Arctic Ocean' is defined here as the marine waters north of the Bering Strait, Greenland, Svalbard, and Franz Josef Land, excluding the Barents Sea. Canada, Denmark/Greenland, Norway, the Russian Federation, and the United States (US) are 'Arctic Ocean coastal states'. These five states and Finland, Iceland, and Sweden are 'Arctic states' by virtue of their membership of the Arctic Council. There are four high seas pockets in the marine Arctic, namely

[1] The term 'Arrangement' is derived from the term 'arrangement' as defined in Article 1(1)(d) of the Fish Stocks Agreement (1995). The main differences between an RFMO's constitutive instrument and an Arrangement are that the latter (a) does not establish an international organization, (b) does not have to be legally binding, and (c) can be bilateral.

the so-called 'Banana Hole' in the Norwegian Sea, the so-called 'Loophole' in the Barents Sea, the so-called 'Donut Hole' in the central Bering Sea, and the so-called 'Central Arctic Ocean'.

5.2 Arctic Fish Stocks, Fisheries, and Climate Change

There are a number of potentially significant commercial fish stocks in the marine Arctic. The ranges of distribution of some of these are confined to the North Pacific or the North Atlantic, while others have a circumpolar distribution. Important North Pacific fish stocks include Alaska pollock (*Theragra chalcogramma*), Pacific cod (*Gadus macrocephalus*), snow crab (*Chionoecetes opilio*), and various Pacific salmon species (*Oncorhynchus* sp.). As regards the North Atlantic, important fish stocks include North-East Arctic cod (*Gadus morhua*), haddock (*Melanogrammus aeglefinus*), Norwegian spring-spawning (Atlanto-scandian (AS)) herring (*Clupea harengus*), Atlantic salmon (*Salmo salar*), and red king crab (*Paralithodes camtschaticus*). Significant circumpolar fish stocks include capelin (*Mallotus villosus*), Greenland halibut (*Reinhardtius hippoglossoides*), and northern shrimp (*Pandalus borealis*). Polar cod (*Boreogadus saida*) and Arctic char (*Salvelinus alpinus*) also have circumpolar distribution, but the former is only marginally targeted by commercial fisheries and the latter is predominantly fished for subsistence purposes (ACIA 2005; AOR 2011; Anchorage Science Meeting Report 2011; Zeller et al. 2011).[2]

While Arctic marine ecosystems have always been highly dynamic and variable (AOR 2011), both qualitatively and quantitatively, the impacts of climate change on the marine Arctic—e.g., increasing water temperature, reduced sea ice coverage and thickness, reduced salinity, and increasing acidification (Anchorage Science Meeting Report 2011)—are likely to make these changes more rapid, more profound, and probably also more difficult to predict. Some existing fish stocks may collapse and never recover, others may become more dominant, and new fish species may successfully invade the marine Arctic. The various assessments and projects currently undertaken within the Arctic Council, such as the Arctic Biodiversity Assessment, the Arctic Change Assessment, and the Arctic Resilience Report, are expected to shed more light on this.

While there are large-scale commercial fisheries in the more southerly waters of the marine Arctic—namely the Bering Sea, Barents Sea, Baffin Bay, and along the coast of east and west Greenland—in the Arctic Ocean there are currently mainly small-scale subsistence fisheries and no significant commercial fisheries, and in the Central Arctic Ocean no fisheries at all. It seems more likely that new fishing opportunities will occur within coastal state maritime zones before occurring on the high seas. According to some commentators, in the short term, it is

[2] See also information at <arcticportal.org/fishing-portlet>.

unlikely that abundance of fish stocks in the high seas portion of the Arctic Ocean will allow for commercially viable fisheries (Hoel 2011). Others disagree while pointing to Polar cod, which has a circumpolar distribution, both inshore and offshore, and may be highly abundant in view of the pivotal role it plays at the bottom of Arctic marine ecosystems.[3] Finally, as reduced ice coverage and thickness will also enable other human activities—most importantly shipping and offshore hydrocarbon activities—these activities may compete with fishing in a spatial sense or affect them by pollution—including noise—and other impacts.

The impact of current and future Arctic fisheries on the marine environment and marine biodiversity in the Arctic is not likely to be fundamentally different from fisheries impacts to the marine environment and biodiversity in other parts of the globe. Arctic fisheries could lead to over-exploitation of target species and a variety of impacts on non-target species, for instance on dependent species due to predator–prey relationships, on associated species due to bycatch, and on benthic species and habitats due to bottom fishing techniques. In view of the broad spatial scope of the marine Arctic, such undesirable effects are without doubt already occurring, even though not necessarily on a very serious scale.

5.3 International Legal and Policy Framework for Arctic Fisheries Management

5.3.1 Interests, Rights, Obligations, and Jurisdiction

The international legal and policy framework for fisheries conservation and management seeks to safeguard the different interests of the international community with those of states that have rights, obligations, or jurisdiction in their capacities as flag, coastal, port, or market states or with respect to their natural and legal persons. While the term 'flag state' is commonly defined as the state in which a vessel is registered and/or whose flag it flies (LOS Convention 1982, art. 91(1)), there are no generally accepted definitions for the terms 'coastal state', 'port state', or 'market state'. For the purposes of this chapter, however, the term 'coastal state' refers to the rights, obligations, and jurisdiction of a state within its own maritime zones over foreign vessels.

The term 'port state' refers to the rights, obligations, and jurisdiction of a state over foreign vessels that are voluntarily in one of its ports. In order to avoid an overlap with jurisdiction by coastal states, this chapter regards port state jurisdiction as relating to fishing by foreign vessels beyond the coastal state's maritime zones as well as over violations of conditions for entry into port (Molenaar 2007).

While there is no universally accepted definition for the term 'market state', this chapter uses a definition that was proposed during the negotiation of the SPRFMO

[3] For additional information see <www.arcodiv.org>.

Convention (2009), but did not make it to the final text. That definition reads "a State [...] which imports, exports, re-exports or has a domestic market for fish or fish products derived from fishing in the Convention Area" (SPRFMO 2008, art. 1(m)).

Both flag and coastal states have, in principle, an interest in the long-term exercise of their entitlements over marine living resources in the various maritime zones. However, as coastal states have exclusive access to marine living resources within areas under their national jurisdiction, their commitment to that objective may often be stronger than that of flag states. A port state will commonly pursue socio-economic interests related to the port and its *hinterland*. States generally have interests, rights, obligations, and jurisdiction in more than one capacity. This commonly leads to a more balanced compromise position, but occasionally also to contradictory positions of the same state within different fora. There is no reason or indication to assume that Arctic states are different in this regard.

The interests of the international community—e.g., sustainable utilization, protection, and preservation of the marine environment and conservation of marine biodiversity—normally overlap with those of states within the various capacities in which they can act, but are usually broader and more general. The interests of some states, however, clearly undermine those of other states and the international community, for instance, by not ensuring that their ships comply with international minimum standards or by allowing foreign vessels in their ports to be non-compliant with international minimum standards. These states, vessels, and ports thereby have a competitive advantage over states, vessels, and ports that *do* comply with international minimum standards. Such 'free riders' clearly benefit from the consensual nature of international law—meaning that a state can only be bound to a rule of international law when it has in one way or another consented to that rule.

5.3.2 Substantive Fisheries Standards

Fisheries conservation and management authorities often make use of the following substantive fisheries standards:

1. Restrictions on catch and effort, for instance, by setting the total allowable catch (TAC) and allocating the TAC by means of national quotas;
2. Designated species for which targeted fishing is prohibited;
3. Minimum size limits for target species;
4. Maximum bycatch limits, for instance, in terms of the number of individuals (e.g., in relation to marine turtles and marine mammals) or as a percentage of the target catch;
5. Gear specifications, for instance, minimum mesh sizes, bycatch mitigation techniques (e.g., turtle excluder devices, bird-scaring lines); and
6. Temporal/seasonal or spatial measures (e.g., closed areas) aimed at avoiding catch of target species (e.g., nursing and spawning areas) or non-target species (e.g., important feedings areas) or avoiding impact on sensitive habitat (e.g., cold water coral reefs).

5.3.3 Global Bodies and Instruments

All the relevant global intergovernmental bodies and instruments discussed in this subsection also apply to the marine Arctic, however defined. The main intergovernmental bodies of relevance to this chapter are the United Nations General Assembly (UNGA) and the United Nations Food and Agriculture Organization (FAO). While the mandate of these bodies would not preclude them from dealing specifically with Arctic fisheries as such, most of the Arctic Ocean coastal states would oppose this (Molenaar 2012a).

The following are the main global legally binding and non-legally binding fisheries instruments:

1. LOS Convention (1982);
2. Fish Stocks Agreement (1995);
3. Compliance Agreement (1993);
4. Port State Measures Agreement (2009);
5. Other FAO fisheries instruments, most importantly, the FAO Code of Conduct for Responsible Fisheries (FAO 1995)—including its Technical Guidelines and international plans of action [IPOAs; e.g., the IPOA-IUU (FAO 2001)], the International Guidelines on Deep-sea Fisheries in the High Seas (FAO 2008), and the International Guidelines on Bycatch Management and Reduction of Discards (FAO 2010); and
6. Certain (parts of) UNGA Resolutions, which have contributed to the phase-out of large-scale pelagic driftnet fishing and imposed innovative restrictions on bottom-fisheries on the high seas.[4] Both initiatives were predominantly aimed at the conservation of non-target species and vulnerable marine ecosystems.

The provisions on marine capture fisheries in the LOS Convention and the Fish Stocks Agreement have a so-called 'framework' character. They contain overall objectives and basic rights and obligations for states, but not the key substantive fisheries standards set out here. Actual fisheries regulation is carried out by states individually or collectively, including through RFMOs (see Sect. 5.3.4).

The key objectives of the LOS Convention are (a) avoidance of overexploitation by means of striving for the maximum sustainable yield (MSY) and setting TACs, and (b) optimum utilization, which obliges coastal states that cannot catch the entire TAC themselves to give other states access to the surplus. The LOS Convention acknowledges or grants rights to coastal states over marine living resources in their maritime zones and to other states on the high seas. These rights are subject to the key objectives just mentioned and many other related obligations, for instance, the obligation to take account of impacts on associated species (e.g., through bycatch) or dependent species (e.g., through predator–prey relationships) and to cooperate with relevant coastal and/or flag states on transboundary stocks/species and discrete high seas stocks (see Table 5.1). The objective of

[4] See, *inter alia*, UNGA Res. 46/215 (1991) and UNGA Res. 61/105 (2006), paras 80–89.

Table 5.1 Categories of fish stocks

Category	Definition
Discrete inshore stocks	Occur exclusively in the maritime zones of one single state
Joint or shared stocks	Occur within the maritime zones of two or more coastal states, but not on the high seas
Straddling stocks	Occur within the maritime zones of one or more coastal states and on the high seas
Highly migratory stocks	The fish species listed in Annex I to the LOS Convention (e.g., tuna)
Anadromous stocks	Spawn in rivers but otherwise occur mostly at sea (e.g., salmon)
Catadromous stocks	Spend greater part of life cycle in internal fresh waters but spawn at sea (e.g., eels)
Discrete high seas stocks	Occur exclusively on the high seas

optimum utilization does not apply to marine mammals and many obligations do not apply to sedentary species or to maritime zones under sovereignty (LOS Convention 1982, arts. 61–72 and 116–120). With respect to anadromous and catadromous stocks, the relevant coastal states have primary responsibility for conservation and management. For catadromous species, this specifically includes ensuring that inbound and outbound migration can take place (LOS Convention 1982, arts. 66–67).

The Fish Stocks Agreement only applies to straddling and highly migratory fish stocks. Its overarching objective is to implement the basic jurisdictional framework of the LOS Convention by means of a modernized and more elaborate and operational regulatory framework. The incorporation of an operationalised precautionary approach and a *de facto* ecosystem approach to fisheries (EAF), the clarification that RFMOs are the primary vehicles for the conservation and management of straddling and highly migratory fish stocks, and the intricate provisions on non-flag state high seas enforcement powers bear witness to that objective.

While the Fish Stocks Agreement retains MSY as a key objective, this is qualified by the need to apply the precautionary approach as operationalised in Article 6 and Annex II as well as a range of ecosystem considerations, which together constitute a *de facto* EAF. These ecosystem considerations require state parties to, among other things, minimize pollution, waste, discards, catch by lost or abandoned gear, catch of non-target species—in particular endangered species—and more generally to protect biodiversity in the marine environment (Fish Stocks Agreement 1994, art. 5).

The Fish Stocks Agreement regards RFMOs as the preferred vehicles for fisheries regulation at the regional level and imposes an obligation on state parties to the Fish Stocks Agreement to cooperate with and through them (art. 8(3)). Of crucial importance in that regard is Article 8(4), which stipulates that access to fisheries is limited to members and cooperating states. New is also the right in Article 8(3) of states with a 'real interest' to become members. Arguably, the duty to cooperate with the relevant RFMO laid down in Article 8(3) is already part of customary international law and thereby entitles the relevant members to take measures against (non-cooperating) non-members that would otherwise be in violation of international law, for instance, imposing trade-related measures. No practices of RFMOs on trade-related measures have at any rate been challenged by means of the establishment of a dispute settlement procedure under the World Trade Organization.

Article 8(5) of the Fish Stocks Agreement stipulates that RFMOs are to be established where these do not exist. This, however, only applies in the case of the presence of "a *particular* straddling fish stock or highly migratory fish stock" (emphasis added). While highly migratory fish stocks currently do not occur in the Central Arctic Ocean, this may be different for straddling fish stocks. Even though Sect. 5.3.4 below concludes that a gap in high seas coverage with RFMOs exists for most of the Central Arctic Ocean, this does not automatically mean that relevant states are obliged to ensure full coverage with RFMOs. There is nevertheless broad support in the international community to ensure that all high seas fisheries fall within the mandate of an RFMO. These developments have among other things led to the 'filling' of gaps for full high seas coverage in the Southern Indian Ocean, South Pacific, and, most recently, the Northern Pacific (Convention on the Conservation and Management of High Seas Fisheries Resources in the North Pacific Ocean 2012). As the discussion in Sect. 5.4 will reveal, several states and entities also support full high seas coverage with RFMOs in the Arctic Ocean.

5.3.4 Regional and Bilateral Fisheries Bodies and Instruments

As regards the regional and bilateral level, the following are the main regional and bilateral fisheries bodies and instruments whose spatial scope overlaps at least to some extent with the marine Arctic:

1. The bilateral (Canada and the US) International Pacific Halibut Commission (IPHC), established by the IPHC Convention (1953);
2. The bilateral (Canada and the US) Yukon River Panel of the bilateral Pacific Salmon Commission (PSC), established by the Pacific Salmon Treaty (1985)[5];
3. The North Pacific Anadromous Fish Commission (NPAFC), established by the NPAFC Convention (1992);
4. The Western and Central Pacific Fisheries Commission (WCPFC), established by the WCPFC Convention (2000);
5. The Conference of Parties (COP) to the CBS Convention (1994);
6. The North Atlantic Salmon Conservation Organization (NASCO), established by the NASCO Convention (1982);

[5] The Yukon River Panel was established by means of the Yukon River Salmon Agreement of 4 Dec 2002, which added Chap. 8 to the Pacific Salmon Treaty.

7. The International Commission for the Conservation of Atlantic Tunas (ICCAT), established by the ICCAT Convention (1966);
8. The Northwest Atlantic Fisheries Organization (NAFO), established by the NAFO Convention (1978);
9. The North East Atlantic Fisheries Commission (NEAFC), established by the NEAFC Convention (1980); and
10. The Joint Norwegian-Russian Fisheries Commission (Joint Commission), established by the bilateral Framework Agreement (1975).

This list can be categorized in several ways. An important distinction exists between NEAFC and the Joint Commission (Nos. 9–10) and all the other bodies, because only the spatial mandates of NEAFC and the Joint Commission indisputably extend to (part of) the Arctic Ocean. Moreover, while the bodies under Nos. 1–5 apply to certain more southerly waters of the marine Arctic in the Pacific, the bodies under Nos. 6–8 apply to certain more southerly waters of the marine Arctic in the Atlantic. The spatial mandates of ICCAT and NASCO do not clearly extend to (part of) the Arctic Ocean, even though some room for interpretation exists. Significant occurrence of tuna and tuna-like species in the Arctic Ocean is not expected in the short- or medium-term, but this may well be different for anadromous species.

As only NEAFC and the Joint Commission have clear mandates in the Arctic Ocean, this warrants some closer attention to them. While Article 1(a) of the NEAFC Convention restricts NEAFC's competence to the North-East Atlantic sector of the Arctic Ocean, the Joint Commission's constitutive instrument does not specify its spatial mandate. Fisheries for species whose distributional range extends into the (Central) Arctic Ocean therefore fall in principle within the Joint Commission's mandate, and this has also been asserted by the Joint Commission on several occasions. It is submitted, however, that this assertion relates first of all to areas of the Arctic Ocean adjacent to the Barents Sea that are part of the maritime zones of Norway and the Russian Federation. If the Joint Commission would actually exercise competence over the Central Arctic Ocean in a similar manner as with regard to the Loophole, this would not be acceptable to the other Arctic Ocean coastal states and other members of NEAFC. With respect to the Loophole, Norway and the Russian Federation have encouraged third states and entities to discontinue, or not to commence, fishing for particular species and thereby not to exercise their entitlements under international law to fish in the high seas and to be involved in high seas fisheries management. In return, they have granted fisheries access to their maritime zones and discontinued withholding benefits such as access to ports (Molenaar 2012a).

Based on the foregoing analysis, it can be concluded that, except for the area covered by NEAFC, the Central Arctic Ocean is a gap in high seas coverage with RFMOs. While Norway and the Russian Federation may not necessarily share this conclusion, other Arctic Ocean coastal states and non-Arctic states and entities do (see Sect. 5.4).

While there is a significant competence-overlap—both spatially and on species—between the two bodies, there seems to be no, or hardly any, actual conflict between their conservation and management measures. Their current relationship can therefore be regarded as complementary. As Norway and the Russian Federation form two-fifths of NEAFC's membership,[6] they are also well-positioned to withstand challenges from the three other members of NEAFC to downsize the role of the Joint Commission and enhance that of NEAFC.[7] Norway and the Russian Federation are also highly unlikely to support broader participation in the Joint Commission, as this would fundamentally alter its nature.

The following list contains other relevant regional, trilateral, and bilateral fisheries arrangements:

1. The meetings of Arctic Ocean coastal states. In addition to the two ministerial meetings held in Ilulissat in May 2008 and in Chelsea in March 2010, dedicated fisheries meetings have taken place at the level of senior officials[8] and at least one meeting of scientific experts (Anchorage Science Meeting Report 2011);
2. The trilateral Loophole Agreement between Iceland, Norway, and the Russian Federation (1999)[9];
3. Bilateral cooperation between Greenland and Norway, pursuant to an Agreement on Mutual Fishery Relations (1992), which is, *inter alia*, implemented through annual bilateral consultations;
4. Bilateral cooperation between Greenland and the Russian Federation pursuant to an Agreement on Mutual Fishery Relations (1992), presumably also implemented through annual bilateral consultations;
5. Bilateral cooperation between the Russian Federation and the US pursuant, *inter alia*, to the bilateral Intergovernmental Consultative Committee, established by an Agreement on Mutual Fisheries Relations (1988);
6. Bilateral cooperation between Canada and Greenland, which is not formalized, even though meetings are held on an annual basis (Molenaar 2012a); and
7. The numerous bilateral and multilateral agreements and arrangements establishing TACs, allocations of fishing opportunities, and mutual access to maritime zones between coastal states (including the European Union (EU)) in the North-East Atlantic (Churchill 2001; Molenaar 2012b).

It is clear from the sheer number of these regional, sub-regional, and bilateral bodies and instruments that they cannot possibly be discussed in a meaningful way in this chapter. More in-depth analyses of some of them are contained in other

---

[6] The other three are Denmark (in respect of the Faroe Islands and Greenland), the EU, and Iceland.

[7] These ratios would change if, for instance, Iceland becomes an EU Member State or Greenland becomes fully independent.

[8] One took place in June 2010, in Oslo, Norway (see the Chair's summary at <www.regjeringen.no/upload/UD/Vedlegg/Folkerett/chair_summary100622.pdf>. Accessed 27 Nov 2012).

[9] This Agreement is complemented by two Protocols between Iceland and Norway and Iceland and the Russian Federation respectively, which are currently in force.

literature (e.g., Barnes 2011; Molenaar 2012a). It is nevertheless important to highlight that many of these bodies adopt conservation and management measures that contain the types of substantive fisheries standards listed in Sect. 5.3.2.

Several of the above-mentioned bodies rely for scientific advice on other bodies, most notably the International Council for the Exploration of the Sea (ICES) and the North Pacific Marine Science Organization (PICES) (Takei 2013).

Finally, reference should be made to the fundamental disagreement that exists between Norway and most other parties to the Spitsbergen Treaty (1920) on the treaty's applicability to the maritime zones of Svalbard. As a consequence, Norway has not established an Exclusive Economic Zone (EEZ) but a Fisheries Protection Zone (FPZ) around Svalbard. Several states enjoy fisheries access to the FPZ and territorial waters of Svalbard as a result of the provisions of equal access laid down in the Spitsbergen Treaty (Molenaar 2012b).

5.3.5 Arctic Council and Arctic Council System

The Arctic Council is a high-level forum established by means of the Ottawa Declaration (1996). The choice for a non-legally binding instrument is a clear indication that the Council was not intended to be an international organization and implies that the Council cannot adopt legally binding decisions or instruments. The Arctic SAR Agreement (2011) was therefore not adopted by the Council, even though it was negotiated under its auspices and the Council's May 2011 Ministerial Meeting was also used as the occasion for its signature.

So far, the Arctic Council has not explicitly involved itself in fisheries management issues; not as the Arctic Council *per se* and not through the Arctic Council System (ACS; see further below). There is nevertheless no juridical obstacle for this; not for the Arctic Council *per se* and also not for the ACS. The mandate of the Arctic Council is very broad and relates to "common Arctic issues" with special reference to "issues of sustainable development and environmental protection in the Arctic" (Ottawa Declaration 1996, art. 1). A footnote nevertheless specifies that the Council "should not deal with matters related to military security".

In spite of this very broad mandate, however, the Council has so far avoided involvement in certain marine mammal issues (Bloom 1999) and at the November 2007 Meeting of the Senior Arctic Officials (SAOs), decided not to become involved in fisheries management issues either. The matter came up because the US drew the meeting's attention to Senate joint resolution No. 17 of 2007 (S.J. Res. 17 2007)[10] "directing the United States to initiate international discussions and take necessary steps with other Nations to negotiate an agreement for managing migratory and transboundary fish stocks in the Arctic Ocean". The ensuing

[10] Passed by the Senate on 4 Oct 2007. The House of Representatives voted in favor of S.J. Res. No. 17 in May 2008 and President George W. Bush signed it on 4 June 2008.

discussion at the SAO's meeting was summarized as follows: "There was strong support for building on and considering this issue within the context of existing mechanisms" (SAO 2007).

Even though the Council has not explicitly reversed its view since then, the issue of international fisheries management has come up in the context of the Arctic Ocean Review (AOR) project that is currently carried out within the Council's Protection of the Arctic Marine Environment (PAME) working group. Phase II of this project is intended to culminate in a final report adopted at the Council's May 2013 Kiruna Ministerial Meeting that will:

> summarize potential weaknesses and/or impediments in the global and regional instruments and measures for [the] management of the Arctic marine environment; outline options to address these weaknesses and/or impediments; and, make agreed recommendations to help ensure a healthy and productive Arctic marine environment in light of current and emerging trends (AOR 2011).

The AOR Phase II draft Report contains a Chap. 4 on 'Marine Living Resources', with Sect. 4.1 (Part A) on 'Fishery Resources'. Its last subsection entitled 'Opportunities' offers various proposals for policy recommendations at its very end, which are copied verbatim into Chap. 9 entitled 'Conclusions and Recommendations' and will be presented to SAOs for negotiation. These proposals consistently use either "Arctic Council States" or "Arctic Council States with coasts on the central Arctic Ocean"—and do not explicitly recommend a role for the Arctic Council as such, but do not rule that out either. Among the options mentioned in the subsection 'Opportunities' that are not specifically retained at the end—and therefore also not in Chap. 9—are a Ministerial Declaration or a statement (AOR 2013).

As alluded to above, the Council could also pursue certain options through the ACS. The concept of the ACS has been introduced by the present author (Molenaar 2012c) to clarify that legally binding instruments such as the Arctic SAR Agreement—and their institutional components—can be part of the Council's output even though they are not—and in fact could not be—formally adopted by it.

The ACS concept consists of two basic components. The first component is made up of the Council's constitutive instrument—the Ottawa Declaration, other Ministerial Declarations, other instruments adopted by the Arctic Council—for instance its Arctic Offshore Oil and Gas Guidelines (PAME 2009), and the Council's institutional structure. The second component consists of instruments 'merely' negotiated under the Council's auspices and their institutional components. The Arctic SAR Agreement and the Meetings of the Parties envisaged under its Article 10 belong to this category. Expansion of this category will occur at the 2013 Kiruna Ministerial Meeting, which will also be used as the occasion for the signature of the Agreement on Cooperation on Marine Oil Pollution Preparedness and Response in the Arctic.

While the section 'Opportunities' in Sect. 4.1 (Part A), Chap. 4 of the AOR Phase II draft Report highlights that the Arctic Council "has been the catalyst" for the adoption of the abovementioned two treaties, it does not explicitly identify this among the various 'modes of delivery' further down that could be used for a treaty on Arctic fisheries (AOR 2013). But in light of the objections by at least one Arctic Council member to the text relating to the catalyst-role of the Arctic Council—and many other parts of Sect. 4.1—it is not even certain if it will be included in the final report.[11]

5.4 National Regulation and Policy

Within the context of this chapter it is not possible to give a comprehensive overview of national regulation and policy by Arctic and non-Arctic states and entities (most notably the EU, but also Taiwan) on the conservation and management of target species and the regulation of the impacts of fishing on non-target species within the marine Arctic. A choice has therefore been made to focus on Arctic Ocean coastal states—in particular the US—and the EU.

In some parts of the marine Arctic, for instance, the North Atlantic, national regulation and policy is expected to be extensive, tailor-made, and related to all or most of the relevant capacities in which states can exercise jurisdiction, namely as flag, coastal, port, and market states and with regard to their natural and legal persons. For other parts of the marine Arctic, however, the presence of ice for most of the year may have rendered tailor-made national fisheries regulation and policy unnecessary. But as diminishing ice-coverage will attract fishing vessels looking for possible new fishing opportunities, all relevant states and entities must ensure that new Arctic fisheries in their own maritime zones and/or by their vessels, comply with applicable global and regional fisheries standards or, where these do not exist, are not conducted in such an unregulated manner that this would be "inconsistent with State responsibilities for the conservation of living marine resources under international law" (FAO 2001, para. 3.3.2).

US action with respect to its EEZ off Alaska in the Arctic Ocean clearly precludes this because it prohibits commercial fishing in the EEZ "until information improves so that fishing can be conducted sustainably and with due concern to other ecosystem components" (NPFMC 2009, sec. E.S.1.2). This proactive and precautionary action by the US is consistent with its Senate joint resolution No. 17 of 2007 mentioned earlier. The 2009 Arctic Region Policy of the US (NSPD-66 2009) contains a few paragraphs on fisheries but not a separate section (secs. III(H)(4) and (6)(b) and (c)).

Norway's laws and regulations relating to fisheries in Svalbard's maritime zones allow 'unregulated fisheries'—but not necessarily in the sense discussed above—to continue or develop, unless explicitly prohibited (Molenaar 2012b). Canada's 2001 'New and Emerging Fisheries Policy'[12] differs from both

[11] These written comments on Sect. 4.1 are on file with the author.

[12] The New and Emerging Fisheries Policy is one of Canada's Fisheries Management Policies and is available at <www.dfo-mpo.gc.ca>. Accessed 8 Jan 2013.

approaches as it neither freezes expansion nor allows unregulated fisheries to develop, but requires licenses for each of its three stages (feasibility, exploratory, and commercial) (Ridgeway 2010).[13] Other Canadian policies would apply to new Arctic fisheries as well.[14]

The precautionary approach as such is probably contained in the legal and policy frameworks of most Arctic Ocean coastal states and other key states and entities, even though not necessarily directly in relation to (new) Arctic fisheries. The approaches of Canada and the US are both precautionary and proactive, even though in different ways. The Kingdom of Denmark's 'Strategy for the Arctic' acknowledges that illegal, unreported, and unregulated fishing is a serious threat, that the lack of knowledge of fish stocks and fishing opportunities calls for the application of the precautionary approach, and that fisheries should not commence where a conservation and management system is not available, while explicitly mentioning the Central Arctic Ocean in this regard (Kingdom of Denmark 2011). Conversely, while Norway's 'The High North. Visions and Strategies' (Norway 2011) emphasizes the need for sustainable and science-based fisheries management and the application of the precautionary approach in a general sense, no attention is devoted to new fisheries.

As regards the EU, several of its policy statements in recent years devote specific attention to Arctic fisheries. Sect. 3.2 of the European Commission's Arctic Communication (European Commission 2008) is specifically devoted to 'Fisheries' and is among other things supportive of a temporary ban on new fisheries. Both the Council of the European Union's conclusions on Arctic issues (Council of the European Union 2009) and a 2011 European Parliament resolution (European Parliament 2011) stressed the need to avoid unregulated fishing as well.

The European Commission's 2008 Arctic Communication still viewed extension of the spatial mandate of NEAFC as the preferred option for addressing the gap in high seas coverage with RFMOs in the Central Arctic Ocean. However, the Council conclusions on Arctic issues do not contain a preferred option, but mention the possibility of extending the mandate of existing RFMOs "or any other proposal to that effect agreed by the relevant parties". The most recent EU policy document (European Commission and High Representative 2012) mentions that RFMOs "could in principle extend their geographical scope". While Iceland may prefer extending the spatial scope of NEAFC, none of the Arctic Ocean coastal states seem supportive of this option or the extension of the spatial scope of other existing RFMOs (Molenaar 2009; IAFS 2010).

---

[13] Note also the May 2010 Report of Canada's Standing Senate Committee on Fisheries and Oceans on 'The Management of Fisheries and Oceans in Canada's Western Arctic' (available at <www.parl.gc.ca>), which recommends Canada to adopt an approach for the Canadian part of the Beaufort Sea that is similar to the United States' Arctic FMP (Recommendation No. 12).

[14] Particularly relevant seem to be the 'Policy for Managing the Impacts of Fishing on Sensitive Benthic Areas' and the 'Policy on New Fisheries for Forage Species' (both available at <www.dfo-mpo.gc.ca>, accessed 8 Jan 2013).

It is not clear if other key high seas fishing states and entities such as China, Iceland, Japan, South Korea, or Taiwan have policies on (new) Arctic fisheries.

These practices and policy statements will function as points of departure in ongoing and future international discussions on the international regime for Arctic Ocean fisheries. The contrast between many of these policy statements—even the more explicit ones—and the US action with respect to its EEZ off Alaska in the Arctic Ocean is nevertheless obvious. Whereas US action already actually constrains fisheries by its nationals, most of the policy statements by other states or entities merely advocate or envisage similar action.

The action by the US is also noteworthy in light of the consensual nature of international law. Rather than awaiting agreement at the international level, the US proactively adopted more stringent domestic regulation unilaterally and thereby essentially created a competitive disadvantage for itself. As some of the fish stocks that occur in the EEZ off Alaska in the Arctic Ocean may be transboundary and in view of a more general preference for a level playing field, the US must have hoped, and must continue to do so, that other states and entities follow with similar actions or actions with similar effectiveness; both for the maritime zones of Arctic Ocean coastal states as well as for the Central Arctic Ocean (IAFS 2010). All states and entities must at any rate ensure that new Arctic Ocean fisheries are conducted in compliance with applicable international law. When policy statements are inadequate for this purpose—which is often the case—they must implement them into domestic regulation, acting in all relevant capacities; for instance, as coastal states, flag states, port states, market states, or with regard to their natural or juridical persons.

5.5 Gaps in the International Legal and Policy Framework and National Regulation and Options for Addressing Them

The existing international legal and policy framework and national regulation for Arctic fisheries contains the following main gaps:

1. Science-based and ecosystem-based fisheries management cannot be ensured due to lack of data;
2. Regulation by Arctic Ocean coastal states and other states and entities may not be adequate;
3. Gaps in Arctic Ocean coastal state fora and instruments; and
4. Gap in high seas coverage with RFMOs.

As a comprehensive discussion of all potential options for addressing these gaps is not possible in this chapter, only some comments and observations are offered here. As regards the lack of data identified in gap No. 1, some progress has already been made among Arctic Ocean coastal states (Anchorage Science Meeting Report 2011) and within ICES and PICES, among other factors due to a joint request to ICES by Norway and the Russian Federation (Molenaar 2012a; Takei 2013). Potential future options include a one-off assessment on Arctic fisheries—within or outside the Arctic Council—or the establishment of a new permanent scientific assessment and advisory body on Arctic fisheries, either self-standing, within the Arctic Council, ICES, or the International Arctic Science Committee (IASC), or established jointly by ICES/PICES. If an assessment will indeed be undertaken, it should as a minimum include plausible future scenarios and take account of the impacts of fisheries intensification and expansion on Arctic indigenous peoples. ICES might be the most likely forum for such initiatives to take place, judging by the fact that, at its 100th Meeting in October 2012, the ICES Council agreed to give its Science Committee a mandate to promote science activities in Arctic waters related to various issues, including expansion of distribution/migration of ranges of commercial fish species (ICES 2012).

As regards the regulatory gap identified in No. 2, reference is made to Sect. 5.4, which contains various options to address gaps. Gaps in Arctic coastal state fora and instruments identified in No. 3 could be addressed by formalizing existing informal cooperation, for instance that between Canada and Greenland (see Sect. 5.3.4). New fora and instruments could relate to the conservation and management of shared, straddling, or anadromous fish stocks or provide a framework for mutual fisheries access and exchange of fishing opportunities. Such fora and instruments could be bilateral, trilateral, or involve all Arctic Ocean coastal states.

The gap in RFMO coverage of the high seas identified in No. 4 has undoubtedly generated the most debate, some of which is already covered in Sect. 5.4. As noted there, insufficient support exists for extending the spatial mandate of NEAFC. A full-fledged RFMO does not seem the most likely option either, due to considerations of cost-effectiveness in light of the fact that significant commercially viable fisheries are not expected in the short term. The negotiation-process on the CBS Convention eventually decided not to push for a full-fledged RFMO for similar reasons (Balton 2001). An Arrangement[15]—whether legally binding or non-legally binding—is therefore a more likely option.

As noted in Sect. 5.3.5, there is no obstacle for such an Arrangement to be adopted by the Arctic Council—in case of a non-legally binding instrument—or through the ACS approach as previously described. If the instrument also related to the Central Arctic Ocean—where the freedom of fishing applies—its effectiveness would benefit from support by key non-Arctic states and entities. Such support could be ensured through a format or mechanism that allows them to participate in the instrument's negotiation as well as to express their consent to be bound. Involving only the current non-Arctic state observers would not work. This group consists of six EU Member States[16]—which have transferred

[15] See *supra* note 1.

[16] Namely France, Germany, the Netherlands, Poland, Spain, and the United Kingdom.

most of their competence on marine capture fisheries to the EU—and therefore does not include any key high seas fishing states and entities such as China, the EU, Japan, South Korea, and Taiwan. While all these except Taiwan have applied for observer status, it is by no means clear when and if these applications will be approved (Molenaar 2012c).[17] If all applications *are* approved, however, involving observers could be an important component of the aforementioned format or mechanism.

An alternative to proceeding through the Arctic Council or the ACS would be to adopt the Arrangement as a stand-alone instrument. If the instrument's spatial scope, and thereby the measures it contains, were limited to the Central Arctic Ocean—as advocated by some[18] -, similar measures or measures with similar effectiveness need to be adopted by Arctic Ocean coastal states for their own maritime zones. The need for compatibility is particularly evident as it is likely that new fishing opportunities will arise in coastal state maritime zones before arising in the high seas. An exception may nevertheless be granted in furtherance of the rights and interests of Arctic indigenous peoples.

While the assumption is that other states and entities besides the Arctic five would be allowed to participate in the negotiation of a stand-alone fisheries instrument on the Arctic Ocean or the Central Arctic Ocean, and eventually become parties thereto, this assumption could of course be proven wrong. A coastal states-only *inter se* approach would not necessarily be inconsistent with international law. This would only occur if the exercise of the right to engage in high seas fishing by other states and entities was interfered with in ways that were not consistent with international law. At-sea high seas enforcement would be an obvious example. Presumably, however, the Arctic five prefer to avoid these issues as well as the lack of legitimacy that is associated with coastal states-only approaches.[19] In a worst-case scenario, such lack of legitimacy might even prompt high seas fishing states or entities to engage in high seas fishing in the Central Arctic Ocean in order to assert their right as such, even if such fishing was not commercially viable. Fortunately, the clear commitment to peace, order, and cooperation that underlies the Ilulissat Declaration implies also a commitment by the Arctic Ocean coastal states to avoid such a scenario.

---

[17] Other non-Arctic state applicants are India (application submitted on 6 Nov 2012; information provided by N. Buvang to the author by email on 7 Feb 2013), Italy, and Singapore.

[18] For instance the Pew Environment Group's 'Oceans North' campaign, which also led to the submission of a letter signed by a large number of scientists to the International Polar Year Conference in Montréal, Canada (22–27 Apr 2012) (info at <oceansnorth.org/international>).

[19] Reference can in this context be made to the controversial 'Galapagos Agreement' (Framework Agreement for the Conservation of the Living Marine Resources on the High Seas of the Southeast Pacific, Santiago, 14 Aug 2000. Not in force, *Law of the Sea Bulletin*, 70–78, No. 45 (2001)), which never entered into force and has now essentially been replaced by the SPRFMO Convention (2009).

5.6 Potential for EU–US Cooperation

A discussion on the potential for EU–US cooperation on Arctic fisheries must acknowledge at the outset that whereas the US is an Arctic Ocean coastal state, the EU cannot rely on such a *de facto* capacity. Whereas Denmark is an Arctic Ocean coastal state with respect to Greenland and an Arctic coastal state with respect to the Faroe Islands, Denmark's EU Membership does not extend to Greenland or the Faroe Islands (TFEU 2008, arts. 204 and 355(5)(a)). But the EU can still act in various other *de facto* capacities; for instance as a flag state—including pursuant to the freedom of fishing on the high seas -, port state, market state, or with respect to the natural and legal persons of its Member States.

The "conservation of marine biological resources under the common fisheries policy" is one of the five areas listed in Article 3(1) of the TFEU (2008) in which the EU has exclusive competence, subject to some exceptions (Churchill and Owen 2010). The consequential external competence of the EU in the sphere of fisheries implies that the EU represents its Member States, for instance in negotiations with non-EU Member States and in RFMOs. In some cases, however, EU Member States can still become members to RFMOs alongside the EU. One of these exceptions relates to 'overseas countries and territories' and enables for instance Denmark to become a member of RFMOs alongside the EU in respect of the Faroe Islands, Greenland, or both; for instance, in relation to NEAFC (for both).

The differences between the EU and the US as regards their involvement in marine capture fisheries worldwide must be acknowledged as well. It is widely known that—compared to US fishing vessels—fishing vessels flying flags of EU Member States operate to a much larger extent on the high seas and within maritime zones of third states (non-EU Member States) in not only the Atlantic Ocean and adjacent seas, but also the Indian Ocean, Pacific Ocean, and Southern Ocean. The fishing fleet of the EU Member States is therefore much more a distant water fishing fleet compared to the US fishing fleet. Furthermore, it is also widely known that natural and legal persons with the nationality of EU Member States—in particular Spain—are more extensively involved as beneficial owners of fishing vessels flying the flag of third states (i.e., non-EU Member States), compared to natural and legal persons with US nationality. For these reasons, the international community views the EU in the domain of marine capture fisheries primarily as a flag state and the US primarily as a coastal state. In view of the widespread tendency to hold flag states responsible—even though not necessarily always justifiably—for failures in regional fisheries management, the US may well be quite critical of the track-record and ability of the EU and its Member States in ensuring a high standard of flag state performance.

These observations should not be interpreted as suggesting that there is no potential for cooperation between the EU and the US on Arctic fisheries. Despite their different entitlements to fisheries resources in the marine Arctic, the EU and the US share common interests in avoiding over-exploitation of target species and undesirable impacts of fisheries on non-target species, including marine mammals. Moreover, as noted in Sect. 5.4, both the EU and the US take the view that unregulated fishing in the Central Arctic Ocean must be avoided.

A suitable domain for cooperation between the EU and the US is research on the marine Arctic in general and Arctic ecosystems, commercial fish species, and fisheries in particular. Such cooperation could take place within the context of existing bodies like ICES, PICES, and the Arctic Council. As regards the Arctic Council, it seems that the success of US–EU cooperation would benefit from US support for EU observer status. The US could also consider advocating that Arctic Ocean coastal states expand participation in their fisheries science meetings with representatives of the EU.

In light of their shared concerns and positions on unregulated fishing in the Central Arctic Ocean, cooperation in that regard could be considered as well. The fact that the EU is not an Arctic Ocean coastal state and incapable of acting in such a *de facto* capacity, requires the US and the EU to proceed carefully in order not to antagonize Arctic Ocean coastal states that feel strongly about the lead role they should have in any intergovernmental consultations on international regulation of Arctic Ocean fisheries. Joint EU–US initiatives therefore run the risk of being counterproductive. Individual action in support of shared concerns and positions could avoid this. The EU could express public support for a position aligned with that of the US, for instance as laid down in the latter's Senate joint resolution No. 17 of 2007. Moreover, both the EU and the US could take individual proactive steps, for example by adopting regulations that impose a temporary ban on fishing by their vessels in the Central Arctic Ocean until such time as fishing would be permitted pursuant to an international instrument.

Even though the US and the EU should proceed carefully, their cooperation on unregulated fishing in the Central Arctic Ocean offers clear opportunities as well. The effectiveness of a future international instrument on Central Arctic Ocean fisheries would benefit from support by key non-Arctic Ocean coastal states and entities (see Sect. 5.5), and early EU support would be crucial for fostering support among them. The US could make an important contribution in this regard by advocating the inclusion of the notion of compatibility—between fisheries regulation on the high seas and fisheries regulation in coastal state maritime zones—in the future instrument. While the future instrument could also be made applicable to the Arctic Ocean as a whole, there is no indication that the US or even the EU would find this desirable; even though not necessarily for similar reasons.

Support by the EU and other key non-Arctic Ocean coastal states and entities for a future international instrument on Central Arctic Ocean fisheries also depends on their ability to participate in a meaningful way in the instrument's negotiation. Such participation would enhance the legitimacy and credibility of the negotiation-process and thereby also the instrument adopted by it. The US could contribute to these ends by undertaking efforts to ensure that preparatory consultations among Arctic Ocean coastal states have not advanced to such an extent that key non-Arctic Ocean coastal states and entities are essentially presented with a *fait accompli*. Improvements in the EU's and its Member States' performance as *de facto* flag states and coastal states in fisheries management and conservation more broadly, would also help to secure support for broader participation in negotiations. If it is decided to conduct such negotiations within the Arctic Council or by means of the ACS approach, the success of US-EU cooperation would also benefit from US support for EU observer status.

5.7 Conclusions

The unprecedented pace of change that the Arctic is currently experiencing makes it very difficult to argue that the current international legal and policy framework for Arctic fisheries conservation and management is adequate for responding to the huge challenges that lie ahead. This chapter identifies a number of gaps in the international framework as well as in national regulation and policy, and suggests various options for addressing these gaps.

Most of the suggested options to address identified gaps also offer opportunities for cooperation between the EU and the US. Cooperative initiatives could be undertaken to strengthen scientific research, 'domestic' fisheries regulation and the international regime for Arctic Ocean fisheries. The fact that the US is an Arctic Ocean coastal state and the EU is not, and is also incapable of acting in such a *de facto* capacity, requires them to proceed carefully but offers clear opportunities as well. Joint EU–US initiatives could antagonize Arctic Ocean coastal states who feel strongly about the lead role they should have in any intergovernmental consultations on international regulation of Arctic Ocean fisheries. Individual action in support of shared concerns and positions could avoid this.

Early EU support for a future international instrument on Central Arctic Ocean fisheries would be crucial for fostering support among other key non-Arctic states and entities and could thereby contribute to the future instrument's effectiveness, legitimacy, and credibility. The US could play a crucial role in this regard by advocating the inclusion of the notion of compatibility in the future instrument and by undertaking efforts to ensure that the EU and key other non-Arctic Ocean coastal states and entities can participate in a meaningful way in the instrument's negotiation.

References

ACIA (2005) Arctic Climate Impact Assessment. Cambridge University Press, New York

Agreement on Mutual Fisheries Relations between the United States and Union of Soviet Socialist Republics (1988) Agreement between the Government of the United States of America and the Government of the Union of Soviet Socialist Republics on Mutual Fisheries Relations, 31 May 1988, Treaties and other International Acts Series 11,422. Entered into force 28 Oct 1988

Agreement on Mutual Fishery Relations between Denmark and Greenland (1992) Agreement between the Government of the Kingdom of Denmark and the Local Government of

Greenland, on the one hand, and the Government of the Russian Federation, on the other hand, concerning Mutual Fishery Relations between Greenland and the Russian Federation, 7 Mar 1992, 1719 U.N.T.S. 89. Entered into force provisionally on 7 Mar 1992 and definitively on 16 Oct 1992

Agreement on Mutual Fishery Relations between Greenland/Denmark and Norway (1992) Agreement between Greenland/Denmark and Norway concerning Mutual Fishery Relations, Copenhagen, 9 June 1992, 1829 U.N.T.S. 223. Entered into force provisionally with retroactive effect from 24 Sept 1991 and definitively on 4 Mar 1994

AMAP (1997) Arctic Pollution Issues: A State of the Arctic Environment Report. Arctic Monitoring and Assessment Programme, Oslo

Anchorage Science Meeting Report (2011) Report of a Meeting of Scientific Experts on Fish Stocks in the Arctic Ocean. Anchorage, Alaska, 15–17 June 2011; on file with the author

AOR (2011) Phase I Report (2009–2011) of the Arctic Ocean Review (AOR) project. <www.pame.is>. Accessed 29 Jan 2013

AOR (2013) The Arctic Ocean review project. Phase II Report 2011–2013, consolidated version of 9 Feb 2013, on file with the author

Arctic SAR Agreement (2011) Agreement on Cooperation on Aeronautical and Maritime Search and Rescue in the Arctic, 12 May 2011, 50 I.L.M. 1119 (2011). Entered into force on 19 Jan 2013

Balton D (2001) The Bering Sea Doughnut Hole convention: Regional Solution, Global Implications. In: Stokke OS (ed) Governing High Seas Fisheries: the Interplay of Global and Regional Regimes. Oxford University Press, Oxford, pp 143–177

Barnes R (2011) International Regulation of Fisheries Management in Arctic Waters. German Yearbook Int Law 54:193–230

Bloom ET (1999) Establishment of the Arctic Council. Am J Int Law 93:712–722

CBS Convention (1994) Convention on the Conservation and Management of Pollock Resources in the Central Bering Sea, 16 June 1994, 34 I.L.M (1995). Entered into force 8 Dec 1995

Churchill RR (2001) Managing Straddling Fish Stocks in the North-East Atlantic: A Multiplicity of Instruments and Regime Linkages—but How Effective a Management? In: Stokke OS (ed) Governing High Seas Fisheries: The Interplay of Global and Regional Regimes. Oxford University Press, Oxford, pp 235–272

Churchill RR, Owen D (2010) The EC Common Fisheries Policy. Oxford University Press, Oxford

Convention on the Conservation and Management of High Seas Fisheries Resources in the North Pacific Ocean (2012) Tokyo, 24 Feb 2012. Not in force; on file with author

Compliance Agreement (1993) Agreement to Promote Compliance with International Conservation and Management Measures by Fishing Vessels on the High Seas, 24 Nov 1993, 33 I.L.M. 969 (1994). Entered into force 24 Apr 2003

Council of the European Union (2009) Council conclusions on Arctic issues, 2985th Foreign Affairs Council meeting, 8 Dec 2009

European Commission (2008) Communication from the Commission to the European Parliament and the Council. The European Union and the Arctic Region, 20 Nov 2008, COM(2008) 763

European Commission and High Representative (2012) Joint Communication to the European Parliament and the Council. Developing a European Union Policy towards the Arctic Region: progress since 2008 and next steps. 26 June 2012, JOIN(2012) 19 final

European Parliament (2011) A sustainable EU policy for the High North. European Parliament resolution of 20 Jan 2011 on a sustainable EU policy for the High North (2009/2214(INI)), 20 Jan 2011, P7_TA(2011)0024

FAO (1995) Code of Conduct for Responsible Fisheries. Adopted by the Twenty-eight Session of the FAO Conference, 31 Oct 1995

FAO (2001) International Plan of Action to Prevent, Deter and Eliminate Illegal, Unreported and Unregulated Fishing. Adopted by consensus by FAO's committee on Fisheries on 2 Mar 2001 and endorsed by the FAO Council on 23 June 2001

FAO (2008) International Guidelines for the Management of Deep-Sea Fisheries in the High Seas, 29 Aug 2008 (contained in Appendix F to the Report of the Technical Consultation on

International Guidelines for the Management of Deep-Sea Fisheries in the High Seas, 4–8 Feb and 25–29 Aug 2008 (FAO Fisheries and Aquaculture Report No. 881))

FAO (2010) International Guidelines on Bycatch Management and Reduction of Discards, Rome, 10 Dec 2010 [contained in Appendix E to the Report of the Technical Consultation to Develop International Guidelines on Bycatch Management and Reduction of Discards, 6–10 Dec 2010 (FAO Fisheries and Aquaculture Report No. 957)]

Fish Stocks Agreement (1995) Agreement for the Implementation of the Provisions of the United Nations Convention on the Law of the Sea of 10 Dec 1982 relating to the Conservation and Management of Straddling Fish Stocks and Highly Migratory Fish Stocks, 4 Aug 1995, 2167 U.N.T.S. 3. Entered into force 11 Dec 2001

Hoel AH (2011) Living marine resources in the Arctic. Trends & Opportunities. Presentation dated 20 Sept 2011 at the 2nd Arctic Ocean review (AOR) project workshop, Reykjavik, on file with author

IAFS (2010) In: Proceedings of the International Arctic Fisheries Symposium. Managing Resources for a Changing Arctic, pp 23–32 (2010). <www.nprb.org/iafs2009>. Accessed 29 Jan 2013

ICCAT Convention (1966) International Convention for the Conservation of Atlantic Tunas, 14 May 1966, 673 U.N.T.S. 63. Entered into force 21 Mar 1969. As amended by Protocols adopted in 1984 and 1992, which both entered into force

ICES (2012) Report of the 100th ICES Council Meeting, 24–25 Oct 2012. Copenhagen: ICES Headquarters. <www.ices.dk>. Accessed 29 Jan 2013

IPHC Convention (1953) Convention for the Preservation of the Halibut Fishery of the North Pacific Ocean and the Bering Sea, 2 Mar 1953, 22 U.N.T.S. 78 (1955). Entered into force 28 Oct 1953. Exchange of Notes Constituting an Agreement to Amend the [IPHC Convention], Washington, 29 Mar 1979, 1168 U.N.T.S. 380 (1980). Entered into force 29 Mar 1979

Kingdom of Denmark (2011) Denmark, Greenland and the Faroe Islands: Kingdom of Denmark Strategy for the Arctic 2011–2020. <www.nanoq.gl>. Accessed 29 Jan 2013

LOS Convention (1982) United Nations Convention on the Law of the Sea, 10 Dec 1982, 1833 U.N.T.S. 396. Entered into force 16 Nov 1994

Molenaar EJ (2007) Port State Jurisdiction: Toward Comprehensive, Mandatory and Global Coverage. Ocean Dev Int Law 38:225–257

Molenaar EJ (2009) Arctic Fisheries Conservation and Management: Initial Steps of Reform of the International Legal Framework. Yearbook Polar Law 1:427–463

Molenaar EJ (2012a) Arctic Fisheries and International Law. Gaps and Options to Address Them. Carbon Climate Law Rev 1:63–77

Molenaar EJ (2012b) Fisheries Regulation in the Maritime zones of Svalbard. Int J Marine Coastal Law 27:3–58

Molenaar EJ (2012c) Current and Prospective Roles of the Arctic Council System within the Context of the Law of the Sea. Int J Marine Coastal Law 27:553–595

NAFO Convention (1978) Convention on Future Multilateral Cooperation in the Northwest Atlantic Fisheries, 24 Oct 1978, 1135 U.N.T.S. 369. Entered into force 1 Jan 1979. 2007 Amendment, Lisbon, 28 Sept 2007. Not in force (at 27 Nov 2012), NAFO/GC Doc. 07/4

NASCO Convention (1982) Convention for the Conservation of Salmon in the North Atlantic Ocean, 2 Mar 1982, 1338 U.N.T.S. 33 (1983). Entered into force 1 Oct 1983

NEAFC Convention (1980) Convention on Future Multilateral Cooperation in the North-East Atlantic Fisheries, 18 Nov 1980, 1285 U.N.T.S. 129. Entered into force 17 Mar 1982. 2004 Amendments (Art. 18bis), London; 12 Nov 2004. Not in force (at 27 Nov 2012), but provisionally applied by means of the 'London Declaration' of 18 Nov 2005. 2006 Amendments, London (Preamble, Arts 1, 2 and 4), 11 Aug 2006. Not in force, but provisionally applied by means of the 'London Declaration' of 18 Nov 2005

Norway (2011) The High North. visions and Strategies, 18 Nov 2011

Norwegian-Russian Bilateral Framework Agreement (1975) Agreement between the Government of the Kingdom of Norway and the Government of the Union of Soviet Socialist Republics on Cooperation in the Fishing Industry, 11 Apr 1975, 983 U.N.T.S. 1975. Entered into force 11 Apr 1975

NPAFC Convention (1992) Convention for the Conservation of Anadromous Stocks in the North Pacific Ocean, 11 Feb 1992. 22 Law Sea Bull 21 (1993). Entered into force 16 Feb 1993

NPFMC (2009) Arctic Fishery Management Plan. United States North Pacific Fishery Management Council. Adopted 5 Feb 2009, effective 3 Dec 2009. 50 C.F.R. 679

NSPD-66 (2009) National Security Presidential Directive and Homeland Security Presidential Directive. Arctic Region Policy, 9 Jan 2009, NSPD-66/HSPD-25

Ottawa Declaration (1996) Declaration on the Establishment of the Arctic Council, 19 Sept 1996, 35 I.L.M. 1382 (1996)

Pacific Salmon Treaty (1985) Treaty between the Government of Canada and the Government of the United States of America Concerning Pacific Salmon, 28 Jan 1985. 1469 U.N.T.S. 358 (1987). Entered into force 18 Mar 1985. As amended

PAME (2009) Arctic Offshore Oil and Gas Guidelines. Last updated 29 Apr 2009

Port State Measures Agreement (2009) Agreement on Port State Measures to Prevent, Deter and Eliminate Illegal, Unreported and Unregulated Fishing, 22 Nov 2009. Not in force (at 27 Nov 2012)

Ridgeway L (2010) Issues in Arctic Fisheries Governance: A Canadian Perspective. In: Nordquist MH, Heidar TH, Moore JN (eds) Changes in the Arctic Environment and the Law of the Sea. Martinus Nijhoff Publishers, Canada, pp 409–446

SAO (2007) Report of the November 2007 SAOs Meeting. <www.arctic-council.org>. Accessed 29 Jan 2013

S.J. Res. 17 (2007) A joint resolution directing the United States to initiate international discussions and take necessary steps with other Nations to negotiate an agreement for managing migratory and transboundary fish stocks in the Arctic Ocean, 110th Congress, 2007

Spitsbergen Treaty (1920) Treaty concerning the Archipelago of Spitsbergen, 9 Feb 1920, 2 L.N.T.S. 7. Entered into force 14 Aug 1925

SPRFMO (2008) Draft Convention on the Conservation and Management of High Seas Fishery Resources in the South Pacific Ocean. Doc. SP/06/WP1, Revision 4 (Oct 2008). Text at <www.southpacificrfmo.org>

SPRFMO Convention (2009) Convention on the Conservation and Management of High Seas Fishery Resources in the South Pacific Ocean, 14 Nov 2009. Entered into force 24 Aug 2012

Takei Y (2013) Marine Scientific Research in the Arctic. In Molenaar EJ, Oude Elferink AG, Rothwell DR (eds) The law of the Sea and Polar Regions: Interactions between Global and Regional Regimes. Martinus Nijhoff Publishers

TFEU (2008) Treaty on the Functioning of the European Union. Consolidated Version. O.J. C 115/47

Trilateral Loophole Agreement between Iceland, Norway, and the Russian Federation (1999) Agreement between the Government of Iceland, the Government of Norway and the Government of the Russian Federation Concerning Certain Aspects of Cooperation in the Area of Fisheries, 15 May 1999. 41 Law Sea Bull 53 (1999). Entered into force 15 July 1999

WCPFC Convention (2000) Convention on the Conservation and Management of Highly Migratory Fish Stocks in the Western and Central Pacific Ocean, 5 Sept 2000, 2275 U.N.T.S. 2004 (2007) Entered into force 19 June 2004

Zeller D, Booth S, Pakhomov E, Swartz W, Pauly D (2011) Arctic fisheries catches in Russia, USA, and Canada: baselines for neglected ecosystems. Polar Biol 34:955–973

Chapter 6
Status and Reform of International Arctic Shipping Law

Erik J. Molenaar

Abstract The last years have witnessed a continuous decrease in Arctic sea ice coverage and thickness and a significant expansion in the volume of shipping traffic. This chapter assesses the adequacy of the current international legal and policy framework for Arctic shipping in view of the likelihood that these trends continue in coming years. An overview of the international legal and policy framework is provided, the main gaps therein are identified, and options for addressing these are suggested. Options could be undertaken within the International Maritime Organization (IMO) but also outside the IMO, for instance within the Arctic Council or among ad hoc groupings of states. Separate attention is devoted to the potential for cooperation between the European Union and the United States in this regard.

6.1 Introduction

As future trends in Arctic marine shipping depend to a significant extent on sea ice coverage and thickness, it is important to note that both have been steadily declining. The summer 2012 Arctic sea ice extent was the lowest on record, the

Writing this chapter was facilitated by funding from the Netherlands Polar Programme and the EU's COST Action IS1105 'NETwork of experts on the legal aspects of MARitime SAFEty and security (MARSAFENET)'. The author is grateful for comments by Carien Droppers, Henrik Ringbom, Greg O'Brien, and others on an earlier version of the chapter. This chapter builds on Molenaar EJ, Corell R (2009). Background Paper. Arctic Shipping. Arctic TRANSFORM.

E. J. Molenaar (✉)
Netherlands Institute for the Law of the Sea (NILOS), Utrecht University, Utrecht, The Netherland
Faculty of Law, University of Tromsø, Tromsø, Norway
e-mail: E.J.Molenaar@uu.nl

E. Tedsen et al. (eds.), *Arctic Marine Governance*,
DOI: 10.1007/978-3-642-38595-7_6,

December 2012 sea ice was the second lowest on record, and—perhaps even more important—multi-year sea ice has declined even further after the enormous loss in the summer of 2007. For several summers now, both the Northwest Passage and the Northern Sea Route have been open.[1]

Intra- and trans-Arctic marine shipping can be interesting alternatives for the much longer routes using the Panama and Suez Canals, or for Arctic routes that are partly terrestrial and partly marine. But even though summers without sea ice in much or all of the Arctic Ocean may occur in the not too distant future, sea ice will still be widespread in winter. While much or most of this will be first-year sea ice, there may be other factors that could adversely affect shipping conditions and care must therefore be taken not to overestimate the potential growth of Arctic marine shipping (Kraska 2007; Brigham 2010; ICS 2012). However, even a limited expansion of intra- and trans-Arctic marine shipping requires an assessment of the adequacy of national and international regulation. The 1989 *Exxon Valdez* disaster has not been forgotten and the 2010 crisis with the *Deepwater Horizon* in the Gulf of Mexico painfully exposed human failure on numerous counts, including a tendency to underestimate risks. In the summer of 2010, this tendency was also illustrated by two groundings in Canadian waters: a tanker carrying fuel, close to Pangnirtung, and the cruise-vessel *Clipper Adventurer*, just east of Kugluktuk.[2]

This chapter assesses the adequacy of the current international legal framework for Arctic shipping in light of the current and expected impacts of global climate change on the marine Arctic. The focus is predominantly on the impacts of Arctic marine shipping on the Arctic marine environment and its biodiversity. The maritime safety dimension is covered where regulation serves a significant subsidiary purpose of pollution prevention. Delimiting the scope of this chapter in this way, however, is not meant to suggest that ensuring maritime safety in the marine Arctic is not as important or urgent, particularly in view of the continued and increasing interest in Arctic sea-borne tourism.

After providing some context and background information on current and future Arctic marine shipping in Sect. 6.2, an overview of the international legal and policy framework for Arctic marine shipping is provided in Sect. 6.3. Gaps in the international legal and policy framework and options for addressing them are covered in Sect. 6.4 and the potential for EU-US cooperation is examined in Sect. 6.5. Some conclusions are offered in Sect. 6.6.

For the purposes of this chapter, Arctic marine shipping is regarded as the shipping that occurs or could occur in the marine Arctic. This chapter uses the same definitions for the terms 'marine Arctic', 'Arctic Ocean', 'Arctic states', and 'Arctic Ocean coastal states' as throughout this book. It is worth noting that

---

1 Information obtained from <nsidc.org/arcticseaicenews>, <www.arctic.noaa.gov/reportcard>, and <www.climatewatch.noaa.gov> on 11 Jan 2013.

2 See the press releases and other information at <barentsobserver.com>, <www.arcticmonitor.net>, and <www.institutenorth.org>.

the definition of marine Arctic is a broader area than the 'Arctic waters' under the International Maritime Organization (IMO) Polar Shipping Guidelines (2009), as described in Chap. 1.

Arctic marine shipping can be intra-Arctic or trans-Arctic. Trans-Arctic marine shipping can take place by means of various routes and combinations of routes. Two of these routes are the Northwest Passage and the Northern Sea Route. The official Northern Sea Route encompasses all routes across the Russian Arctic coastal seas from Kara Gate (at the southern tip of Novaya Zemlya) to the Bering Strait (Tymchenko 2001). The Northwest Passage is the name given to the marine routes between the Atlantic and Pacific Oceans along the northern coast of North America that span the straits and sounds of the Canadian Arctic Archipelago. Pharand identified seven main routes, with minor variations (Pharand 2007). An alternative to all these routes is the Central Arctic Ocean Route, which runs straight across the middle of the Central Arctic Ocean.

6.2 Current and Future Arctic Marine Shipping

While the current volume of Arctic marine shipping is still very modest, recent years have seen steady—and sometimes significant—increases in many types of shipping, including transits through the Northwest Passage and the Northern Sea Route, destinational traffic associated with offshore resource activity, and Arctic sea-borne tourism. As regards transits, various records and first-evers have recently occurred; for instance: back-to-back crossings of both the Northwest Passage and the Northern Sea Route in a single summer by yachts; the transit of the Northwest Passage by *The World* in the summer of 2012; and—in relation to the Northern Sea Route—tankers carrying oil, gas condensate, and liquid natural gas (LNG); the first ferry (the *Georg Ots*); the first non-Russian bulk-carrier (the *MV Nordic Barents*) carrying iron-ore concentrate without stopping at a Russian port; and the first cargo vessel (the *Monchegorsk*) to transit without ice-breaker assistance. However, regular container ship operations through the Northern Sea Route have not proven commercially viable.[3]

Indications are that traffic in the Northwest Passage will grow much less compared to the Northern Sea Route (AMSA 2009; AOR 2013). This is to some extent caused by projections about the presence of sea ice and natural restraints (e.g., shallowness) in parts of the Northwest Passage. Another important factor is that Russia is much more interested in developing Arctic marine shipping than Canada, and has made large efforts in support of development (Emmerson 2011). In addition to substantial investments in infrastructure, vessels, and equipment, and reforms in domestic legislation and institutional arrangements (Solski 2013),

---

[3] See the press releases and other information at <barentsobserver.com>, <www.arcticmonitor.net>, and <www.institutenorth.org>.

reference can be made to the strategic agreement adopted in 2010 between the Sovcomflot Group and the China National Petroleum Corporation, which envisions increased usage of the Northern Sea Route.[4] The 'shale gas revolution' has also meant that Russian gas originally intended to be shipped to the east coast of the United States (US), is now shipped through the Northern Sea Route to China and other Asian states. This is a good example of the role of key variables, uncertainties, or 'wildcards' in the scenarios developed as part of the Arctic Marine Shipping Assessment (AMSA). Examples of wildcards are an accelerated Arctic meltdown, major Arctic shipping disasters, and technology breakthroughs (AMSA Scenarios 2008). The risk assessments of classification societies and the marine insurance industry are also likely to be a crucial factor for the economic viability of all Arctic marine shipping.

Due to the accelerated melting of Arctic sea ice, the Central Arctic Ocean Route may soon be an option as well (Humpert and Raspotnik 2012). The most suitable course of this latter route may vary from year to year and lead to various combinations of the Central Arctic Ocean Route on the one hand and the Northwest Passage and Northern Sea Route on the other hand. Some of the routes of which the Northern Sea Route consists already pass through the high seas area of the Central Arctic Ocean. It is finally important to note that all trans-Arctic marine shipping must pass through the Bering Strait.

Marine shipping has the following actual and potential impacts on the marine environment and marine biodiversity:

1. Shipping practices and incidents leading to accidental discharges of polluting substances (cargo or fuel) or physical impact on components of the marine ecosystem (e.g., on the benthos and larger marine mammals);
2. Operational discharges (cargo residues, fuel residues (sludge), (incineration of) garbage and sewage), and emissions;
3. Introduction of alien organisms through ballast-water exchanges or attachment to vessel hulls (e.g., in crevices); and
4. Other navigation impacts (noise pollution and other forms of impacts on, or interference with, marine species potentially causing, for instance, disruption of behaviour, abandonment, or trampling of the young by fleeing animals or displacement from normal habitat).

All these actual and potential impacts are also relevant for Arctic marine shipping. The likelihood for some of these impacts—for instance shipping incidents—to occur is higher in some parts of the marine Arctic due to the presence of ice(bergs), lack of accurate charts, and in the case of insufficient experience in navigating in ice-covered areas. In addition, cold temperatures may affect machinery, and icing can create additional loads on the hull, propulsion systems, and appendages (VanderZwaag et al. 2008). The remoteness of much of the marine Arctic, the limited available maritime safety information data, and the challenges

[4] See the press release of 22 Nov 2010 at <www.sovcomflot.ru>.

of navigating therein moreover mean that once shipping incidents do occur, a response will take a relatively long time and may even then be inadequate to address impacts on the marine environment and marine biodiversity.

6.3 International Legal and Policy Framework for Arctic Marine Shipping

6.3.1 Interests, Rights, Obligations, and Jurisdiction

The international legal and policy framework for vessel-source pollution seeks to safeguard the different interests of the international community as a whole with those of states that have rights, obligations, or jurisdiction in their capacities as flag, coastal, or port states or with respect to their natural and legal persons. While the term 'flag state' is commonly defined as the state in which a vessel is registered and/or whose flag it flies (LOS Convention 1982, art. 91(1)), there are no generally accepted definitions for the terms 'coastal state' or 'port state'. For the purposes of this chapter, however, the term 'coastal state' refers to the rights, obligations, and jurisdiction of a state within its own maritime zones over foreign vessels.

The term 'port state' refers to the rights, obligations, and jurisdiction of a state over foreign vessels that are voluntarily in one of its ports. In order to avoid an overlap with jurisdiction by coastal states, this chapter regards port state jurisdiction as relating to illegal discharges by foreign vessels beyond the coastal state's maritime zones as well as over violations of conditions for entry into port (Molenaar 2007).

The balance in the above-mentioned framework is first of all between the socioeconomic interests of flag states in unimpeded navigation and a minimum of globally uniform international regulation, and the environmental interests of the coastal state. The port state commonly seeks to balance its local environmental interests and the broader environmental interests that 'its' coastal state has over its maritime zones, against the socioeconomic interests of the port and its *hinterland*. States generally have interests, rights, obligations, and jurisdiction in more than one capacity. This commonly leads to a more balanced compromise position, but occasionally also to contradictory positions of the same state within different fora. There is no reason or indication to assume that Arctic states are different in this regard.

The interests of the international community—e.g., sustainable utilization, protection, and preservation of the marine environment, and conservation of marine biodiversity—normally overlap with those of flag, coastal, and port states, but are usually broader and more general. The interests of some states, however, clearly undermine those of other states and the international community, for instance, by not ensuring that their ships comply with international minimum standards or by allowing foreign vessels in their ports to be non-compliant with international minimum standards. These states, vessels, and ports thereby have a competitive advantage over states, vessels, and ports that *do* comply with international minimum standards. Such 'free riders' clearly benefit from the consensual nature of international law—meaning that a state can only be bound to a rule of international law when it has in one way or another consented to that rule.

6.3.2 Substantive Shipping Standards

The categories of substantive shipping standards below are based on the substantive focus of this chapter (see Sect. 6.1), the LOS Convention's (1982) jurisdictional framework for vessel-source pollution, and practice within IMO so far. The categories are:

1. Discharge and emission standards, including standards relating to ballast water exchange;
2. Construction, design, equipment, and manning (CDEM) standards, including fuel content specifications and anti-fouling and ballast water treatment standards;
3. Navigation standards, in the form of ships' routeing measures, ship reporting systems (SRSs), and vessel traffic services (VTS);
4. Contingency planning and preparedness standards; and
5. Liability, compensation, and insurance requirements.

This categorization is merely meant to facilitate the discussion below, however. It does not capture the entire spectrum of types of standards or requirements developed within IMO or applied by individual states acting in their various capacities. An Arctic Ocean coastal state may for instance require use of ice-breaker assistance and the payment of fees for such services.

6.3.3 Global and Regional Bodies

International regulation of vessel-source pollution by merchant ships is primarily done by global bodies. While IMO is the most prominent, the International Labour Organization (ILO) and the International Atomic Energy Agency (IAEA) adopt relevant regulation as well. The pre-eminence of global bodies is a direct consequence of the global nature of international shipping and the interest of the international community in globally uniform international minimum regulation (VanderZwaag et al. 2008; Chircop 2009; Molenaar 2010). The LOS Convention safeguards this by allowing unilateral coastal state prescription in only a few situations. These exceptions are explained in Sect. 6.3.4, which also devotes attention to so-called 'residual port state jurisdiction'.

As unilateral coastal state prescription and residual port state jurisdiction can also be exercised collectively (in concert), regional regulation of marine shipping by regional bodies or *ad-hoc* groupings of states is not inconsistent with the

LOS Convention, and thereby also not with the primary role accorded to the IMO by the LOS Convention (Stokke 2012). Regional regulation can also be pursued on an *inter se* basis to ships flying the flag of parties—for instance, Annex IV on 'Prevention of Marine Pollution' to the Antarctic Treaty's (1959), Protocol on Environmental Protection (1991).

Several IMO instruments also allow or encourage regional implementation. This has led the Arctic Council to undertake efforts to implement IMO's SAR (Search and Rescue) Convention (1979) by means of the Arctic SAR Agreement (2011), and IMO's OPRC 90 (1990) by means of the future Arctic MOPPR (Marine Oil Pollution Preparedness and Response) Agreement (2013), which is scheduled to be signed at the Arctic Council's Ministerial Meeting in Kiruna in May 2013.

Regional action within the broad spectrum of monitoring, surveillance, inspection, and enforcement is consistent with the LOS Convention as well, even though the Convention does not explicitly encourage it. IMO has frequently encouraged such regional action, for instance by means of the 1991 IMO Assembly Resolution A.682(17) 'Regional Co-operation in the Control of Ships and Discharges', which triggered the creation of a global network of regional arrangements on port state control (PSC) modelled on the then already almost decade-old Paris MOU Convention (1982).

The IMO bodies of most relevance to this chapter are the Marine Environment Protection Committee (MEPC), the Maritime Safety Committee (MSC), and the latter's Sub-Committee on Navigation (NAV) and its Sub-Committee on Design and Equipment (DE). Amendments to MARPOL 73/78 (1973/1978) are adopted by the MEPC and amendments to SOLAS 74 (1974) by the MSC. The MEPC has a coordinating role in relation to particularly sensitive sea areas (PSSAs) and the MSC has the authority to adopt mandatory SRSs and VTS pursuant to the SOLAS 74 and COLREG 72 (1972). Proposals for many of the associated protective measures (APMs) that are made applicable within PSSAs are first discussed in the NAV. The ongoing process to develop a mandatory Code for Shipping in Polar Waters (Polar Code) takes predominantly place within DE.

Other bodies relevant to Arctic marine shipping are the International Association of Classification Societies (IACS)—in particular on account of its Unified Requirements concerning Polar Class -, the Association of Arctic Expedition Cruise Operators (AECO), the Arctic Council—in particular through the efforts of its Protection of the Arctic Marine Environment (PAME) and Emergency, Prevention, Preparedness, and Response (EPPR) working groups -, and the OSPAR Commission.

6.3.4 Global and Regional Instruments

LOS Convention

Most of the LOS Convention's provisions on vessel-source pollution are laid down in its Part XII, entitled 'Protection and Preservation of the Marine Environment'. This part begins with Chap. 1, entitled 'General Provisions' and applies to all sources of pollution. Its first provision, Article 192 lays down the general obligation for all states—in whatever capacity therefore—"to protect and preserve the marine environment". This is elaborated in Article 194 with regard to measures to prevent, reduce, and control pollution of the marine environment; aimed specifically at vessel-source pollution in paragraph (3)(b). Other relevant general obligations relate to rare or fragile ecosystems and the habitat of endangered species (art. 194(5)), introduction of alien species (art. 196), cooperation on a global or regional basis (art. 197), contingency plans against pollution (art. 199), monitoring of the risks or effects of pollution (art. 204), and assessment of potential effects of activities (art. 206). Sections 5 and 6 of Part XII contain separate provisions on prescription and enforcement for all each of the sources of pollution.

The jurisdictional framework relating to vessel-source pollution laid down in the LOS Convention is predominantly aimed at flag and coastal states. Apart from one explicit provision (art. 218), port state jurisdiction is only implicitly dealt with (see 'Port state jurisdiction' below).

Prescriptive jurisdiction by flag and coastal states is linked by means of rules of reference to the notion of 'generally accepted international rules and standards' (GAIRAS). These refer to the technical rules and standards laid down in instruments adopted by regulatory bodies, in particular IMO. It is likely that the rules and standards laid down in legally binding IMO instruments that have entered into force can at any rate be regarded as GAIRAS (Molenaar 1998).

The basic duty for flag states to exercise effective jurisdiction and control over ships flying their flag laid down in Article 94 of the LOS Convention is further specified in Article 211(2), which stipulates that flag state prescriptive jurisdiction over vessel-source pollution is mandatory and must at least have the same level as GAIRAS. It is therefore up to flag states to require their vessels to comply with more stringent standards than GAIRAS, but this will of course impact their competitiveness.

This mandatory minimum level of flag state jurisdiction established by the LOS Convention is balanced by according all states the following navigational rights:

1. The right of innocent passage—suspendable or non-suspendable -, in territorial seas, archipelagic waters outside routes normally used for international navigation or—if designated—archipelagic sea lanes, internal waters pursuant to Article 8(2) of the LOS Convention, and certain straits used for international navigation;
2. The right of transit passage in straits used for international navigation;
3. The right of archipelagic sea lanes passage within routes normally used for international navigation or—if designated—archipelagic sea lanes; and
4. The freedom of navigation within exclusive economic zones (EEZs) and on the high seas.

Coastal state prescriptive jurisdiction over vessel-source pollution is optional under the LOS Convention but, if exercised, cannot be more stringent than the level of GAIRAS (LOS Convention 1982, arts. 21(2), 39(2) and 211(5)). This is the general rule even though it is subject to some exceptions that are discussed further below.

Straits Used for International Navigation

The general rule just mentioned is also applicable to marine areas where the right of transit passage applies (LOS Convention 1982, arts. 41 and 42(1)(a) and (b)). This regime was developed for narrow international straits that would no longer have a high seas corridor once strait states would extend the breadth of their territorial seas to 12 nautical miles (nm). The applicability of the regime of transit passage is nevertheless dependent on various conditions.

One of these is laid down in Article 37 and stipulates that the regime of transit passage only applies to "straits which are used for international navigation". Diverging views exist on the words "are used", where the normal meaning points to 'actual' and not 'potential' usage. The latter is adhered to by the US, which takes the view that "the term 'used for international navigation' includes all straits capable of being used for international navigation" (Sen. Exec. Rep. 110-9 2007). Conversely, Canada and the Russian Federation take the view that the words refer to actual usage, and most commentators embrace this interpretation as well (e.g., Rothwell 1996; Pharand 2007). Close reading of the ICJ's Judgment in the *Corfu Channel* case[5]—from which the phrase originates—nevertheless reveals that it also touches on potential usage (cf. McDorman 2010).

Consistent with its above view on potential usage, the US regards the Northwest Passage and parts of the Northern Sea Route as straits used for international navigation subject to the regime of transit passage (NSPD-66 2009). None of the European Union's (EU) Arctic policy statements in recent years contain a position on the issue, even though the EU Council of the European Union's Conclusions on Arctic issues (Council of the European Union 2009) mention transit passage. However, one would assume that at least some states with large fleets engaged in international shipping or with a special interest in Arctic shipping—for instance China, Japan, Norway, South Korea, and several EU Member States—share the view of the US.

Consistent with its above view on actual usage, Canada does not regard the Northwest Passage as a strait used for international navigation. Canada combines this position with two other positions. First, the waters within its Arctic archipelago enclosed by its 1985 straight baselines[6] are internal waters based on historic title (CYIL 1987, 1988). As a corollary, the right of innocent passage pursuant to Article 8(2) of the LOS Convention does not apply (Lalonde 2004). Both the US and the then European Community (EC) Member States lodged diplomatic protests against the 1985 straight baselines, regarding them as inconsistent with international law and explicitly rejecting that historic title could provide an adequate justification (Roach and Smith 1996). The second position that Canada combines

---

[5] *Corfu Channel* case (United Kingdom of Great Britain and Ireland v. Albania), Judgment on the Merits of 9 Apr 1949, *ICJ Reports* 1949, p. 1.

[6] Territorial Sea Geographical Coordinates (Area 7) Order, S.O.R./85-872; effective on 1 Jan 1986.

with its view on the Northwest Passage is that the transit passage regime is trumped by Article 234 of the LOS Convention (see Section 'Unilateral Coastal State Prescription').

Despite their bilateral Agreement on Arctic Cooperation (1988), the dispute between Canada and the US on the legal status of the Northwest Passage and the applicable regime of navigation remains unresolved. The broad saving-clause included in its Sect. 6.4 indicates that the agreement should above all be regarded as an 'agreement-to-disagree'. The 2010 debates within IMO on Canada's mandatory Northern Canada Vessel Traffic Services (NORDREG) Regulations (SOR/2010-127)—which focus predominantly on Article 234 of the LOS Convention, however—are further proof that their disputes remain unresolved (see 'Unilateral coastal state prescription' below).

The position of the Russian Federation vis-à-vis the Northern Sea Route seems largely similar to that of Canada and consists of combined positions on actual usage, internal waters included within straight baselines pursuant to historic title, and transit passage being trumped by Article 234 (Brubaker 1999; Brubaker 2001).

General Exceptions

The above-mentioned restriction on coastal state jurisdiction applies only in relation to pollution of the marine environment, as defined in Article 1(1)(4) of the LOS Convention. Once coastal state jurisdiction is not exercised for that purpose but, for instance, for the conservation of marine living resources instead, the general rule on coastal state jurisdiction over vessel-source pollution does not apply either. As regards anchoring, this view is supported by the practice of the US and—more recently—the Netherlands on regulating anchoring beyond the territorial sea without seeking IMO approval and apparently without any objection by other states. As regards ballast water discharges, the above view is supported by the fact that, instead of an Annex to MARPOL 73/78, IMO decided to deal with ballast water management in a stand-alone treaty, namely the BWM Convention (2004). Moreover, the BWM Convention allows states individually or in concert to regulate more stringently above the minimum ballast water exchange level laid down in the Convention (BWM Convention 2004, arts. 2(3) and 13(3) and Section C of the Annex).

More stringent standards can also be adopted for special areas pursuant to Article 211(6) of the LOS Convention. But as this requires at any rate IMO approval, it gives coastal states no unilateral prescriptive authority. The PSSA Guidelines (IMO 2005) developed by IMO also implement Article 211(6) and are clearly inspired by, and consistent with, that provision (IMO 2005, para. 7.5.2.3(iii)). PSSA status is not a precondition for obtaining the majority of possible APMs as, for instance, ships' routeing measures, SRSs, or VTS can also be made applicable to the maritime zones of a coastal state upon its request by means of IMO approval.

Unilateral Coastal State Prescription

There are three exceptions to the abovementioned general rule that coastal state prescription cannot be more stringent than GAIRAS. First, as general international law does not grant foreign vessels any navigational rights in internal waters—apart from a minor exception laid down in Article 8(2) of the LOS Convention—coastal state jurisdiction is in principle unrestricted. The observations on port state jurisdiction below applies therefore *mutatis mutandis* to internal waters.

Second, a coastal state is entitled to prescribe more stringent (unilateral) standards for the territorial sea, provided they "shall not apply to the design, construction, manning or equipment of foreign ships unless they are giving effect to generally accepted international rules or standards" (LOS Convention 1982, art. 21(2)). Unilateral discharge, navigation, and ballast water management standards are, among others, allowed. The rationale of this provision is to safeguard the objective of globally uniform international minimum regulation, which would be undermined if states unilaterally prescribe standards that have significant extra-territorial effects.

A third exception is laid down in Article 234 of the LOS Convention. It is entitled 'Ice-covered areas' and provides:

> Coastal States have the right to adopt and enforce non-discriminatory laws and regulations for the prevention, reduction and control of marine pollution from vessels in ice-covered areas within the limits of the exclusive economic zone, where particularly severe climatic conditions and the presence of ice covering such areas for most of the year create obstructions or exceptional hazards to navigation, and pollution of the marine environment could cause major harm to or irreversible disturbance of the ecological balance. Such laws and regulations shall have due regard to navigation and the protection and preservation of the marine environment based on the best available scientific evidence.

Article 234 was included in the LOS Convention as a result of in particular the efforts of Canada, which sought to ensure that its Arctic Waters Pollution Prevention Act (AWPPA 1970) and underlying regulations and orders would no longer be regarded as inconsistent with international law. The negotiations on Article 234 were predominantly conducted by Canada, the Soviet Union, and the US and were closely connected to what eventually became Article 211(6) on special areas (McRae 1987; Huebert 2001; Bartenstein 2011).

While Article 234 contains a number of ambiguities—not unlike many other provisions in the LOS Convention, and in fact many treaties—the basic purpose is to provide a coastal state with broader prescriptive and enforcement jurisdiction in ice-covered areas than in maritime zones elsewhere. In particular, in contrast with Article 211(6) on special areas, Article 234 does not envisage a role for the 'competent international organization' (IMO) in case the coastal state takes the view that more stringent standards than GAIRAS are needed.

As the wording of Article 234 indicates, however, jurisdiction is subject to several restrictions and can only be exercised for a specified purpose. One such restriction follows from the words "for most of the year". Decreasing ice coverage will mean that, gradually, fewer states will be able to rely on Article 234 in fewer areas. As regards the phrase "within the limits of the exclusive economic zone", it is submitted that the better interpretation is that this is merely meant to indicate the outer limits of the EEZ but not to exclude the territorial sea (Molenaar 1998).

The purpose for which jurisdiction can be exercised pursuant to Article 234 is "the prevention, reduction and control of marine pollution from vessels". Even though 'navigation' is mentioned twice in Article 234, it does not explicitly grant jurisdiction for the purpose of ensuring maritime safety. It is nevertheless submitted that Article 234 allows regulations that have environmental protection as a primary purpose and maritime safety as a secondary purpose as well as regulations for which both purposes are more or less equally important.

The LOS Convention does not explicitly address the scenario of waters that are both ice-covered and subject to the regime of transit passage, but many commentators argue that the inclusion of the stand-alone Article 234 in the separate Sect. 8 of Part XII supports the dominance of Article 234 over transit passage (Hakapää 1981; McRae 1987; Pharand 2007). While the International Chamber of Shipping (ICS) supports the opposite view (ICS 2012), the US does not seem to have ever explicitly and publicly stated that transit passage trumps Article 234, even though this might well be its position (Roach and Smith 1996; *contra* McDorman 2010). There may be several reasons for this, including the fact that the US is not a party to the LOS Convention, awareness that its position is not very strong, and a preference for a cooperative rather than a confrontational stance.

The following states would currently be entitled to exercise jurisdiction pursuant to Article 234: Canada, Denmark (in relation to Greenland), Norway [in relation to Svalbard but subject to the Spitsbergen Treaty (1920)], the Russian Federation, and the US. So far only Canada and the Russian Federation have actually exercised such jurisdiction (Franckx 1993; Brubaker 2005; VanderZwaag et al. 2008; Solski 2013). The Kingdom of Denmark's 'Strategy for the Arctic' (2011) refers to Denmark's willingness to invoke Article 234 when adequate standards cannot be adopted within IMO.

The consistency of the national laws and regulations of Canada and the Russian Federation with international law has been questioned from time to time. For instance: the applicability of certain CDEM standards to foreign warships and other governmental vessels (re Canada); discriminatory navigation requirements, icebreaker fees, and insurance requirements; lack of transparency; and high levels of bureaucracy (primarily re Russian Federation, even if not stated) (Molenaar et al. 2010; ICS 2012; Solski 2013).

The consistency of Canada's NORDREG Regulations with Article 234 of the LOS Convention was debated within IMO's NAV (56th Session)[7] and MSC (88th Session)[8] in 2010 (McDorman 2012). Canada introduced the voluntary NORDREG system in 1977 but decided to make it mandatory as a consequence of Canada's

[7] IMO doc. NAV 56/20, of 31 Aug 2010, at paras 19.21–19.24.

[8] IMO docs MSC 88/11/2, of 22 Sept 2010; MSC 88/11/3, of 5 Oct 2010; MSC 88/26, of 15 Dec 2010, at paras 11.28–11.39 and Annexes 27 and 28.

Northern Strategy (Minister of Indian Affairs and Northern Development 2009). The NORDREG Regulations became mandatory on 1 July 2010 within the extended (200 nm) scope of the AWPPA, and therefore have a much wider scope than the Northwest Passage. The cornerstone of the NORDREG Regulations is the requirement for prescribed vessels—whether domestic or foreign—to submit, prior to entering the NORDREG Zone, certain information and to obtain clearance.[9] Contravention of these requirements could lead to the vessel's detention and the imposition of a fine and/or imprisonment (c.f. Canada Shipping Act 2001, sec. 138), but none of these seem to have been imposed so far. The NORDREG Regulations are enacted pursuant to the Canada Shipping Act (2001), whose objectives include marine environmental protection (sec. 6).

At MSC 88, the debate centred mainly around the question whether or not Canada was required to seek IMO approval before imposing the NORDREG Regulations on foreign vessels. The US argued that IMO approval was necessary because in its view SOLAS 74 and associated guidelines do not provide an adequate basis for imposing the NORDREG Regulations unilaterally. The US made no references to Article 234 or even the international law of the sea, even though it made the latter references at NAV 56 and its diplomatic notes to Canada (McDorman 2013). The NORDREG Regulation's requirement to obtain clearance is probably the most troublesome for the US, among other things because it essentially amounts to the need for prior authorization and could have precedent-setting effects for other waters that the US regards as straits used for international navigation.

The US was in particular supported by interventions from Germany and Singapore at MSC 88. While the former closely followed the US position at MSC 88, the latter explicitly viewed Canada's actions as inconsistent with the LOS Convention. Prior to MSC 88, France, Germany, and the United Kingdom—and presumably other states as well—had sent *Notes Verbales* to Canada. Before the United Kingdom issued its *Note Verbale*, it approached the European Commission to verify if the Commission would be willing to issue a *Note Verbale*. The Commission declined, in part because it felt that it was not evident that Canada's actions warranted a diplomatic protest and in part also due to concerns that a diplomatic protest could compromise the EU's more important interests in cooperation with Arctic states within and outside the Arctic Council.[10]

Canada—supported among others by Norway and the Russian Federation—took the view that IMO approval was unnecessary as Article 234 provided an adequate basis. While the debates at NAV 56 and MSC 88 were inconclusive and did not resurface within IMO, they illustrate that many more states than just the US

---

[9] Cf. sec. 4 of the NORDREG Regulations (SOR/2010-127); Canada Shipping Act 2001, sec. 126(1)(a); and IMO doc. SN.1/Circ.291, of 5 Oct 2010, 'Information on the Mandatory Canadian Ship Reporting System in Canada's Northern Waters (NORDREG)'.

[10] Based on communications between the author and officials from Germany, France, the United Kingdom, and the Commission in late 2010 and early 2011.

are concerned about navigational rights and coastal state jurisdiction over shipping in ice-covered areas and potential precedent-setting effects for straits used for international navigation.

Port State Jurisdiction

As ports lie wholly within a state's territory and fall on that account under its territorial sovereignty, customary international law acknowledges that a port state has wide discretion in exercising jurisdiction over its ports. This was explicitly stated by the International Court of Justice (ICJ) in the *Nicaragua* case where it observed that it is "by virtue of its sovereignty, that the coastal state may regulate access to its ports".[11] While there may often be a presumption that access to port will be granted, customary international law gives foreign vessels no general right of access to ports (Lowe 1977). Articles 25(2), 211(3), and 255 of the LOS Convention implicitly confirm the absence of a right of access for foreign vessels to ports as well as the port state's wide discretion in exercising jurisdiction under customary international law. This so-called 'residual' jurisdiction is also recognized in several IMO instruments and has on some occasions been exercised by the US and the EU. Nevertheless, some exceptions apply—for instance in case of *force majeure* and distress—and uncertainties exist—for instance on the implications of international trade law. International law only very rarely authorizes port states to impose enforcement measures that are more stringent than denial of access or use of port (services) for extra-territorial behaviour (Molenaar 2007). Article 218 of the LOS Convention is one of these instances. This innovative provision gives port states enforcement jurisdiction over illegal discharges beyond their own maritime zones, namely the high seas and the maritime zones of other states.

IMO Instruments

In view of the focus of this chapter (see Sect. 6.1) and the main categories of substantive shipping standards listed in Sect. 6.3.2, the following are the most important legally binding IMO instruments:

1. MARPOL 73/78;
2. SOLAS 74;
3. STCW 78 (1978);
4. The Anti-Fouling Convention (2001);
5. The BWM Convention (2004);
6. COLREG 72;

---

[11] Case concerning Military and Paramilitary Activities In and Against Nicaragua (Nicaragua v. United States of America), Judgment of 27 June 1986, *ICJ Reports* 1986, p. 14.

7. OPRC 90 and its HNS Protocol (2000);
8. The various IMO instruments relating to liability, compensation, and insurance, e.g., the Civil Liability Convention (1969), the Fund Convention (1971) (each modified by several protocols), the HNS Convention (1996), and the Bunker Oil Convention (2001).

In addition, the following are the most important non-legally binding IMO instruments:

1. The General Provisions on Ships' Routeing (IMO 1985);
2. The MARPOL 73/78 Special Area Guidelines (IMO 2001);
3. The PSSA Guidelines (IMO 2005);
4. The Arctic Shipping Guidelines (2002); and
5. The Polar Shipping Guidelines (2009).

Apart from the Arctic and Polar Shipping Guidelines, all these legally binding and non-legally binding instruments have a global scope of application and therefore apply in principle to the entire marine Arctic. As is illustrated below, however, many IMO instruments allow for the adoption of more stringent measures in specified geographical areas. It should also be noted that where Arctic states are not parties to certain legally binding IMO instruments, they will not implement them in any capacity, including as a coastal state.

The remainder of this subsection will elaborate further on (a) discharge and emission standards, (b) CDEM standards, (c) navigation standards, (d) PSSA Guidelines, and (e) other standards.

Discharge and Emission Standards

MARPOL 73/78 and the BWM Convention are the only IMO instruments that contain discharge and emission standards. The Annexes to MARPOL 73/78 contain discharge standards for oil (Annex I), noxious liquid substances (Annex II), sewage (Annex IV), and garbage (Annex V), and emission standards for ozone depleting substances, nitrogen oxides (NOx), sulphur oxides (SOx), and volatile organic compounds (VOCs) (Annex VI). Annexes I, II, IV, and V allow the designation of so-called 'special areas' where more stringent discharge standards apply. Annex VI allows the designation of so-called 'Emission Control Areas' for SOx, particulate matter, and NOx. A substantial number of special areas and Emission Control Areas are currently in effect, but none of these apply to the marine Arctic.

The BWM Convention stipulates that vessels using the ballast water exchange method should not discharge ballast water within 200 nm from the nearest land or in waters less than 200 metres deep and must meet an efficiency of at least 95 % volumetric exchange (Regulations B-4 and D-1). Also, as was noted above, the BWM Convention allows states individually or in concert to regulate more stringently above this minimum level.

CDEM Standards

CDEM standards are contained in many of the main legally binding IMO instruments, in particular SOLAS 74 and STCW 78. The well-known double-hull standard—which was triggered by the *Exxon Valdez* disaster in 1989—is nevertheless laid down in Annex I to MARPOL 73/78. Arguably, the fuel content requirements in Annex VI to MARPOL 73/78 (within and beyond Emission Control Areas) and the ballast water treatment requirements in the BWM Convention must be regarded as, or treated analogous with, CDEM standards. A similar argument could be made for prescriptions on the use of certain paints or coatings pursuant to the Anti-Fouling Convention. A recent amendment to the STCW Code connected to STCW 78 concerns the inclusion of a new (voluntary) Section B–V/g on 'Guidance regarding training of Masters and officers for ships operating in Polar waters'.

As the Polar Shipping Guidelines are envisaged to be replaced by the Polar Code, it will only be briefly mentioned here. The Polar Shipping Guidelines are more elaborate and extensive than the Arctic Shipping Guidelines, for instance in relation to life-saving appliances. The Polar Shipping Guidelines contain the definition of 'ship' used in SOLAS 74 and apply to all voyages in Antarctic waters but as regards Arctic waters only to international voyages. They contain mostly CDEM standards and strong links with the IACS Unified Requirements concerning Polar Class.

The Polar Code is still under development and at least two more years beyond its original target of 2012 are needed until its adoption. The Code's maritime safety component is in a more advanced stage than the marine pollution component.[12] Slow progress on the Code is at least in part caused by complexities on the linkage between the Code and other IMO instruments, and a failure to select one of the available options early on. Apart from SOLAS 74 and MARPOL 73/78, there also seems to be support for including linkages with other IMO 'pollution' instruments such as the Anti-fouling Convention and the BWM Convention. At MSC 91 in November 2012, it was confirmed that all relevant IMO instruments "should be amended to mandate the associated provisions of the future Polar Code, as opposed to making it mandatory under the SOLAS Convention only or developing a stand-alone new Convention".[13] It is submitted that, if pursued, this approach would make the Polar Code the first genuine regional legally binding IMO instrument. Despite its being bi-polar, it would clearly be much broader and more comprehensive than, for instance, special areas and Emission Control Areas under MARPOL 73/78, or the packages of APMs under PSSAs. The preferred approach may also imply that the special area designations for Antarctica under the Annexes to MARPOL 73/78 will eventually be transferred to the Polar Code.

---

[12] The most recent publicly available version seems to be contained in IMO doc. DE 56/WP.4, of 16 Feb 2012.

[13] IMO doc. MSC 91/WP 1, of 29 Nov 2012, at para. 8.2.

Navigation Standards

As noted in Sect. 6.3.2, the category of navigation standards includes ships' routeing measures, SRSs, and VTS. These navigation standards can be adopted by the MSC based on their authority under SOLAS 74 and COLREG 72. As regards ships' routeing measures, reference should be made to the General Provisions on Ships' Routeing. Examples of routeing measures are: traffic separations schemes, deep-water routes, precautionary areas, areas to be avoided, and no anchoring areas. Apart from the regulation of anchoring for the purpose of the conservation of marine living resources, the LOS Convention does not authorize coastal states to adopt mandatory navigation standards seaward of its territorial sea without IMO approval.

While several IMO navigation standards currently apply within the marine Arctic, there is no comprehensive mandatory or voluntary IMO ships' routeing system for the Arctic Ocean or a large part thereof.

PSSA Guidelines

Designation of an area as a PSSA pursuant to the PSSA Guidelines does not bring about regulation of shipping within that area as such. This requires adoption of one or more APMs. Attention can in this context be drawn to the possibility to have special discharge standards within PSSAs (other than by means of designation as special area under MARPOL 73/78) and "other measures aimed at protecting specific sea areas against environmental damage from ships, provided that they have an identified legal basis" (IMO 2005, para. 6.1.3). Innovative standards are therefore not ruled out.

Other Standards

Reference should also be made to IMO Assembly Resolution A.999(25), 'Guidelines on voyage planning for passenger ships operating in remote areas' (IMO 2008), that was adopted a week after the tragic sinking of the *MS Explorer*—a purpose-built, ice-strengthened tourist vessel originally named *MS Lindblad Explorer*—on 23 November 2007 in Antarctic waters. IMO Assembly Resolution A.999(25) complements the more general IMO Assembly Resolution A.893(21), 'Guidelines for voyage planning' (IMO 1999). Resolution A.999(25) refers, *inter alia*, to the need to take account of shortcomings in available hydrographic data, the presence of places of refuge, and the need of experience in navigating in ice-covered areas. As regards places of refuge, IMO Assembly Resolution A.949(23), of 5 December 2003, 'Guidelines on Places of Refuge for Ships in Need of Assistance', adopted in the aftermath of the disaster with the *Prestige* in 2002, is also relevant. Finally, mention should be made of Regulation V/5 of SOLAS 74 on 'Meteorological services and warnings', Regulation V/6 on 'Ice Patrol Service', and Chapter V's Appendix on 'Rules for the management, operation and financing of the North Atlantic Ice Patrol'.

Regional PSC Arrangements

Regional PSC Arrangements for merchant shipping were established to enhance compliance with internationally agreed standards by means of commitments to carry out harmonized and coordinated inspections and to take predominantly corrective enforcement action (i.e., detention for the purpose of rectification). The instruments in which these internationally agreed standards are contained are commonly referred to as the 'relevant instruments' and include all the main IMO instruments. A participating Maritime Authority must only apply standards that are not just in force generally but also for that Maritime Authority.[14] Some applicability gaps therefore remain unavoidable.

The Arrangements are non-legally binding and, rather than states as such, maritime authorities are parties to them (Molenaar 2007). Saving-clauses have nevertheless been incorporated to ensure that nothing in them affects residual port state jurisdiction, which would include the right to take more onerous enforcement measures.[15]

The expansion in participation in the Paris MOU and the creation and expansion of eight new regional PSC Arrangements since then means that almost complete global coverage has now been achieved. However, no Arrangements have been adopted specifically for the Arctic Ocean/region or the Southern Ocean/ Antarctic region. Some advantages and disadvantages of an Arctic Ocean/region MOU will be discussed under Sect. 6.4.2, among other things in view of the likelihood that practically all ships engaged in either intra- or trans-Arctic marine shipping will make use of ports subject to either the Paris MOU or Tokyo MOU Tokyo (1993). None of the other Arrangements seem therefore relevant for Arctic marine shipping. However, when considering amendments to the Paris MOU it is,—in light of the EU's Directive on Port State Control (2009) and the need of convergence between the Directive and the Paris MOU[16]—essential to obtain prior agreement within the EU.

The maritime authorities of the following 27 states currently participate in the Paris MOU:

Belgium	Estonia	Ireland	Norway	Spain
Bulgaria	Finland	Italy	Poland	Sweden
Canada	France	Latvia	Portugal	United Kingdom
Croatia	Germany	Lithuania	Romania	
Cyprus	Greece	Malta	Russian Federation	
Denmark	Iceland	Netherlands	Slovenia	

[14] Cf. Paris MOU, sec. 2.3, and Tokyo MOU (1993), sec. 2.4.

[15] E.g., of the Paris MOU, secs. 1.7 and 9.1.

[16] See the 13th preambular paragraph of the Directive.

The participation by the Danish Maritime Authority extends to Greenland as well. Moreover, even though the US Coast Guard merely has observer status, it has been cooperating with the Paris MOU since at least 1986—when it first attended meetings within the Paris MOU—and its PSC system is more or less compatible with that of the Paris MOU.[17]

The Paris MOU does not contain a provision that explicitly defines its spatial coverage. However, Sect. 9.2 stipulates that adherence is open for "A Maritime Authority of a European coastal state and a coastal state of the North Atlantic basin from North America to Europe". This has facilitated the participation or cooperation of the Maritime Authorities of all Arctic states, even though the description is not intended to encompass the entire marine Arctic.

As the Maritime Authorities of both Canada and the Russian Federation also participate in the Tokyo MOU (see below)—and, in addition, the Maritime Authority of the Russian Federation also participates in the Black Sea MOU (2000)-, clarity is needed as to which of their ports are subject to which Arrangement. In 2009, Canada decided to also subject its Pacific ports to the Paris MOU. The Pacific ports of the Russian Federation are currently still subject to the Tokyo MOU, even though some Paris MOU requirements, for instance on training for PSC officers, are applicable throughout the Russian Federation.[18]

The Maritime Authorities of the following states currently participate in the Tokyo MOU:

Australia	Fiji	Republic of Korea	Philippines	Thailand
Canada	Hong Kong, China	Malaysia	Russian Federation	Vanuatu
Chile	Indonesia	New Zealand	Singapore	Vietnam
China	Japan	Papua New Guinea	Solomon Islands	

Sections 1.2 and 8.2 of the Tokyo MOU and Sect. 1.1 of its Annex 1, entitled 'Membership of the Memorandum', stipulate that the Tokyo MOU applies to the "Asia–Pacific region", a term that is not further defined. The US Coast Guard has observer status with the Tokyo MOU and cooperates in a similar way as with the Paris MOU.

Output of the Arctic Council

The Arctic Council is a high-level forum established by means of the Ottawa Declaration (1996). The choice for a non-legally binding instrument is a clear indication that the Council was not intended to be an international organization

[17] Information provided to the author by C. Droppers, Paris MOU Secretariat, on 25 Jan 2013.

[18] Information provided to the author by C. Droppers, Paris MOU Secretariat, on 25 Jan 2013.

and implies that the Council cannot adopt legally binding decisions or instruments. The Arctic SAR Agreement was therefore not adopted by the Council, even though it was negotiated under its auspices and the Council's May 2011 Ministerial Meeting was also used as the occasion for its signature.

The mandate of the Arctic Council is very broad and relates to "common Arctic issues" with special reference to "issues of sustainable development and environmental protection in the Arctic" (Ottawa Declaration 1996, art. 1). A footnote nevertheless specifies that the Council "should not deal with matters related to military security". Marine shipping falls squarely under this broad mandate and this is also subscribed by the fact that the Arctic Council has produced output that relates specifically to marine shipping as well less specific or more indirectly relevant output.

The latter includes the 2004 Arctic Marine Strategic Plan (AMSP; Arctic Council 2004), which was developed under PAME and is currently under revision, with adoption scheduled for the 2014 Deputy Ministerial Meeting (PAME 2012b). Both the Arctic SAR Agreement and the future Arctic MOPPR Agreement—scheduled to be signed at the 2013 Kiruna Ministerial Meeting—could be regarded to belong in this category as well. While both implement global IMO instruments—namely the SAR Convention and OPRC 90 -, the Arctic SAR Agreement also implements the ICAO Convention (1994), and neither deals exclusively with shipping incidents, but also with incidents relating to air traffic and offshore installations. Finally, much of the output of EPPR belongs in this category as well, as evidenced by EPPR's role in the negotiation process of the future Arctic MOPPR Agreement by developing Operational Guidelines that will be appended to the Agreement (EPPR 2012).

The most important Arctic Council output that focuses specifically on Arctic marine shipping is the AMSA, completed by PAME in 2009. The AMSA contains a considerable number of recommendations categorized under the headings 'Enhancing Arctic Marine Safety', 'Protecting Arctic People and the Environment', and 'Building the Arctic Marine Infrastructure'. At least three of these recommendations have already been implemented, namely Recommendation I(B), which includes support for the updating and mandatory application of the Arctic Shipping Guidelines; Recommendation I(E), which supports the negotiation of an Arctic SAR instrument; and Recommendation III(C), which supports, *inter alia*, the development of circumpolar agreements on environmental response capacity.

As Recommendation I(B) eventually shaped to a considerable extent—in addition to actions undertaken within the Antarctic Treaty System (ATS)—the decision to develop the IMO Polar Code, it is a good example of the Arctic Council's so-called 'decision-shaping' function (Molenaar 2012). This function continues to be relevant through PAME's continuous monitoring of progress with the implementation of the AMSA Recommendations.

New Arctic Council initiatives in the domain of Arctic marine shipping will probably arise from the Arctic Ocean Review (AOR) project that is currently carried out within PAME. Phase II of this project is intended to culminate in a final report adopted at the Council's May 2013 Kiruna Ministerial Meeting that will:

> summarize potential weaknesses and/or impediments in the global and regional instruments and measures for [the] management of the Arctic marine environment; outline options to address these weaknesses and/or impediments; and, make agreed recommendations to help ensure a healthy and productive Arctic marine environment in light of current and emerging trends (AOR 2011).

The AOR Phase II draft Report (AOR 2013) contains a Chap. 3 on 'Arctic Marine Operations and Shipping', with specific opportunities (A-L) in Sect. 3.4.3, some of which build on the AMSA recommendations. Section 6.4 below incorporates some of these.

As the Polar Code will ultimately be adopted by the IMO, it will be regarded as that body's output and not as the Council's. The connection between the Polar Shipping Code and the Council is clearly very different from the connection between the Council and the Arctic SAR Agreement and the future Arctic MOPPR Agreement. This author has introduced the concept of the Arctic Council System (ACS) to clarify that legally binding instruments such as the Arctic SAR Agreement and the future Arctic MOPPR Agreement—and their institutional components—can be part of the Council's output even though they are not—and in fact could not be—formally adopted by it (Molenaar 2012).

The ACS concept consists of two basic components. The first component is made up of the Council's constitutive instrument—the Ottawa Declaration, other Ministerial Declarations, other instruments adopted by the Arctic Council—for instance its Arctic Offshore Oil and Gas Guidelines (PAME 2009), and the Council's institutional structure. The second component consists of instruments 'merely' negotiated under the Council's auspices and their institutional components. The 2011 Arctic SAR Agreement and the Meetings of the Parties envisaged under its Article 10 belong to this category and the Arctic MOPPR Agreement and, if included, its institutional component, will soon be as well.

The AOR Phase II draft Report proposes recourse to the ACS approach in order to amend or complement the future Arctic MOPPR Agreement to ensure coverage of other pollutants—in particular noxious liquid substances—, as well as on the collection and sharing of Arctic marine traffic data (AOR 2013).

Acts of the OSPAR Commission

The spatial competence of the OSPAR Commission extends to the 'OSPAR Maritime Area', which includes areas within and beyond national jurisdiction (OSPAR Convention 1992, art. 1(a)). The OSPAR Maritime Area roughly overlaps with the Atlantic sector of the marine Arctic, but about half extends further south. Nothing in the OSPAR Convention or the Acts of the OSPAR Commission challenges IMO's primacy in the regulation of international merchant shipping, but also does not entirely preclude action in relation to merchant shipping. Article 4(2) of

Annex V to the OSPAR Convention stipulates that Members of the OSPAR Commission can raise the need for regulatory action within IMO and requires them to cooperate among other things on regional implementation of IMO instruments. An example of action by the OSPAR Commission in the domain of shipping is the voluntary interim application of certain standards of the BWM Convention adopted in 2007. In 2012, this action was replaced by joint action between the OSPAR Commission and the regional seas bodies for the Baltic and Mediterranean Seas.[19]

6.4 Gaps in the International Legal and Policy Framework and Options for Addressing them

6.4.1 Gaps

Before identifying gaps in the international legal and policy framework for Arctic marine shipping, it is only fair to note that much progress has been made in addressing such gaps in the last five years or so. IMO managed to adopt the Polar Shipping Guidelines in 2009 and is working hard on the adoption of the Polar Code. The Arctic Council finalized AMSA in 2009 and has made good progress in implementing several of its recommendations, most notably culminating in the signature of the Arctic SAR Agreement in 2011 and the future Arctic MOPPR Agreement, scheduled to be signed in May 2013.

As the Polar Code is still to be adopted and enter into force, the following appear to be the main gaps in the international legal and policy framework for Arctic marine shipping:

1. Insufficient participation in relevant international instruments, for instance the LOS Convention—to which the US is not a party—and instruments such as the BWM Convention and the Arctic SAR Agreement, which are still to enter into force;
2. No dedicated, legally binding IMO standards for the marine Arctic, for instance in relation to discharge, emission, or ballast water exchange standards and CDEM (including fuel content (e.g., heavy fuel oil (HFO), anti-fouling, and ballast water treatment) standards;
3. No comprehensive mandatory or voluntary IMO ships' routeing system for all or part of the Arctic Ocean; and
4. No dedicated pan-Arctic mechanisms on monitoring, surveillance, inspection, and enforcement.

[19] Joint Notice to Shipping from the Contracting Parties of the Barcelona Convention, OSPAR and HELCOM on: 'General Guidance on the Voluntary Interim Application of the D1 Ballast Water Exchange Standard by Vessels Operating between the Mediterranean Sea and the North-East Atlantic and/or the Baltic Sea' (Annex 17 to 2012 OSPAR Summary Record).

6.4.2 Options

This subsection contains various options for addressing the gaps identified in the previous subsection. They are discussed under the following subsections: options for action within IMO; options for action outside IMO; and options for PSC initiatives. Some of the options for action outside IMO also highlight a potential role for the Arctic Council or the ACS approach.

Options for Action Within IMO

As the Polar Code is still under negotiation, the most obvious option for action within IMO is inclusion of the commonly used IMO standards mentioned under Gap No. 2 in Sect. 6.4.1 above. The desirability of restrictions on the use and carriage of HFO in the marine Arctic is under consideration within PAME as well. The negotiations on the Polar Code may also provide opportunities to include new types of standards, for instance compulsory pilotage and ice-breaker or tug assistance.

So far, there are no indications that navigation standards (ships' routeing measures, SRSs, and VTS) will be included in the Polar Code. Navigation standards could be adopted on a case-by-case basis as stand-alone standards, for instance a SRS or VTS for the Bering Strait, or speed restrictions for certain areas in order to avoid ship strikes of marine mammals or to reduce emissions.[20] An alternative would be to develop a comprehensive ships' routeing system for part or all of the Arctic Ocean, which may be desirable in view of the continuous expansion of Arctic marine shipping. As the main shipping routes described in Sect. 6.2 resemble somewhat archipelagic sea lanes established pursuant to Article 53 of the LOS Convention, the procedure laid down in Article 53—implemented by Annex 2 to the IMO's General Provisions on Ships' Routeing—may be suitable as a model for developing 'Arctic Sea Lanes'. It is not a problem if one or more of these sea lanes would be partially on the high seas, as consensus-based IMO approval reflects support by the entire international community.

Designating one or more PSSAs with APMs in the marine Arctic is also an option that could be pursued in parallel with the Polar Code. Many of the above-mentioned standards could be adopted as APMs. Area-based measures for the marine Arctic are also under consideration by PAME as part of the implementation of AMSA Recommendations II(A), (C), and (D) and the AOR project. It is disappointing—but to some extent understandable—that PAME decided to limit its efforts on PSSAs and MARPOL 73/78 special areas to the high seas of the Arctic Ocean (PAME 2012a; AOR 2013).

[20] See IMO doc. MEPC 60/4/24, of 15 Jan 2010, at para. 8, in the context of black carbon emissions. The MSC has adopted at least one speed restriction—albeit recommendatory and primarily for the purpose of human safety—, namely in the TSS "Between Korsoer and Sprogoe" (see IMO doc. MSC 78/26/Add.2, of 4 June 2004, at Annex 21, p. 8, n. 3 to the TSS).

Options for Action Outside IMO

As was pointed out in Sect. 6.3.3, the LOS Convention does not preclude action outside IMO.

1. All states—in their capacities as flag states, port states, coastal states, or with regard to their tour operators—can encourage self-regulation by the (cruise) shipping industry;
2. All states, whether individually or collectively, can in their capacities as flag states impose standards on their vessels for shipping in the marine Arctic that are more stringent than GAIRAS, for instance special discharge, emission, and ballast water exchange standards or HFO standards. Such proactive steps can also be taken in anticipation of the entry into force of the Polar Code and other relevant IMO instruments and standards;
3. Arctic Ocean coastal states and other (Arctic) states—whether within or outside the scope of the Arctic Council or by pursuing the ACS approach—can, in their capacities as port states, develop a collective multifaceted strategy on port state jurisdiction for Arctic marine shipping. This strategy could consist of the following elements:
   a. PSC initiatives (see 'Options for PSC initiatives' further below in this subsection);
   b. Coordinated and optimized use of port state jurisdiction, for instance by implementing Article 218 of the LOS Convention in concert, exercising 'departure state jurisdiction', or using criminal or administrative law to impose charges such as furnishing false information or obstruction of inspection in connection with behaviour prior to entry into port (Molenaar 2007); and
   c. Exercise of port state residual jurisdiction in concert in case the Polar Code takes too long to enter into force or its stringency level is deemed insufficient.
4. Arctic Ocean coastal states or Arctic states—whether within or outside the scope of the Arctic Council or by pursuing the ACS approach—could, in their capacities as coastal states, collectively
   a. Amend or complement the future Arctic MOPPR Agreement to ensure coverage of other pollutants, in particular noxious liquid substances (AOR 2013);
   b. Develop a new regional instrument on the collection and sharing of Arctic marine traffic data (AOR 2013);
   c. Ensure regional implementation of IMO's Guidelines on Places of Refuge;
   d. Develop a regional mechanism for coordinated aerial and satellite surveillance of intentional and accidental marine pollution;
   e. Harmonize relevant domestic laws, regulations, and policies, including in relation to enforcement; and
   f. Take other action consistent with international law, including by relying on Article 234 of the LOS Convention, in case the Polar Code takes too long to enter into force or its stringency level is deemed insufficient.
5. Canada, the Russian Federation, and key flag states could convene multilateral consultations on Arctic marine shipping in order to exchange views and address concerns on navigation in the Northwest Passage and the Northern Sea Route.

Options for PSC Initiatives

PSC initiatives can either be undertaken within existing regional PSC Arrangements or by establishing a new Arrangement, namely an Arctic Ocean/region MOU.

As regards possible initiatives on Arctic marine shipping within existing Arrangements, one approach would be to bring as much Arctic marine shipping as possible under the scope of the Paris MOU. This would be based on the assumption that the stringency level and performance of the Paris MOU is the highest of all the regional PSC Arrangements. Accordingly, the Russian Federation could follow Canada's example (see Sect. 6.3.4 'Regional PSC Arrangements') by subjecting all its Pacific ports to the Paris MOU. The Paris MOU would thereby cover all intra-Arctic shipping and a sizeable part of trans-Arctic shipping, in particular if relatively much use would be made of transhipment ports in the high North Atlantic and the high North Pacific.

Further initiatives could be developed within the Paris MOU as well. As highlighted earlier, these would not relate to the prescription of new standards, but rather be concerned with harmonized and coordinated inspection, and corrective enforcement action with respect to existing standards. Initiatives should be specifically tailored to ships that have engaged in Arctic marine shipping since their last port visit and those that will do so before their next port visit. As regards the Paris MOU, adjustments could for instance be made to one or more Port State Control Committee Instructions (e.g., 'Guidance on Type of Inspections') to include special guidance/instructions for inspections of ships that have engaged or will engage in Arctic marine shipping, as well as specific requirements for the qualification and training of PSC officers in that regard. Such guidance could also be developed by, and made applicable to, a sub-set of the maritime authorities that participate in, or cooperate with, the Paris MOU.

But unless trans-Arctic shipping would make extensive use of transhipment ports in the high North Pacific, departure or destination ports in the Asia–Pacific region would still constitute a significant gap. Similar dedicated guidance/instructions on Arctic marine shipping could therefore be developed within the Tokyo MOU.

An alternative to developing initiatives under the Paris and Tokyo MOUs is the development of an Arctic Ocean/region MOU. As participation in regional PSC Arrangements is always reserved for maritime authorities of the region's coastal states, it follows that the maritime authorities from the following states would be participants: Denmark (Greenland), Canada, Norway, the Russian Federation, the US and—especially in case ships involved in Arctic marine shipping are expected to make extensive use of Icelandic ports—Iceland.

As noted in the discussion on Regional PSC Arrangements in Sect. 6.3.4, the maritime authorities from these states either already formally participate in, or cooperate with, both the Paris and Tokyo MOUs (Canada, the Russian Federation,[21] and the US) or just the Paris MOU (Denmark (Greenland), Iceland, and Norway).

---

[21] See also *supra* note 64 and accompanying text. Memorandum of Understanding on Port State Control in the Black Sea Region, Istanbul, 7 Apr 2000. In effect 19 Dec 2000, as regularly amended. Most recent text at <www.bsmou.org>.

While the cost-effectiveness of all regional PSC Arrangements as a whole would not necessarily be negatively affected by further overlaps in participation, the six maritime authorities will have to weigh the costs of participating in, or cooperating with, yet another MOU against the benefits that its establishment would bring. This would seem to depend, among other things, on their views as to the need and urgency of dedicated PSC initiatives for Arctic marine shipping; the extent to which Arctic marine shipping is expected to be composed of intra-Arctic shipping and ships using transhipment ports in the high North Atlantic and the high North Pacific; and the prospects of adopting satisfactory dedicated PSC initiatives for Arctic marine shipping within the Paris or Tokyo MOUs (Stokke 2012).

6.5 Potential for EU–US Cooperation

A discussion on the potential for EU-US cooperation on Arctic marine shipping must acknowledge at the outset that whereas the US is an Arctic Ocean coastal state, the EU cannot rely on such a *de facto* capacity. Denmark is an Arctic Ocean coastal state with respect to Greenland and an Arctic coastal state with respect to the Faroe Islands, but Denmark's EU Membership does not extend to Greenland or the Faroe Islands (TFEU 2008, arts. 204 and 355(5)(a)). However, the EU can still act in various other *de facto* capacities; for instance as a flag state—pursuant to the various navigational rights applicable within the marine Arctic -, port state, or with respect to natural and legal persons of its Member States.

'Transport' and 'environment' are among the areas listed in Article 4(2) of the TFEU where the EU and its Member States share competence. This shared competence in shipping is among things reflected in the fact that the EU is not a member of IMO. The European Commission nevertheless has observer status with IMO.

Both the US and several EU Member States—in particular Cyprus, Denmark, Germany, Greece, and Malta—are important actors in the domain of international merchant shipping on account of the number of vessels and cumulative deadweight tonnage registered under their flags or in terms of (beneficial) ownership over such vessels (UNCTAD 2011). More in general, seaborne trade is vital to the economies of the US and EU Member States and they also have a range of other interests that are closely associated with shipping (Raspotnik and Rudloff 2012). The high priority accorded to safeguarding navigational rights from undue interference also results from the naval capability of the US and several EU Member States and concerns that restrictions on merchant shipping may spill over to warships and other government ships. It is worth noting that the US Arctic Region Policy NSPD-66 (2009) discusses navigation rights and interests in Section III(B), entitled 'National Security and Homeland Security Interests in the Arctic'.

As regards merchant shipping in the marine Arctic, the US seems so far mainly involved in intra-Arctic traffic. Among the EU Member States, Denmark, Finland, and Germany are involved in both intra- and trans-Arctic shipping. Other EU Member States may also have an interest in Arctic marine shipping on account or their ports, for instance the Netherlands on account of Rotterdam.

In addition to their shared shipping interests, the US and the EU share interests and obligations in relation to the Arctic marine environment and its marine biodiversity, including marine mammals. All these rights and interests of the EU and the US related or associated with Arctic shipping are reflected in various recent policy statements including the US Arctic Region Policy, the European Commission's Arctic Communication (European Commission 2008) and the EU Council conclusions on Arctic issues Council of the European Union (2009). For the purpose of this section, they are grouped together below under three headings, followed by various opportunities for bilateral cooperation derived in some cases from the options identified in Sect. 6.4.2:

1. Protection and preservation of the Arctic marine environment and its marine biodiversity:
   a. Joint and coordinated engagement and support for the negotiations on the IMO Polar Code;
   b. Joint efforts to ensure that the US and EU Member States have a reputation as responsible and high-performance shipping states, including by ensuring compliance with international obligations in all their capacities; and
   c. Joint pro-active steps in their (*de facto*) capacities as flag states (see No. 2 under 'Options for action outside IMO' in Sect. 6.4.2).
2. Safeguarding navigational rights from undue interference:
   a. Cooperation on monitoring the laws, regulations, and practices of Canada, the Russian Federation, and other Arctic states to verify consistency with the international law of the sea; and
   b. Joint or coordinated diplomatic protests in case laws, regulations, or practices are not consistent with the international law of the sea.
3. Promoting multilateral regulation of Arctic marine shipping:
   a. Joint and coordinated engagement within relevant international bodies, including the IMO, the Paris and Tokyo MOUs, and the Arctic Council;
   b. Joint actions to initiate multilateral consultations on the Northwest Passage and the Northern Sea Route (see No. 5 under 'Options for action outside IMO' in Sect. 6.4.2);
   c. Joint action to discourage unnecessary reliance on Article 234 of the LOS Convention, among other things by working towards a high stringency level of the IMO Polar Code and its speedy entry into force; and
   d. Joint action to encourage regional harmonization of relevant laws, regulations, and practices, including in relation to enforcement.

6.6 Conclusions

In coming years, Arctic sea ice coverage and thickness are highly likely to gradually decrease and Arctic marine shipping to increase. In view of these trends, the international legal and policy framework for Arctic marine shipping cannot be assumed to be adequate. This is in fact broadly acknowledged as much progress has been made in addressing gaps in this framework during the last five years or so. The IMO managed to adopt the Polar Shipping Guidelines in 2009 and is working hard on the adoption of the Polar Code, even though this will not be achieved before 2014. The Arctic Council finalized AMSA in 2009 and has made good progress in implementing several of its recommendations, most notably culminating in the signature of the Arctic SAR Agreement in 2011 and the future Arctic MOPPR Agreement, scheduled to be signed in May 2013.

Many of the options suggested in Sect. 6.4.2 to address the gaps identified in Sect. 6.4.1 also offer opportunities for cooperation between the EU and the US. These opportunities can be grouped together under headings that reflect the rights and interests of the EU and the US related or associated with Arctic shipping in line with their recent policy statements, namely: protection and preservation of the Arctic marine environment and its biodiversity; safeguarding navigational rights from undue interference; and promoting multilateral regulation of Arctic marine shipping.

References

Agreement on Arctic Cooperation (1988) Agreement between the Government of Canada and the Government of the United States of America on Arctic Cooperation. Canada Treaty Series 1988, No. 29. Entered into force 11 Jan 1988

AMSA (2009) Arctic Marine Shipping Assessment 2009 Report. Arctic Council, Apr 2009

AMSA Scenarios (2008) The Future of Arctic Marine Navigation in Mid-Century. Scenario Narratives, Prepared for the Arctic Marine Shipping Assessment. May 2008

Antarctic Treaty (1959) The Antarctic Treaty, 1 Dec 1959, 402 U.N.T.S. 71. Entered into force 23 June 1961

Anti-Fouling Convention (2001) International Convention on the Control of Harmful Anti-fouling Systems on Ships, 5 Oct 2001, IMO Doc. AFS/CONF/26, of 18 Oct 2001. Entered into force 17 Sept 2008

AOR (2011) Phase I report (2009–2011) of the Arctic Ocean Review (AOR) project. <www.pame.is>. Accessed 1 Feb 2013

AOR (2013) The Arctic Ocean Review Project. Phase II report 2011–2013, consolidated version of 9 Feb 2013. On file with author

Arctic Council (2004) Arctic Marine strategic Plan. 24 Nov 2004. Available at <http://www.pame.is/images/stories/AMSP_files/AMSP-Nov-2004.pdf>. Accessed 5 Feb 2013

Arctic MOPPR Agreement (2013) Agreement on Cooperation on Marine Oil Pollution Preparedness and Response in the Arctic, scheduled to be signed at the Arctic Council's Ministerial Meeting in Kiruna, May 2013

Arctic SAR Agreement (2011) Agreement on Cooperation on Aeronautical and Maritime Search and Rescue in the Arctic, 12 May 2011, 50 I.L.M. 1119 (2011). Entered into force on 19 Jan 2013

Arctic Shipping Guidelines (2002) Guidelines for Ships Operating in Arctic Ice-Covered Waters, IMO MSC/Circ. 1056, MEPC/Circ. 399, 23 Dec 2002

AWPPA (1970) Arctic Waters Pollution Prevention Act 1970, R.S.C. 1985, Canada

Bartenstein K (2011) The "Arctic Exception" in the Law of the Sea Convention: a Contribution to Safer Navigation in the Northwest Passage? Ocean Dev Int Law 42:22–52

Black Sea MOU (2000) Memorandum of Understanding on Port State Control in the Black Sea Region, Istanbul, 7 Apr 2000. Entered into force 19 Dec 2000, as regularly amended

Brigham LW (2010) Think again: The Arctic. Foreign Policy, Sept/Oct. 2010

Brubaker RD (1999) The Legal Status of the Russian Baselines in the Arctic. Ocean Dev Int Law 30:191–233

Brubaker RD (2001) Straits in the Russian Arctic. Ocean Dev Int Law 32:263–287

Brubaker RD (2005) The Russian Arctic Straits. Martinus Nijhoff, Leiden

Bunker Oil Convention (2001) International Convention on Civil Liability for Bunker Oil Pollution Damage, 23 Mar 2001, LEG/CONF.12/19. Entered into force 21 Nov 2008

BWM Convention (2004) International Convention for the Control and Management of Ships' Ballast Water and Sediments, 13 Feb 2004. Not in force, IMO Doc. BWM/CONF/36, of 16 Feb 2004

Canada Shipping Act (2001) S.C. 2001, c. 26

Chircop A (2009) The Growth of International Shipping in the Arctic: Is a Regulatory Review Timely? Int J Marine Coastal Law 24:355–380

Civil Liability Convention (1969) International Convention on Civil Liability for Oil Pollution Damage, Brussels, 29 Nov 1969, 9 I.L.M. 45 (1970). Entered into force 19 June 1975. Replaced and entered into force 30 May 1996

COLREG 72 (1972) Convention on the International Regulations for Preventing Collisions at Sea, 20 Oct 1972, 1050 U.N.T.S. 16. Entered into force 15 July 1977

Council of the European Union (2009) Council Conclusions on Arctic issues, In: 2985th Foreign Affairs Council Meeting, 8 Dec 2009

CYIL (1987) Canadian Yearbook of International Law, Vol. 25. University of British Columbia Press, Victoria

CYIL (1988) 26 Canadian Yearbook of International Law, Vol. 26. University of British Columbia Press, Victoria

Directive on Port State Control (2009) Directive 2009/16/EC of the European Parliament and of the Council of 23 Apr 2009 on port State control, 2009 O.J. (L 131/57) (23 Apr 2009)

Emmerson C (2011) Russia's Arctic Opening. Foreign Policy 30 Mar 2011. <http://www.foreignpolicy.com/articles/2011/03/30/russias_arctic_opening>

EPPR (2012) EPPR Progress report to Senior Arctic Officials; ACSAO-SE03 Haparanda, Doc 6.1

European Commission (2008) Communication from the Commission to the European Parliament and the Council. The European Union and the Arctic Region, 20 Nov 2008, COM(2008) 763

Franckx E (1993) Maritime Claims in the Arctic: Canadian and Russian Perspectives. Martinus Nijhoff Publishers, Dordrecht

Fund Convention (1971) International Convention on the Establishment of an International Fund for Compensation for Oil Pollution Damage, 18 Dec 1971, 11 I.L.M. 284 (1972). Entered into force 16 Oct 1978

Hakapää K (1981) Marine Pollution in International Law: Material Obligations and Jurisdiction. Suomalainen Tiedeakatemia, Helsinki

HNS Convention (1996) International Convention on Liability and Compensation for Damage in Connection with the Carriage of Hazardous and Noxious Substances by Sea, 3 May 1996, 35 I.L.M. 1406. Not in force

HNS Protocol (2000) Protocol on Preparedness, Response and Co-operation to Pollution Incidents by Hazardous and Noxious Substances, 15 Mar 2000, IMO Doc. HNS-OPRC/CONF/11/Rev.1, of 15 Mar 2000. Entered into force 14 June 2007

Huebert R (2001) Article 234 and Marine Pollution Jurisdiction in the Arctic. In: Oude Elferink AG, Rothwell DR (eds) The Law of the Sea and Polar Maritime Delimitation and Jurisdiction. Martinus Nijhoff Publishers, New York, pp 249–267

Humpert M, Raspotnik A (2012) The Future of Arctic Shipping Along the Transpolar Sea Route. Arctic Yearbook 2012:281–307

ICAO Convention (1994) Convention on International Civil Aviation, Chicago, 7 Dec 1944, 15 U.N.T.S. 295 (1948). Entered into force 4 Apr 1947

ICS (2012) Position Paper on Arctic Shipping. International Chamber of Shipping. <www.ices-shipping.org>. Accessed 1 Feb 2013

IMO (1985) General Provisions on Ships' Routening. IMO Resolution A.572(14). Adopted on 20 Nov 1985, as amended
IMO (1999) Guidelines for voyage planning. IMO Doc. A 21/Res.893, 4 Feb 2000. Adapted on 25 Nov 1999
IMO (2001) Guidelines for the Designation of Special Areas under MARPOL 73/78', as set out in Annex 1 to IMO Assembly Resolution A.927(22) 2001, to be revoked in 2013 by means of an IMO Assembly Resolution (see IMO Doc. 63/23/Add.1, 14 Mar 2012, Annex 27)
IMO (2005) IMO Assembly Resolution A. 982(24) of 1 Dec 2005, 'Revised Guidelines for the Identification and Designation of Particularly Sensitive Sea Areas' (IMO doc. A 24/Res.982, 6 Feb 2006)
IMO (2008) Guidelines on voyage planning for passenger ships operating in remote areas. IMO Doc. A 25/Res.999, 3 Jan 2008. Adapted on 29 Nov 2007
Kingdom of Denmark (2011) Denmark, Greenland and the Faroe Islands: Kingdom of Denmark Strategy for the Arctic 2011–2020. <www.nanoq.gl>. Accessed 29 Jan 2013
Kraska J (2007) The Law of the Sea Convention and the Northwest Passage. Int J Marine Coastal Law 22:257–282
Lalonde S (2004) Increased Traffic Through Canadian Arctic Waters: Canada's State of Readiness. Revue Juridique Themis 38:49–124
LOS Convention (1982) United Nations Convention on the Law of the Sea, 10 Dec 1982, 1833 U.N.T.S. 396. Entered into force 16 Nov 1994
Lowe AV (1977) The Right of Entry into Maritime Ports in International Law. San Diego Law Rev 14:597–622
MARPOL 73/78 (1973/1978) International Convention for the Prevention of Pollution from Ships, 2 Nov 1973, 2 I.L.M. 1319 (1973), as modified by the 1978 Protocol Relating to the International Convention for the Prevention of Pollution from Ships (17 Feb 1978, 17 I.L.M. 546 (1978)) and the 1997 Protocol to Amend the International Convention for the Prevention of Pollution from Ships (26 Sept 1997) and as regularly amended. Entry into force varies for each Annex. At the time of writing Annexes I-VI were all in force
McDorman TL (2010) The Northwest Passage: International Law, Politics and Cooperation. In Nordquist MH, Heidar TH, Moore JN (eds) Changes in the Arctic Environment and the Law of the Sea, Martinus Nijhoff Publishers, Dordrecht, pp 227–250
McDorman TL (2012) National Measures for the Safety of Navigation in Arctic Waters: NORDREG, Article 234 and Canada. In Nordquist MH, Moore JN, Soons AHA, Kim H (eds) The Law of the Sea Convention. US Accession and Globalization. Martinus Nijhoff Publishers, Leiden
McDorman TL (2013) Canada, the United States and International Law of the Sea in the Arctic Ocean. In VanderZwaag D, Stephens T (eds) Polar Oceans Governance in an Era of Environmental Change. Edward Elgar, Cheltenham
McRae D (1987) The Negotiation of Article 234. In: Griffiths F (ed) Politics of the Northwest Passage. McGill-Queen's University Press, Kingston and Montreal, pp 98–114
Minister of Indian Affairs and Northern Development (2009) Canada's Northern Strategy: Our North, Our Heritage, Our Future, Ottawa
Molenaar EJ (1998) Coastal State Jurisdiction Over Vessel-Source Pollution. Kluwer Law International, The Hague
Molenaar EJ (2007) Port State Jurisdiction: Toward Comprehensive, Mandatory and Global Coverage. Ocean Dev Int Law 38:225–257
Molenaar EJ (2010) Arctic Marine Shipping. Overview of the International Legal Framework, Gaps and Options. J Trans Law Policy 18:289–325
Molenaar EJ (2012) Current and Prospective Roles of the Arctic Council System within the Context of the Law of the Sea. Int J Marine Coastal Law 27:553–595
Molenaar EJ, Hodgson S, VanderZwaag D, Heidar Hallsson H, Henriksen T, Holm-Peterson L, Vladimirovich Korel'skiy M, Kraska J, Már Magnússon B, Rolston S, Serdy A (2010) Legal Aspects of Arctic Shipping. Report commissioned by the European Commission, Directorate-General Maritime Affairs & Fisheries, Feb 2010. On file with author

NSPD-66 (2009) National Security Presidential Directive and Homeland Security Presidential Directive. Arctic Region Policy, 9 Jan 2009, NSPD-66/HSPD-25

OPRC 90 (1990) International Convention on Oil Pollution Preparedness, Response, and Cooperation, 30 Nov 1990. 30 I.L.M. 733 (1991). Entered into force 13 May 1995

Ottawa Declaration (1996) Declaration on the Establishment of the Arctic Council, 19 Sept 1996, 35 I.L.M. 1382

PAME (2009) Arctic Offshore Oil and Gas Guidelines. Last updated 29 Apr 2009

PAME (2012a) PAME Working Group Meeting Report No: I-2012. 26–27 Mar 2012, Sweden

PAME (2012b) Record of Decisions and Follow-up Actions. PAME II-2012. 18–20 Sept 2012

Paris MOU (1982) Memorandum of Understanding on Port State Control, Paris, 26 Jan 1982. Entered into force 1 July 1982, as regularly amended. This chapter uses the version that includes the 34th amendment and entered into force on 1 July 2012

Pharand D (2007) The Arctic Waters and the Northwest Passage: a Final Revisit. Ocean Dev Int Law 78:3–69

Polar Shipping Guidelines (2009) Guidelines for ships operating in polar waters, IMO Assembly Resolution A.1024(26), 2 Dec 2009

Protocol on Environmental Protection (1991) Protocol on Environmental Protection to the Antarctic Treaty. Annexes I–IV, 4 Oct 1991. Entered into force 14 Jan 1998; Annex V (adopted as Recommendation XVI-10), 17 Oct 1991. Entered into force 24 May 2002; Annex VI (adopted as Measure 1(2005)), 14 June 2005. Not in force

Raspotnik A, Rudloff B (2012) The EU as a shipping actor in the Arctic. characteristics, interests and perspectives. Working Paper FG 2, 2012/Nr. 4 Dec 2012, SWP Berlin

Roach JA, Smith RW (1996) United States Responses to Excessive Maritime Claims, 2nd edn. Martinus Nijhoff Publishers, The Hague

Rothwell DR (1996) The Polar Regions and the Development of International Law. Cambridge University Press, Cambridge

SAR Convention (1979) International Convention on Maritime Search and Rescue, 27 Apr 1979. 1405 U.N.T.S. 118. Entered into force 22 June 1985

Sen. Exec. Rep. 110-9 (2007) United States Senate Executive Report 110-9: Convention on the Law of the Sea. In: 110th Congress, 19 Dec 2007

SOLAS 74 (1974) International Convention for the Safety of Life at Sea, 1 Nov 1974, 1184 U.N.T.S. 278. Entered into force 25 May 1980

Solski J (2013) New Legal Developments in the Russian Regulation of Navigation in the Northern Sea Route. Arctic Rev Law Politics, pp 90–120

Spitsbergen Treaty (1920) Treaty concerning the Archipelago of Spitsbergen, 9 Feb 1920, 2 L.N.T.S. 7. Entered into force 14 Aug 1925

STCW 78 (1978) International Convention on Standards of Training, Certification and Watchkeeping for Seafarers, 1 Dec 1978, 1361 U.N.T.S. 2 Entered into force 28 Apr 1984. As amended and modified by the 1995 Protocol

Stokke OS (2012) Regime interplay in Arctic shipping governance: explaining regional niche selection. Int Environ Agree Politics Law Econ 13(1):65–85

TFEU (2008) Treaty on the Functioning of the European Union. Consolidated Version. O.J. C 115/47

Tokyo MOU (1993) Asia-Pacific Memorandum of Understanding on Port State Control in the Asia-Pacific Region, Tokyo, 1 Dec 1993. Entered into force effect 1 Apr 1994, as regularly amended. This chapter uses the version that includes the 12th amendment and in effect on 1 June 2012

Tymchenko L (2001) The Northern Sea Route: Russian Management and Jurisdiction Over Navigation in Arctic Seas. In: Oude Elferink AG, Rothwell DR (eds) The law of the sea and polar maritime delimitation and jurisdiction. Martinus Nijhoff Publishers, The Hague, pp 269–291

UNCTAD (2011) Review of Maritime Transport. In: United Nations Conference on Trade and Development. doc. UNCTAD/RMT/2011

VanderZwaag D et al (2008) Governance of Arctic Marine Shipping. Report prepared for the AMSA. Marine & Environmental Law Institute, Dalhousie University, Halifax

Chapter 7
Understanding Risks Associated with Offshore Hydrocarbon Development

Kamrul Hossain, Timo Koivurova, and Gerald Zojer

Abstract Arctic sea ice is rapidly reducing due to climatic changes occurring in the region, allowing for easier access to vast amounts of undiscovered oil and gas resources. In recent years, growing interest in exploitation of Arctic hydrocarbon resources has led to an increase in exploration activity. Nevertheless, because of the Arctic's harsh conditions, activities remain costly and are linked to serious environmental risks for vulnerable and unique Arctic ecosystems. Clean-up of potential oil spills would be highly complicated, if not impossible, and routine operational activities connected to hydrocarbon development, such as drilling or increased shipping traffic, have adverse consequences on marine flora and fauna. This chapter examines past, current, and potential future hydrocarbon activities in the Arctic, associated environmental impacts from accidents as well as normal operations, and possible cooperation between the European Union (EU) and United States (US) in mitigating the adverse environmental consequences of oil and gas development. Possibilities for transatlantic cooperation regarding hydrocarbon development in the Arctic are considered, including the use of legal and institutional frameworks to which both the EU and US have commitments.

Based on Koivurova T, Hossain K (2008).

K. Hossain · T. Koivurova · G. Zojer (✉)
Arctic Centre, University of Lapland, Rovaniemi, Finland
e-mail: info@gerald-zojer.com

K. Hossain
e-mail: kamrul.hossain@ulapland.fi

T. Koivurova
e-mail: timo.koivurova@ulapland.fi

E. Tedsen et al. (eds.), *Arctic Marine Governance*,
DOI: 10.1007/978-3-642-38595-7_7,

7.1 Introduction

Offshore hydrocarbon development is anticipated to be a major future economic activity in the Arctic. One of the primary driving factors behind this is climate change, from which access to the Arctic marine area is gradually becoming easier. The Arctic Climate Impact Assessment (ACIA) estimates that there are significant oil and gas reserves in the Arctic marine area, most of which are located in Russian territory, with additional fields in Canada, Alaska (United States (US)), Greenland (Denmark), and Norway (ACIA 2004). The US Geological Survey (USGS) suggests that the area north of the Arctic Circle holds approximately 30 % of the world's undiscovered gas and 13 % of undiscovered oil, with most resources located under less than 500 metres of water (Gautier et al. 2009). Approximately 84 % of this undiscovered oil and gas, representing about 90 billion barrels of technically recoverable oil, is believed to be offshore (Bird et al. 2008). While the Eurasian side of the Arctic has more natural gas reserves, the North American Arctic is more oil-prone: The North American Arctic is estimated to have 65 % of undiscovered Arctic oil, but only 26 % of undiscovered Arctic natural gas (Hong 2012).

Despite the economic opportunities that a surge in offshore hydrocarbon activity may bring, the resulting environmental impacts will contribute to instability in Arctic ecosystems. The end result of the activity, leading to increased use of fossil fuels, will further accelerate climate change, making the Arctic's climate, weather, and ice conditions seemingly less predictable. Although the long-term trends are clear,[1] there will be large variations in ice from year to year, with some seasons colder and having more ice than has been 'normal' in recent years.[2] These variations may have the effect of limiting possibilities for moving oil and gas operations further out to sea, even if the ice edge retreats. Thus, industry cannot necessarily count on areas remaining ice-free and in the case of fixed installations, operators need to set up infrastructure in consideration of maximum extent of ice (Rottem and Moe 2007).

There are four main stages of hydrocarbon development: geological and geophysical survey, exploration, development and production, and decommissioning. Each of these stages has associated environmental impacts. This chapter focuses on the potential risks associated with offshore hydrocarbon development in the Arctic, as well as on potential cooperation between the European Union (EU) and US in the context of hydrocarbon activities.

[1] A recent study by the Arctic Monitoring and Assessment Program (AMAP) suggests that the Arctic Ocean will be seasonally ice-free during the summer within the next thirty to forty years (AMAP 2011).

[2] The National Snow and Ice Data Center announced a new record low in Arctic sea ice extent in Sept 2012, following the previous minimum in summer 2007 (NSIDC 2012).

7.2 Offshore Hydrocarbon Potential in the Arctic

7.2.1 Development of Oil and Gas Extraction in the Arctic

Oil and gas exploration and extraction in the Arctic began in the 1920s, but the second half of the twentieth century has witnessed a rapid growth in activities. In recent years, both energy industries and Arctic states have found oil and gas exploration in the Arctic to be increasingly attractive. In many regions within the circumpolar North, exploitation of oil and gas is already a major economic driver (AMAP 2007). Due to estimated increases in global oil demand[3] and even greater natural gas demand,[4] it is projected that hydrocarbon extraction will continue to expand in the future (IEA 2011a).

A number of reasons are arguably behind this trend: First, the price of oil on the world market is anticipated to remain on a relatively high level, supporting investment in hydrocarbon development, even when extracted in costly regions. Second, resource exploitation may become increasingly feasible in the Arctic with continued advances in ship design and drilling equipment. Third, in comparison to many other hydrocarbon-rich regions, the Arctic can be viewed as a relatively safe region, as there are no ongoing conflicts that would potentially disrupt production.[5] Nonetheless, Arctic offshore drilling is more expensive than in other regions of the world, due to harsh Arctic conditions, which require the use of advanced technologies and enhanced safety measures. The resulting high cost of doing business in the Arctic suggests that only the world's largest oil and gas companies may have the financial, technical, and managerial strength to meet the costs and long lead-times for projects that are dictated by challenging Arctic conditions (Hong 2012). From an investment standpoint, the risks in the Arctic are considered to be far greater than in other regions, which is one reason why hydrocarbon extraction in the Arctic remains in the early phases of development, with limited activity to date. The short drilling season due to sea ice onset, and in some cases tighter environmental regulations than in other regions, could slow the development of new fields.

[3] The International Energy Agency (IEA) projects in a current policies scenario, as well as in a new policies scenario (with more oil use efficiency and switching to other fuels), an absolute global primary oil use increase (e.g., reaching 107 MMb/d in 2035 in the current policies scenario, compared to 84 MMb/d in 2009), even if the share of oil in total primary energy demand is expected to decrease (IEA 2010b).

[4] For natural gas, the IEA expects an absolute increase in demand as well as an increase in the share of total primary energy demand in all scenarios (e.g., in the current policies scenario, the share grows 1.6 % per year, attaining 4.9 tcm in 2035, compared to 3.2 tcm in 2008) (IEA 2010b).

[5] This fear stems from the October 1973 world oil crisis, when Arab members of petroleum producing countries announced a ban on oil shipment to countries supporting Israel in the 1973 Arab–Israeli war (EIA n.d.).

7.2.2 Arctic Resources and Activities

Within the US, more than half of the current Alaskan oil and gas production rate of approximately 0.55 Bb/d[6] originates onshore in the Prudhoe Bay area (AOGCC 2012). The most important *offshore* extracted oil fields[7] in the North of Alaska are the fields of Endicott, Point McIntyre, Oooguruk, and Northstar; from the latter, oil is transported to shore by the first Arctic subsea pipeline, operating since 2001 (Piepul 2001). These fields are, however, near-shore and in shallow waters, using causeways or infrastructure on artificial islands (ADNR 2009), and thus in relatively easy conditions for the Arctic. The results of past exploration, as well as recent surveys, suggest that the chances of discovering large volumes of oil and gas in Alaska's Beaufort and Chukchi Seas are promising. The USGS estimates that 72,766 MMboe of undiscovered oil-equivalent are located in the Alaskan Arctic (Bird et al. 2008). For the US, concerns after the Deepwater Horizon disaster in the Gulf of Mexico in April 2010 led to a temporary moratorium on deepwater offshore drilling in the Outer Continental Shelf and to tighter regulations. In the US Department of Interior's Five Year Outer Continental Shelf Oil and Gas Leasing Program 2012–2017, which was finished in the aftermaths of the Deepwater Horizon accident, new leasing sales for the Beaufort and Chukchi Seas were shifted towards the end of this five-year period. Shell Oil, holding pre-existing leases, was given permission to initiate drilling in 2012, but was forced to postpone completion of wells for another year, after a spill containment dome was damaged during testing in September 2012 (Krauss 2012). In February 2013, Shell Oil announced it would suspend planned offshore drilling in 2013. Under optimistic estimates, assuming high oil and gas prices, development of new oil and gas resources in the offshore US Arctic might compensate for a current decline in Alaskan production (EIA 2012).

The Russian Federation is one of world's leading oil and natural gas producers, producing 10.28 MMb/d oil and 670 Bcm gas in 2011 (Watkins 2012), and holds the world's largest natural gas reserves with approximately 474.6 Tcm. The majority of Russia's estimated 88.2 billion barrels proven oil reserves (BP 2012) are located in Western Siberia. Current production from oil and natural gas fields in Western Siberia is expected to decline, thus in order to maintain high levels of production, "a new generation of higher-cost fields need to be developed, both in the traditional production areas of Western Siberia and in the new frontiers of Eastern Siberia and the Arctic" (IEA 2011a). With an estimated 132,572 MMboe in the West Siberian Basin and 61,755 MMboe in the Eastern Barents Basin (Bird et al. 2008), Russia has strong prospects for maintaining its position as an oil and gas supply leader. Nonetheless, high exploration costs have delayed utilization of many of these resources. For example, even after years of negotiations, the Shtokman project, one of the world's biggest undeveloped gas fields, located in the Barents Sea, was put on hold due to its high development costs (Chazan 2012). On the other hand, some projects have moved forward in the Russian Arctic: The Prirazlomnaya platform in the Nenets Autonomous

[6] Average daily production rate for Sept 2012.

[7] Other offshore fields are extracted from land, using directional drilling technology (ADNR 2009).

Okrug was put in place in August 2011 and, despite several delays, "will be the first ever offshore field in the Russian Arctic put in production" (Pettersen 2012). An agreement between Russia and Norway on the demarcation of their maritime border in the Barents Sea in February 2011 increased stability in the region, opening the door for new exploration in both Norwegian and Russian waters (IEA 2011b).

In 2010, Norway produced 2.16 MMb/d crude oil, and had an estimated natural gas production of 105.9 Bcm in 2009. A share of 3.4 Bcm originated from the Snøhvit LNG plant (IEA 2011b). The Snøhvit gas field is the only field in the Norwegian Arctic where production has taken place. The Snøhvit field is situated in the Barents Sea along with the Goliat field, where, according to the operator Eni, production is scheduled to commence in 2014 (Eni Norge n.d.). Since initial gas field discoveries in the 1970s, hydrocarbon extraction has been the main driver of Norway's economy, with most resources exploited from offshore platforms off Norway's western coast on the continental shelf. Since 2002, however, oil production has declined, and without significant new discoveries, peak Norwegian oil production may have already been reached (IEA 2011b). Such new discoveries are anticipated in Arctic territories, where vast untapped oil and gas reserves are expected. USGS estimates an undiscovered 7,322 MMboe in the Norwegian Margin and 6,704 MMboe in the Barents Platform (Bird et al. 2008).

Combining both conventional and non-conventional oil reserves, Canada is the world's second largest oil-resource holder behind Saudi Arabia with an estimated 267 billion barrels and rising production (IEA 2010a). The Canadian National Energy Board estimates an oil production rate of 3.45 MMb/d for 2012 (NEB 2012a). Canada is also the world's third largest natural gas producer, with an estimated production of 145.7 Bcm in 2012 (NEB 2012b). However, the vast majority of this gas, approximately 98 %, originates *onshore* in the Western Canada Sedimentary Basin and only 2 % is produced—mainly offshore—in Atlantic Canada (IEA 2010a). In the Canadian Arctic, oil and gas activities have been carried out since the 1960s. The first significant discoveries that led to production, such as the Drake Point gas field, were located on Canada's Arctic islands and the Beaufort-Mackenzie area. In total, 65 fields were discovered until the late 1980s, when activities began to decline (McCracken et al. 2007) due to problems with transporting hydrocarbons to the markets, tight environmental regulations, and unsettled land claims (AMAP 2010). Canada's first Arctic offshore wells were drilled in the 1970s in the Beaufort Sea, but no field development took place until the (now ceased) Cohasset Panuke started production in 1992. In recent years, activities have begun increasing again, showing a rise in interest in undiscovered hydrocarbon deposits in the Canadian Arctic. Currently, production takes place off of Nova Scotia (Sable project, 235,067 b/d[8]; production from Deep Panuke was expected to start in 2012), Newfoundland, and Labrador[9] (Hibernia, Terra Nova, White Rose, and its satellite

8 Average of monthly production rates in 2011. CNSOPB. <http://www.cnsopb.ns.ca/pdfs/production_report.pdf.> Accessed 5 July 2012.

9 Following the determination of the Arctic Human Development Report (AHDR), parts of the North Atlantic Ocean, including the Labrador Sea, belong to the Arctic marine area (Young and Einarsson 2004).

North Amethyst; together: 271,791 b/d; CNLOPB 2011), only contributing a moderate share to Canada's overall production. However, USGS estimates potential undiscovered resources of, for example, 17,063 MMboe in West Greenland-East Canada, or 5,108 MMboe in the Amerasian Basin (Bird et al. 2008).

Even though exploration for hydrocarbons in Greenland was first initiated in the early 1970s, to date, no economically feasible amounts of oil or gas have been discovered. Nevertheless, the area offshore of Northwest Greenland is expected to hold 17 BBoe of undiscovered oil and gas resources, and another 31.4 Bboe are predicted to exist in Northeast Greenland (BMP 2011). Greenland's new self-government is active in inviting foreign partners to invest in offshore oil and gas exploration. British Cairn Energy, one of the most prominent of these, finished eight exploration wells in a number of basins from 2010 to 2011 and discovered reservoir-quality sands in the Atammink block (BMP 2012).

7.3 Risks Associated with Hydrocarbon Development in the Arctic

Hydrocarbon development in the Arctic marine area could result in potentially devastating damage to the environment, particularly were a serious accident to occur at any stage of extraction activities, although normal operational activities affect the environment as well. Pollution, such as from oil discharges during drilling operations, releases of drilling mud and chemicals used during the development and production phases, and operational air emissions (e.g., carbon dioxide, nitrous oxide, sulfur dioxide, VOCs, methane, black carbon) all have negative impacts on marine flora and fauna abundance, health, and diversity (Casper 2009).

Oil spills are more dramatic events and cause both short- and long-term adverse effects on the marine environment. Oil spills and releases can occur as the result of blowouts during exploration or production activities, from slow releases of oil from sub-sea pipelines and on-land storage tanks or pipelines travelling to water, or from accidents involving oil transportation vessels or vessels carrying fuel oil (Casper 2009). While intermittent oil spills can occur quickly and remain in specific areas (such as from tanker accidents), persistent oil spills occur mainly in the phase of exploration or production, for example, by means of a blowout or through leaking pipelines. In the case of a persistent oil spill, oil releases continuously and may spread over a larger area if it cannot be embanked in time (Belanger et al. 2010). Spills from drilling platforms normally last for longer periods of time, resulting in immediate and drastic consequences to the environment and wildlife within the marine area, depending on factors such as the type of crude oil spilled, environmental conditions, time of year, currents, and more (Belanger et al. 2010). The impacts of spilled oil, petroleum by-products, and dispersants used for clean-up are of great concern for marine organisms (Muhling et al. 2012).

While true that the Arctic has not experienced any major spills from oil drilling activities to date, it is likewise true that extensive offshore development in the Arctic has yet to commence and would increase the risk of spills. The 1989 *Exxon*

Valdez spill, the largest sub-Arctic oil spill, occurred in significantly more accessible and favourable conditions, but nonetheless left a severe footprint in the region after a huge amount of oil was released in a short time and gradually spread along the coastline (Pew Environment Group n.d.). Following the (non-Arctic) 2010 *Deepwater Horizon* accident, leaking of oil into water column continued persistently for over ninety days and an estimated 4.9 million barrels of oil were discharged into the ocean environment (Muhling et al. 2012), causing shocking consequences to a large marine area. The official estimate suggests that despite clean-up operation efforts—in conditions that were far more accessible than in the Arctic and with greater infrastructure at hand—26 % of residual oil remained in the seawater, whereas 24 % was naturally or chemically dispersed (Maltrud et al. 2010). Should any such accident occur in the Arctic Ocean, the potential impacts would presumably be much greater considering the fact that an effective clean-up operation in Arctic conditions would face greater challenges, not to mention high financial costs.

The main concerns related to offshore development in the Arctic are the potential environmental impacts on its fragile ecosystems. The unique environmental conditions in the Arctic include extended periods of darkness, reduced visibility, ice-covered ocean areas, severe cold, high winds, and extreme storms (Casper 2009). The Arctic marine environment also has a unique seasonal shoreline and oceanographic changes. The Arctic shore consists of ice shelves, glacier margins, ice foot features, and tundra coast; the unique seasonal oceanographic and shoreline changes are due to open water, freeze-up, frozen conditions, and break-up (EPPR 1998). If a large-scale oil spill were to occur in the Arctic, the marine environment would undoubtedly suffer serious adverse impacts and the impacts would be severe for the region's species and ecosystems, causing long-term contamination that would affect populations and ecosystems for decades (Kaczynski and Brosnan 2008; AMAP 2007; Carpenter 2009). For example, twelve years after the (sub-Arctic) *Exxon Valdez* accident, a survey found sixty-one tons of undecayed oil in the subsurface sediments of Prince William Sound's intertidal shorelines and an almost equal amount of only minimally decayed subsurface oil, representing only a 20–26 % per year decay rate. Direct exposure to oil, oil by-products, and dispersants almost certainly results in increased rates of mortality for many organisms. The effects of incorporation of oil into marine food webs are not yet fully understood (Muhling et al. 2012), but would arguably cause contamination. Moreover, increased infrastructure development in the Arctic to facilitate transportation of potential oil and gas, both on land and at sea, will further accelerate adverse consequences on the Arctic environment, which is already vulnerable due to climate change.[10]

The potential environmental impacts of oil and gas production are also relevant for probable costs and insurance issues. Given the difficulty of handling an accident, risk criteria will be set higher than in other offshore areas, such as the North Sea, and thus managing and insuring risk in the offshore Arctic is likely to be costly.

[10] Climate change may allow for increased transport and greater access to Arctic resources (particularly fossil fuels) which would not only create potential environmental consequences, but the burning of extracted fuels to meet global energy demand would further accelerate climate change. See Koivurova T, Hossain K (2008).

Allowing investors without sufficient funds for potential clean-up operations to drill in the Arctic is essentially a risk transfer towards the public sector (Emmerson and Lahn 2012).

7.3.1 Oil Spill Pollution

It is estimated that at present, 80–90 % of petroleum hydrocarbons entering the Arctic come from natural seeps. Increases in oil and gas exploitation, however, will correspondingly heighten the risk of oil contamination from spills and leaks. Oil spills during the exploration, production, and transportation phases of hydrocarbon development are the most serious and direct sources of oil pollution (AMAP 2002). Near-shore facilities and tanker routes near land pose the risk of coastal damage, with spills possibly dispersing over wide stretches (AMAP 2007).

Arctic conditions significantly enhance routine oil spill risks. Ice keels—large pieces of ice that gouge the seafloor—may pummel undersea supply pipes (Wolf 2007). In remote Arctic conditions, a pipeline leak could go undetected for months, leaving extensive pollution behind. Further, oil trapped under ice could be difficult to reach and could potentially travel long distances and towards land, fouling bays, estuaries, and inlets and harming coastal and near-coast species, or travelling further out to sea, placing offshore species at risk and contaminating the sea floor. The persistence of oil, even in small amounts, harms wildlife by reducing species' survival rates, slowing reproduction, and stunting growth (Carpenter 2009).

There are a number of ways in which oil spills could contaminate the Arctic marine area. First, there is a risk of losing well control during the exploration phase, resulting in discharges of oil or releases of gas and creating the potential for fires. Second, other accidental spills and releases may occur from operating activities, both at sea and on land during storage and shipping.[11] Third, oil may be released during offloading in harbours and terminals. Finally, undersea oil pipelines could leak, which poses a special spill risk in the Arctic, as this might not immediately get detected and clean-up operations would be difficult if surface-level sea ice is present (Carpenter 2009).

As mentioned, oil spills are not the only harmful releases that might occur during drilling operations. The offshore oil and gas industry generates huge amounts of dirty water; this water may contain chemicals, and unless it is re-injected, may cause contamination of sea water if discharged into the ocean.[12]

---

[11] Smaller, diffused spills might occur from increased transportation by ships in the Arctic (AMAP 2007).

[12] See OSPAR Recommendation 2001/1 for the Management of Produced Water from Offshore Installations. Under this recommendation, each contracting party was to ensure that the total quantity of oil in produced water discharged into the sea in the year 2006 was reduced by a minimum of 15 % compared to the equivalent discharge in the year 2000. The means used by most of the contracting parties to achieve the goal of a 15 % reduction is the re-injection of produced water (OSPAR Commission 2007).

7.3.2 Clean-Up Challenges

One major reason why the Arctic is considered to be exceptionally vulnerable to oil spills is because of the potentially slow recovery from a spill in cold and highly seasonal ecosystems and the difficulty of clean-up in the remote, cold region, especially where sea ice is present. The ability to cope with spills and to conduct clean-up efforts would vary greatly depending on the location, time of year, weather conditions, volume and characteristics of the oil spilled, and more (Belanger et al. 2010). According to the ACIA report, ice movement, even in the open sea, would hinder some clean-up operations (ACIA 2004). A recent study commissioned by Canada's National Energy Board suggests that clean-up efforts for an offshore oil spill in the Arctic could be impossible at least one day in five because of bad weather or sea ice (Weber 2011).

One of the main methods of clean-up for ocean spills is through the burning of oil slicks. In the Arctic Ocean, these would be subject to high winds that can reach over 10 metres per second, potentially making it impossible to burn the slicks. Even in June, the most favourable Arctic month, weather and ocean conditions would likely prevent conventional clean-up methods from being effectively used about 20 % of the time. These conditions deteriorate over time through the summer until October, when traditional clean-up would likely be impossible 65 % of the time (SLR 2011). Spills in broken sea ice and under sea ice are the most difficult conditions to respond to and cannot be cleaned up effectively (NRC 2003; AMAP 2007). Effective spill response strategies are still being developed for Arctic ice-filled waters (Carpenter 2009). There have been no successful oil spill response tests in ice-covered waters[13] that demonstrate the actual difficulties in clean-up operations.

7.3.3 Oil Spill Impacts on Marine Living Resources

Unlike in ecosystems with less extreme climatic conditions, most organisms in the fragile Arctic depend on limited sources of food supply. The rates of biological factors—such as productive season lengths, generational turnover time, and age of maturity—which determine how quickly an ecosystem would recover from a spill are much slower in the Arctic. Therefore, a single serious oil spill could destroy entire populations and greatly endanger unique species, particularly were the event to overlap with the presence of migratory species, which often congregate

[13] See, e.g., Pedersen (2012), which states that the machinery Shell planned to use for oil spill response has not yet successfully been used in Arctic waters. Wolf (2007) quotes Michael Macrander, a biologist with Shell Oil, as saying that, to date, there are no methods to clean up oil in ice-laden conditions.

in relatively small areas. In addition, flora in Arctic terrestrial environments tends to be much more susceptible than in less extreme environments (Kaczynski and Brosnan 2008).

The living resources in the Arctic that would be most severely affected by oil spills are fish stocks in the embryonic stage, and feathered and fur-bearing animals, which are harmed if they inhale or ingest oil (Carpenter 2009). Seasonal aggregations of animals may be particularly vulnerable (e.g., marine mammals in open water areas in sea ice, seabirds in breeding colonies or feeding sites, or fish at spawning time). Near-coast species are affected by even small oil spills and animals far offshore are at risk if a spill moves out to sea rather than along the coast (Wolf 2007).

Oil spills pollute seawater and adult fish readily take up oil components; however, it is unlikely that high concentrations of these components would accumulate in the fish, which are able to metabolize and excrete them. Massive fish kills caused by oil spills have not been documented in the open sea, although this is largely because toxics concentrations in the wider and deeper open seas seldom reached significant levels. Another reason for this could be avoidance behaviour, whereby adult fish are mobile and can escape contaminated areas, as is witnessed among salmon and cod (Mosbech 2002). Nonetheless, oil spills do directly contribute to fish kills and lead to the gradual reduction of fish stocks. Spawn and fish larvae are particularly sensitive to the effects of petroleum products, as eggs and larvae cannot move to avoid spilled oil, unlike adult fish, and even greater mortality may occur (Mosbech 2002). Hydrocarbons poison the larvae of many aquatic organisms and can kill them during the initial days following an oil spill (Lesikhina et al. 2007). Low levels of dissolved oil hydrocarbons may also slow larval growth rates and affect swimming and feeding behaviours (Muhling et al. 2012). An oil spill in spawning areas could severely reduce that year's recruitment. The risk is greater in the Arctic conditions where effective oil spill clean-up operation is difficult. Dispersants, if used for spill clean-up, could also expose fish eggs and larvae to harmful concentrations of oil components, as dispersants' low evaporation rate increases aquatic exposure.

In ice-free waters, marine mammals may be able to avoid oil, but frozen sea ice may limit open water areas on which marine mammals rely. White whales, narwhals, bowhead whales, ringed seals, walrus, and bearded seals are particularly at risk from oil exposure, as these species' primary habitat is ice-covered waters. Polar bears, seals, and walrus are the most commonly occurring species in the Arctic waters during the icy period. Marine mammals with fur, such as sea otters, polar bears and seals, are more vulnerable to oil spills than other sea mammals, as fur contaminated with oil mats and loses its ability to retain heat and repel water. Oiling may disrupt fur's insulating effect, which species like polar bears depend upon. Oil is, as for other marine species, toxic to the bears and studies suggest that ingestion results in lethal poisoning (Boertmann and Aastrup 2002). Oil can additionally cause irritation to animals' skin and eyes and impede their normal ability to swim (AMAP 1997). Whales and most seals, which rely on blubber rather than fur for insulation, are generally less vulnerable to oiling, but ingested oil can still result in cause gastrointestinal bleeding, renal failure, liver poisoning, and blood pressure disruption. Fumes resulting from the evaporation of oil lead to problems in the respiratory organs of mammals near to, or in the immediate vicinity of, large-scale oil spills. Moreover, oil spills and contamination result in loss of the mammals' food supply (OGP 2002).

Seabirds are among the immediate indicators of wildlife and environmental damage during marine spill events and spilled oil contaminates birds' food supplies, eggs, and habitat (Montevecchi et al. 2012). Birds that feed at sea throughout or for part of the year are considered sensitive to oil spills (Mosbech 2002). Exposure to oil destroys plumage, mats feathers, and causes eye irritation. Oily feathers hinder birds from flying and deprive them of their ability to retain warmth, eventually causing death from hypothermia. Arctic seabirds, which live in cold water, are especially vulnerable to the destruction of the insulating capacity of plumage. Long-term exposure to toxic oil may also hamper bird's reproductive capacity (AMAP 2007).

7.3.4 Other Operational Impacts

Oil and gas activities have a number of other operational impacts on the marine environment. For example, the construction of gravel islands and causeways can impede fish migrations and near-shore water flow. Drill cutting piles accumulating near rigs can disturb bottom-dwelling animals.

The use of ice-breakers can affect ice habitats and also create considerable noise, as can air traffic noise that may occur during transport of logistics supplies to the offshore installations and can frighten animals, causing displacement and disrupting feeding schedules. Large increases in ocean vessel traffic to support hydrocarbon development will raise the number of bird and animal strikes and disturb wildlife (Wolf 2007). Fish and marine mammals both are affected by noise, the effects of which can extend tens of kilometres from the source, particularly by sounds generated from seismic exploration (NRC 2003). For instance, in the Alaskan Beaufort Sea, bowhead whales have been observed to change swimming direction in response to noise sources up to 30 kilometres away. Whale hunters in northern Alaska report that they must travel farther offshore to find whales, a change attributed to the displacement of whales from near-shore areas by industrial noise (AMAP 2007). Species such as whales, walruses, and seals are sensitive to man-made sounds and research shows they move away from industrial noises (AMAP 2007), even though such avoidance behaviour is often temporary. Further, since marine mammals rely on hearing to locate prey, seismic activities could drive animals away from important feeding sites (Wolf 2007).

Hydrocarbon-related transportation and other activities create pressures for improving infrastructure, which may cause fragmentation of both maritime and terrestrial habitats. Many animals have dense seasonal aggregations on breeding grounds, along migratory pathways, or along the ice edges and in open water polynyas in the sea ice, making them temporarily vulnerable to even localised incidents. Rigs, drill ships, and offshore pipelines also tend to impair migration routes. Even without pollution or accidents, oil and gas activities can reduce the wilderness character of a region (AMAP 2007).

Oil and gas development in the Arctic is expected to exacerbate global climate change by increasing the availability of oil and gas to be consumed, contributing to increased greenhouse gas emissions (see, e.g., Casper 2009). Offshore installations and gravel islands, infrastructural development, transportation facilities, and industrial activities related to offshore development will lead to significant new emissions sources, perpetuating the impacts of climate change and Arctic sea ice melt (Hossain 2010). A rough estimate suggests that a barrel of crude oil produces 300 kg of carbon dioxide after refining and combustion processes. If the Arctic's resources comprise 90 billion barrels of technically recoverable oil (based on the USGS 2008 estimate), the region's reserves could eventually produce 27 billion tons of carbon dioxide emissions, an amount comparable to current world total annual emissions, further hindering efforts to mitigate climate change (Greenpeace 2010).

7.4 Potential for EU-US Cooperation

Energy is one of the EU's significant Arctic interests: Although EU member states do not have direct access to Arctic offshore areas, a major share of the EU's energy demands—currently 50 % of total consumption—is imported, 53 % of which comes from the Arctic (Russia and Norway). It has been estimated that over the next twenty years, EU energy imports will rise approximately 65–70 % (Airoldi 2010). Energy development in the Arctic is an EU concern in terms of climate change and sustainability policies as well as energy security. Still, without territorial jurisdiction over the Arctic Ocean, the EU does not have direct authority in regards to hydrocarbon developments in the offshore Arctic. Denmark, which is an EU Member State, does not provide EU territorial jurisdiction over Arctic waters as Greenland withdrew itself from the EU by referendum in 1982. While not coastal, the EU does have Arctic territory through Finland and Sweden, both members of the Arctic Council. Iceland and Norway, coastal Arctic states, are not members of the EU, but are committed to many EU regulations through the EEA Agreement and Iceland has applied for EU membership. Thus, even while the EU lacks direct jurisdiction over offshore activities in the Arctic, it nonetheless has competences and influence that do intersect with its Arctic interests (Jerome et al. 2009).

The US, on the other hand, holds a considerably different position by virtue of its Arctic coastal territory in Alaska. The US has already engaged in Arctic offshore hydrocarbon production and holds notable offshore oil and gas resources within its territorial jurisdiction in accordance with the law of the sea. Further, as a member of the Arctic Council, the US has an effective means of participation and strong voice in any joint decision making processes, including issues directly concerning hydrocarbon development. The US, along with the other Arctic countries and the European Parliament, is also a member country of the Conference of Arctic Parliamentarians. Accompanying its jurisdiction over offshore resources in the Arctic Ocean, the US has legal and political commitments, both domestic and international, to ensure environmental standards in resource development. In addition, the US is bound to observe a number of contingency plans concluded with Russia and Canada. In 1986, the US and Canada, for example, concluded the Canada-United States Joint Marine Pollution Contingency Plan, which was later revised in 2003 (JPC Canada-US 2003). A similar contingency plan—the US-Russia Joint Contingency Plan against Pollution in the Bering and Chukchi Seas—was created between the US and Russia (JPC US-Russia 1989). Moreover, the US also takes part in other regional and multilateral bodies, such as International Regulators Forum for global offshore safety (IRF), that work for the promotion of safe and sustainable offshore hydrocarbon development.

Due to the vast potential hydrocarbon reserves underneath the Arctic Ocean seabed and increasing global demand for energy resources, including in both EU and US markets, extensive offshore exploitation is likely to occur sooner rather than later. The critical factor for EU-US cooperation in this arena will be maintaining high environmental standards so that demand is not met at the expense of environmental sustainability.

Both the EU and the US view the Arctic Council as a high-level forum that can enhance cooperation between states on Arctic issues. The US is a member of the Arctic Council, unlike the EU, which nevertheless supports Arctic Council agendas through various channels such as through financing Arctic research in order to promote science-based knowledge on environmental challenges and sustainable Arctic development. The EU has applied to be an observer to the Arctic Council.

The Arctic Council, through its working groups, has produced a number of scientific documents with a view to promoting Arctic environmental protection and which can support better understanding of offshore oil and gas risks and impacts. In 2009, the PAME (Protection of the Arctic Marine Environment) working group adopted the Arctic Offshore Oil and Gas Guidelines (PAME 2009), a comprehensive document identifying fundamental issues to be taken into account in all stages of offshore oil and gas activities in the Arctic region. These issues include, *inter alia*, environmental impact assessment (EIA), environmental monitoring, safety and environmental management, operating practices, decommissioning, and site clearance. Although the guidelines are not legally binding, they provide important guidance for best practices in offshore activities and have been endorsed by all eight Arctic states, demonstrating wide support. The Arctic Council's EPPR (Emergency Prevention, Preparedness, and Response) working group also supports marine oil pollution preparedness and response.

Currently, a new legally binding agreement on oil spill response is being concluded under the auspices of the Arctic Council in order to implement effective response mechanisms. The new Agreement on Cooperation on Marine Oil Pollution Preparedness and Response (MOPPR) in the Arctic has been drafted and negotiated by a task force, as mandated at the Arctic Council Ministerial Meeting in Nuuk in May 2011, led by Norway, the Russian Federation, and the US, and is intended for completion and signature in May 2013. Following the OPRC Convention (1990), the agreement looks to achieve prompt and effective cooperation amongst Arctic states to undertake actions to minimize any damage or threat of damage resulted from oil spills. Steps taken to combat oil spill in the Arctic marine area are to include strengthened cooperation, coordination, and mutual assistance among the parties and the establishment of national contingency plans and contract points. Voluntary operational guidelines are being developed as well.

In light of the Arctic Council's successful history of cooperation and focus on offshore oil and gas activities, it follows that EU-US cooperation on hydrocarbon development in the offshore Arctic should primarily be conducted through Arctic Council initiatives. Arctic Council initiatives have provided science-based policy documents, founded on extensive research within the auspices of Arctic Council to which the Arctic states have political commitment. Additionally, it may be argued that the Arctic Offshore Oil and Gas Guidelines should be made a legally binding convention, creating legal obligations for actors in Arctic hydrocarbon development.

In addition to the Arctic Council, the EU has engaged with the region's development through other bilateral and multilateral arrangements such as through Barents Euro-Arctic Cooperation (BEAC) and Northern Dimension Policy (NDP). Cooperation between the EU and the US on mitigation of risks associated with offshore hydrocarbon development should additionally be viewed in consideration of these institutional arrangements. Bridging BEAC's initiatives to enhance regional sustainable development, which the EU is a part of, with the Arctic Council's agenda could further enhance EU-US cooperation. Transatlantic partnership between the EU and US could also be developed through the NDP, a common policy of the EU, Iceland, Norway, and Russia that promotes partnership between the EU and other northern non-Member States with regard to sustainable development. Both Canada and the US participate as observers in Northern Dimension initiatives, providing an opportunity whereby the forum could serve as a venue for transatlantic cooperation on sustainable offshore activity.

EU-US cooperation can also address the implementation of existing international legal mechanisms, such as the LOS Convention (1982). Despite not being a party to the LOS Convention, the US has expressed its commitment to following the law of the sea in the Arctic in the Ilulissat Declaration (2008). Implementation of other pertinent regulations, such as IMO (International Maritime Organization) regulations and the OPRC Convention, may contribute to enhancing EU-US cooperation in offshore development as far as these regulations deal with oil discharges from sources such as ships and offshore installations, environmental impacts of offshore development, and preparedness and response in the case of potential oil spills. The EU is not a signatory to the OPRC Convention (while most of its Member States are), but is an observer in the IMO (while all of its 27 Member States are IMO members) as well as a party to the Espoo Convention on Environmental Impact Assessment (1991). The US is a party to the OPRC Convention and is a signatory to but has not ratified the Espoo Convention.

7.5 Conclusion

On balance, there are many promising aspects for EU-US cooperation to help achieve sustainable offshore hydrocarbon development in the Arctic. Both the EU and the US have highlighted the importance of regional knowledge-building in their respective Arctic policies. Sharing information, new knowledge, and industry best practices can facilitate effective management of offshore energy development. The Arctic Council's initiatives to conclude legally binding instruments, such as the Agreement on Cooperation on Aeronautical and Maritime Search and Rescue in the Arctic (Arctic SAR Agreement 2011) and forthcoming oil spill prevention and response agreement, are major developments to which both the EU and US have shown strong support and commitment. In addition, both the EU and US, because of their commitments to sustainable hydrocarbon development in the Arctic, support the types of measures suggested by the guidance provided by the Arctic Council Offshore Oil and Gas Guidelines. The EU and US therefore could further cooperate to support implementation of and adherence to the Guidelines, perhaps even including a push to make the guidelines legally binding. Additionally, cooperation could lend support to effective utilization of existing mechanisms, such as implementation of relevant provisions of the LOS Convention, the OPRC Convention within the auspices of IMO, the Espoo Convention, and more. While cooperation on offshore hydrocarbon development may best be promoted under the auspices of the Arctic Council, the EU and US could also utilize available institutional arrangements, such as cooperation through BEAC and NDP initiatives.

References

ACIA (2004) Impacts of a Warming Arctic: Arctic Climate Impact Assessment. Cambridge University Press, Cambridge

ADNR (2009) Beaufort Sea area wide oil and gas lease sale: Final finding of the Director. Alaska Department of Natural Resources. 9 Nov 2009

Arctic SAR Agreement (2011) Agreement on Cooperation on Aeronautical and Maritime Search and Rescue in the Arctic, 12 May 2011, 50 I.L.M. 1119 (2011). Entered into force on 19 Jan 2013

Airoldi A (2010) The European Union and the Arctic. Main developments July 2008–2010. Nordic Council of Ministers, Copenhagen

AMAP (1997) Arctic Pollution Issues: a State of the Arctic Environment Report. Arctic Monitoring and Assessment Programme, Oslo

AMAP (2002) Arctic Pollution Issues: a State of the Arctic Environment Report. Arctic Monitoring and Assessment Programme, Oslo

AMAP (2007) Arctic Oil and Gas 2007. Arctic Monitoring and Assessment Programme, Oslo

AMAP (2010) Assessment 2007: Oil and Gas Activities in the Arctic: Effects and Potential Effects, vol 1. Arctic Monitoring and Assessment Programme, Oslo

AMAP (2011) Snow, Water, Ice and Permafrost in the Arctic (SWIPA). Executive Summary. Oslo: Arctic Monitoring and Assessment Programme. <http://amap.no/swipa/SWIPA2011ExecutiveSummaryV2.pdf>. Accessed 6 Aug 2011

AOGCC (2012) Alaska Average Daily Oil and NGL Production Rates. Alaska Oil and Gas Conservation Commission, Alaska Department of Administration. <http://doa.alaska.gov/ogc/ActivityCharts/Production/2012_09-ProdChart.pdf>. Accessed 8 Nov 2012

Arctic MOPPR Agreement (2013) Agreement on Cooperation on Marine Oil Pollution Preparedness and Response in the Arctic, scheduled to be signed at the Arctic Council's Ministerial Meeting in Kiruna, May 2013

Belanger M, Tan L, Askin N, Wittnich C (2010) Chronological effects of the Deepwater Horizon Gulf of Mexico oil spill on regional seabird casualties. J Marine Animals Ecology 3(2):10–14

Bird KJ, Charpentier RR, Gautier DL, Houseknecht DW, Klett TR, Pitman JK, Moore TE, Schenk CJ, Tennyson ME, Wandrey CJ (2008) Circum-Arctic Resource Appraisal: Estimates of Undiscovered Oil and Gas North of the Arctic Circle. U.S. Geological Survey, Fact Sheet 2008–3049

BMP (2011) Report to Inatsisartut on mineral resource activities in Greenland. Bureau of Minerals and Petroleum. <http://www.bmp.gl/images/stories/about_bmp/publications/Report_to_inatsisartut_on_mineral_reousrce_activities_in_2011.pdf>. Accessed 13 June 2012

BMP (2012) Report to Inatsisartut, the Parliament of Greenland, concerning mineral resources activities in Greenland. Bureau of Minerals and Petroleum

Boertmann D, Aastrup P (2002) Impacts on Mammals. In A. Mosbech (Ed.), Potential environmental impacts of oil spills in Greenland. An assessment of information status and research needs.(pp 113–117). NERI (National Environmental Research Institute), Technical Report 415

BP (2012) Statistical Review of World Energy, June 2012. <http://www.bp.com/statisticalreview>. Accessed 27 July 2012

JPC Canada-US (2003) Canada-United States joint marine pollution contingency plan, Revised 2003. <http://www.nrt.org/production/NRT/NRTWeb.nsf/AllAttachmentsByTitle/A-403CANUSJCPEnglish/$File/CANUS%20JCP%20English.pdf?OpenElement>

JPC US-Russia (1989) Agreement between the government of the United States of America and the government of the union of soviet socialist republics concerning cooperation in combating pollution in the Bering and Chukchi Seas in emergency situations, with joint contingency plan against pollution in the Bering and Chukchi. Agreement, Moscow, 11 May 1989. Plan, London, 17 Oct 1989. *TIAS* 11446

Carpenter B (2009) Warm is the new cold: global warming, oil, UNCLOS Article76, and how an Arctic treaty might stop a new Cold War. Environ Law Rev 39(1):215–252

Casper KN (2009) Oil and Gas Development in the Arctic: Softening of Ice Demands Hardening of International Law. Nat Res J 49(3/4):825–882

Chazan G (2012) 31 Aug Gazprom puts Shtokman project on ice. Financial Times. <http://www.ft.com/intl/cms/s/0/604b9b38-f359-11e1-9ca6-00144feabdc0.html>. Accessed 3 Oct 2012

CNLOPB (2011) Annual Report 2010/2011. Canada-Newfoundland and Labrador Offshore Petroleum Board

OPRC Convention (1990) International Convention on Oil Pollution Preparedness, Response, and Cooperation, 30 Nov 1990. 30 I.L.M. 733 (1991). Entered into force 13 May 1995

EIA (2012) Annual Energy Outlook 2012. US Energy Information Administration. DOE/EIA-0383(2012)

EIA (n.d.) 25th Anniversary of the 1973 oil Embargo. US Energy Information A dministration. <http://www.eia.doe.gov/emeu/25opec/anniversary.html>. Accessed 5 Feb 2013

Emmerson C, Lahn G (2012) Arctic Opening: Opportunity and Risk in the High North. Chatham House, Lloyd's. <http://www.lloyds.com/news-and-insight/risk-insight/reports/arctic-report-2012>. Accessed 6 Feb 2013

Eni Norge (n.d.) Field development facts. <http://www.eninorge.com/en/Field-development/Goliat/Facts>. Accessed 26 Aug 2012

EPPR (1998) Field Guide for Oil Spill Response in Arctic Water. <http://eppr.arctic-council.org/content/fldguide/index.html>. Accessed 09 Aug 2011

Espoo Convention (1991) Convention on Environmental Impact Assessment in a Transboundary Context, 25 Feb 1989, 1989 U.N.T.S. 309. Entered into force 10 Sept 1997

Gautier DL, Bird KJ, Charpentier RR, Grantz A, Houseknecht DW, Klett TR, Moore TE, Pitman JK, Schenk CJ, Schuenemeyer JH, Sørensen K, Tennyson ME, Valin ZC, Wandreym CJ

(2009) Assessment of Undiscovered Oil and Gas in the Arctic. Science 324(59321):1175–1179. doi:10.1126/science.1169467

Greenpeace (2010) The risks and potential impacts of oil exploration in the Arctic. Media Briefing, Apr 23. <http://www.greenpeace.org.uk/files/pdfs/climate/arctic_briefing_gp.pdf>. Accessed 09 Aug 2011

Hong N (2012) The energy factor in the Arctic dispute: A pathway to conflict or cooperation? J World Energy Law Bus 5(1):13–26. doi:10.1039/jwelb/jwr023

Hossain K (2010) International Governance in the Arctic: The Law of the Sea Convention with a Special Focus on Offshore Oil and Gas. In G. Aflredsson, T. Koivurova (Ed.), Yearbook of Polar Law, 2, 139–169

IEA (2010a) Oil and Gas Security. Emergency Response of IEA countries: Canada. <http://www.iea.org/papers/security/canada_2010.pdf>. Accessed 13 June 2012

IEA (2010b) World Energy Outlook 2010. International Energy Agency, Paris OECD/IEA

IEA (2011a) World Energy Outlook 2011. Executive Summary. Paris: OECD/IEA. <http://www.iea.org/Textbase/npsum/weo2011sum.pdf>. Accessed 13 June 2012

IEA (2011b) Oil and Gas Security. Emergency Response of IEA Countries: Norway. <http://www.iea.org/papers/security/Norway_2011.pdf>. Accessed 13 June 2012

Ilulissat Declaration (2008) Arctic Ocean Conference. Ilulissat, Greenland. 27 May 2008, 48 I.L.M. 382 (2009)

Jerome D, Hossain K, Koivurova T (2009) Canadian Arctic Offshore Oil and Natural Gas and European Union Energy Diversification: Towards a New Perspective? In T. Koivurova, A. Chircop, E. Franckx, E.J. Molenaar and D.L. VanderZwaag (Eds.), Understanding and strengthening European union-Canada relations in law of the sea and ocean governance. Juridica Lapponica, 35 (227–251), University of Lapland Printing Centre, Rovaniemi

Kaczynski V, Brosnan M (2008) Management of Arctic Resources: Economic, Environmental, Legal and Policy Considerations. In: Proceedings of 4th international conference on globalization, energy and environment, Warsaw School of Economics, May 29–30, 2008

Koivurova T, Hossain K (2008) Background Paper: Offshore Hydrocarbon—Current Policy Context in the Marine Arctic. Arctic transform

Krauss C (2012) Shell Delays Arctic Oil Drilling Until 2013. New York Times, 17 Sept. <http://www.nytimes.com/2012/09/18/business/global/shell-delays-arctic-oil-drilling-until-next-year.html>. Accessed 31 Oct 2012

Lesikhina N, Rudaya I, Kireeva A, Krivonos O, Kobets E (2007) Offshore Oil and Gas Development in Northwest Russia: Consequences and Implications. Bellona. <http://www.bellona.org/reports/report/russian_arctic_shelf>. Accessed 15 Nov 2012

Maltrud M, Peacock S, Visbeck M (2010) On the possible long-term fate of oil released in the Deepwater Horizon incident, estimated using ensembles of dye release simulations. Environ Res Lett 5(3):035301. doi:10.1088/1748-9326/5/3/035301

McCracken AD, Poulton TP, Macey E, Monro Gray JM, Nowlan GS (2007) Arctic Oil and Gas. Geological Association of Canada. <http://www.gac.ca/PopularGeoscience/factsheets/ArcticOilandGas_e.pdf>. Accessed 4 July 2012

Montevecchi W, Fifield D, Burke C, Garthe S, Hedd A, Rail JF, Robertson G (2012) Tracking long-distance migration to assess marine pollution impact. Biol Lett 8(2):218–221. doi:10.1098/rsbl.2011.0880

Mosbech A (2002) Impacts of oil spill on fish. In: A. Mosbech (ed.), Potential environmental impacts of oil spills in Greenland. An assessment of information status and research needs. (pp 79–92) NERI (National Environmental Research Institute), Technical Report 415

Muhling BA, Roffer MA, Lamkin JT, Ingram GW, Upton MA, Gawlikowksi G, Muller-Karger F, Habtes S, Richards WJ (2012) Overlap between Atlantic bluefin tuna spawning grounds and observed Deepwater Horizon surface oil in the Northern Gulf of Mexico. Mar Pollut Bull 64:679–687. doi:10.1016/j.marpolbul.2012.01.034

NEB (2012a) Estimated Production of Canadian Crude Oil and Equivalent. National Energy Board. <http://www.neb.gc.ca/clf-nsi/rnrgynfmtn/sttstc/crdlndptrlmprdct/stmtdprdctn-eng.html>. Accessed 14 Nov 2012

NEB (2012b) Marketable Natural Gas Production in Canada. National Energy Board. <http://www.neb.gc.ca/clf-nsi/rnrgynfmtn/sttstc/mrktblntrlgsprdctn/mrktblntrlgsprdctn-eng.html>. Accessed 14 Nov 2012

NRC (2003) Cumulative Environmental Effects of Oil and Gas activities on Alaska's North slope. National Research Council: National Academies Press, Washington, DC

NSIDC (2012) Press Release: Arctic sea ice shatters previous low records; Antarctic sea ice edges to record high. National Snow and Ice Data Center. <http://nsidc.org/news/press/20121002_MinimumPR.html>. Accessed 30 Oct 2012

OGP (2002) Oil and gas exploration and production in arctic offshore regions. Guidelines for environmental protection. Report No. 2.84/329. International Association of Oil and Gas Producers

OSPAR Commission (2007) Annual Report of the OSPAR Commission 2006/07

PAME (2009) Arctic Offshore Oil and Gas Guidelines. Last updated 29 Apr 2009

Pedersen S (2012) Shell's Arctic Drilling Mission Begins Without Oil Recovery Vessel. The international, 15 July. <http://www.theinternational.org/articles/221-shells-arctic-drilling-mission-begins-wi>. Accessed 13 Aug 2012

Pettersen T (2012) More delays at Prirazlomnoye. Barents Observer, 24 May. <http://barentsobserver.com/en/energy/more-delays-prirazlomnoye>. Accessed 16 June 2012

Pew Environment Group (n.d.) Oil Spills. Oceans North. <http://www.oceansnorth.org/oil-spills>. Accessed 6 Feb 2013

Piepul R (2001) Northstar field begins producing through first subsea Arctic pipeline. Oil Gas J. <http://www.ogj.com/articles/2001/11/northstar-field-begins-producing-through-first-subsea-arctic-pipeline.html>. Accessed 27 Oct 2012

Rottem SV, Moe A (2007) Climate Change in the North and the Oil Industry. Input to Strategic Impact Assessment. Barents region 2030. Fridtjof Nansens Institute Report 9/2007

SLR (2011) Spill Response Gap Study for the Canadian Beaufort Sea and the Canadian Davis Strait. SL Ross Environmental Research Limited, National Energy Board. <https://www.neb-one.gc.ca/ll-eng/livelink.exe?func=ll&objId=702903&objAction=Open>. Accessed 10 Nov 2012

Watkins E (2012) CGES: Russia's 2011 output sets new post-Soviet production record. Oil Gas J 110(1b), 17–18

Weber B (2011) Arctic oil spill cleanup impossible one day in five: energy board report. The globe and mail. <http://www.theglobeandmail.com/news/national/arctic-oil-spill-cleanup-impossible-one-day-in-five-energy-board-report/article2116242/>. Accessed 9 Aug 2011

Wolf E (2007) Oil and water. The Arctic Seas Face Irreversible Damage. Earth I J, 22(2)

Young OR, Einarsson N (2004) Introduction. In: AHDR (Arctic Human Development Report. Steffanson Arctic Institute, Akureyri, pp 15–26

Part III
Improving Marine Governance

Chapter 8
Impact Assessments and the New Arctic Geo-Environment

Pamela Lesser and Timo Koivurova

Abstract This chapter discusses impact assessment on both sides of the Atlantic, focusing on its current use in, and potential for, the Arctic region. There is a large body of information regarding the use of impact assessments, primarily as a tool for environmental analysis, but also in a lesser capacity for multinational collaboration. As a result of climate change, regardless of whether impact assessment as a regulatory mechanism becomes legally enforceable, this instrument has an opportunity for an increasingly useful role in pan-Arctic cooperation. The chapter concludes with a discussion on how impact assessments can be used as a mechanism to view the Arctic and address environmental problems in a new 'geo-environmental' context.

8.1 An Overview of Environmental Impact Assessment and the EU and US

According to the International Association for Impact Assessment (IAIA), 'impact assessment', simply defined, is the process of identifying the future consequences of a current or proposed action (IAIA 2013). The requirements for, and implementation of, impact assessments vary widely from country to country. Much of the Arctic is considered to be the sovereign territory of the eight Arctic states, and therefore, it is primarily national and sub-national authorities that implement environmental impact assessments (EIAs).

P. Lesser (✉)
Ecologic Institute, Berlin, Germany
e-mail: pamela.lesser@ecologic.eu

T. Koivurova
Arctic Centre, University of Lapland, Rovaniemi, Finland
e-mail: timo.koivurova@ulapland.fi

E. Tedsen et al. (eds.), *Arctic Marine Governance*,
DOI: 10.1007/978-3-642-38595-7_8,

The European Union's (EU) EIA Directive (1985) provides the legal framework for EIAs in the EU. These assessments may be either mandatory—when projects fall into a designated category automatically triggering preparation of an EIA—, or discretionary —when project impacts are not yet known but are considered likely to result in significant impacts and where Member States determine whether an EIA is needed. The Directive has a limited role for the marine Arctic. Of the Arctic states, Denmark, Finland, and Sweden are EU Member States and are clearly subject to the Directive. Iceland and Norway, as European Free Trade Association (EFTA) countries and parties to the European Economic Area (EEA) agreement are also bound to the EIA Directive. Denmark, via Greenland, has Arctic coastal waters, however, Greenland withdrew from the European Economic Community (EEC) in 1985 and is therefore not subject to the EIA Directive. With respect to the marine Arctic, the EIA Directive's reach is limited due to Greenland's withdrawal from the EEC, the fact that neither Finland nor Sweden have Arctic coastal waters, and that only the mainland of Norway (not the Svalbard Islands which are excluded from the EEA agreement) is subject to the Directive.

In general, preparation and implementation of EIAs are left to the individual EU Member States as well as to sub-national authorities (Koivurova 2008). The European Commission has issued Guidelines for EIAs and Environmental Impact Statements (EIS) with the "aim…to provide practical help to those involved in [stages of screening, scoping, and review of the EIA process] drawing upon experience from around Europe and worldwide" (European Commission 2001). The Guidelines note that while "designed to be useful across Europe [they] cannot reflect all the specific requirements and practice of EIA in different countries [and] cannot substitute for Member State guidance on EIA which should always be referred to first."

While such guidelines are helpful, they remain precisely that—guidelines–, having little to no regulatory power behind them and often, at best, are used in a piecemeal fashion, and at worst, may be forgotten. In terms of the Arctic, the Guidelines for Environmental Impact Assessment (AEPS 1997) adopted as part of the Arctic Environmental Protection Strategy (AEPS 1991) are legally non-binding; as a result, they have not been fully incorporated into the national legislation of the eight Arctic states (Koivurova 2008).

The United States (US) has federal regulations for impact assessments and several individual states also have their own regulations applicable to projects that complement the federal regulations. The National Environmental Policy Act (NEPA 1969) requires federal agencies to undertake an assessment of the environmental effects of proposed actions prior to making decisions. Federal government agencies have a responsibility to implement NEPA and are required to determine if their proposed actions have significant environmental effects and to consider the environmental and related social and economic effects of the proposed actions (CEQ 2007). The range of actions considered is broad and includes issuing regulations, providing permits for private actions, funding private actions, making federal land management decisions, constructing publicly-owned facilities, and more. US state environmental impact assessment laws require additional EIA requirements for state-level agencies, often adding more substantive, rather than procedural, requirements. It should be noted, however, that Alaska (the only US state with Arctic territory) does not have a state-level environmental impact assessment law in place. While the information that must be included in analyses under these laws is specific, the rigor with which that analysis has been undertaken is far from consistent, the result being that the quality of the assessments is often challenged by opponents of a particular project with the courts becoming the final arbiters of what qualifies as an adequate environmental assessment.

8.2 Environmental Assessments in the Marine Environment

Although there are marked differences between marine and terrestrial ecosystems, the most obvious being the prevalence of water versus air, there is also an important legal difference in that the marine environment is generally common property, rather than a freehold or owned by an individual, as on land; nonetheless, the conventional approach to EIA for both is still typically the same. Some have argued that using the same analytical framework for both actually diminishes the influence EIA has to control development and to protect the environment (Smith 2008). There is no conclusive evidence to validate this assertion, and there are of course many similar pressures on both marine and terrestrial environments, but it is worth noting that marine environments, and in particular the marine Arctic, may benefit from more tailored EIAs that account for the many differences.

There are few developments that have either a positive or neutral impact on the marine environment. EIAs are most often used in the marine environment for offshore wind farms, hydrocarbon production, aquaculture, dredging, laying cable lines, hydropower projects, and for permitting desalination plants (OSPAR 2010). While there has been some good work regarding legislation, planning, and application of the EIA process to the marine environment, there is still much additional work that could be done to ensure accurate evaluation of impacts and effective implementation of mitigation measures (Smith 2008).

National guidance has been—and is being—developed in many countries to assist developers and regulators in adapting the environmental impact assessment process to the marine environment. One example is the German *Standards for Environmental Impact Assessments of Offshore Wind Turbines on the Marine Environment* (BSH 2007). The United Kingdom has also been very active in this area, having prepared several guidance documents, including *Offshore Wind-Farms—Guidance Note for Environmental Impact Assessment In Respect of FEPA and CPA requirements* (CEFAS 2004) and the *Nature Conservation Guidance on Offshore Wind Farm Development* (Defra 2005).

8.3 The Role of Impact Assessments (EIAs, SEAs, and TEAs) in the Arctic

8.3.1 Environmental Impact Assessments

For over 20 years, environmental protection in the Arctic has been a catalyst for multilateral cooperation among the eight Arctic states (Canada, Denmark/Greenland, Finland, Iceland, Norway, Sweden, Russia, and the US). The Arctic Environmental Protection Strategy (AEPS), signed in 1991, was followed by, and eventually enveloped under, the Arctic Council which was established five years later. One of the Arctic Council's key objectives has been to promote sustainable development, which is described in the Terms of Reference for the Sustainable Development Program (Arctic Council 2000) as including opportunities to protect and enhance the environment and the economies, culture, and health of indigenous communities and of other inhabitants of the Arctic, as well as to improve the environmental, economic, and social conditions of Arctic communities as a whole. In order to begin implementing this goal, the Guidelines for Environmental Impact Assessment in the Arctic ('Arctic EIA Guidelines'; AEPS 1997) were developed. Finland volunteered to take on the lead role, and the Arctic EIA Guidelines take an arguably holistic approach to looking at the Arctic's unique natural, historical, and cultural environment.

The objectives of an Arctic EIA, following the Arctic EIA Guidelines, are similar to those of non-Arctic environmental assessments: i.e., the nature and likelihood of environmentally damaging events should be accounted for to provide a basis for decision making; the location, technical solutions, construction, operation, and decommissioning aspects of a project should be identified if they are likely to cause adverse environmental effects; alternative options to the project should be identified with the goal of balancing environmental protection and the conservation of natural resources with other social, health, and economic considerations; and to devise and implement remedial measures for eliminating or minimizing undesirable impacts. A unique facet of the Arctic EIA Guidelines is the objective "to provide for the incorporation of traditional knowledge and consultations with the developer, the public, regulatory, and non-regulatory authorities to guide decision making." Later in this chapter, case studies illustrating the role of indigenous peoples in Arctic impact assessments show how traditional knowledge can fuse with contemporary science.

While there is genuine opportunity to enhance environmental protection by using the Arctic EIA Guidelines as a tool for evaluation, they are legally nonbinding and therefore have not been directly incorporated into national legislation (Koivurova 2008). That said, all Arctic countries also have national EIA procedures that may mirror some of the procedures in the Arctic EIA Guidelines, despite not having directly adopted them (Koivurova 2012).

The role of the Arctic EIA Guidelines is to offer a framework for preparing EIAs, but, as explained, as they are not mandatory, it is often national EIA laws that determine the information contained in an EIA. In the marine Arctic, there have been relatively few EIAs, and of those, the comprehensiveness of topics covered and analytical robustness vary considerably. Nevertheless, guidelines generally can be useful, an example of which is the Arctic Offshore Oil and Gas Guidelines (PAME 2009), prepared by the Arctic Council's Protection of Arctic Marine Environment (PAME) working group. These, too, are not mandatory, but as they have already been revised two times, most recently in 2009, this demonstrates that the Arctic Offshore Oil and Gas Guidelines continue to be a living document and presumably have some continuing utility, in contrast to the Arctic EIA Guidelines.

Box 8.1 Example of EIAs in the Arctic: Oden Arctic Technology Research Cruise

The Swedish Polar Research Secretariat (SPRS) is a government agency promoting and coordinating Swedish polar research, including planning research and development and organizing and leading research expeditions to the Arctic and Antarctic regions. The Norwegian University of Science and Technology (NTNU) hosts a centre for research-based innovation, Sustainable Arctic Marine and Coastal Technology (SAMCoT), which works with the development of technologies necessary for the sustainable exploration and exploitation of the Arctic region (Swedish Polar Research Secretariat 2012).

In 2012, the NTNU and the SPRS established a collaboration known as the 'Nordic Cooperation in Polar Research'. The first step in this collaboration was performing a research cruise to the waters northeast of Greenland on the Swedish icebreaker *Oden* in the summer of 2012. The cruise, called the 'Oden Arctic Technology Research Cruise 2012' (OATRC), was a SAMCoT-associated project with financial support from Statoil (Swedish Polar Research Secretariat 2012).

Oden's first scientific expedition was actually carried out in 1991 and an EIA prepared in 1993 to respond to concerns about the effects of underwater noise, interference with marine mammals, and exhaust emissions (Swedish Polar Research Secretariat 2012). The EIA verified that the environmental impacts for *Oden* during polar operations would be minor and any adverse effects were likely to be the result of the use of fossil fuels and resultant air emissions. Emissions were expected to be transitory and in negligible concentrations. All waste would be retrograded back to Sweden for proper disposal. *Oden* would not visit any terrestrial areas, and the cumulative impacts also were considered negligible and to have less than minor or transitory impacts (Swedish Polar Research Secretariat 2012).

The main modification of *Oden* since the 1993 EIA was the installation of a multi-beam echo sounder with an integrated sub-bottom profiler, to be operated during transits and in the survey areas on the continental shelf in the Arctic Ocean. Debate over underwater sound and its effect on marine mammals became the catalyst for another EIA. Similar to the 1993 EIA, the EIA undertaken in 2012 noted that the unavoidable environmental impacts from the OATRC would be transitory and less than significant. At most, minor or transitory impacts would result from logistic activities. Thus, from an environmental point of view, there were no reasons not to perform OATRC, assuming the expedition was conducted within the framework described in the EIA (Swedish Polar Research Secretariat 2012). Interestingly, the actual EIA noted that "[p]resently there is a lack of coherent policy framework for evaluating the environmental impacts of underwater noise pollution. There are no regulations that specifically address the operations of sonar or other ship-borne transmissions of sound in the Arctic" (Swedish Polar Research Secretariat 2012). Given the lack of either a policy or regulatory framework against which to evaluate the project's impact, i.e., underwater noise pollution, the conclusions of significance may be seen as challenging to evaluate and perhaps arbitrary.

8.3.2 Strategic Environmental Assessments

Strategic environmental assessments (SEAs) are often referred to as the environmental impact process applied to policies, plans, and programmes. Although the Arctic EIA Guidelines only briefly address SEAs, with their utility described as a regional and sectoral planning tool, useful in facilitating identification of general sustainability issues and setting the context for more specific project EIAs (AEPS 1997), SEAs are becoming increasingly relevant with respect to EIAs in the region.

SEA Protocol to the Espoo Convention

The SEA Protocol (2011) of the Espoo Convention (1991) entered into force in July 2010 (see Sect. "Espoo Convention" below on the Espoo Convention). The impetus for adding the Protocol stemmed from the need to include environmental and health issues in the preparation of plans, programmes, policies and legislation. Transboundary environmental assessments (TEAs) are only discretionary under the Espoo Convention; therefore, a specific protocol on SEA was developed (Azcarate et al. 2011). The new SEA Protocol is directly linked to TEAs in that the SEA Protocol contains rules requiring a transboundary SEA in certain cases of transboundary environmental effects.

The Protocol focuses on creating national SEA procedures and augments the Espoo Convention by ensuring that individual parties integrate environmental assessment into their plans and programmes at the earliest stage, thus helping to lay the groundwork for sustainable development. SEA is undertaken much earlier in the decision making process than project EIAs, and is therefore seen as a key tool for sustainable development. The SEA Protocol also provides for extensive public participation in government decision making processes. In practice, the applicability of the SEA Protocol to the Arctic marine area is limited given the fact that of the Arctic states, only Finland, Denmark, Norway, and Sweden have ratified it; Canada, Iceland, Russia, and the US are not even signatories.

EU SEA Directive

Under the EU's SEA Directive (2001), SEA is mandatory for plans and programmes (but not policies) which are prepared for agriculture, forestry, fisheries, energy, industry, transport, wastewater management, telecommunications, tourism, town and country planning, or land use and which set the framework for future development consent or have been determined to require an assessment under the EU's Habitats Directive. In general, for plans and programmes not included in the above, Member States must carry out a screening procedure to determine whether they are likely to have significant environmental effects; if there are significant effects, an SEA is needed.

The European Commission summarizes the SEA procedure as follows: An environmental report is prepared in which the likely significant effects on the environment and the reasonable alternatives of the proposed plan or programme are identified and public and environmental authorities are then consulted on the draft documents. Regarding plans and programmes likely to have significant effects on the environment of another Member State, the Member State in whose territory the plan or programme is being prepared must consult the potentially affected Member State(s). The environmental report and the results of consultations are taken into account before adoption. Once the plan or programme is adopted, the environmental authorities and public are informed and relevant information made available. In order to identify unforeseen adverse effects at an early stage, significant environmental effects of the plan or programme are to be monitored (see European Commission 2013).

The EU's SEA Protocol was largely inspired by the SEA Directive and the influence can be seen when comparing the Protocol and Directive, particularly where each requires consultation with potentially affected states/Member States. The Protocol does, however, go beyond the scope of the Directive in that it proposes SEAs be discretionally applied to legislation and policies (Azcarate et al. 2011). Commonalities to both include a provision on transboundary consultations and both explicitly apply to offshore hydrocarbon exploitation.

Box 8.2. Example of SEAs in the Arctic: Dreki maritime area

In 2007, a SEA for the development of Iceland's Dreki maritime area for oil and gas exploitation was undertaken. The proposed plan entailed granting exclusive licenses for exploration and production of oil and gas in the northern part of the Dreki area, which covers approximately 42,700 km^2,. An agreement between Iceland and Norway on the continental shelf between Iceland and Jan Mayen applies to almost 30 % of this area (Ministry of Industry 2007). The possible environmental effects associated with issuing licenses for prospecting, exploration, and production of oil and gas were evaluated along with the risk of accidents. As part of the analysis of the plan, the SEA included a review of the existing legal framework, health, safety, and environmental issues (geology, biota, and climate in the Dreki area), an evaluation of potential hydrocarbon resources in the Dreki area, and also a gap analysis. No major obstacles were discovered to initiating exploration and production of hydrocarbons in the area by granting exclusive licenses.

The SEA made note of its limited scope by stating that "[i]t is important to keep in mind that a strategic environmental assessment of a plan is never as exhaustive as an environmental impact assessment of the environmental effects of construction that may be required because of particular aspects of these activities," the implication being that an EIA would be required for the actual exploratory and production drilling activities in accordance with Icelandic law on EIA (Ministry of Industry 2007).

Iceland's government provisionally awarded Faroe Petroleum the offshore exploration licenses under the country's second licensing round, without requiring any mitigation measures (Offshore 2012). In January 2013, the final licenses were issued. It is interesting to note that the SEA states that the main environmental impact during the prospecting phase is noise from an air gun used during measurements of the ocean floor taken with the help of seismic surveys, finding that "research has shown that whales avoid the noise and alter their diving pattern for a distance from the source of up to 20 kilometres. Since this involves a limited and demarcated activity, no special measures are deemed necessary because of this" (Ministry of Industry 2007). As mentioned previously in this chapter, the analytical rigor between impact assessments can vary greatly and decisions on whether or not to impose mitigation measures remain highly subjective.

The analysis and determination of alternatives are also often highly subjective. For example, the Dreki SEA lists three possible alternatives: to offer licenses throughout the northern part of the Dreki area (42,700 km^2) in accordance with the plan; to restrict the size of the licensing area from what the plan provides to the 5,000–10,000 km^2 considered most promising for finding oil and gas; and not to offer any licenses in the area—either temporarily or for the indefinite future (Ministry of Industry 2007). None of these alternatives were specifically analyzed, but the conclusion was that even without issuing any licenses: the impacts of climate change will result in greater pollution (i.e., more oil tankers transiting the Arctic will increase exposure to oil pollution and emissions); Norway might begin exploring for oil in the neighbouring region, thus exposing the Dreki ecosystem to adverse impacts; and anthropogenic climate change has a large impact on the Arctic region that will likely cause other diverse changes in the environment to occur. While these conclusions are technically not incorrect, an analysis of alternatives could have been conducted in a more robust manner. Impact assessments in general, when sufficiently rigorous, can be valuable tools for marine governance, but can also be used merely as procedural tools to justify political and business priorities.

8.3.3 Transboundary Environmental Assessments

The Arctic EIA Guidelines address the need for looking at transboundary issues, but there is little discussion of actual methods, or of their use as a possible tool for implementing environmental strategies. Rather, they point to the Espoo Convention (see below Sect. "Espoo Convention") as the best framework in which to address transboundary issues, suggesting that these be later coupled with either bilateral or multilateral agreements. Rather than focusing on implementation, the discussion centres more on building trust (especially with indigenous peoples); although, the linkage is later made in the statement that building trust would also, over the long term, produce better EIAs and ultimately a better environmental state. The Arctic EIA Guidelines note that this would be a particularly relevant and important aspect of TEAs as it is sovereign states that need most to cooperate. For any TEA to be successful, "it is essential that individual states understand it will be a long process, that the need to discuss issues early is essential, and that the knowledge of likely transboundary impacts is built together and is followed up together" (AEPS 1997).

Espoo Convention

The main international instrument on transboundary EIA is the Espoo Convention (1991). Adopted in 1991 and entered into force in 1997, the Convention now has a total of 30 signatories and 45 parties.[1] Of the Arctic states, Canada, Denmark, Finland, Norway, and Sweden have all ratified the Convention. Iceland, the Russian Federation, and the US are signatories. While this provides a good foundation for a pan-Arctic legal framework for transboundary EIA, it is not a

[1] As of 10 Feb 2013. See <http://www.unece.org/env/eia/ratification.html>.

comprehensive, multilateral legal framework. Still, because transboundary issues are especially prevalent in the Arctic, the influence of the Espoo Convention as a catalyst for Arctic cooperation, in particular during the stage of the AEPS, should be acknowledged (Koivurova 2009).

Activities requiring a TEA include development activities that are expected to cause significant impacts to the environment of another state; however, TEAs for strategic actions above the project planning level are discretionary, as described in Sect. 8.3.3. TEA processes follow the standard procedures of EIAs and SEAs, but the transboundary issues dealt with in TEAs add administrative, political, and regulatory complexities (Azcarate et al. 2011). The origin state is only obliged to notify a potentially affected state if the planned activity (e.g., offshore hydrocarbon extraction) is likely to cause significant adverse transboundary environmental impacts. If the concerned states disagree on the likelihood of impacts, Appendix IV of the Convention provides for an inquiry commission procedure. It should be noted that the Espoo Convention does not apply to cases of potential harm to global commons, such as the high seas, but only when proposed activities are likely to cause harm to the environment located in another state's maritime zone (Koviurova and Molenaar 2009).

Box 8.3 Example of transboundary environmental assessments in the Arctic: Tornio Stainless Steel Works and Tornio Fairway

The Tornio Stainless Steel Works plant in Finland is situated at the northern end of the Gulf of Bothnia and at the mouth of the river Tornionjoki. The plant has produced ferrochrome since 1968 and steel since 1976. The first EIA for the facility was prepared from 1996 to1997, and the second in 2005. The procedures of the two were identical with the exception that the latter required the draft EIA report to both be submitted for public comment in Finland and submitted to the affected party—in this case, Sweden.

The project to enlarge the Steel Works plant was multi-faceted and included an increase in capacity, enlargement of the nearby harbour, sulphuric acid installation, and a recycling installation for metal dust. According to national Finnish legislation, the proposed project required preparation of an EIA. The enlargement project triggered the requirements of the Espoo Convention as it was determined that transboundary impacts from chemical installation and waste-disposal installation of dangerous wastes could easily impact Sweden, only 2 km away.

Procedures under the second EIA encountered some difficulty as the initial draft of the EIA report sent to Sweden was only partially translated; Sweden asserted that the translated version did not include all of the information that had been stated at an earlier stage during the scoping process. The additions were subsequently provided, which then allowed Finland to receive

comments from Sweden and transmit the document to the Lapland Regional Environmental Centre (Rantakallio 2005).
Public involvement in both projects was substantial. For the steel works component, a combined public information meeting for Sweden and Finland with interpretation was provided and the developer also arranged a separate meeting for the media. In the case of the Tornio fairway project, public information meetings were held separately in Sweden and Finland (Rantakallio 2005). Cooperation with the authorities in both projects consisted of a steering group and a follow-up group with relevant authorities from both sides of the border, including the Lapland Regional Environmental Centre, the Swedish provincial environmental administration, the towns of Tornio (Finland) and Haaparanta (Sweden), the Swedish fisheries administration, and fishermen's associations from Sweden and Finland (for the fairway project).

Lessons learned included that procedures should be well-planned in advance, responsibilities should be clear with training regional competent authorities and providing written guidance, and meetings should be held with points of contacts during the process to agree on practicalities and improve procedure (Rantakallio 2005).

8.3.4 Other Application Under International Law

In addition to the Espoo Convention and SEA Protocol, a duty to conduct an EIA is found under other international treaties such as the LOS Convention (1982) (art. 206) and the Convention on Biological Diversity (1992) (art. 14). Principle 17 of the Rio Declaration (1992) calls for an EIA to be undertaken for "proposed activities that are likely to have a significant adverse impact on the environment".

Transboundary EIA under the LOS Convention is of particular interest for the Arctic marine area. Under the Convention, when there are reasonable grounds for believing that planned activities within the jurisdiction or control of a state may cause substantial pollution of or significant harmful changes to the marine environment, states must assess the potential effects of the activities on the marine environment. States are required to conduct an assessment of the effects of activities taking place within their maritime jurisdiction on the marine environment located in other states' jurisdiction as well as on areas beyond national jurisdiction. Unfortunately, it is unlikely that an assessment of transboundary impacts on the marine environment located in another state's jurisdiction would be evaluated in a systematic manner given the fact that there is no guidance on how potentially affected states can contribute to an assessment. More importantly, the duty of assessment is qualified by the phrase "as far as practicable", giving the origin state a fair amount of discretion. The results of assessments are to be communicated to competent international organizations "which should make them available to all states", and it is in this manner that a potentially affected state can obtain information.

The International Court of Justice (ICJ) has recognised a requirement under customary international law to carry out assessments where there is a risk that a proposed industrial activity may have a significant adverse impact in a transboundary context, in particular, on shared resources. In the *Pulp Mills on the river Uruguay* case,[2] the ICJ held that conducting an EIA can be considered a requirement under general international law and is part of exercising due diligence; this judgment does not necessarily extend to a requirement for SEAs. The ICJ did not specify the requisite content of assessments, but did clarify that content be determined "having regard to the nature and magnitude of the proposed development and its likely adverse impact on the environment" (para. 205), that impact assessments be carried out prior to implementation of activities, and that continuous monitoring is needed. As a rule of customary international law, this development may require additional clarification. The ICJ also suggested that the content of this obligation might evolve over time.

In its Advisory Opinion for the *Nauru* case,[3] the International Tribunal on the Law of the Sea with its specialised section for the settlement of seabed disputes elaborated on aspects of due diligence with regard to EIAs. It concluded that there exists an obligation of conduct—not, however, an obligation of result—and understood due diligence as a variable concept. It also remains to be seen if the transfer of this principle to areas beyond national jurisdictions will be confirmed in the future.

8.3.5 Do Any of These Impact Assessments and Guidelines Matter and are They Effective?

Although specific and easy to understand, in practice, the Arctic EIA Guidelines are often not applied due to both a lack of awareness of their existence coupled with the fact they are not mandatory (Koivurova 2012). While many project-level EIAs have been prepared, there is little consistency regarding methodology and rigor of analysis, or even what substantively is to be covered.

A similar lack of methodology is also a persistent problem in SEAs. In theory, by having adopted various regulations for EIA and SEA within national legal systems, the Arctic states are obligated to carry out environmental assessments for projects, plans, and programmes, and, in some cases, for policies that could potentially harm the environment. However, as variations exist between the environmental assessment systems of the different Arctic countries, and between rigor of analysis

[2] Pulp Mills on the River Uruguay (Argentina v. Uruguay), Judgment of 20 Apr 2010, *ICJ Reports* 2010.

[3] Responsibilities and obligations of States sponsoring persons and entities with respect to activities in the Area (Request for Advisory Opinion submitted to the Seabed Disputes Chamber), Case No. 17, Advisory Opinion, 1 Feb 2011, see paras. 110ff., 117.

and implementation, the application of EIA and SEA has varied considerably in the circumpolar region (Azcarate et al. 2011).

Regarding TEAs, planned activities for the Arctic have not yet taken place on a large scale, and where activities have occurred, they have been far away from national borders (Koivurova 2008). Moreover, there seems to be a lack of capacity and knowledge on how to implement TEAs, demonstrated by the low number of national assessments and TEAs that have been implemented in the Arctic (Azcarate et al. 2011).

Thus, while most of the Arctic is covered by national provisions on EIA and SEA, and international treaties such as the Espoo Convention and its SEA Protocol require certain Arctic states to carry out TEAs, gaps still remain, suggesting that some activities with environmental impacts in the Arctic are not assessed.

8.4 Climate Change in the Arctic and the EIA process

With climate change, Arctic glaciers have drastically receded, but it is the Arctic Ocean itself that is most changed. From the 1970s to the 1990s, the minimum extent of polar pack ice fell by around 8 % per decade. In 2007, record sea ice lows left the Northwest Passage ice-free for the first time in memory (see Astill 2012). Scientists found that in 2007 every natural variable, including warm weather, clear skies, and warm currents, had lined up to reinforce the seasonal melt. But in 2012, under less 'primed' conditions, sea ice reached a new record low. Past climate models predicted that the Arctic Ocean could be ice-free in summer by the end of this century; an analysis published in 2009 in *Geophysical Research Letters* Wang and Overland (2009) suggested it might happen as early as 2037. Many now think it will be sooner.

8.4.1 The Arctic Climate Impact Assessment

During the US chairmanship of the Arctic Council from 1998 to 2000, the Council and the International Arctic Science Committee (IASC) commenced work on the Arctic Climate Impact Assessment (ACIA), and their findings were released in a publicly accessible format in 2004 (ACIA 2004). The ACIA concluded that ocean warming and loss of ice is expected to accelerate, exacerbating the major physical, ecological, social, and economic changes already underway in the Arctic marine environment.

Even though a 2001 report by the Intergovernmental Panel on Climate Change (IPCC) had already noted that warming is more intense in the Arctic, the Arctic Council-sponsored ACIA report established the Arctic as an early warning region for climate change observation. In addition to revealing serious impacts on the environment, ecosystems, and local communities, the ACIA led to important changes in the way the Arctic is perceived. The ACIA findings demonstrated that the Arctic is a region undergoing dramatic transformation. The public view developed from an understanding of the Arctic as being naturally guarded from human activity to an image of the region as dynamic and rich with economic potential, and therefore requiring stricter governance measures.

8.4.2 Climate Change and a Stronger Arctic Council

Until recently, Arctic cooperation has functioned for over fifteen years in a fairly consistent mode of operation with work primarily conducted by the six working groups focused on environmental protection and sustainable development (see Chap. 1). But in response to alarming climate change, and the new changes and challenges that are rapidly manifesting in the region, the Council has recently strengthened the way it functions. Given the enormity of challenges posed by climate change in the Arctic, the Arctic Council has gradually started to support the adoption of binding legal responses, rather than sticking only to traditional soft law regulation through guidelines, best practices, and manuals. These soft law measures serve their purpose in some policy arenas, especially since the region's indigenous peoples' organizations can participate in their drafting and some soft law guidance has likely made its way into practice, although this is often difficult to verify. One case that indicates the limitations of soft law regulation, particularly where Arctic Council member states have viewed guidance as an intrusion upon their sovereignty, is the EIA Guidelines, which as noted, appear to have been gradually forgotten (Koivurova 2012).

8.4.3 Analyzing Climate Change in EIAs

There are no written guidelines for the Arctic that provide a framework for assessing climate change in impact assessments. However, factors to consider when analyzing climate change effects in the Arctic include rising temperatures, increased extreme weather, loss of permafrost affecting facilities and transportation design and performance, environmental and social changes as a result of climate changes, and effects arising from increased pressure to develop resources given easier Arctic access.

8.4.4 The Canadian Example

At the prompting of the IAIA and the Canadian Environmental Assessment Agency, the *Practitioner's Guide to Incorporating Climate Change into the*

Environmental Impact Assessment Process—ClimAdapt EIA Guide (Bell et al. 2003) was developed in 2003 with the intent to provide the following:

- An understanding of the implications of climate change in relation to the preparation of an EIA;
- On a project-specific basis, to give direction on how to determine whether or not climate change should be considered;
- Sources of information for use in assessing climate change implications; and
- Guidance in incorporating climate change consideration into the EIA process.

Part of the guidance includes reversing the normal thought process, which looks at the impacts of the project on the environment. Conversely, the question was asked—*what are the effects of the environment on the project*? To answer this question, issues such as the following were addressed: human health and safety; how the operation and productivity of a project may be adversely affected; how the cost of development may rise and project design need to be modified; how maintenance requirements could increase; and the increasing importance of monitoring (Bell et al. 2003).

It also promoted the consideration of cumulative effects within the context of climate change, involving issues such as the increased transport of physical or chemical constituents; an increase or decrease in habitat area for a species or species group that is already affected by the project; and the secondary effects related to climate change modification to the environment or its effects on the project (Bell et al. 2003). Six existing EIAs in Canada were used as case studies, for example:

Gahcho Kue Diamond Mine—2007: The MacKenzie Valley Review Board has jurisdiction for EIAs in a large portion of Canada's Northwest and was faced with the proposal for a new proposed Debeers diamond mine. Given scientific consensus that the North is particularly vulnerable to impacts from a changing climate, the Board found that the EIS must examine and evaluate the development as a potential greenhouse gas contributor, as well as examine potential climate change effects on the proposed development (Bell et al. 2003).

Beaufort Sea Gas Development Project (Canadian Arctic)—1982: A detailed EIS described the impacts of a project to recover Arctic oil and gas and deliver it to southern markets through either a MacKenzie River pipeline or tanker. While current and historic climate conditions were reviewed in detail in the EIA, there was no consideration or integration of climate change into the assessment. The decision making panel did, however, recommend the possibility that climate change be considered in the design and construction of a pipeline and other fixed facilities in areas of permafrost (Bell et al. 2003).

Voisey's Bay Nickel Mine (Canadian province of Newfoundland and Labrador near the bay of Voisey)—1997: The project was located in Labrador where climate change predictions at the time were for reductions in temperatures rather than increases. The assessment revealed that Valued Environmental Components warranted ongoing monitoring and observation, and the preparation of a risk assessment that looked at caribou, birds, Arctic char, sea ice with respect to shipping and its interaction with local coastal activity, and changes to social activities of local aboriginal peoples. The last two are possibly the most likely to involve interactions of the project and the effects of climate change together (Bell et al. 2003).

Although the Guide was prepared in 2003, it appears that little has changed in EIA practice. The emphasis on addressing climate change impacts remains focused on the reduction of greenhouse gas emissions and the potential effects of change on the project. As noted by Collins (2010), the question should be whether we can agree that climate change is the largest global cumulative impact that we need to address. While effort is increasing to mainstream climate change into existing sustainable development management processes, integration in EIA has been largely limited to greenhouse gas contributions and effects on the project. There is a world of difference between consideration of greenhouse gas reductions and climate change contributions, and planning for and taking into consideration the impacts of climate change.

8.5 Progressiveness in Using Traditional Knowledge

One important aspect of impact assessments in the Arctic is the role and contribution of indigenous peoples. As explained the Arctic EIA Guidelines emphasize the inclusion of indigenous peoples into the EIA process, reflecting the status of indigenous peoples as permanent participants in the Arctic Council.

Comparing impact assessments between indigenous and non-indigenous peoples provides a revealing contrast in terms of environmental approaches and resultant analyses. For example, the Sámi perceive landscape in a cultural way, as part of the community's history, language, myths, values, and livelihoods. In short, the Sámi cultural landscape, heritage, and biodiversity are inextricably linked (see Chap. 4).

As a result of this cultural landscape having been formed through traditional livelihoods, traditional ecological knowledge (TEK) and respect for the land can result in a more holistic approach in terms of environmental assessment. In Canada, this can be seen in a number of instances, and through the following example of the Baffinland project, it is clear that TEK helps provides a foundation for a more integrated approach.

8.5.1 Baffinland Project

The Mary River Project is a proposed iron ore mine on North Baffin Island, in the Nunavut Settlement Area (NSA) of the Mackenzie Valley, in conjunction with the development of a railway that is planned to transport 18 million tons per year of ore from an open pit mine to an all-season deepwater port. The project would be the largest in Nunavut history, involving technological challenges such as the construction of a railway on permafrost and cumulative impact concerns such as land use changes, mining, land transportation, and marine shipping.

The Nunavut Impact Review Board (NIRB) is responsible for the environmental assessment of project proposals in the NSA [Collins N (2003)]. An EIA was requested by the NIRB in June 2008, designed to include preparation of a climate change assessment addressing project impacts to permafrost and soils with high ice content, the hydrological regime, marine ice flow regimes, and the long-term impacts of such changes on the project. Multiple impact assessment scenarios spanning the range of possible future climates were to be designed and applied, and the long-term effects of climate change were to be discussed up to the projected closure phase of the project (Collins 2010).

The NIRB's guidelines (NIRB 2008) took a more holistic view that required not only a detailed evaluation of climate change impacts, as discussed above, but in addition, that an *ecosystem-based approach* to EIA be adopted, that socioeconomic issues—such as the economic development within the region—be included, that past and potential future environmental, economic, and social trends be taken into account, and that the well-being of residents of Canada outside of the Nunavut Settlement Area also be considered (NIRB 2008; see also Collins 2010).

In addition to seeing how climate change impacts are addressed in both the Canadian and Nunavut cases, what is interesting about the Baffinland example is the extent to which TEK underlay many of the other environmental baseline assumptions against which impacts were later measured. For example, five local communities were consulted regarding their knowledge of terrestrial wildlife, caribou in particular. Traditional knowledge of caribou habitat and abundance cycles was then collected and summarized in a spatial database. The traditional knowledge and survey data were ultimately used as a baseline to conduct an impact assessment to determine practical options to mitigate the effects of the mine, road, and railway on caribou habitat and movement (Environmental Dynamics n.d.).

The need to survive extremely harsh conditions for generations has built a repository of local knowledge that easily melds the understanding of how local and regional ecosystems operate with the more specific interrelationships between animals, plants, and the physical environment. In fact, it has been documented that in remote parts of the world, the ecological knowledge of indigenous peoples is often 'geographically and temporally more extensive than scientific knowledge' (Ferguson and Messier 1997).

The final public hearings on the Mary River Project by the NIRB occurred in July 2012. The first phase of the mine is expected to last 21 years and generate revenues of 3–5 billion CND while creating 950 jobs. Baffinland representatives at the hearing stated that they are committed to hiring Inuit and creating economic opportunities (Murphy 2012).

In spite of these professed benefits, the NIRB, Nunavut Tunngavik Inc., and Qikiqtani Inuit Association repeatedly voiced concerns regarding Baffinland's preparedness in the event of a spill. Baffinland insisted the risk of fuel spills associated with the project's transportation system are low and that emergency clean-up plans are in place in the "unlikely" event that one will occur; nonetheless representatives admitted their current techniques for dealing with a spill are "not optimal for recovery on the ice" (Murphy 2012).

8.6 Toward 'Geo-Environmental' Cooperation in the Arctic'

The Espoo Convention initially held much promise as the best hope for transboundary, pan-Arctic environmental cooperation in the Arctic region. Since the late 1990s, however, that promise has waned as Iceland, Russia, and the US have been unwilling to accept the Convention as legally binding; however, recent developments indicate that Russia may be moving closer to acceptance of the Convention. Iceland's reasons for not ratifying the Convention are said to be the impracticality of TEA procedures due to the country's location and the fact that priorities need to be established for a country with such a small civil service (Koivurova 2008). The US, while a signatory, is not interested in becoming a party. Even so, the Espoo Convention is still generally cited as being a standard for TEA in the Arctic, as exemplified by the voluntary application by Finland of Espoo procedures towards Russia in the hopes of inducing Russia to finally ratify the Convention.

It has been more than 15 years since Espoo has come into force, and in the interim, the ramifications of climate change, and the projections of future adverse impacts from climate change, continue to grow. Could this changing reality be the catalyst for a new type of environmental cooperation—one that takes climate change fully into account and addresses not only traditional environmental issues, but also, in an integrated fashion, those relating to environmental security and socioeconomic considerations? Could this new reality be the catalyst for a new form of Arctic cooperation (i.e., the 'geo-environmental' dimension) where geopolitics—including relationships based on shipping and trade, offshore oil and gas development, and energy security—are increasingly shaped by environmental issues? Perhaps transboundary environmental assessments will build stronger relations among the Arctic nations by encouraging environmental cooperation in the recognition of a common goal: sustainable use of the Arctic.

Perhaps most important of all is the potential for new cooperation in Arctic governance. Whether or not an EIA or SEA with actual legal authority will ever be a reality, can the Arctic, in all its complexity, be looked at from a geo-environmental perspective, and if not, what are the ramifications for the Arctic and the world?

The key question is: Can new and better approaches to impact assessments in the Arctic help to inform and shape a more effective approach for the region's environment? The limits to this will likely be set by Arctic politics and sovereignty concerns, but the demand for more rigorous and enforceable EIA guidelines will almost certainly grow over the coming years. The growing significance of Arctic natural resources has a direct impact on events in the global economy and politics, and with climate change, globalization, and technological developments, the

Arctic region is no longer seen as a peripheral area, but is gaining a place at the centre of international affairs. Arctic change is inevitable, and still largely unpredictable, but of huge—'geo-environmental'—dimensions.

References

ACIA (2004) Impacts of a Warming Arctic: Arctic Climate Impact Assessment. Cambridge University Press, Cambridge

AEPS (1991) Arctic Environmental Protection Strategy, 14 Jan 1991, 30 I.L.M. 1624

AEPS (1997) Guidelines for Environmental Impact Assessment (EIA) in the Arctic. Sustainable Development and Utilization, Arctic Environment Protection Strategy. Finnish Ministry of the Environment, Finland

Arctic Council (2000) Framework Document (Chapeau) for the Sustainable Development Programme, Arctic Council, 13 Oct 2000

Astill J (2012) The vanishing north, The Economist, 16 June

Azcarate J, Balfors B, Destouni G, Bring A (2011) Shaping a Sustainability Strategy for the Arctic. Royal Institute of Technology and Stockholm University. IAIA11 Conference on Impact Assessment and Responsible Development for Infrastructure, Business and Industry

Bell A, Collins N, Young R (2003) Practitioner's Guide to Incorporating Climate Change into the Environmental Impact Assessment Process: Clim Adapt EIA Guide

BSH (2007) Standards for Environmental Impact Assessments of Offshore Wind Turbines on the Marine Environment. Bundesamt für Seeschifffahrt und Hydrographie, Feb 2007

CEFAS (2004) Offshore Wind-Farms—Guidance Note for Environmental Impact Assessment in Respect of FEPA and CPA requirements (United Kingdom), Version 2 June 2004. Prepared by the Centre for Environment, Fisheries and Aquaculture Science on behalf of the Marine Consents and Environment Unit

CEQ (2007) A Citizen's Guide to the NEPA: Having your voice heard. Council on Environmental Quality, Executive Office of the President. Dec 2007

Collins N (2003) Incorporating Climate Change into Cumulative Effects Assessment. Presentation for the 2003 IAIA Calgary Conference, CEF Consultants Ltd., Halifax

Collins N (2010) Climate Change and Canada's North. Presentation for the IAIA conference in Washington DC, CEF Consultants Ltd., Halifax

Convention on Biological Diversity (1992) Convention on Biological Diversity, 5 June 1992, 1760 U.N.T.S. 79. Entered into force 29 Dec 1991

Defra (2005) The Nature Conservation Guidance on Offshore Wind Farm Development—A guidance note on the implications of the EC Wild Birds and Habitats Directives for developers undertaking offshore wind-farm developments. Mar 2005

EIA Directive (1985) Council Directive 85/of 27 June 1985 on the assessment of the effects of certain public and private project son the environment, 27 June 1985, O.J. (L 175/40), as amended as amended by Directives 97/11/EC and 2003/35/EC

Environmental Dynamics (n.d.) Mary River Terrestrial Wildlife Baseline and Impact Assessment. <http://www.edynamics.com/assets/files/Baffinland_MaryRiver_Project.pdf>

Espoo Convention (1991) Convention on Environmental Impact Assessment in a Transboundary Context, 25 Feb 1989, 1989 U.N.T.S. 309. Entered into force 10 Sept 1997

European Commission (2001) Guidance on EIA—EIS review. Prepared by Environmental Resources Management, June 2001

European Commission (2013) Additional tools. Strategic Environmental Assessment—SEA. <http://ec.europa.eu/environment/eia/sea-legalcontext.htm>. Accessed 20 Feb 2013

Ferguson M, Messier F (1997) Collection and Analysis of Traditional Ecological Knowledge about a Population of Arctic Tundra Caribou. Arctic 50(1):17–28

IAIA (2013) International Association for Impact Assessment. <www.iaia.org>. Accessed 8 Feb 2013

Koivurova T (2008) Transboundary environmental assessment in the Arctic. Impact Assess Project Appraisal 26(4):265–275

Koivurova T (2009) Transboundary EIA in the Arctic. UNECE Seminar. <http://www.unece.org/fileadmin/DAM/env/eia/documents/WG12_may2009/seminar_arctic.pdf>. Accessed 10 Feb 2013

Koivurova T (2012) New ways to Respond to Climate Change in the Arctic. insights. American Society of International Law, Insights, 16(33)

Koivurova T, Molenaar EJ (2009) International Governance and Regulation of the Marine Arctic: Overview and Gap Analysis. WWF International Arctic Programme, Oslo

LOS Convention (1982) United Nations Convention on the Law of the Sea, 10 Dec 1982, 1833 U.N.T.S. 396. Entered into force 16 Nov 1994

Ministry of Industry (2007) Strategic Environmental Assessment (SEA) of the proposed plan to offer exclusive exploration and production licenses for oil and gas in the Dreki (Dragon) area on the Jan Mayen Ridge, Northeast of Iceland. Ministry of Industry, Iceland Mar 2007

Murphy D (2012) Baffinland reps field questions on risk, socio-economics, July 18. Nunatsiaq Online. <http://www.nunatsiaqonline.ca/stories/article/65674baffinland_reps_field_questions_on_risk_socio-economics/>. Accessed 11 Feb 2013

NEPA (1969) National Environmental Policy Act of 1969, 42 U.S.C. 4321 et seq

NIRB (2008) Guide to the NIRB Review Process. Guide 5. Nunavut Impact Review Board, Sept 2008. <http://www.nirb.ca/04-GUIDES/NIRB-F-Guide%205-The%20NIRB%20Review%20Process-OT5E.pdf>. Accessed 11 Feb 2013

Offshore (2012) Iceland issues offshore licenses, 4 Dec <http://www.offshore-mag.com/articles/2012/12/iceland-issues-offshore-licenses.html>. Accessed 20 Feb 2013

OSPAR (2010) Quality Status Report 2010. Assessment report on the environmental impact of offshore wind farms. Publication No. 385/2008

PAME (2009) Arctic Offshore Oil and Gas Guidelines. Last updated 29 Apr 2009

Rantakallio S (2005) Practical application of Espoo convention in Finland. Presentation from the Finnish Ministry of the Environment. UNECE Conference, Stockholm

Rio Declaration (1992) Rio Declaration on Environment and Development, UN Doc. A/CONF.151/26 (vol. I), 31 I.L.M. 874 (1992)

SEA Directive (2001) Directive 2001/42/EC of the European Parliament and of the Council of 27 June 2001 on the assessment of the effects of certain plans and programmes on the environment,21 July 2001, 2001 O.J. (L 197/30)

SEA Protocol (2011) Protocol on Strategic Environmental Assessment to the Convention on Environmental Impact Assessment in a Transboundary Context, 21 May 2001, UNECE Document ECE/MP.EIA/2003/2. Entered into force 11 July 2010

Smith A (2008) Impact assessment in the marine environment—the most challenging of all. IAIA08—Impact assessment in the marine environment. <http://www.iaia.org/iaia08perth/pdfs/themeforums/TF4-1_sector_marine_Smith.pdf>. Accessed 8 Feb 2013

Swedish Polar Research Secretariat (2012) Environmental Impact Assessment (EIA) for OATRC 2012, Dnr 2012-160, Swedish Polar Research Secretariat, Stockholm, Sweden

Wang M Overland JE (2009) A sea ice free summer Arctic within 30 years? Geophys Res Lett 36, L07502. doi:10.1029/2009GL037820

Chapter 9
Pan-Arctic Marine Spatial Planning: An Idea Whose Time Has Come

Charles N. Ehler

Abstract Driven by outside economic forces and the effects of climate change, the Arctic, its ecosystems, and its people are all faced with substantial change ranging from the loss of ice-dependent species, more intense human uses of the Arctic, and the loss of natural services provided by Arctic ecosystems. In addition to business opportunities, these changes represent new risks to the Arctic's unique natural environment and to the people who now live and work in the Arctic. Once new human activities begin in the Arctic Ocean, it will be difficult for policymakers and planners to put limits on them. This paper explores a new approach to the integrated management of human activities—marine spatial planning (MSP). MSP seeks to reduce conflicts among human activities and balance the conservation of ecologically important areas with the sustainable development of marine resources in the Arctic. With the exception of Norway, most Arctic governments have been slow to advance marine spatial planning. A way to advance MSP in the Arctic would be to explicitly recognize the importance of moving beyond sole reliance on the initiatives of national governments and towards a pan-Arctic approach to guide the future of the region. Networks and partnerships of non-governmental actors, including indigenous peoples, environmental NGOs, academia, and private industry, all of whom have influence over governmental policies and actions, could be used to initiate MSP across the Arctic.

9.1 Introduction

The Arctic Ocean—all 14 million square kilometres of it—is one of the most pristine, yet vulnerable, ecosystems in the world. Protected by its historic inaccessibility, harsh environment, relatively small human population, and slow rate of economic development, the Arctic Ocean has been relatively less affected by human activity than most other marine regions of the earth.

C. N. Ehler (✉)
Ocean Visions Consulting, Paris, France
e-mail: charles.ehler@mac.com

E. Tedsen et al. (eds.), *Arctic Marine Governance*,
DOI: 10.1007/978-3-642-38595-7_9,

This is about to change. Driven by outside economic forces and the effects of climate change, the Arctic, its ecosystems, and its people are all faced with substantial change ranging from the loss of ice-dependent species, particularly polar bears and walruses, more intense human uses of the Arctic (e.g., oil and gas exploration and development, commercial fishing, and marine transport), and the loss of natural services provided by Arctic ecosystems. As the Arctic warms, its ice melts, and its ecosystems change, and as technology improves and the demand for natural resources increases, opportunities open up for industry—shorter shipping routes, virgin fishing grounds, new areas of oil and gas exploration and development, and new places for commercial tourism. Several scientists have predicted that the Arctic will be ice-free by the 2030s. As well as business opportunities, these changes represent new risks to the Arctic's unique natural environment and to the people who now live and work in the Arctic. *Once these activities begin in the Arctic Ocean, it will be difficult for policymakers and planners to put limits on them.*

International and national interest in mitigating and adapting to these changes has led to increased calls to manage human activities through an ecosystem-based approach. Marine spatial planning has emerged as an operational approach to translate this relatively vague concept into management practice in many marine areas around the world.

Marine spatial planning is already in place in one Arctic country. In 2006, the Norwegian Parliament approved an integrated management plan, including spatial and temporal management measures, for the Norwegian part of the Barents Sea and the Sea Areas off the Lofoten Islands (Norwegian Ministry of the Environment 2005). The plan integrates previously separate management regimes for fisheries, shipping, oil and gas, and nature conservation. The Barents Sea plan was revised in 2010. The Norwegian Parliament also approved an integrated management plan for the Norwegian Sea in 2010 (Norwegian Ministry of the Environment 2009).

Canada has developed an integrated management plan for its part of the Beaufort Sea that included development of a marine spatial plan as one of its future actions (Beaufort Sea Partnership 2009). However, while the Canadian Beaufort Sea plan was 'supported' by the Minister of Fisheries and Oceans in 2010, no funds have been allocated to its implementation.

In the United States (US), the federal government is currently "implement[ing] comprehensive, integrated, ecosystem-based coastal and marine spatial planning and management in the United States", including its Beaufort, Chukchi, and Bering seas (National Ocean Council 2012). However, planning for the Arctic Ocean lags behind other marine regions of the US.

Russian legislation and regulations currently have no references to integrated management or marine spatial planning. Governmental organizations of the Russian Federation that have responsibilities for marine management operate on a sector-by-sector basis. However, existing legislation does contain areas in which marine spatial planning could be an important instrument for addressing problems of Russian marine management. Marine planning is under discussion now only in the Russian academic community. Russia has identified biologically important areas (Spiridonov et al. 2011). Still, only a few marine areas have been designated as protected areas. For example, Wrangel Island and its surrounding waters between the East Siberian Sea and the Chukchi Sea are designated as a national 'Zapovednik' (strict nature preserve) and listed as the only World Heritage marine site in the entire Arctic Ocean. Some coastal terrestrial areas have been designated as protected areas, a few of which extend a kilometre into the territorial sea.

While the European Union (EU) has no authority for marine (or maritime) spatial planning in the Arctic, it recently committed to pursue its involvement within existing international frameworks on Arctic issues such as biodiversity, ecosystem-based management, and marine protected areas (European Commission and High Representative 2012). The EU has been a strong advocate for maritime spatial planning—one of the pillars of its Maritime Strategy (European Commission 2010). The EU is planning to issue a directive on coastal and maritime spatial planning to Member States during 2013.

9.2 Why is Marine Spatial Planning Needed?

Before the last century, the oceans of the world were used mainly for two purposes: marine transportation and fishing. Conflicts between uses were few and far between except around some ports. Fisheries were managed separately from marine transportation, which over the last century was managed separately from offshore oil and gas development and other emerging marine activities, despite increasing real conflicts between and among these uses.

Single-sector management has often failed to resolve conflicts among users of marine space, rarely dealing explicitly with trade-offs among uses, and even more rarely dealing with conflicts between the cumulative effects of multiple uses and the marine environment. New uses of marine areas including offshore wind, ocean energy, offshore aquaculture, and marine tourism, as well as the demand for new marine protected areas, have only exacerbated the situation. Single-sector management has also tended to reduce and dissipate the effect of enforcement at sea because of the scope and geographic coverage involved and the environmental conditions in which monitoring and enforcement have to operate. In sharp contrast to the land, little 'public policing' of human activities takes place at sea—and is even less likely in the Arctic Ocean due to its inaccessibility.

As a consequence, marine ecosystems around the world are in trouble. Both the severity and scale of impacts on marine ecosystems from overfishing, habitat loss and fragmentation, pollution, invasive species, and climate change are increasing, with virtually no corner of the world left untouched.

Awareness is growing that the ongoing degradation in marine ecosystems is in large part a failure of governance (Crowder et al. 2006). Many scientists and policy analysts have advocated reforms centred on the idea of 'ecosystem-based management' (EBM). To date, however, a practical method for translating this concept into operational management practice has not emerged (Arkema et al. 2006).

Many recommendations for an ecosystem-based approach to marine management in the Arctic already exist. In fact, most Arctic countries are working to implement ecosystem-based management of their marine areas. The Arctic Council has repeatedly called for an ecosystem-based approach to marine management. For example, in its 2004 Arctic Marine Strategic Plan, EBM is defined as an approach that "[...] requires that development activities be coordinated in a way that minimizes their impact on the environment and integrates thinking across environmental, socio-economic, political and sectoral realms" (Arctic Council 2004). The key features of this approach include consideration of multiple scales, a long-term perspective, the recognition that humans are an integral part of ecosystems, an adaptive management perspective, and a concern for sustaining production and consumption potential for goods and services. The Best Practices in Ecosystem-based Oceans Management in the Arctic (BePOMAr) Project of the Arctic Council has summarized the practices that Arctic countries have used to apply an ecosystem-based management approach to marine management (Hoel 2009). The conclusions of that project include:

- Flexible application of effective ecosystem-based oceans management;
- Decision making must be integrated and science based;
- National commitment is required for effective management;
- Area-based approaches and transboundary perspectives are necessary;
- Stakeholder and Arctic resident participation is a key element; and
- Adaptive management is critical.

An integrated, ecosystem-based management approach has been identified in many marine places as an appropriate evolution to address problems caused by today's incremental, single-sector approach to marine management. However, examples of practical applications of an ecosystem-based approach are elusive. How to begin is the first challenge. A step in the right direction is the increasing worldwide interest in 'marine spatial planning' as an operational process through which to implement EBM.

9.3 What is Marine Spatial Planning?

Marine spatial planning (known as *maritime* spatial planning in Member States of the EU) or MSP is a practical way to create and establish a more rational organization of the use of marine space and the interactions between its uses, to balance demands for development with the need to protect marine ecosystems, and to achieve social and economic objectives for marine regions in an open and planned way.

MSP is a public process of analyzing and allocating the spatial and temporal distribution of human activities in marine areas to achieve ecological, economic, and social goals and objectives that are usually specified through a political process (Ehler and Douvere 2007). Its characteristics include:

- *Integrated* across economic sectors and governmental agencies, and among levels of government;

- *Strategic* and *future-oriented,* focused on the long-term;
- *Participatory,* including stakeholders actively in the entire process;
- *Adaptive,* capable of learning by doing;
- *Ecosystem-based,* balancing ecological, economic, social, and cultural goals and objectives toward sustainable development and the maintenance of ecosystem services; and
- *Place-based or area-based,* i.e., integrated management of all human activities within a spatially defined area identified through ecological, socioeconomic, and jurisdictional considerations (Ehler and Douvere 2007).

It's important to remember that we can only plan and manage human activities in marine areas, not marine ecosystems or components of ecosystems. We can allocate human activities to specific marine areas by objective, e.g., development or preservation areas, or by specific uses, e.g., offshore energy, offshore aquaculture, or sand and gravel mining.

9.4 Why is Marine Spatial Planning Needed?

Most countries already designate or zone marine space for a number of human activities such as maritime transportation, oil and gas development, offshore renewable energy, offshore aquaculture, and waste disposal. However, the problem is that usually this is done on a sector-by-sector, project-by-project basis without much consideration of effects either on other human activities or the marine environment. Consequently, this situation has led to two major types of conflict:

- Conflicts among human uses (user–user conflicts); and
- Conflicts between human uses and the marine environment (user-nature conflicts).

These conflicts weaken the ability of the ocean to provide the necessary ecosystem services upon which humans and all other life on Earth depend. Furthermore, decision makers in this situation usually end up only being able to react to events, often when it is already too late, rather than having the choice to plan and shape actions that could lead to a more desirable future of the marine environment.

By contrast, MSP is a future-oriented process. It offers a way to address both these types of conflict and select appropriate management measures to maintain and safeguard necessary ecosystem services. It is the missing piece that can lead to truly integrated planning from coastal watersheds to marine ecosystems.

When effectively put into practice, MSP can be used to:

- *Set priorities*—to enable significant inroads to be made into meeting the development objectives of marine areas in an equitable way, it is necessary to provide a rational basis for setting priorities, and to manage and direct resources to where and when they are needed most;
- *Stimulate opportunities* for new users of marine areas, including ocean energy;

- *Coordinate actions and investments* in space and time to ensure positive returns from those investments, both public and private, and to facilitate complementarities among jurisdictions and institutions;
- *Provide a vision and consistent direction,* not only of what is desirable, but what is possible in marine areas;
- *Protect nature,* that has its own requirements that should be respected if long-term sustainable development is to be achieved and if large-scale environmental degradation is to be avoided or minimized;
- *Reduce fragmentation of marine habitats,* i.e., when ecosystems are divided by human activities and prevented from functioning properly;
- *Avoid duplication of effort* by different public agencies and levels of government in planning, monitoring, and permitting; and
- *Achieve higher quality of service* at all levels of government, e.g., by ensuring that permitting of human activities is streamlined when proposed development is consistent with a comprehensive spatial management plan for the marine area.

9.5 Why are Space and Time Important?

Some areas of the ocean are more important than others—both ecologically and economically (Crowder and Norse 2008). Species, habitats, populations of animals, oil and gas deposits, sand and gravel deposits, and sustained winds or waves—are all distributed in various places and at various times. Successful marine management needs planners and managers that understand how to work with the spatial and temporal diversity of the sea. Understanding these spatial and temporal distributions and mapping them is an important aspect of MSP. Managing human activities to enhance compatible uses and reduce conflicts among uses, as well as to reduce conflicts between human activities and nature, are important outcomes of MSP. Examining how these distributions might change due to climate change and other long-term pressures (e.g., overfishing) on marine systems is another important step of MSP.

9.6 What is the Principal 'Driver' of MSP in the Arctic?

Pressures from human activities have often led to initiatives to better manage marine areas. For example, the Santa Barbara (US) oil spill from an oil well blowout in 1969 resulted in numerous pieces of legislation focused on marine pollution. In the 1970s, the threat of offshore oil and gas development and phosphate mining lead to initial efforts to protect the Great Barrier Reef. More recently, particularly in western Europe, MSP has been driven by national policies to develop offshore wind energy in Belgium, the Netherlands, and Germany (all of whom have developed and implemented marine plans), and the United Kingdom (England began development of marine spatial plans for two sub-regions of its marine area in 2011) and the requirement to designate more marine protected areas under directives of the European Commission. These 'new' uses, including offshore aquaculture, have had to compete with traditional users for scarce ocean space.

However, most of the Arctic Ocean is different. One of the major issues in the Barents Sea was the potential expansion of oil and gas activities into areas of the Barents used by fisheries and living marine resources. MSP is at the core of the plan, identifying particularly valuable and vulnerable areas, either from ecological and/or human perspectives. Within the plan, access to specific areas for human activities is carefully managed, for example, by moving shipping lanes outside Norwegian territorial waters (12-nautical miles), limiting trawling in sensitive areas, not opening particularly valuable and vulnerable areas to petroleum activities, including the ice edge, and extending marine protected areas and fishery closure areas to protect spawning aggregations, fish eggs and larvae, and juvenile fish and shellfish.

The level of human use of the Arctic marine environment remains relatively undeveloped. This situation is rare for marine areas. Marine spatial planners usually have to fit new uses into a heavily-used mosaic of other uses. The opportunity to shape the future of the Arctic in a socially desirable way is also a principal driver.

9.7 What are the Benefits of Marine Spatial Planning?

When developed effectively, marine spatial planning can have significant economic, social, and environmental benefits. The following table identifies some of the most important benefits of marine spatial planning (Table 9.1).

Table 9.1 Examples of benefits of marine spatial planning

Economic	Greater certainty of access to desirable areas for new private sector investments, frequently amortized over 20–30 years
	Identification and early resolution of conflicts between incompatible uses through planning instead of litigation
	Streamlining and transparency in permit and licensing procedures
	Improved capacity to plan for new and changing human activities, including emerging technologies and their associated effects
Environmental	Identification of biologically and ecologically important areas as a basis for space allocation
	Establish context for planning a network of marine protected areas
	Identification and reduction of the cumulative effects of human activities on marine ecosystems
Social	Improved opportunities for local community and citizen participation
	Identification of effects of decisions on the allocation of ocean space (e.g., closure areas for certain uses, protected areas) on communities
	Identification and preservation of social, cultural, and spiritual values related to use of ocean space
Administrative	Improve speed, quality, accountability, and transparency of decision making, and reduction of regulatory costs
	Improve consistency and compatibility of regulatory decisions
	Improve information collection, storage and retrieval, access, and sharing

9.8 What are the Key Steps of Marine Spatial Planning?

The development and implementation of MSP involves a number of steps, including:

- Identifying need and establishing *authority*;
- Obtaining *financial support*;
- Organizing the process through *pre-planning;*
- Organizing *stakeholder participation;*
- Defining and analyzing *existing conditions;*
- Defining and analyzing *future conditions;*
- Preparing and approving the *spatial management plan;*
- *Implementing and enforcing* the spatial management plan;
- *Monitoring and evaluating performance;* and
- *Adapting* the marine spatial management process (Ehler and Douvere 2009).

These ten steps are not simply a linear process that moves sequentially from step to step. Many feedback loops should be built into the process. For example, goals and objectives identified early in the planning process are likely to be modified as costs and benefits of different management measures are identified later in the planning process. Analyses of existing and future conditions will change as new information is identified and incorporated into the planning process. Stakeholder participation will change the planning process as it develops over time. Planning is a dynamic process and planners and stakeholders have to be open to accommodating changes as the process evolves over time.

Comprehensive MSP provides an integrated framework for management that provides a guide for, but does not replace, single-sector management. For example, MSP can provide important contextual information for guiding marine protected area management or for fisheries management, but does not replace it.

MSP answers four simple questions:

1. *Where are we today?* What are the baseline conditions?
2. *Where do we want to be?* What are the alternative spatial scenarios of the future? What is the desired vision?
3. *How do we get there?* What spatial management measures move us toward the desired future?
4. What *have we accomplished?* Have the spatial management measures moved us in the direction of the desired vision? If not, how should they be adapted in the next round of planning? (Ehler and Douvere 2009)

9.9 What are Examples of MSP Management Measures?

Management measures are at the heart of any management plan. They are the collective actions that will be implemented to achieve the overall management goals and objectives of the plan.

However, marine spatial planning cannot do it all. An integrated management plan for a marine area will have many management measures that will be applied to the important sectors of human activities (e.g., fisheries, marine transport, minerals extraction, and oil and gas) that use the resources of the marine area. There are four categories of management measures:

- *Input management measures*: measures that specify the inputs of human activities in marine areas;
- *Process management measures*: measures that specify the nature of the production process of human activities in marine areas;
- *Output management measures*: measures that specify the outputs of human activities in marine areas; and
- *Spatial/temporal management measures*: measures that specify where in space and when in time human activities can occur in marine areas (see Table 9.2).

Table 9.2 Examples of MSP management measures

Human activity	Spatial/temporal management measures
Oil and gas exploration	Restrict seismic operations when marine mammals are present in the marine area
	Restrict oil and gas activities in areas of subsistence access and harvest
	Identify areas where oil and gas activities should be prohibited at any time
	Identify areas where oil and gas activities should be prohibited by season, e.g., during marine mammal migrations
	Restrict oil and gas activities to winter months to reduce effects on biota and habitats, e.g., ice roads instead of permanent roads
	Select supply routes, frequency, and timing to avoid effects on biota or harvesting of wildlife by residents
	Prohibit discharges of drilling cuttings and produced water in sensitive marine areas
	Require reconditioning or recycling of drilling fluids in sensitive marine areas
	Prohibit discharges of solid waste into marine environment
	Install, operate, and maintain pipelines to minimize disturbance of seafloor habitats and other uses of the seafloor
Marine transportation	Designate ecologically and biologically sensitive Arctic areas as 'special areas' or 'particularly sensitive sea areas' (PSSAs)
	Require double-hulled and ice-strengthened vessels in Arctic waters
	Manage shipping to reduce marine mammal strikes
	Develop a traffic information system to improve monitoring of vessel traffic in Arctic waters
	Improve information on the use of Arctic marine environment by indigenous communities to avoid conflicts with marine transportation
	Identify areas of heightened ecological and cultural significance in light of changing climate conditions and increasing marine use to protect these areas from impacts of Arctic shipping
	Identify 'areas to be avoided' (ATBAs) based on navigation hazards and biological characteristics
	Improve hydrographic, oceanographic, and meteorological information about the Arctic marine environment to support safe navigation

(*Continued*)

Table 9.2 (Continued)

Human activity	Spatial/temporal management measures
Commercial fishing	Prohibit commercial fishing in Arctic waters until more information is available to support sustainable management (the US and Denmark have employed such a ban, Canada, Russia, and Norway have not)
	Improve baseline data on living marine resources, including potential changes in their composition in the Arctic
	Improve knowledge base of hydrographic and oceanographic features of benthic areas (habitat, communities, and species) and effects of fishing in those areas
	Increase transparency and participation of stakeholders, including indigenous communities, in international fisheries management planning
Tourism	Limit the negative effects of visits to any special area and local cultures
	Ensure the tourism activities do not conflict with nature conservation efforts
	Limit the size of visiting groups according to nature and wildlife vulnerabilities of special areas at any time
Nature conservation	Designate ecologically and biologically sensitive areas (EBSAs), e.g., nesting and feeding habitats of seabirds
	Designate marine protected areas
	Designate marine reserves (no take areas)
Indigenous peoples	Identify and designate subsistence hunting areas
	Identify and designate culturally important areas

9.10 What are the Outputs of Marine Spatial Planning?

The principal output of MSP is a comprehensive spatial plan for a marine area or ecosystem. The plan moves the whole marine system toward a 'vision for the future'. It sets out priorities for the area and—more importantly—defines what these priorities mean in time and space. Typically, a comprehensive spatial management plan has a 10–20-year horizon and reflects political and social priorities for the area. The comprehensive marine spatial plan is often implemented through a zoning map, zoning regulations, and/or a permit system similar to a comprehensive regional plan on land. Individual permit decisions made within individual sectors (for example, the fisheries, oil and gas, or tourism sectors) should then be based on the zoning maps and regulations. A strategic environmental assessment (SEA) or programmatic environmental impact statement (PEIS) is usually required before a marine spatial plan is approved by government (see Chap. 8).

MSP does not replace single-sector planning and decision making. Instead it aims to provide guidance for a range of decision makers responsible for particular sectors, activities, or concerns so that they have the means to make decisions confidently in a more comprehensive, integrated, and complementary way.

9.11 Why is Stakeholder Participation Critical to Marine Spatial Planning?

Involving key stakeholders, including indigenous peoples in the Arctic, in the development of MSP is essential for a number of reasons. Of these, the most important is because MSP aims to achieve multiple objectives (social, economic, and ecological) and should therefore reflect as many expectations, opportunities, or conflicts that are occurring in the MSP area. The scope and extent of stakeholder involvement differs greatly from country to country and is often culturally influenced. The level of stakeholder involvement will largely depend on the legal or cultural requirements for participation that often exist in each country.

Generally speaking, all individuals, groups, or organizations that are in one way or another affected, involved, or interested in MSP can be considered stakeholders. However, involving too many stakeholders at the wrong moment or in the wrong form can be very time consuming and can distract resources from the expected or anticipated result. To involve stakeholders effectively (e.g., leading toward expected results) and efficiently (e.g., producing expected results at least-cost), three questions should be asked:

- *Who* should be involved?
- *When* should stakeholders be involved?
- *How* should stakeholders be involved?

Where no legal obligations exist, it is important to define what type of stakeholder participation will be most suitable for a successful result. For instance, involving indigenous people in MSP efforts may not be a legal requirement, but they could however be greatly affected (positively or negatively) by MSP management measures, and should therefore participate.

Wide-ranging and innovative approaches to stakeholder participation and proactive empowerment should be used in the MSP process. Stakeholder participation and involvement in the process should be early, often, and sustained throughout the process. Stakeholder participation and involvement encourages "ownership" of the plan and can engender trust among the various stakeholders. Different types of stakeholder participation should be encouraged at various stages of the MSP process. The key stages at which stakeholders should be involved in the process include:

- *The planning phase*: Stakeholders need to be involved and contribute to the setting of goals and objectives of MSP. They also need to be involved in the evaluation and choice of specific management measure options and the consequences of these choices on their areas of interest;
- *The implementation phase*: Stakeholders should be involved in the actual implementation of MSP and its management measures. For example, an approach to enforcement may be identified that would involve local communities in the regulatory and enforcement process. When the local communities understand the problems and benefits of taking action—and agree upon the management

measures to be taken—they will be part of the enforcement process, at least to the extent of encouraging compliance; and

- *The monitoring and evaluation (post-implementation) phase*: Stakeholders should be involved in the evaluation of the overall effectiveness of MSP in achieving goals and objectives. The post-evaluation effort should involve all stakeholders in a discussion to identify plan results, evaluate results against objectives, and plan for the next round of planning (Pomeroy and Douvere 2008).

9.12 How Can Pan-Arctic Marine Spatial Planning Advance?

While most existing MSP efforts have been focused at the national level, especially in Norway, several current initiatives have a pan-Arctic perspective. For example, the Arctic Council has several spatial data initiatives that are developing regional or circumpolar datasets or to provide a framework that would allow integration, access, and coordination of spatial data on the Arctic. It has discussed the possibility of developing a common interface for access to spatial data. However, existing efforts of the Arctic Council to address cartography, geographic information systems, and spatial analysis have been conducted in isolation with no attempt at harmonization or integration. The working group on Emergency Prevention, Preparedness, and Response (EPPR), the working group on the Conservation of Arctic Flora and Fauna (CAFF), and the Arctic Monitoring and Assessment Program (AMAP) working group have discussed the possibility of developing a common interface for access to spatial data, the first tentative steps toward providing a framework to allow for data standardization and integration within the Arctic Council.

Preliminary work on MSP has begun through the Arctic Council's Protection of the Arctic Marine Environment (PAME) working group Work Plan for 2009–2011 (PAME 2009). It is developing pilot projects to make the large marine ecosystem (LME) assessment approach operational (in the Canadian/US Beaufort Sea and the US/Russian Federation West Bering Sea). However, no current plan exists to develop an integrated marine spatial plan for the pan-Arctic region. Given the working arrangement for PAME (two meetings a year) and the level of investment in its work plan (about $140,000 per year (see PAME 2009)), little progress on a pan-Arctic approach to MSP can be expected over the next few years without a substantial infusion of new resources.

In May 2011, the Arctic Council Ministers established an Expert Group on Arctic ecosystem-based management, recognizing the "...planning and management of human activities on a cross-sectoral basis can assist in reducing conflict among activities and in supporting the conservation and sustainable use of natural resources" (Arctic Council 2012). The Expert Group has met two times and developed a definition of ecosystem-based management and a set of principles. Its work is now focused on compiling EBM best practices, identifying conservation standards, and identifying ecological objectives. The Expert Group will present its findings to the Arctic Council Ministers in 2013.

The International Union for Conservation of Nature (IUCN) and the Natural Resources Defense Council (NRDC) have completed a cooperative Arctic Marine Ecosystem-Based Management Project to explore ways of advancing implementation of ecosystem-based management, and to begin the process of identifying specific ecologically significant and vulnerable marine areas that should be considered for enhanced protection in any new management arrangements (Speer and Laughlin 2011). A workshop held in November 2010 convened 34 scientists and representatives of indigenous peoples with expertise in various aspects of Arctic marine ecosystems and species to identify biologically or ecologically significant or vulnerable habitats using internationally accepted criteria developed under the auspices of the Convention on Biological Diversity (CBD 1992). The seven CBD criteria were: uniqueness; life history importance; importance to endangered/threatened species; vulnerable/fragile/slow recovery areas; areas of high productivity; areas of high diversity; and 'naturalness'. Importance of an area for subsistence or cultural heritage was also considered.

The workshop produced a set of preliminary maps depicting 77 Arctic marine EBSAs based on the CBD criteria. Thirteen 'Super EBSAs' were also identified, i.e., areas that all or almost all of the CBD criteria at a global level of significance. The area around the Bering Strait, for example, was identified as a 'Super EBSA'. Caveats and limitations of the results, including lack of data on some species, were identified in the report of the workshop.

Another way to advance MSP in the Arctic would explicitly recognize the importance of moving beyond sole reliance on the initiatives of national governments and toward a pan-Arctic approach to guide the future of the development. Networks and partnerships of non-governmental actors including indigenous peoples, environmental NGOs, academia, and private industry, all of which have influence over governmental policies and actions, could be used to initiate MSP.

Precedents exist. In Belgium, the University of Ghent laid the groundwork for MSP that was later implemented by the national government. In the US, NGOs and the private sector, particularly new private users of ocean space, e.g., wind farms and offshore aquaculture, have been particularly influential in the development of the national MSP framework. Local indigenous peoples (Coastal First Nations) have led by example the MSP process in British Columbia from the 'bottom up'.

Similarly, indigenous peoples from the Arctic could take the initiative to develop an Arctic-wide approach to MSP through a network of their organizations including the Aleut International Association, the Arctic Athabaskan Council, the Gwich'in Council International, the Inuit Circumpolar Council, the Russian Association of Indigenous Peoples of the North (RAIPON), and the Sámi Council. If capacity building to begin MSP is needed, technical advice could be sought from the Coastal First Nations Planning Office in Vancouver, British Columbia, and the Beaufort Sea Planning Office in Inuvik, Northwest Territories, Canada. Initially the Arctic Council Indigenous Peoples Secretariat in Copenhagen,

Denmark could provide coordination of this initiative. Alternatively, the initiative could be self organizing, relying on the interests and initiative of a few indigenous organizations and their leadership.

Leadership through a MSP initiative by indigenous peoples could provide the basis for other stakeholders, e.g., the business and NGO communities, to collaborate in the planning process. Eventually the Arctic Council and national governments would participate, particularly in the implementation of many spatial and temporal management measures.

References

Arctic Council (2004) Arctic Marine Strategic Plan. 24 Nov 2004. Available at <http://www.pame.is/images/stories/AMSP_files/AMSP-Nov-2004.pdf>. Accessed 5 Feb 2013

Arctic Council (2012) Report to senior Arctic officials. The 2nd Arctic Council Ecosystem-based Management Experts Group Meeting, Gothenburg, Sweden, 16–18 Apr 2012

Arkema KK, Abramson SC, Dewsbury BM (2006) Marine ecosystem-based management: from characterization to implementation. Front Ecol Environ 4(10):525–532

Beaufort Sea Partnership (2009) Integrated Ocean Management Plan for the Beaufort Sea. Beaufort Sea Planning Office, Inuvik

Convention on Biological Diversity (1992) Convention on Biological Diversity, 5 June 1992, 1760 U.N.T.S. 79. Entered into force 29 Dec 1991

Crowder L, Norse E (2008) Essential ecological insights for marine ecosystem-based management and marine spatial planning. Marine Policy 32:772–778

Crowder LB, Osherenko G, Young OR, Airamé S, Norse EA, Baron N et al (2006) Resolving mismatches in US ocean governance. Science 313(5787):617–618

Ehler C, Douvere F (2007) Visions for a Sea Change. Report of the first international workshop on marine spatial planning. Intergovernmental Oceanographic Commission and Man and the Biosphere Programme. IOC Manual and Guides, 46: ICAM Dossier, 3. UNESCO, Paris

Ehler C, Douvere F (2009) Marine Spatial Planning: a step-by-step approach toward ecosystem-based management. Intergovernmental Oceanographic Commission and Man and the Biosphere Programme. IOC Manual and Guides No. 53, ICAM Dossier No. 6. UNESCO, Paris

European Commission (2010) Communication from the Commission to the European Parliament, the Council, the European Economic and Social Committee and the Committee of the Regions: Maritime Spatial Planning in the EU—Achievements and Future Development, COM (2010) 771 final. 12 Dec 2010

European Commission and High Representative (2012) Joint Communication to the European Parliament and the Council. Developing a European Union Policy towards the Arctic Region: progress since 2008 and next steps. JOIN(2012) 19 final. 26 June 2012

Hoel AH (2009) Best practices in ecosystem-based oceans management in the Arctic. Report No. 129 of the Norwegian Polar Institute, Polar Environmental Centre

National Ocean Council (2012) Draft National Ocean Policy Implementation Plan, 12 Jan 2012. <http://www.whitehouse.gov/sites/default/files/microsites/ceq/national_ocean_policy_draft_implementation_plan_01-12-12.pdf>. Accessed 5 Feb 2013

Norwegian Ministry of the Environment (2005) Integrated Management of the Marine Environment of the Barents Sea and the Sea Areas Off the Lofoten Islands. Report No. 8 to the Storting

Norwegian Ministry of the Environment (2009) Integrated Management of the Marine Environment of the Norwegian Sea. Report No. 37 to the Storting

PAME (2009) PAME Work Plan 2009–2011. <http://www.pame.is/pame-work-plan/work-plan-2009-2011>. Accessed 5 Feb 2013

Pomeroy R, Douvere F (2008) The engagement of stakeholders in the marine spatial planning process. Marine Policy 32:816–822

Speer L, Laughlin T (2011) IUCN/NRDC Workshop to identify areas of ecological and biological significance or vulnerability in the Arctic marine environment. Workshop report. 7 Apr 2011

Spiridonov VA, Gavrilo MV, Krasnova ED, Nikolaeva NG (2011) Atlas of marine and coastal biological diversity of the Russian Arctic. WWF Russia, Moscow

Chapter 10
Marine Protected Areas as a Tool to Ensure Environmental Protection of the Marine Arctic: Legal Aspects

Ingvild Ulrikke Jakobsen

Abstract The sensitive Arctic marine environment, with its fragile ecosystems and habitats, may be threatened by increasing human activities and melting of sea ice due to climate change. Marine protected areas (MPAs) are an environmental tool that, by ensuring a higher level of protection within a defined geographical area, can benefit conservation and sustainable use of marine biological diversity. To ensure conservation of marine biological diversity, a number of states have committed themselves to establishing a coherent network of MPAs within 2012, an objective endorsed also by the Arctic states. In this chapter, political developments with regard to the establishment of MPAs within and outside of national jurisdiction and legal obligations applicable to the marine Arctic are examined. Despite having both legal obligations and political initiative, there is currently no network of MPAs in the marine Arctic. To ensure implementation of global obligations in the Arctic region, there is a need for further operationalization and clarification of scientific criteria for the selection and designation of sites that may contribute to a future network of MPAs. To fulfill the objective of an ecologically coherent network of MPAs in the marine Arctic, stronger efforts must also be made by the Arctic states through the Arctic Council.

10.1 Introduction

The marine environment in the Arctic is vulnerable with its unique ecosystems and habitats, and numerous rare and threatened species (CAFF 2010). As a result of climate change, increasing sea temperatures and melting sea ice have been witnessed in the Arctic. These climatic changes have extensive consequences for the marine environment and Arctic biodiversity, but also provide increased opportunities for human activities and economic development (ACIA 2005). Due to this

I. U. Jakobsen (✉)
University of Tromsø, Tromsø, Norway
e-mail: ingvild.jakobsen@uit.no

E. Tedsen et al. (eds.), *Arctic Marine Governance*,
DOI: 10.1007/978-3-642-38595-7_10, © Springer-Verlag Berlin Heidelberg 2014

development and the expansion of activities—such as for fishing, shipping, oil and gas, and tourism—in combination with climate change, the sensitive Arctic marine environment may face new and increasing threats.

The Convention on Biological Diversity (CBD 1992), along with a number of other international agreements, obliges states to protect and conserve marine biodiversity (hereby ecosystems, habitats, and species). To help meet this obligation, states have committed themselves to establishing representative networks of marine protected areas (MPAs), consistent with international law and based on scientific information, by 2012 (WSSD 2002, para. 32). With climatic changes and increasing impacts from human activities in the marine Arctic, the need for networks of protective MPAs within the Arctic is increasingly apparent (CAFF 2010).

The aim of this chapter is to provide an introduction to MPAs as a legal environmental management tool to ensure protection of the marine environment and conservation of biological diversity and ecosystems in the marine Arctic. The chapter examines existing global and regional obligations applicable to the marine Arctic on the conservation and sustainable use of biological diversity and ecosystems, to be met through the use of MPAs (see AOR 2011). MPAs have been on the global political agenda for the past two decades, during which time the legal regime for MPAs has further developed through interactions between political and legal instruments. This chapter therefore reviews political developments with regard to MPAs as well.

The maritime areas of the Arctic are subject to different legal regimes, ranging from territorial seas—subjected to the sovereignty of the coastal states -, to Exclusive Economic Zones (EEZs)—where coastal states enjoy sovereign rights over natural resources and other states enjoy the right of navigation -, to the high seas—where the principle of the freedom of the seas applies.[1] Even though most of the marine Arctic is within the maritime zones of the Arctic coastal states, there are four high seas pockets in the marine Arctic.[2] Living resources that are important for the conservation of biological diversity and ecosystems move between areas both within and beyond the limits of national jurisdiction. Analyses here thus include the possibilities of establishing MPAs in areas beyond national jurisdiction.

10.2 MPAs as an Environmental Tool

10.2.1 Defining MPAs

There is no formal definition of 'MPA'. The term is defined in different ways in a variety of international and regional instruments, thus, there are various types as

[1] As discussed in Chap. 1, there are many different definitions for areas constitute the marine Arctic. Regarding conservation of biological diversity, there are good reasons for adopting a wide definition of the 'marine Arctic'. In this chapter, the same definition as established in Chap. 1 and throughout this book, which includes the Arctic Ocean and its adjoining seas, is applied.

[2] The 'Banana hole' in the Norwegian Sea, the 'Loop Hole' in the Barents Sea, the 'Donut Hole' in the Bering Sea, and the 'Central Arctic Ocean'.

well as varying definitions of MPAs. The International Union for Conservation of Nature (IUCN) has adopted a definition that is often applied and has served as the basis for definitions of MPAs within certain legal instruments:

> Any area of inter tidal or sub tidal terrain, together with its overlaying water and associated flora and fauna, historical and cultural features, which has been reserved by law or other effective measures to protect part or all of the enclosed environment (IUCN/WCMC 1994).

Under the CBD, 'protected areas' is defined as a "geographically defined area which is designated or regulated and managed to achieve specific conservation objectives" (CBD 1992, art. 2). Under the OSPAR Convention (1992) MPAs are defined as:

> [...] an area within the maritime area for which protective, conservation, restorative or precautionary measures, consistent with international law have been instituted for the purpose of protecting and conserving species, habitats, ecosystems or ecological processes of the marine environment (OSPAR Commission 2003a, para. 1.1.).

Generally, MPAs can be characterized as geographically limited areas that are protected to achieve one or more conservation objectives through regulations and prohibitions on human activities which may threaten or damage the environment (Koivurova and Molenaar 2009). A prohibition on all human activities is not necessary for classification as an MPA. Rather, the level of restrictions or prohibitions that apply in an MPA can vary from a total ban on all human activities to more moderate limitations for sustainable use of natural resources (IUCN 2011).

The focus of this chapter is on integrated, cross-sectoral MPAs. These are MPAs where all activities that may threaten conservation objectives may be regulated or prohibited. Related concepts that do not cover all activities, but aim to protect a specific area against one particular human activity, such as areas closed for fisheries or areas where shipping is regulated (e.g., MARPOL 'special areas' and particularly sensitive sea areas (PSSAs)) are not addressed here (see Tanaka 2012; see also Chap. 6).

10.2.2 Potential Benefits of Using MPAs

MPAs help to ensure a higher level of protection of a defined geographical area from the environmental impacts of human activities, to ensure conservation and sustainable use of ecosystems, habitats, and sensitive and vulnerable marine species. Conservation of Arctic marine biological diversity is important not only for the Arctic region, but also for global biological diversity. Many species, such as migratory sea birds and mammals, use the marine Arctic seasonally, as Arctic habitats provide vital resources (WWF 2012). As explained, climate change is emerging and identified as a significant threat to the Arctic's biodiversity (CAFF 2010). Even though the establishment of MPAs may not address sources of greenhouse gas emissions, MPAs may nonetheless serve as tool for mitigation and adaptation to climate change. Through the establishment of protected areas, the total pressures on ecosystems may be controlled, thus strengthening ecosystems and making them more adaptive and resilient in the face of climate change. MPAs have also been emphasised as a mitigation tool, as MPAs may be designed to protect marine ecosystems that serve as sinks and reservoirs of greenhouse gases. The CBD has emphasised the role of MPAs in this respect.[3]

MPAs are, however, accompanied by certain limitations. Not all activities that cause environmental damage take place within areas that are protected, and therefore these activities may not be regulated within an MPA. For example, pollution from land-based activities is considered a major threat to the marine environment and biological diversity in the Arctic, but may fall outside of protected areas. Additionally, marine species move across the geographical limits of MPAs and between different legal regimes. It is therefore important that MPAs are used in addition to other traditional management structures and regulation of activities that may cause damage to the marine environment. To achieve ecosystem-based management of the marine Arctic, MPAs must be established and managed together with other tools that provide for integrated management of human activities, such as marine spatial planning.

10.3 International Framework for MPAs

10.3.1 Introduction

In this section, the competence to establish MPAs in maritime zones within and beyond national jurisdiction is reviewed. Next, the political development of MPAs is examined, followed by a review of significant global obligations and developments. Finally, regional obligations and cooperation between the Arctic states are investigated.[4]

10.3.2 The Establishment and Management of MPAs Under the LOS Convention

The competence to establish and manage MPAs is based on the United Nations Law of the Sea Convention (LOS Convention 1982). MPA regimes are closely linked to the jurisdiction and powers that states can exercise in maritime zones

[3] See CBD COP Decision VII/5, para. 8, Seventh Ordinary Meeting of the Conference of the Parties to the Convention on Biological Diversity, 9–20 Feb 2004, Kuala Lumpur. CBD COP Decision IX/33, para. 8, Ninth meeting of the Conference of the Parties to the Convention on Biological Diversity, 19–30 May 2008, Bonn.

[4] See Jakobsen (2012).

under the LOS Convention. Whereas states enjoy full sovereignty over land-based territory, the situation is different in the sea.

In the territorial sea, coastal states enjoy sovereignty and therefore may, in this zone, adopt any regulations on human activities to protect marine biodiversity, including the creation of MPAs. However, MPAs can only be adopted subject to the rights of other states to enjoy innocent passage (LOS Convention 1982, art. 17). As coastal states enjoy sovereignty in the territorial sea, this provides jurisdictional competence to regulate innocent passage, such as for purposes of environmental protection (LOS Convention 1982, arts. 21 and 22).

In the EEZ, coastal states enjoy sovereign rights over marine resources and jurisdiction over the marine environment (LOS Convention 1982, art. 56(1)(a), (b)(iii)). A coastal state may therefore, for the purposes of conservation of living resources and environmental protection, adopt MPAs within its EEZ.[5] Other states, however, continue to enjoy the freedom of navigation and the navigational rights of other states must be respected. Due to jurisdiction over the marine environment, coastal states have the authority to adopt regulations on shipping for the purpose of protecting the marine environment against marine pollution. These are, pursuant to Article 211(5) of the LOS Convention, limited to regulations that are "…conforming to and giving effect to generally accepted international rules and standards established through the competent international organization…". The competent international organization referred to here is the International Maritime Organization (IMO).

On the continental shelf, coastal states have, as set out in Article 77, sovereign rights "…for the purpose of exploring it and exploiting its natural resources". Natural resources consist of mineral and other non-living resources together with "…living organisms belonging to sedentary species…" (LOS Convention 1982, art. 77(4)).[6] The wording of "exploring and exploiting" is wide enough to include a right for coastal states to adopt measures for management and conservation of natural resources (Frank 2007). Thus, coastal states may establish MPAs to restrict fishing for sedentary species or other activities, such as mining and oil and gas extraction, that may threaten sedentary species (Frank 2007).

Such MPAs may also be adopted for the outer continental shelf, where coastal states have delineated the outer limits on the basis of the requirements of the Commission on the Limits of the Continental shelf (LOS Convention 1982, art. 76(8)). Regulations adopted for MPAs on the continental shelf must not, however, according to Article 78(2), "…infringe or result in any unjustifiable interference with navigation and other rights and freedoms of other States…" provided for by the LOS Convention. In relation to this, it can be questioned to what extent a

---

[5] For a discussion of the competence to establish MPAs in the EEZ, see Frank (2007), Lagoni (2003), and Jakobsen (2010).

[6] Sedentary species are defined in the provision as "…organisms which, at the harvestable stage, either are immobile on or under the seabed or are unable to move except in constant physical contact with these a-bed or the subsoil".

coastal state may regulate fishing activities on the outer continental shelf, where sedentary species are not the target species, but rather where the regulated activity may threaten sedentary species. On the one hand, it is natural that coastal states—when exercising their sovereign rights—may create an MPA where activities that may threaten sedentary species, such as bottom trawling, are restricted. This may, however, easily come into conflict with the freedoms of the high seas. Therefore, to be legitimate, such regulations would have to be based on a careful consideration of these conflicting interests (Frank 2007). A more appropriate way for coastal states may be to approach the relevant Regional Fisheries Management Organization (RFMO) with competence over the destructive fishing activities and call upon the adoption of appropriate regulations.

The principle of freedom of the high seas means that all states enjoy certain freedoms in high sea areas, such as the freedom of navigation, over flight, to lay submarine cables and pipeline, fishing, or scientific research (LOS Convention, Article 87). As no states have sovereignty or sovereign rights over marine resources on the high seas, no states are entitled to unilaterally define and designate high seas areas as MPAs or to adopt regulations on foreign vessels or other activities. To date, there are no legal regimes that provide for MPAs on the high seas.[7] If some states, for instance the Arctic states, were to agree to take measures to conserve an ecosystem in the high seas through the establishment of MPAs, the resulting restrictions could not apply and be binding on states that were non-parties to the agreement. Even so, the creation of MPAs on the high seas is controversial due to the lack of a clear legal basis in international law. To ensure legitimacy of MPAs established on the high seas on the basis of an agreement between states, careful consideration must be given both to the selection of areas and to compatibility with the freedoms of the high seas (Tanaka 2012). In addition to cooperation between states, MPAs on the high seas require both coordination and cooperation between different competent organizations (Molenaar and Oude Elferink 2009).

The 'Area' is defined in the LOS Convention Article 1(1) as the "…seabed and ocean floor and subsoil thereof beyond the limits of national jurisdiction". The Area and its resources are the common heritage of mankind (LOS Convention 1982, art. 136) When an MPA is established in an area beyond national jurisdiction, it must be determined whether an activity that desired for regulation is subject to the legal regime of the high seas, or to the legal regime of the Area. It is, for instance, not clear which legal regime that is applicable to the living resources of the Area. Questions related to the relationship between these two legal regimes will not, however be dealt with any further here.[8]

---

7 Note that while MPAs on the high seas are established under the OSPAR Convention, none of these are within the maritime areas of the Arctic.

8 For a discussion of this, see Molenaar and Oude Elferink (2009).

10.3.3 Global Political Developments with Regard to MPAs

Agenda 21 and the WSSD Plan of Implementation

Agenda 21, the United Nations (UN) Programme of Action, was adopted at the Rio Conference in 1992 and encourages states to take a more holistic approach to ocean management that is to be "...integrated in content and are precautionary and anticipatory in ambit" (Agenda 21 1992, para. 17.1). MPAs are emphasised as a significant tool and a core element of this new approach. Agenda 21 stresses the importance of protecting species, habitats, and ecologically sensitive areas, both within and beyond areas of national jurisdiction (Agenda 21 1992, para. 17.46(e) and (f), 17.75(e) and (f)).

The significance of adopting an integrated ecosystem-based management system of the oceans and of establishing MPAs was confirmed in the Word Summit on Sustainable Development (WSSD) Plan of Implementation, adopted in Johannesburg in 2002. In particular, the WSSD Plan of Implementation declares that states should:

> ...maintain the productivity and biodiversity of important and vulnerable marine and coastal areas, including in areas within and beyond national jurisdiction.

and

> ...develop and facilitate the use of diverse approaches and tools, (...) including (...) the establishment of marine protected areas, consistent with international law and based on scientific information, including representative networks by 2012... (WSSD 2002, para. 32c).

These stated objectives have had significance for the legal development of MPAs. As a response to the Plan's commitment, the objective to establish MPAs has been endorsed in legal instruments such as the CBD and OSPAR Convention, as well as within EU law.[9] Further, to achieve the objective, it has been transformed into more specific goals within the frameworks of instruments such as through the adoption of decisions, recommendations, and guidelines adopted by the Conference of Parties (COP) of CBD and the OSPAR Commission.

United Nations General Assembly

While the LOS Convention does not contain any specific references to the use of MPAs, the UN General Assembly (UNGA) has in recent years addressed the use of MPAs as a tool to protect oceans and to ensure conservation of biological diversity (UNGA 2008, para. 212–135). The UNGA has reaffirmed the need to develop and facilitate tools for conserving and protecting vulnerable marine ecosystems, including the establishment of MPAs and the development of representative networks of

---

[9] See, for instance, the EU's Marine Strategic Framework Directive (2008).

marine protected areas within 2012 (UNGA 2008, para. 134). In the latest resolution of significance here, UNGA encourages states to:

> ...further progress towards the 2012 target for the establishment of marine protected areas, including representative networks, and calls upon States to further consider options to identify and protect ecologically or biologically significant areas, consistent with international law and on the basis of the best available scientific information (UNGA 2011, para. 178).

The UNGA has also addressed conservation of biological diversity in areas beyond national jurisdiction, reaffirming its central role. At the same time that the UNGA reaffirmed its role in this area, it made note of the work performed under the CBD (UNGA 2011, para. 172). The UNGA's latest resolution recalls that the COP has adopted scientific criteria for identifying ecologically or biologically significant marine areas in need of protection in open-ocean waters and deep-sea habitats in addition to scientific guidance for selecting areas to establish a representative network of marine protected areas, including in open-ocean waters and deep-sea habitats (UNGA 2011, para. 179).

In 2004, the UNGA requested the appointment of an Ad Hoc Open-ended Informal Working Group to study issues relating to the conservation and sustainable use of marine biological diversity. At its fourth meeting in 2011, the Working Group adopted by consensus a set of recommendations, requesting a process to be initiated by the UNGA with a view to the "possible development of a multilateral agreement" under the LOS Convention to ensure conservation and sustainable use of the biological diversity in areas beyond national jurisdiction (IISD 2012). It was also recommended that the process address conservation and sustainable use of biological diversity in areas beyond national jurisdiction (by the use of MPAs, for instance).[10] At the group's fifth meeting, the issue of an implementation agreement under the LOS Convention was discussed. The EU called for an implementing agreement that provides for a global mechanism for establishing MPAs in areas beyond national jurisdiction (IISD 2012). While the Working Group achieved consensus with regard to the recommendation of a "possible development of a multilateral agreement under" the LOS Convention, it did not yet manage to agree on a process forward to develop the legal framework (IISD 2012). At the 2012 Rio + 20 Conference, states gave support to the development of a new implementation agreement and agreed to initiate, as soon as possible, the negotiation of an implementing agreement to the LOS Convention that addressed the conservation and sustainable use of marine biodiversity in areas beyond national jurisdiction (UNGA 2012, para. 181; United Nations 2012).

Following these developments, there appears to be broad support in the international community for creating such an implementation agreement. This is significant for the Arctic Ocean, as parts of the marine Arctic are beyond national jurisdiction. Moreover, with the melting of sea ice, large areas in the Central Arctic Ocean will become ice free, at least in the summer months, and will be part of the high seas. With the prospect of increased human activity in these areas, it

[10] These recommendations were endorsed in UNGA (2011) This obligation is reasonable, para. 167.

will be necessary to identify sensitive or valuable areas that need protection and that may constitute a part of a global network of MPAs. This requires clarification of the legal basis and competence to establish and manage MPAs on the high seas.

10.3.4 Global Legal Obligations with Regard to MPAs

The LOS Convention and the Obligation to Protect and Preserve the Marine Environment

The LOS Convention aims to establish a "legal order for the oceans" (preamble). The LOS Convention part XII includes legal obligations to protect and conserve the marine environment that apply to all states in all maritime zones. In the Ilulissat Declaration (2008), the five Arctic coastal states declared that the law of the sea is the relevant legal regime for the management of the Arctic and also acknowledged that the Arctic Ocean is a unique ecosystem that they have a particular responsibility to protect. The states expressed a commitment to:

> ...take steps in accordance with international law both nationally and in cooperation among the five states and other interested parties to ensure protection and preservation of the fragile marine environment of the Arctic Ocean (Ilulissat Declaration 2008).

In the light of these statements and interest in the development of a framework for MPAs, it is of interest to examine whether the general obligations of the LOS Convention also contain obligations to use this particular tool or to provide for its use.

Does the obligation of LOS Convention Article 192 contain a duty for states to adopt MPAs in order to ensure conservation of biological diversity within their maritime zones? It follows from the Article that states "have the obligation to protect and preserve the marine environment". This obligation is reasonable to understand as providing states with the discretion to decide which measures and tools to adopt to comply with it. A legal duty to adopt MPAs would thus mean constraining the discretion of the states when it comes to the choice of means to fulfil this obligation (Jakobsen 2010).

'Biological diversity' and 'ecosystems' are relatively new legal terms which were introduced to international law through the CBD in particular. A relevant question in relation to LOS Convention Article 192 and the use of MPAs is whether the wording "marine environment" also covers the concept of 'biological diversity'. The term 'marine environment' is not defined in the LOS Convention, but is arguably wide enough to also cover biological diversity, e.g., the diversity within species, between species, and of ecosystems.[11] States are also, according to

[11] Biological diversity is defined in the CBD as "the variability among living organisms from all sources including, *inter alia,* terrestrial, marine and other aquatic ecosystems and the ecological complexes of which they are part: this includes diversity within species, between species and of ecosystems" (CBD 1992, art. 2).

Article 194(5), required to adopt measures "…necessary to protect and preserve rare or fragile ecosystems as well as the habitat of depleted, threatened or endangered species and other forms of marine life." A contextual interpretation of Article 192 implies that the 'marine environment' includes protection of 'ecosystems', 'fragile habitats', and 'marine life'.

Following the adoption of the LOS Convention, there has been development from the protection of the environment against single activities, such as fishing and pollution, to a broader approach to protection of the environment, whereby ecosystems and biological diversity are conserved and protected (Birnie et al. 2009). Subsequent instruments, such as the CBD and Agenda 21, spell out the need to take a holistic ecosystem-based approach to the regulation of human activities together with an understanding of the natural variations of species and ecosystems. According to the Vienna Convention (1969)·, the term 'marine environment' must be interpreted in light of these 'new' developments (art. 31(3) (c)). It follows from the Vienna Convention that when a "rule" is interpreted, "any relevant rules of international law applicable in the relations between the parties" shall be applied as a source of interpretation. Such dynamic interpretation also finds support in international case law,[12] although the suggested interpretation must, however, remain within the wording and objective of the convention (Boyle 2006). An interpretation where 'marine environment' covers the newer legal term 'biological diversity' is acceptable within the wording of the LOS Convention.[13]

An MPA is a cross-sectoral measure where all potentially damaging activities may be restricted to protect biological diversity. Related is the question of whether Article 192 requires states to protect the marine environment against all human activities. It can be held that Part XII of the LOS Convention deals primarily with marine pollution. On the other hand, the wording of 'protection' and 'preservation' implies that states must adopt a broader approach in order to comply with the obligation (Nordquist and Rosenne 1991). Article 193 suggests that the conservation of living resources is part of protection of the marine environment, as is also confirmed in international case law.[14] Formulations in the preamble to the LOS Convention such as "the problems of the ocean space are closely interrelated and need to be considered as a whole" and that the aim is to establish "a legal order for the oceans", indicate that Article 192 is an obligation to protect the marine environment against all activities that may cause damage, not only pollution. This could include, for example, a duty to ensure that coral reefs are not damaged by shipping or oil and gas activities.

---

[12] See Namibia Advisory Opinion, *ICJ Reports* 1971, 16: 31; Case concerning the Gabcikovo-Nagymaros Project (Hungary v. Slovakia), *ICJ Reports* 1997, para. 140; and the arbitration case between Belgium and the Netherlands: *The Arbitration Regarding the Iron Rhine*, The Permanent Court of Arbitration, 2005, para. 59.

[13] The relationship between the CBD and the LOS Convention is addressed in CBD Article 22.

[14] Southern bluefin cases (New Zealand v. Japan; Australia v. Japan), 38 *ITLOS Report* 1724 (1999), para. 170.

Consequently, the obligation in Article 192 does not give any direction regarding which tools states shall apply to achieve protection of the marine environment and states are free to choose which measures and tools they want to apply to fulfil the obligation, including MPAs. Article 194(5) specifically deals with measures to protect the marine environment against pollution, although an interpretation of Article 192 in the light of Article 194(5) suggests that there is a duty for states to adopt measures to protect a sensitive ecosystem that is endangered by pollution or other threats. Even though Article 192 does not contain a general duty to establish MPAs, the use of this tool is consistent with legal and political developments, and is an appropriate way to comply with the obligation. Article 192 is not an absolute obligation and it does not require that all activities that may cause damage to the marine environment be prohibited.

The CBD and MPAs

All of the Arctic states are parties to the CBD, with the exception of the United States (US).[15] The objective of the Convention is to ensure the conservation of biological diversity, sustainable use of its components, and the fair and equitable sharing of the benefits arising from genetic resources (CBD 1992, art. 1).

CBD Article 8 includes different measures to ensure in situ conservation of biological diversity. In situ conservation is defined in Article 1 as:

> …the conservation of ecosystems and natural habitats and the maintenance and recovery of viable populations of species in their natural surroundings and, in the case of domesticated or cultivated species, in the surroundings where they have developed their distinctive properties.

Several of the obligations in Article 8 deal with protected areas (CBD 1992, art. 8(a), (b), (c) and (e)). States are, according to Article 8(a), required to "as far as possible and as appropriate" establish a system of protected areas. The wording "system of protected areas" can be understood as a 'network', and implies that states should establish protected areas in a systematic way, as part of a wider plan for conservation of biodiversity. The obligation to establish protected areas applies, pursuant to Article 4(a), in maritime zones within national jurisdiction.[16] The obligation to establish and manage MPAs must, however, be implemented so that it does not infringe with the rights and duties of other states pursuant to the LOS Convention, as follows from CBD Article 22(2).[17]

It follows from Article 8 that states "shall" establish protected areas. This indicates a legal duty to establish protected areas—not something states may choose to do

[15] For an overview of the member states, see <www.cbd.int/convention/parties/list>.

[16] As for areas outside national jurisdiction, it follows from Article 4(b) that the CBD applies in the case of "…processes and activities, regardless of where their effects occur, carried out under its jurisdiction or control, within the area of its national jurisdiction or beyond the limits of national jurisdiction".

[17] Regarding Article 22, see Wolfrum and Matz (2000) and Jakobsen (2010).

when or if they find it "possible" and "appropriate".[18] Arguably, "as far as possible" points to the level of implementation of this duty. States are committed to comply with the obligation in accordance with their conditions and capabilities (Burhenne-Guilmin and Casey-Lefkowitz 1992). This is consistent with the principle of differentiated responsibility.[19] The other element of the clause, "as appropriate", is reasonable to understand as a reference to the manner of implementation, providing states with discretion as to how and where MPAs should be established.[20] This is supported by Article 8(b) which leaves it to the states to develop guidelines for "the selection, establishment and management of protect areas…".

Article 8 does not provide guidance as to what areas that should be selected, what human activities should be prohibited or regulated, or how strict regulations adopted to protect and conserve marine biodiversity should be. Restrictions do apply however, and regulations within protected areas must be designed to meet the objectives of Article 8 and of the CBD.

Requirements as to the content of MPAs may be understood through interpretation of relevant sources, such as the Article's objectives, relevant principles such as the precautionary principle, principle of sustainable development, and the ecosystem approach, and a general legal obligation to ensure conservation and sustainable use of biological diversity (Jakobsen 2010). Work within the CBD regime is also of significance.

The CBD COP has adopted several legally non-binding decisions of importance for MPAs, elaborating on criteria for the selection and management of MPAs.[21] The COP has adopted scientific criteria for identifying ecologically or biologically significant marine areas in need of protection in open ocean waters and deep sea habitats.[22] With regard to management of MPAs, the COP has established that the MPAs should be "…effectively managed, ecologically based […] where human activities are managed […] to maintain the structure and functioning of the full range of marine and coastal ecosystems".[23]

Article 8(l) is also important when defining legal requirements for MPAs. According to Article 8(l) "…where a significant adverse effect on biological diversity has been determined pursuant to Article 7…" the states shall as far as possible and as appropriate "…regulate or manage the relevant processes and categories of activities…". This implies a duty for states to take actions when activities have "significant adverse effects". Within an MPA, which is protected due to its sensitive or vulnerable ecosystems, activities may have a negative effect on

[18] The interpretation and application of the clause "as far as possible and as appropriate" is discussed in Jakobsen (2010).

[19] Regarding the principle of differentiated responsibility, see Birnie et al. (2009).

[20] This interpretation is also adopted by the expert committee who proposed the Norwegian Nature Management Act of 2009, see NOU (2004):28, p. 160.

[21] The COP decisions are available at <http://www.cbd.int/>.

[22] CBD COP Decision IX/20. Ninth meeting of the Conference of the Parties, 19–30 May 2008, Bonn.

[23] CBD COP Decision VII/5, *supra* note 3, para. 18.

biological diversity and therefore states should adopt appropriate regulations for these activities. The duty in Article 8(l) can, for instance, imply that a state must approach the IMO for approval to adopt traffic regulation measures such as a sea lane to direct the traffic outside a sensitive area in the EEZ, if shipping is found to have a "significant adverse effect" on biological diversity. This is also set out in a decision by the COP urging states to address threats to achieve an effective management of MPAs.[24]

A reading of Articles 4(b), 8, and 22 suggests that the obligation to establish MPAs under the CBD only applies in maritime zones within national jurisdiction. The need to establish MPAs on the high seas has nevertheless been emphasised under the CBD.[25] The COP acknowledges that the CBD has a key role in supporting the work of the UNGA to ensure conservation of marine biological diversity in areas beyond national jurisdiction.[26]

10.3.5 Regional Obligations and Cooperation Between the Arctic States on MPAs

General

The section analyzes how global obligations to protect and conserve biological diversity through MPAs are implemented in the marine Arctic. These obligations are generally to be implemented by states at the national level; however, marine ecosystems are large and species migrate across the maritime zones of different states. Further, many threats to biological diversity are transboundary in nature. This raises jurisdictional issues and requires that, in order for protection to be successful, states in the Arctic cooperate in establishing MPAs.

There is no comprehensive regional agreement for the marine Arctic. The OSPAR Convention (1992) applies only to the marine environment in the North East Atlantic (art. 1). The Arctic states of Norway, Denmark, Iceland, Sweden, and Finland are together with Belgium, France, Ireland, Luxembourg, the Netherlands, Portugal, Spain, Switzerland, and the United Kingdom, as well as the European Community, contracting parties to the Convention. Russia, however, is not a contracting party, which means that not all of the European marine Arctic is covered.

Without a comprehensive regional agreement in the marine Arctic, and due to the fact that not all of the Arctic states are parties to relevant global conventions, political cooperation on environmental protection becomes especially important. The Arctic Council was established in 1996 with protection of the environment as

---

[24] CBD COP Decision VII/5, ibid, para. 26.

[25] CBD COP Decision VII/24, para. 35–47, Seventh Ordinary Meeting of the Conference of the Parties, 9–20 Feb 2004, Kuala Lumpur.

[26] CBD COP Decision IX/20, *supra* note 23.

one of its main objectives and serves as a forum for high level political cooperation between the Arctic states (Ottawa Declaration 1996). The Arctic Council does not possess the competence to adopt legally binding regulations and has been described as a consensus-based and project-driven apparatus, rather than an operational body (Koivurova and Molenaar 2009). In recent years, however, the Arctic Council has begun contributing to the development of legally binding regulations and agreements.[27] The Arctic Council also serves a role in implementing the global obligations of states at the regional level (Henriksen 2010). On this basis, political cooperation under the Arctic Council to conserve biological diversity through the use of MPAs is included in this analysis.

The OSPAR Convention

The OSPAR Convention contains principles and obligations to ensure protection of the marine environment and marine biodiversity in the northeast Atlantic. The Convention applies to parts of the Arctic Ocean both within and beyond national jurisdiction (OSPAR Convention 1992, art. 1(a)). Under Article 2 of the Convention, states are under a general obligation to take "all possible steps to prevent and eliminate pollution" and furthermore to take "the necessary measures to protect the maritime area against the adverse effects of human activities" to safeguard the humans health and to conserve marine ecosystems.

In 1998, OSPAR Convention Annex V on the protection and conservation of the ecosystems and biological diversity was adopted. Article 2(a), Annex V declares an obligation whereby states shall take "the necessary measures to protect and conserve the ecosystems and the biological diversity of the maritime area". This wording may cover the use of MPAs.

The OSPAR Commission is under a duty "to develop means, consistent with international law, for instituting protective, conservation, restorative or precautionary measures related to specific areas or sites or related to particular species or habitats" (OSPAR Convention 1992, Annex V, art. 3). On this basis, the Commission has adopted guidelines for the establishment of MPAs, resolving as a goal to establish a representative network of marine protected areas within 2012 in the OSPAR maritime area (OSPAR Commission 2003/3). The recommendation of the Commission is consistent with the obligations of CBD Article 8(a) and the objectives of the WSSD. Even though the recommendation on MPAs is not legally binding, it is significant for the interpretation of Article 2, Annex V.[28] All of the contracting parties to the OSPAR Convention are also parties to the CBD. It is explicit in Article 2 that states shall take necessary measures to protect and conserve ecosystems and biological diversity to meet their obligations under the OSPAR Convention, *as well as* to to

[27] An example of this is the agreement on search and rescue which is negotiated under the auspices of the Arctic Council (Arctic SAR Agreement 2011). See also Chap. 1.

[28] The Commission has the competence to adopt both legally binding decisions and non-legally binding recommendations (OSPAR Convention 1992, art. 10(3) and 13).

fulfil their obligations under the CBD to "to develop strategies, plans or programmes for the conservation and sustainable use of biological diversity". Consequently, the obligation in Article 2, Annex V must be interpreted in light of the CBD and also the obligation to adopt protected areas under CBD Article 8(a). It is reasonable to understand the wording of "take the necessary measures" in Article 2, Annex V to allow states to retain discretion in choosing the means of compliance; nevertheless, there is strong political will under OSPAR to establish MPAs that are part of an ecologically coherent network of MPAs in the OSPAR maritime area. Even though Article 2, Annex V does not include a general duty to establish MPAs, MPAs could be considered as part of states' obligations.

The OSPAR Commission has, in order to assist the contracting parties in establishing MPAs in their domestic maritime zones, adopted recommendations on identifying and selecting areas that may be included in the network as well as recommendations on MPA management (OSPAR Commission 2003b, c). Moreover, the Commission has adopted a guidance document for interpreting the concept of a "ecologically coherent network" (OSPAR Commission 2006). The contracting parties may take this document into account when selecting sites for the OSPAR MPA network.

When it comes to the management and regulation of activities within MPAs, the OSPAR Convention takes a broad, holistic approach, as it aims to protect the marine environment and ecosystems against human activities. There are, however, explicit exceptions made for fisheries management and shipping (OSPAR Convention, preamble and Annex V, art. 4). This means that the Commission cannot adopt recommendations and decisions where restricting these activities, as questions relating to the management of fisheries are to be regulated under international and regional agreements dealing specifically with such questions, and authority is similarly ceded to the IMO as the competent international body for international shipping regulation. Nevertheless, these activities are included in assessments of the quality and status of the maritime area (Molenaar and Oude Elferink 2009). Moreover, the precautionary principle is part of the Convention and OSPAR takes an ecosystem approach to the protection of the marine environment, suggesting that when adopting MPAs, states are obliged to cooperate through relevant organizations such as the IMO and RFMOs to ensure appropriate protection of MPAs. One may argue that a duty could be interpreted under the OSPAR Convention for a state to cooperate with the IMO and obtain approval for a traffic regulation, such as a sea lane in the EEZ, to ensure protection of an MPA against shipping activities.[29]

OSPAR has made progress on the establishment of MPAs on the high seas. The recommendations by the Commission on MPAs also address MPAs on the high seas. In 2009, the Commission identified eight areas as potential high seas MPAs and in 2010, six areas were established as MPAs in areas beyond national jurisdiction

---

[29] In the Guidelines for the Management of Marine Protected Areas in the OSPAR Maritime Area, 2003/18, both fisheries and shipping are mentioned as examples of activities that may require regulation.

(OSPAR Commission 2011). None of these MPAs are, however, established within the marine Arctic areas.

Overall, progress on the OSPAR network is underway, but the objective of an ecologically coherent network of MPAs has not yet been achieved.

The Arctic Council

The establishment of a network of protected areas was emphasised at the outset of the Arctic Council through the Arctic Environmental Protection Strategy (AEPS 1991).

The work carried out by Arctic Council's Conservation of Arctic Flora and Fauna (CAFF) working group has provided critical knowledge about biological diversity in the Arctic. On this basis, the Circumpolar Protected Areas Network (CPAN), a strategy and action plan for the establishment of a circumpolar network of MPAs, was adopted in 1998.[30] The use of protected areas is recognized by CPAN as an effective and necessary tool to ensure conservation and sustainable use of biological diversity (CAFF 1996a). The goal of CPAN is to establish "an adequate and well managed network of protected areas that has a high probability of maintaining the dynamic biodiversity of the Arctic region" (CAFF 1996a). The network shall, as far as possible, cover the large variation of ecosystems in the Arctic (CAFF 1996a). CPAN does not include any legal obligations, but aims to provide a common framework for states for the selection and management of protected areas to ensure protection of significant areas at the national, regional, and circumpolar level (CAFF 1996a). Under CPAN, guidelines for the selection and designation of protected areas were adopted in 1996 (CAFF 1996b). It is emphasised in the CPAN strategy and action plan that an additional goal is to contribute to states' obligation pursuant to the CBD Article 8(a).

The Arctic Marine Strategic Plan was developed by the working group Protection of the Marine Environment (PAME) and adopted by the Arctic Council in 2004 (Arctic Council 2004). The objective of the strategy is to protect the marine environment, marine biological diversity, ecosystems, and their functions. The use of MPAs as an environmental tool is recognized and emphasised in the plan. The objective of the WSSD of establishing MPAs is repeated here, stating that the Arctic Council shall:

> Promote WSSD actions related to the marine and coastal environment, including the application of an ecosystem approach and establishment of marine protected areas, including representative networks (Arctic Council 2004).

Work towards establishing protected areas that are part of a circumpolar network is documented through reports from the states to CAFF in 1997 and 2004 (CAFF 1997, 2004). In spite of the legal obligations of the states in the Arctic

---

[30] Information and relevant documents and publications about CPAN are available at <http://www.caff.is/protected-areas-cpan>. See also Koivurova (2009).

and that the significance of establishing protected areas was emphasised as early as 1996 with the establishment of the Arctic Council, few marine protected areas have been adopted in the region.[31] As the current work plans for CAFF and PAME do not include action on MPAs, this indicates that MPAs are not currently a priority area for the Arctic Council (CAFF 2009, 2011; PAME 2011).

A CPAN task at the national level is to identify gaps in national networks of protected areas and select candidate sites for further action (CAFF 1996a). It is also emphasised that the guidelines for selection and identification of protected areas are to be used by each Arctic state within its own legislative framework (CAFF 1996b). The guidelines are broad and do not provide clear criteria for site selection. Although MPAs must be implemented at the national level, this suggests that the Arctic states have not been willing to commit to a common understanding on the selection and management of sites for Arctic network of MPAs. Stronger commitments from the Arctic states when it comes to selection and designation of marine areas would strengthen the chances of achieving the objective of a circumpolar protected area network.

10.4 Conclusions

Despite legal obligations and political initiative at the global and regional levels, there is currently no network of MPAs in the marine Arctic. Given the transboundary character of biological diversity and of environmental threats, states must cooperate in order to successfully protect the marine environment. The work carried out under the CBD in elaborating and operationalizing obligations to establish MPAs is significant for achieving the objective of a global network of MPAs. Moreover, work under the CBD and UNGA to develop an implementation agreement is crucial for creating the necessary legal basis for conservation of marine biodiversity in areas beyond national jurisdiction.

Global obligations must be successfully implemented by the Arctic states at the regional and national levels. To achieve a network of MPAs in the Arctic, as part of a wider global network, there is a need for further operationalization of global and regional obligations by clarifying scientific criteria for the selection and management of areas as potential MPAs in this network. This would assist Arctic states in overcoming jurisdictional and sectoral challenges that may hamper the designation and effective management of MPAs in the region. Given the lack of a comprehensive, overarching regional agreement, it is natural that the Arctic Council would play an active role in this.

No state alone can ensure the conservation of the Arctic marine biodiversity. It is unlikely that the Arctic states will develop and adopt a legally binding instrument on conservation of marine biological diversity and the establishment of

[31] Possible explanations are provided in Lalonde (2010).

MPAs, but the efforts made by the Arctic states under the Arctic Council are essential for realizing an ecological network of MPAs in the Arctic.

References

ACIA (2005) Arctic Climate Impact Assessment. Cambridge University Press, New York

AEPS (1991) Arctic Environmental Protection Strategy, 30 I.L.M. 1624 14 Jan 1991

Agenda 21 (1992) Agenda 21: Programme of Action for Sustainable Development. Conference on Environment and Development. U.N. GAOR, 46th Sess., Agenda Item 21, UN Doc A/Conf.151/26

AOR (2011) Phase I Report (2009-2011) of the Arctic Ocean Review (AOR) project. <www.pame.is>. Accessed 1 Feb 2013

Arctic Council (2004) Arctic Marine Strategic Plan. 24 Nov 2004. <http://www.pame.is/images/stories/AMSP_files/AMSP-Nov-2004.pdf>. Accessed 5 Feb 2013

Arctic SAR Agreement (2011) Agreement on Cooperation on Aeronautical and Maritime Search and Rescue in the Arctic, 12 May 2011, 50 I.L.M. 1119 (2011). Entered into force on 19 Jan 2013

Birnie P, Boyle A, Redgwell C (2009) International law and the environment, 3rd edn. Oxford University Press, Oxford

Boyle A (2006) Further development of the 1982 convention on the law of the sea. In: Barnes R, Freestone D, Ong D (eds) The law of the sea: Progress and prospects. Oxford University Press, Oxford

Burhenne-Guilmin F, Casey-Lefkowitz S (1992) The Convention on Biological Diversity: A hard won global achievement. Yearb Int Environ Law 3(1):43–59

CAFF (1996a) Circumpolar Protected Areas Network: Strategy and Action Plan. CAFF Habitat Conservation Report No. 6

CAFF (1996b) Circumpolar Protected Area Network (CPAN): Principles and Guidelines. CAFF Habitat Conservation Report No. 4

CAFF (1997) CPAN Progress Report 1997. CAFF Habitat Conservation Report No. 7

CAFF (2004) CPAN Country Updates Report 2004. CAFF Habitat Conservation Report No. 11

CAFF (2009) CAFF WORK PLAN 2009–2011 Ministerial Period. CAFF International Secretariat, Akureyri

CAFF (2010) Arctic Biodiversity Trends 2010—Selected indicators of change. CAFF International Secretariat, Akureyri. <http://www.arcticbiodiversity.is/images/stories/report/pdf/Arctic_Biodiversity_Trends_Report_2010.pdf>

CAFF (2011) Conservation of Arctic flora and fauna (CAFF) Workplan. CAFF Administrative Series Report No. 1. CAFF International Secretariat, Akureyri

Convention on Biological Diversity (1992) Convention on Biological Diversity, 5 June 1992, 1760 U.N.T.S. 79. Entered into force 29 Dec 1991

Frank V (2007) The European community and marine environmental protection in the international law of the sea: Implementing global obligations at the regional level. Utrecht University, Leiden

Henriksen T (2010) Conservation and sustainable use of Arctic marine biodiversity. Arctic Rev Law Politics 2(1):249–278

Ilulissat Declaration (2008) Arctic Ocean Conference. 48 I.L.M. 382 (2009) Ilulissat, Greenland. 27 May 2008

IISD (2012) Earth negotiations bulletin: Summary of the fifth meeting of the working group on marine biodiversity beyond areas of national jurisdiction, vol 25. P 83. 7–11 May 2012

IUCN (2011) Guidelines for applying the IUCN Protected Area Management Categories to Marine Protected Areas (supplementary to the 2008 guidelines), Second Draft. June 2011

IUCN/WCMC (1994) Guidelines for Protected Area Management Categories. IUCN, Gland and Cambridge

Jakobsen IU (2010) Marine protected areas in international law: A Norwegian perspective. PhD Thesis, University of Tromsø. <http://hdl.handle.net/10037/4409>. Accessed 11 Feb 2013
Jakobsen IU (2012) Marine verneområder i Arktis: Kyststatens forpliktelser og rettigheter. In: Henriksen T, Ravna Ø (eds) Juss i nord: Hav, fisk og urfolk. Gyldendal, Oslo
Koivurova T (2009) Governance of protected areas in the Arctic. Utrecht Law Rev 5(1):44–60
Koivurova T, Molenaar EJ (2009) International governance and regulation of the marine arctic: Overview and Gap Analysis. Report prepared for the WWF International Arctic Programme, Oslo
Lagoni R (2003) Marine protected areas in the exclusive economic zone. In Kirchner A (ed) International marine environmental law: Institutions, implementation and innovations. Kluwer Law International, The Hague
Lalonde S (2010) A Network of Protected Areas in the Arctic: Promises and Challenges. In: Nordquist M, Moore J, Heidar T (eds) Changes in the Arctic Environment and the Law of the Sea. Martinus Nijhoff, Leiden, pp 131–142
LOS Convention (1982) United Nations Convention on the Law of the Sea, 10 Dec 1982, 1833 U.N.T.S. 396. Entered into force 16 Nov 1994
Marine Strategy Framework Directive (2008) Directive 2008/56/EC of the European Parliament and the Council of 17 June 2009 establishing a framework for community action in the field of marine environmental policy, June 17 2009, 2008 O.J. (L 164)
Molenaar EJ, Oude Elferink AG (2009) Marine protected areas in areas beyond national jurisdiction: The pioneering efforts under the OSPAR Convention. Utrecht Law Rev 5(1):5–20
Nordquist MH, Rosenne S (1991) United Nations convention on the Law of the Sea 1982: A commentary, vol IV. Martinus Nijhoff, Dordrecht
OSPAR Convention (1992) Convention for the Protection of the Marine Environment of the North-East Atlantic, 22 Sept 1992, 32 I.L.M. 1072 (1993). Entered into force 25 Mar 1998
OSPAR Commission (2003a) Guidelines for the identification and selection of marine protected areas in the OSPAR maritime area, 2003/17
OSPAR Commission (2003b) Guidelines for the management of marine protected areas in the OSPAR maritime area, 2003/18
OSPAR Commission (2003c) Recommendation on a network of marine protected areas, 2003/3, amended by Recommendation 2010/2
OSPAR Commission (2006) Guidance on developing an ecologically coherent network of OSPAR marine protected areas, 2006/3
OSPAR Commission (2011) Status report on the OSPAR network of marine protected areas
Ottawa Declaration (1996) Declaration on the Establishment of the Arctic Council, 35 I.L.M. 1382 19 Sept 1996
PAME (2011) PAME Work Plan 2011–2013. <http://www.pame.is/images/PAME_NEW/PAME%20Work%20Plan/PAME_Work_Plan_2011-2013.pdf>. Accessed 5 Feb 2013
Tanaka Y (2012) The international law of the sea. Cambridge University Press, Cambridge
UNGA (2008) Oceans and the Law of the Sea, A/Res/63/111
UNGA (2011) Oceans and the Law of the Sea, A/Res./66/231, 2011
UNGA (2012) Oceans and the Law of the Sea, A/Res./67/78
United Nations (2012) The Future We Want: Outcome document adopted at the Rio + 20 Conference. <http://www.un.org/en/sustainablefuture>. Accessed 8 Feb 2013
Vienna Convention (1969) The Vienna Convention on the Law of Treaties, 23 May 1969, 1155 U.N.T.S. 331. Entered into force 27 Jan 1980
Wolfrum R, Matz N (2000) The Interplay of the United Nations Convention the Law of the Sea and the Convention on Biological Diversity. In: Bogdandy AV, Rüdiger W (eds) Max Planck yearbook of United Nations law, Max Planck Institute, The Hague
WSSD (2002) Plan of Implementation of the World summit on Sustainable Development. United Nations World Summit on Sustainable Development, Johannesburg, 26 Aug to Sept 2002
WWF (2012) The Circle. World Wildlife Fund. World Wildlife Magazine 3

Part IV
Opportunities for Transatlantic Cooperation

Chapter 11
EU–US Cooperation to Enhance Arctic Marine Governance

Elizabeth Tedsen and Sandra Cavalieri

Abstract Enhanced transatlantic cooperation can help leverage emerging opportunities to improve protection of the Arctic marine environment. Through both international and domestic action, the European Union (EU) and United States (US) can work together to promote environmental leadership and shape agendas and policy making. This chapter examines the shared objectives and interests of the EU and US in the marine Arctic. With this common ground in mind, the chapter examines areas of opportunity for addressing shortcomings in the international legal and policy framework of the marine Arctic through transatlantic action. The need to support a more integrated, ecosystem-based approach to governance is highlighted as a necessary way forward, although sector-based gaps and policy options for the areas of fisheries, shipping, and offshore hydrocarbon development are also considered. The role of indigenous peoples in Arctic decision making and governance is emphasized. Finally, the chapter considers the need for joint action in tackling Arctic governance's greatest underlying challenge: climate change.

11.1 A Changing Environment and the Need for Policy Action

Climate change is occurring more rapidly in the Arctic than in any other region of the world, with sea ice retreating at a pace that exceeds even the most dramatic predictions of scientists. Access to newly accessible Arctic waters is creating

Based on: Best A. et al. (2009a, b).

S. Cavalieri (✉)
Ecologic Institute, Pfalzburger Strasse 43–44, 10717 Berlin, Germany
e-mail: sandra.cavalieri@ecologic.eu

E. Tedsen
Ecologic Institute, 1630 Connecticut Ave NW, Suite 300, 20009 Washington, D.C., USA
e-mail: elizabeth.tedsen@ecologic.eu

E. Tedsen et al. (eds.), *Arctic Marine Governance*,
DOI: 10.1007/978-3-642-38595-7_11, © Springer-Verlag Berlin Heidelberg 2014

economic opportunities for the fishing, shipping, energy, and tourism industries, all of which are expected to expand in both scope and intensity. These changes bring with them new challenges. Increased human activity in the Arctic marine area will require effective policies and international cooperation if the world hopes to protect fragile Arctic ecosystems as well as to safeguard the rights and interests of indigenous peoples.

11.2 Recent Policy Developments

The sense of urgency initiated by rapidly melting Arctic sea ice has contributed to new international action on Arctic governance. Political dynamics are changing nearly as fast as the ice. The world took notice of changing Arctic geopolitics in the summer of 2007 when the Russian flag was planted on the North Pole seabed and the lowest Arctic sea ice extent to that time, since surpassed, was recorded. Since then, Canada, Denmark, Finland, Iceland, Sweden, Russia, the United States (US), and the European Union (EU) have all released new and revised Arctic policies. An increasing number of non-Arctic states and other political entities, such as the EU, have begun developing Arctic policies and seeking a role in Arctic governance, with increasing interest from non-Arctic states and entities in the Arctic Council. Arctic governance is no longer solely of interest to Arctic states, indigenous peoples, academics, and non-governmental organizations (NGOs), but rather, it has become a major item on the policy agenda for a range of international actors.

Emerging global powers are vying to become permanent observers at the Arctic Council and in the process shifting the forum's internal dynamics. A number of non-Arctic states, including China, Italy, Japan, Korea, Singapore, and India, as well as the EU, have applied for permanent observer status with the Arctic Council. The European Commission announced in its November 2008 Arctic Communication (European Commission 2008) that it intended to seek permanent observer status at the Arctic Council, marking a change from its previous policy of participating as an ad-hoc observer, along with the European External Action Service (EEAS) and other EU agencies. At the Arctic Council's April 2009 Ministerial Meeting, a decision on permanent observer status was delayed for all applications, including the European Commission. Although the EU was denied membership—perhaps due to Russia's reluctance as well as conflict with Canada over a proposed seal import ban—the strongest and most comprehensive policy statements from the group of non-Arctic players have come from the EU. The EU updated its application after the Arctic Council's adoption of criteria for observers in May 2011 and reiterated its interest in a Joint Communication in June 2012 (European Commission and High Representative 2012). A decision on observer applications will be taken at the Council's upcoming[1] May 2013 Ministerial Meeting in Kiruna. However, even if the EU is granted observer

[1] As of time of writing, March 2013.

status, it remains unclear what role it would actually play in the Arctic Council as the role of observers is currently limited; for example, observers may only submit written statements at ministerial meetings.

Although climate change in the Arctic is indeed an issue of global concern, the Arctic coastal states are reluctant to have the region viewed—and governed—as a global 'commons'. As such, the basis for the EU and other non-Arctic states defining and pursuing interests in the Arctic may not be as self-evident as supposed (Best et al. 2009a, b).

Influenced by the rate of Arctic change, the Arctic Council itself is undergoing shifts so as to better address new governance needs. The Council does not have the competence to impose legally binding obligations of any kind on its members, permanent participants, or observers. The most it can do directly from a governance perspective is to issue policy recommendations, such as the one commissioning the Arctic Climate Impact Assessment (ACIA), and to adopt guidelines and recommendations. The Arctic Council is an influential contributor to policy making in the Arctic, although largely through issuing non-binding guidelines and recommendations. However, the Council's role has recently evolved to serve as a forum under the auspices of which Arctic states negotiate agreements, which are then signed and transposed by the respective national procedures to become legally binding decisions. At the May 2011 Ministerial Meeting, the Arctic coastal states adopted a legally binding Arctic Search and Rescue agreement (Arctic SAR Agreement 2011). An agreement on Marine Oil Pollution Preparedness and Response (Arctic MOPPR Agreement 2013) is scheduled to be signed by the coastal states during the May 2013 Ministerial Meeting. Also, following a recommendation by the Arctic Environmental Ministers, an agreement on reductions of black carbon emissions could be negotiated under the new Canadian chairmanship (Ministry of the Environment 2013). Thus, while the Arctic Council itself remains limited in its mandate and ability to adopted binding measures, it has nonetheless strengthened its role as an institution that both influences and helps support new policies and governance measures.

Alongside these trends, dynamics among the established actors in the Arctic Council have also undergone a shift in recent years. In part, this change was marked by a meeting of the five Arctic Ocean coastal states in Ilulissat, Greenland in May 2008. Perceiving that the Arctic Ocean was on the brink of crossing a significant threshold, they declared that "[b]y virtue of their sovereignty, sovereign rights and jurisdiction in large areas of the Arctic Ocean the five coastal states are in a unique position to address these possibilities and challenges" (Ilulissat Declaration 2008). The Arctic Ocean coastal states also announced their intention to protect the Arctic environment and the interests of indigenous peoples and local inhabitants:

> Climate change and the melting of ice have a potential impact on vulnerable ecosystems, the livelihoods of local inhabitants and indigenous communities…. The Arctic Ocean is a unique ecosystem, which the five coastal states have a stewardship role in protecting. Experience has shown how shipping disasters and subsequent pollution of the marine environment may cause irreversible disturbance of the ecological balance and major harm to the livelihoods of local inhabitants and indigenous communities.

The Arctic Ocean coastal states also expressed their opinion therein that there is "no need to develop a new comprehensive international legal regime to govern the Arctic Ocean" because:

> Notably, the law of the sea provides for important rights and obligations concerning the delineation of the outer limits of the continental shelf, the protection of the marine environment, including ice-covered areas, freedom of navigation, marine scientific research, and other uses of the sea. We remain committed to this legal framework and to the orderly settlement of any possible overlapping claims. This framework provides a solid foundation for responsible management by the five coastal States and other users of this Ocean through national implementation and application of relevant provisions (Ilulissat Declaration 2008).

Despite the fact that Denmark, which is an Arctic state by virtue of Greenland, had earlier insisted that Arctic Ocean coastal state cooperation should not compete with the Arctic Council (SAO 2007), the meeting in Ilulissat produced friction among the Arctic Council members. Iceland expressed the greatest concern among the three non-Arctic Ocean coastal states (the other two being Finland and Sweden). It had already expressed reservations about strengthened Arctic Ocean coastal state cooperation at a 2007 meeting of Senior Arctic Officials (SAOs)[2] and reiterated its concern during the August 2008 Conference of Arctic parliamentarians (Conference of the Parliamentarians 2008). The meeting in Ilulissat in May 2008 also provoked a reaction from one of the strongest Arctic Council permanent participants, the Inuit Circumpolar Council (ICC) and national Inuit leaders, who issued a statement[3] outlining, *inter alia*, their concerns:

> Concern was expressed among us leaders gathered in Kuujjuaq that governments were entering into Arctic sovereignty discussions without the meaningful involvement of Inuit, such as the May, 2008 meeting of five Arctic ministers in Ilulissat, Greenland. The Kuujjuaq summit noted that while the Ilulissat Declaration asserts that it is the coastal nation states that have sovereignty and jurisdiction over the Arctic Ocean, it completely ignores the rights Inuit have gained through international law, land claims, and self-government processes. Further, while the ministers strongly supported the use of international mechanisms and international law to resolve sovereignty disputes, it makes no reference to those international instruments that promote and protect the rights of indigenous peoples (ICC 2008).

---

[2] In the discussions at the Narvik SAO meeting in 2007, Iceland expressed concerns that: "separate meetings of the five Arctic states, Denmark, Norway, US, Russia and Canada, on Arctic issues without the participation of the members of the Arctic Council, Sweden, Finland and Iceland, could create a new process that competes with the objectives of the Arctic Council. If issues of broad concern to all of the Arctic Council Member States, including the effect of climate change, shipping in the Arctic, etc. are to be discussed, Iceland requested that Denmark invite the other Arctic Council states to participate in the ministerial meeting. Permanent participants also requested to participate in the meeting. Denmark responded that the capacity of the venue may be an issue" (SAO 2007).

[3] Similar sentiments were expressed in the ICC's subsequent 2009 'Circumpolar Inuit Declaration on Sovereignty in the Arctic' that referenced the previous statement and Ilulissat meeting (ICC 2009).

But the ICC and the Inuit leaders were also critical of current Arctic governance structures:

> We recognized the value of the work of the Arctic Council…We further noted the meaningful and direct role that indigenous peoples have at the Arctic Council, while at the same time expressing concern that the Council leaves many issues considered sensitive by member states off the table, including security, sovereignty, national legislation relating to marine mammal protection, and commercial fishing (ICC 2008).

Clarifying their position regarding the creation of any new governance arrangements, the ICC made the following statement:

> We called upon Arctic governments to include Inuit as equal partners in any future talks regarding Arctic sovereignty. We insisted that in these talks, Inuit be included in a manner that equals or surpasses the participatory role Inuit play at the Arctic Council through ICC's permanent participant status (ICC 2008).

Several years later, the Illulisat Declaration may continue to be the best indicator of which way the political winds are blowing. A central purpose of the meeting of the five Arctic Ocean coastal states in Greenland was to demonstrate to the international community and the media that there would not be a scramble for resources in the region, but rather an orderly process governed by the law of the sea. They embraced the rhetoric of environmental conservation, stating "the Arctic Ocean is a unique ecosystem, which the five coastal states have a stewardship role in protecting".

11.3 Transatlantic Cooperation and Common Ground

The EU and US hold notably different positions in the marine Arctic—the US, unlike the EU, possesses Arctic coastal territory—however, both share common interests in promoting environmental protection and sustainable development in the Arctic region and, as global leaders, have particular capacity for realizing these aims. Further, while none of the Arctic EU Member States (Denmark, Finland, and Sweden) have direct influence on Arctic waters, the EU is, by virtue of its Member States, among the largest maritime powers in the world and as such can significantly contribute to the discussion on environmental governance in the marine Arctic. In addition, the EU accession of candidate state Iceland might add a territorial perspective to EU marine governance in the Arctic in the mid-term future that goes beyond the already close economic cooperation under the European Economic Area (EEA) agreement.

Both the EU and the US have released policy statements regarding their Arctic policies. In November 2008, the European Commission issued its Arctic Communication (European Commission 2008) which laid out EU policy objectives in a number of different areas, including environmental protection, indigenous peoples, sustainable use of resources, and international governance options. In June 2012, a Joint Communication on the EU's Arctic Policy (European Commission and High Representative 2012) set out the case for a refined policy and increased EU engagement on Arctic issues based on knowledge, responsibility to achieve sustainable development, and engagement with Arctic states, indigenous peoples, and other partners. Like the 2008 Communication, it highlighted the issues of climate change, research, indigenous peoples, maritime safety, sustainable economic development, and multilateral cooperation. The US's Presidential Directive on Arctic Region Policy (NSPD-66 2009) outlined a similar set of issues, with the notable addition of US security interests, focusing on Arctic security needs, protection of the environment and resources, sustainable development, international cooperation, indigenous communities' participation in decision making, and scientific monitoring and research.

These policy statements have a noteworthy level of agreement, with clear areas of opportunity for potential cooperation. Both affirm commitments to the existing law of the sea framework. Both indicate a preference for working within existing institutions and frameworks rather than creating a new overarching governance regime, while indicating a willingness to modify these frameworks to fit unique Arctic conditions. Both highlight the Arctic Council as a forum for continued cooperation in the region. Both recognise the threats posed to indigenous communities by rapid environmental change and poorly regulated economic expansion and the importance of including indigenous people in Arctic decision making. Both indicate a commitment to greater cooperation in scientific research and monitoring. Both point to agreement that Arctic governance should be informed by principles of ecosystem-based management.

Likewise, in their approaches to marine governance, the EU and US also find common ground. The EU has adopted a system of integrated and holistic maritime policies and recognised the need for an integrated policy approach to the Arctic Ocean in its 2007 Integrated Maritime Policy (IMP) (European Commission 2007). The EU's Marine Strategy Framework Directive (2008) states that marine policies will use an ecosystem-based approach to the management of human activities. The US, through its National Ocean Policy (Exec. Order No. 13547 2010), has also adopted an ecosystem-based management approach. The National Ocean Policy's draft implementation plan singles out the Arctic Ocean as a priority area for ocean policy action (National Ocean Council 2012).

Importantly, the EU and US have also both emphasized the need for international cooperation in approaching Arctic environmental protection. The European Commission made international cooperation one of three key policy objectives in its 2008 Communication and cooperation—internationally and with Arctic stakeholders—was again emphasized in the 2012 Joint Communication, including regarding the sustainable management of marine resources. The US's 2009 Arctic Region Policy highlights the need for international cooperation in responding effectively to environmental challenges, as well as for international scientific and shipping cooperation.

With this common basis in mind, transatlantic cooperation between the EU and US can take advantage of emerging opportunities for improving protection of the Arctic marine environment. Through formal cooperation and informal channels, and both international and domestic action, the EU and US can promote environmental leadership and provide resources in shaping agendas and policy making.

11.4 Addressing Governance Shortcomings

The Arctic marine area is governed by a complex array of legal instruments, including bilateral and multilateral agreements, supra-national, national, and subnational legislation, as well as soft law arrangements. Arctic governance also involves institutions that are national, regional, and global in scope, and that possess mandates ranging from the provision of scientific advice and issuance of recommendations to the prescription of legally binding obligations (see Chap. 3).

The principle challenges with the Arctic's environmental legal and policy framework today are, first, gaps and shortcomings in adherence to global instruments and their implementation; second, insufficient coordination and integration between existing instruments; and third, rapid environmental, socioeconomic, and geopolitical changes that may outpace the ability of current institutions and agreements to adapt to and address new governance challenges.

At the international level, certain challenges stem from the fact that international conventions provide general frameworks that are not specific to Arctic conditions and not all Arctic states are parties to relevant instruments (Best et al. 2009b). At all levels, many of the gaps in Arctic governance today are the result of an incremental, sector-based approach to management that has historically been static in nature and failed to consider interlinkages between systems. Relevant regional data and scientific knowledge suffer from similar gaps, owing both to the complexity of Arctic marine ecosystems and that many scientific efforts to date have been directed toward specific issues, with less attention paid to interdependencies and cause-and-effect relationships present in Arctic ecosystems.

In addition to gaps between different sectoral governance regimes, there are also gaps within these regimes as they apply to the Arctic. Additional discussion on the areas of fishing, shipping, and offshore hydrocarbon extraction can be found in Chaps. 5, 6, and 7.

The changes taking place in the Arctic pose immense challenges for maintaining environmental and cultural sustainability in the region. Many of these are complex, in that they entail a number of ecosystem components that are affected by multiple drivers of change. Additionally, ecosystems sometimes span territorial boundaries and often involve a broad range of stakeholders. Furthermore, approaches to governance in marine environments are often less developed than in terrestrial environments. Implementation of natural resource management in marine ecosystems is arguably more difficult than in terrestrial ecosystems due to the lack of visible boundaries between marine ecosystems and the vast areas of international waters. It is, however, important to understand linkages between terrestrial and marine ecosystems. For instance, the Arctic Ocean receives more river runoff than any other global ocean (AMAP 1998).

To address these challenges, there is a need for flexible and adaptive management approaches in the Arctic that recognize cultural and governmental/legal differences, apply an integrated and interdisciplinary approach to understanding and managing ecosystems, and, ultimately, maintain the resilience of Arctic ecosystems and communities.

Integrated, cross-sectoral governance strategies, taking into account both natural systems and human activities in a holistic and integrated manner, should thus be a key aim of regional and global policies. Cross-sectoral policy options can be distinguished from those of a more narrow focus by their substantive scope and level of participation. An integrated, ecosystem-based management approach is increasingly recognized as a superior approach to marine management, rather than the status quo sectoral approach that is prevalent in the Arctic and elsewhere (see e.g., Ehler 2011). Such an approach can help to maintain ecosystem resilience while also recognizing community and stakeholder needs, maintaining cultural traditions, allowing for economic opportunities, and encouraging flexibility and adaptability. Further, both scientific and traditional knowledge can be incorporated in integrated processes, and multiple interests—e.g., environmental, social, and economic—can be addressed in conjunction. Flexible management approaches can allow governance systems to adapt to changing environmental and climatic conditions as well as new scientific knowledge. As highlighted in EU and US policy statements, the existing governance framework should continue to be modified to better address changing Arctic conditions.

As Arctic challenges are both local and global in nature, there is need for coordination and integrated governance both among Arctic states and at the international level to successfully manage pressures in the Arctic region. Suggestions for improving the current framework range from new sector-based agreements to a comprehensive 'Arctic treaty'; however, as stated in the Ilulissat Declaration, the Arctic coastal states themselves have rejected the idea of a comprehensive new regime.

It is of importance to note that no governing body has a mandate to develop legally binding rules for the entire Arctic region. While it has taken a more active role in policy development, providing a forum for Arctic states to negotiate legally binding instruments, the Arctic Council nonetheless remains limited in its mandate and without regulatory powers of its own: The Ottawa Declaration on the Establishment of the Arctic Council (1996) does not impose binding obligations on its participants and neither is the Arctic Council empowered to do so. The law of the sea provides a general governance framework for the marine environment, which Arctic states agreed to abide by in the Ilulissat Declaration, but relies on additional institutions to implement its provisions.

Recent years have witnessed a surge of interest in Arctic governance and in efforts to improve the existing framework so as to better meet new changes. These include the Arctic SAR agreement (2011), designed to fill gaps in coverage as Arctic marine traffic increases, and the forthcoming Arctic MOPPR Agreement, which was drafted in anticipation of a rise in Arctic offshore hydrocarbon extraction. Will new efforts such as these be sufficient to fill Arctic marine governance gaps? Can they be flexible enough to meet continuing changes? And will they be inclusive in allowing participation by indigenous peoples?

There are several strategies by which a cross-sectoral system of governance in the marine Arctic could be implemented, carrying varying degrees of political support from different Arctic players:

- Relevant actors, including non-Arctic states, indigenous peoples' organizations, industry, and NGOs, could establish *new* issue- or sector-specific instruments and institutions that are complimentary in nature;

- Relevant actors, including non-Arctic states, indigenous peoples' organizations, industry, and NGOs, could participate in multilateral negotiations within the context of *existing* institutions and instruments in order to *modify* them in a coordinated fashion.
- The *Arctic Council* could serve as a forum and coordinator in efforts to *supplement or modify* existing frameworks to function in a more integrated and comprehensive fashion. Observers could play a key role in supporting this work through stronger participation in the Arctic Council working groups and by being given a more active voice in the Council's deliberations. Further, observers could support a strengthened Arctic Council in other fora in which they play a more prominent role themselves.
- State actors, with the involvement of other relevant actors, such as, indigenous peoples' organizations, industry, and NGOs, could negotiate an *overarching*, legally binding regional instrument specifically tailored to address the unique conditions and challenges of the Arctic.

Given the need for a flexible governance regime, and the wide range of sovereign actors involved in the Arctic, the utility of soft law instruments should not be underestimated. Existing international bodies such as the Arctic Council and legal instruments with institutional components may be well situated to create and update guidelines and best practices for the region, although the non-legally binding nature of soft law instruments may produce weaker commitments and less certain outcomes.

11.5 Transatlantic Support for Integrated Management

Regardless of the implementation strategy—or possibly a mix of several strategies—taken, the following outlines certain approaches that could provide a basis for cross-sectoral governance measures and frameworks:

- *Ecosystem-based management* (EBM)—EBM is as a comprehensive, integrated approach to management of human activities that offers flexibility and is widely regarded as a best practice of environmental governance. As illustrated in Chap. 3, EBM has the potential serve as an *organizing framework for integrated management* by balancing a variety of human activities, interests, and priorities while supporting ecosystem sustainability.
- *Marine spatial planning* (MSP)—MSP organizes marine space and uses, balancing development demands with ecosystem protection, to achieve ecological, economic, and social objectives in a planned process. As Chap. 9 describes, MSP is an *operational process through which to implement EBM*. MSP supports managing human activities to enhance compatible uses and reduce conflicts among uses.
- *Marine protected areas* (MPAs)—MPAs often serve as an important *component* of EBM and MSP and can be a helpful tool for implementing the precautionary principle in management by setting aside spaces for protection as conditions

and uses change. As Chap. 10 explains, currently, most protected areas are for terrestrial rather than marine environments and little of the Arctic marine area is designated as MPAs.

EU and the US statements regarding their Arctic policies point to broad areas of agreement, as outlined in Sect. 11.3. In addition to the synergies identified relating to indigenous peoples, the environment, and international cooperation, there also appears to be agreement that marine Arctic governance should be informed by principles of EBM. The EU's 2008 Arctic Communication states that holistic, ecosystem-based management of human activities should complement any efforts to mitigate and adapt to changes in the Arctic caused by climate change, and the 2012 Joint Communication highlights steps for the EU to take on Arctic ecosystem management. Similarly, the US's 2009 Arctic Region Policy states that relevant executive agencies should pursue marine ecosystem-based management in the Arctic.

Both the EU and the US have experience with EBM within their own maritime zones and could push for wider application in transboundary, cross-sectoral Arctic marine governance. The EU has been a strong advocate for MSP, an important component of the IMP, and is planning to issue a directive on coastal and maritime spatial planning in 2013. In the US, the federal government has begun implementing its ocean policy, applying integrated management and MSP, including in Alaska's Beaufort, Chukchi, and Bering Seas. The US National Ocean Policy provides for a strategic plan for EBM in the Arctic, and the draft implementation plan highlights EBM as a guiding theme.

The EU and US support the Arctic Council as a primary forum for cooperation and governance in the Arctic region. As such, supporting the Arctic Council's work may be one of the best areas for EU and US cooperation on promoting integrated management approaches. The Arctic Council has undertaken and continues to lead important work on EBM. As described in Chap. 3, past projects of significance include the PAME (Protection of the Arctic Marine Environment) working group's work on Large Marine Ecosystems (LMEs) and the 2009 Best Practices in Ecosystem-based Oceans Management in the Arctic (BePoMAR) report (Hoel 2009). Currently, an expert group on EBM for the Arctic environment, composed of government experts from Arctic States and representatives from the Arctic Council's permanent participants and working groups, is developing recommendations for advancing EBM in Arctic ecosystems, which will be considered by the Ministers before the end of the Swedish Chairmanship in May 2013.

In practice, EBM approaches are largely implemented at the national level, although regional and international support and coordination are important and can ensure coordination between jurisdictions as well as sharing of best practices. The Arctic Council can help facilitate coordination between states and continue to support the development of EBM principles and practices. In addition to Arctic states, the Arctic Council can also help bring non-governmental actors such as indigenous peoples, NGOs, and industry to the table. Stakeholder participation is an important part of EBM and MSP planning processes and is critical for achieving social, economic, and environmental objectives. The US and EU can support the continuation of this work financially and by appointing experts from multiple cross-cutting agencies to contribute to the working groups and take tasks forward.

Arctic coastal states should designate MPAs in the Arctic, either independently or as part of a larger EBM framework, before a further increase in economic activity might lead to the entrenchment of interests in certain areas. MPAs could help to protect sensitive and unique Arctic ecosystems and increase resilience in the face of changing conditions, as could be especially promoted by a planned network of protected areas, perhaps involving co-management agreements with indigenous communities. MPAs through a network, rather than individually, could cover a wider area, including species' habitat and migratory pathways that cross state boundaries. Thus, working together, states can do more than alone and cooperation can create a transboundary network of protected areas of greater ecological value and resilience. The EU and US could jointly advocate for the creation of an EBM-based network of MPAs.

In its 2012 Joint Communication on the Arctic, the EU outlined the following steps for pursuing Arctic ecosystem management:

- Work through the OSPAR Convention to establish a network of Arctic MPAs and assess measures to manage oil and gas extraction activities in extreme climatic conditions;
- Contribute to work under the Arctic Council's PAME working group; and
- Promote biodiversity protection in areas beyond national jurisdiction in UN bodies such as the LOS Convention.

While the US is neither a contracting party to the OSPAR Convention, nor a party to the LOS Convention, and the EU's application for an observer status at the Arctic Council has yet to be decided upon, the latter two options suggest areas for US as well as EU action. The Arctic Council may present some of the best opportunities for dual action and the EU and US can offer support within the Council, as well as in other international bodies, for environmental priorities and promoting integrated management approaches. Their cooperation in existing regulatory bodies is critical to ensuring that environmental goals remain at the top of the agenda.

11.6 Toward Integrated Management

To successfully implement integrated management approaches in Arctic marine area, key knowledge gaps must first be filled. A commonly identified problem among Arctic policymakers is a lack of information. Good environmental governance must be supported by an understanding of ecosystem conditions, baselines, processes, and changes. Both Arctic and non-Arctic states alike can, via the Arctic Council and other international scientific institutions, continue to improve coordination among scientific research initiatives. The following represent example areas of actions that can support the design, implementation, and adaptation of successful governance practices for the Arctic marine area:

11.6.1 Research and Monitoring

Additional research and data is needed on Arctic systems in order to inform EBM initiatives, as much Arctic research has traditionally had a narrow, issue-based focus. Information is lacking on baseline Arctic ecosystem conditions, thus making the measurement of changes challenging, and on new changes and pressures. Better monitoring and observations are needed to continue to assess changing conditions, activities, interactions, and management needs. To meet adaptation needs in the Arctic, it is necessary to improve observational knowledge and have an understanding of integrated impacts and processes in the region and associated risks (AMAP 2012). Traditional knowledge of indigenous communities should be incorporated into these efforts.

Opportunities for cooperation in Arctic marine science exist between the EU and US, as well as between Arctic and non-Arctic states. Both the EU and US strongly support international scientific cooperation in the Arctic, as noted in their respective Arctic policies. In addition to its 2009 Arctic Policy, the US's Arctic Research and Policy Act (ARPA 1984) provides for a comprehensive national policy dealing with national research needs and objectives in the Arctic. Both the EU and US have devoted significant resources to Arctic research and have helped to promote shared research platforms and collaboration. The US has identified the need to establish a science framework to support science-based EBM implementation as a key action for implementing EBM, noting that "[s]ustainably managing human uses of an ecosystem requires a robust understanding of the nature of the dynamically interacting biological, physical, chemical, and geological components and processes; the effects of human and natural forces; and the results of management efforts" (National Ocean Council 2012).

The Arctic Council's EBM Task Force has recognized the need for supporting EBM through increased knowledge in areas such as ecosystem services, monitoring, data-sharing, and improving understanding of ecosystem interactions of cumulative effects, while, however, noting that "[i]n most cases the missing piece for implementing EBM is not the science but an effective process or organizational structure; without some means to translate the science into a meaningful management approach that meets certain agreed-upon objectives, EBM is just a series of interesting reports" (SAO 2012a, b).

11.6.2 Defining and Assessing Ecosystems

One approach to help distinguish priority areas for policy action is the Large Marine Ecosystem (LME) concept, built on general principles of ecosystem management. LME boundaries are becoming widely used at the international scale to distinguish highly productive areas around the globe for marine ecosystem management.[4] LMEs encompass relatively large areas of approximately 200,000 km^2 or greater and have distinct bathymetry, hydrography, productivity, and trophically-dependent populations (Sherman 1994). They can be evaluated with respect to their productivity, fisheries, pollution, ecosystem health, socioeconomic conditions, and governance (Juda and Hennessey 2001). In addition, LMEs draw attention to the need to understand complex changes in multiple species interactions and the need to manage for resilience rather than composition or structure. LMEs provide a practical basis to evaluate shipping, fishing, and tourism at the regional level.

The Arctic Council PAME working group has developed a map of LMEs in the Arctic that was adopted by the Arctic Council Ministers.[5] Both the Arctic Marine Shipping Assessment (AMSA 2009) and Arctic Oil and Gas Assessment used LMEs as a basis for analysis (IUCN/NRDC 2010), and the Arctic EBM Task Force has suggested using PAME's LME map to "inform EBM implementation" (SAO 2012b). Beyond the Arctic Council, several Arctic Ocean coastal states, notably Canada, Norway, and the US, have organised their national Arctic governance regimes around the concept of LMEs. However, LMEs often cross national borders and there is as yet no established framework for coordinating LME regulatory activities at the bilateral or international level. The EU and US could, for example, continue such efforts through working bilaterally on a comprehensive Arctic Ocean Assessment, drawing on the work of the United Nations (UN) on global reporting and assessment of the marine environment.[6]

An Arctic Ocean Assessment could complement the LME work already taking place and better harmonise governance approaches to issues common to multiple ecosystems. Furthermore, it could build on existing assessments, e.g., environmental impact assessments (EIAs) and strategic environmental assessments (SEAs), to better understand the potential environmental and social impacts of proposed activities and programmes on the Arctic environment. For climate change, EIAs not only help to assess and understand the interactions between a changing environment and human activities, but can also help identify—and mitigate—the climate change impacts of the activity. Both the US and EU are engaged in the UN's global marine assessment and both have their own procedures in place for EIA.

While terrestrial protected areas have increased globally, there remains a need to identify and protect biologically important marine areas, including in the Arctic. An important component of EBM, as well as for the establishment of MPAs, is the identification of ecologically, biologically, or culturally significant and vulnerable areas that should be considered for protection (SAO 2012a; Speer and Laughlin 2010).

---

[4] LMEs are used, among others, by the United Nations Environment Programme, United Nations Development Programme, the World Bank, the US National Oceanic and Atmospheric Association, and the Arctic Council.

[5] See <http://www.pame.is/arctic-large-marine-ecosystems-lme-s>.

[6] See <http://www.un.org/Depts/los/global_reporting/global_reporting.htm>.

These areas can serve as a basis for MPAs as well as for MARPOL 'special areas' or PSSAs (see Sect. 11.8; see also Chap. 6). Efforts have taken place to begin identifying Arctic ecologically and biologically significant areas [EBSAs; as defined by the Convention of Biodiversity (1992)], at the state level - with efforts by Norway, the US, Canada, and Greenland (SAO 2012a), in the AMSA (2009), and by environmental NGOs (Speer and Laughlin 2010).

The identification of EBSAs should be a dynamic process. Many ecologically or biologically and culturally significant areas are tied to Arctic sea ice, which is undergoing dramatic changes (AOR 2012; Speer and Laughlin 2010). The EU and US can support efforts to improve and continue identification of EBSAs as well as to move from the recognition of areas in need of protection to the implementation of protective measures through, *inter alia*, efforts within the IMO, Arctic Council, or bilateral cooperation.

11.7 Transatlantic Policy Options for Fisheries Management

While warmer areas of the Arctic marine area have supported commercial fishing activities for decades, until recently there was little or no major fishing activity in the colder areas of the Arctic, with ice-covered regions completely cutting off access to fishing. The retreat of Arctic sea ice is opening up new parts of the Arctic Ocean to fishing vessels, and there are already signs that certain fish species are migrating north to warming ocean waters.

In light of these changes, gaps exist in both international and national legal and policy frameworks for regulating Arctic fisheries, and the expansion of marine capture fisheries in the Arctic may necessitate adjustments. Any such process would benefit from a needs assessment drawing on basic fisheries research and an evaluation of likely future scenarios regarding, for example, habitats, migration patterns, impacts on target and non-target species, and fishing techniques. For certain Arctic fisheries that have been commercially fished for years, policymakers have access to a wealth of information. In other areas, almost nothing is known. For instance, new fishing opportunities on the Pacific side of the Arctic Ocean may remain located primarily in the maritime zones of coastal states for the near future, whereas fishing opportunities on the Atlantic side may soon extend to areas in the high seas that were previously not fished. There is a pressing lack of scientific data to understanding of Arctic Ocean ecosystems and for use in developing science-based fisheries management.

To address information deficits, individual or collective initiatives geared towards developing mechanisms or procedures similar to an EIA or a SEA for new fisheries in the Arctic marine area could be performed. The EU and US can support further basic research on fisheries, Arctic ecosystems, fish species, and climatic impacts, and for the development of potential scenarios. Assessments could be carried out in the framework of the Arctic Council through its CAFF

(Conservation of Arctic Flora and Fauna) working group or in other forums, such as ICES, where some progress has already been made (see Chap. 5).

Until scientific information and understanding improve, one policy option is to support a freeze on the expansion of commercial fishing in the Arctic, such as the one enacted by the North Pacific Fishery Management Council (NPFMC) in 2009 for fishing in the Alaskan EEZ, until there is sufficient information and scientific understanding of the ecosystem and climatic impacts on fisheries. This freeze followed a 2007 Congressional Joint Resolution (S.J. Res. 17 2007) directing the US to initiate international discussions and take steps to negotiate an agreement for managing migratory and transboundary fish stocks in the Arctic Ocean.

While not having Arctic Ocean coastal waters of its own, the EU likewise "advocate[s] a precautionary approach whereby, prior to the exploitation of any new fishing opportunities, a regulatory framework for the conservation and management of fish stocks should be established for those parts of the Arctic high seas not yet covered by an international conservation and management system" (European Commission and High Representative 2012). Such a statement is made possible since the EU has—according to Articles 3(1)(d) and 38(1) of the Treaty on the Functioning of the European Union (TFEU 2008)—the exclusive competence for the conservation of marine biological resources under its Common Fisheries Policy (CFP), while competence on other fisheries-related issues is shared with EU Member States (however, Member States are only competent to fill gaps in the EU's legislation, see TFEU (2008), art. 2(2)).

Together, both the EU and US could advocate for wider international support of a precautionary position, halting new fisheries activity until scientific understanding and adequate management approaches are established and supporting underlying research and policy measures. Moreover, both could support a declaration that the relevant general principles of the Fish Stocks Agreement (1995), UN Resolutions in relation to vulnerable marine ecosystems and destructive fishing practices (for instance, A/RES/61/105[7]), and relevant conservation and management measures drawn from RFMOs would apply to new and existing fisheries in the Arctic marine area. Following the NPFMC ban, this declaration could stipulate that there shall be no expansion of commercial fishing in the Arctic until adequate assessments of the impacts on target and non-target species and livelihoods of indigenous peoples have been carried out.

Both Arctic Ocean coastal states and other states can adopt individual regulations on fishing activities in the Arctic marine area within their own maritime zones or for their natural and legal persons. The EU and the US could coordinate efforts in this regard, acting independently on the basis of shared concerns, with the EU acting as a flag rather than coastal state in the Arctic Ocean, and thereby expanding the geographic scope and relevance of any adopted regulations. Over time, such transatlantic regulations could serve as a model for international rule making.

[7] See <http://www.un.org/Depts/los/general_assembly/general_assembly_resolutions.htm> for more UN Resolutions on sustainable fisheries.

11.8 Transatlantic Policy Options for Arctic Shipping

With sea ice melting, new intra- and trans-Arctic shipping routes are opening to industry and tourism. Trans-Arctic shipping is increasing during summer months already—with an increase from 34 vessels in 2011 to 46 in 2012 (Pettersen 2012)—and it is increasingly important to address the safety and environmental risks associated with a rise in shipping activity.

Significant progress has been made in recent years towards addressing gaps in the international legal and policy framework for Arctic marine shipping. In 2009, for instance, the International Maritime Organisation (IMO) adopted the Polar Shipping Guidelines (2009) and is currently developing new mandatory guidelines for 'Ships Operating in Arctic Ice-Covered Waters', also known as the 'Polar Code'. At the Arctic Council's Seventh Ministerial Meeting, the Arctic SAR Agreement was signed by the Arctic states. In 2013, the new Arctic MOPPR Agreement is planned for signature. Thus, while gaps still remain (see Chap. 6), significant progress has been made in recent years through multilateral cooperation.

For the remaining gaps, there are various options available for modifying the current international framework to account for the risks presented by shipping to Arctic marine ecosystems and human safety. The EU and the US should consider coordinating a joint and harmonised approach towards supporting or initiating various unilateral, regional, and global shipping options. Relevant international bodies in this regard include the IMO, Arctic Council, and the Paris and Tokyo MOUs on port state control. In considering the suitability of regional and global options in the sphere of shipping vis-à-vis individual options, particular account should be taken of the function of competent international organizations like the IMO and the need for uniformity in the international regulation of shipping.

The EU and US share interests in the protection and preservation of the marine environment and marine biodiversity, as well as in the continued exercise of navigational rights and freedoms for their flagged vessels. In recent policy statements both have advocated for work through the IMO on strengthening Arctic navigation standards (such as through the development of the Polar Code), for freedom of navigation in the Arctic, and for promoting navigational safety. Even though the EU cannot act in a capacity comparable to that of an Arctic Ocean coastal state, it can act in a capacity comparable to that of a flag state, a port state, a market state, or with regard to its natural and legal persons. As flag states, and as major shipping powers, or the EU and US could, for instance, impose requirements on vessels that are more stringent than generally accepted international rules and standards (GAIRAS) (e.g., related to special discharge, emission, and ballast water exchange standards), taking forward-looking steps in advance of the forthcoming Polar Code.

In particular, the EU and US can strengthen cooperation to improve shipping regulations by working together in the IMO to support negotiations on the Polar Code and ensure inclusion of stringent standards This could, as suggested in Chap. 6, include new standards such as for ship routeing measures, compulsory pilotage, and ice-breaker or tug assistance. Also through the IMO, the EU and US could jointly propose and support the designation of the marine Arctic (or parts thereof) as a MARPOL special area or PSSA, accompanied by a comprehensive package of associated protective measures (APMs) consisting of one or more of the above standards.

11.9 Transatlantic Policy Options for Offshore Oil and Gas Development

Under scenarios of retreating sea ice, technological advances, and rising oil prices, the exploration, production, and shipping of Arctic oil and gas are increasingly viable and attractive. However, challenges persist under harsh regional conditions, raising the environmental and safety risks of drilling for both the environment and resource-dependent communities. In addition to risks from oil spills, oil and gas development creates significant operational impacts: For example, fuel combustion for onsite power generation, well testing, gas flaring, and operational leaks release black carbon, methane, NOx, SO_2, VOCs, and CO_2 emissions into the Arctic atmosphere.

The positions of the EU and US in relation to offshore hydrocarbons activities in the Arctic are fundamentally different. As a coastal state, the US is directly involved in offshore hydrocarbon extraction, with significant reserves—possibly 30 % of total Arctic reserves—off the coast of Alaska (Bird et al. 2008). By contrast, the EU does not have any coastal state jurisdiction in the Arctic Ocean.

There is no instrument providing comprehensive global regulation of offshore hydrocarbon activities, nor is there any global regulatory or governance body with such a mandate. There are, however, a number of instruments with broader scope that also apply to offshore hydrocarbon activities, including those taking place in the Arctic. Among global instruments, the LOS Convention (1982) sets out the basic rules on access to and control over offshore hydrocarbon resources and the mandate of the International Seabed Authority (ISA). Other instruments with more limited applicability to offshore hydrocarbon activities include MARPOL (1973/1978), OPRC (1990), the OSPAR Convention (1992), and the Espoo Convention (1991).

The Arctic Council's Arctic Offshore Oil and Gas Guidelines (PAME 2009), adopted in 1997 and most recently updated in 2009, can go a long way toward addressing regulatory gaps and deficiencies if put into practice. The Guidelines provide recommendations on standards, technical and environmental best practices, management, impact assessment, emergencies, decommissioning, and regulatory control for Arctic offshore oil and gas operations, and recommend following the precautionary approach, the polluter-pays principle, the principle of sustainable development, and the principle of continuous improvement. While providing an important starting point for regulating offshore oil and gas activities in the Arctic, the Guidelines are, however, non-legally binding and leave coastal states with a wide margin of discretion in their implementation.

The EU's 2008 Arctic Communication stated in its proposals for action to "press for the introduction of binding international standards, building *inter alia* on the guidelines of the Arctic Council and relevant international conventions" (European Commission 2008). Later, in 2009, the 'Council conclusions on Arctic issues' (Council of the European Union 2009), invited the European Commission and Member States to examine the possibilities to endorse the Guidelines. The EU's most recent Arctic policy document, the 2012 Joint Communication, does not address the Guidelines, but references the EU Commission's October 2011 proposed new regulation[8] on the safety of offshore oil and gas prospection, exploration, and production activities. While the proposal is still in inter-institutional procedures, its direct impact on Arctic offshore drilling is doubtful. Norway's offshore activities in Arctic waters would only be covered if the finalized regulation or directive fell inside the scope of the EEA agreement, which has been challenged by Norway (EEA 2012).

It has been suggested that in order to ensure effective protection, the Arctic Offshore Oil and Gas Guidelines should be made legally binding (see Chap. 7). A legally binding agreement, however, could face political challenges from coastal states with established national practices.

Still, the fact that all Arctic eight states have formally endorsed the Guidelines demonstrates at least an initial level of support and agreement on governance of offshore hydrocarbon development. The EU and US can publicly support application of the Guidelines and good management practices. As an overarching recommendation, the Arctic Ocean Review (AOR) draft report has urged for agreement on *non*-binding internationally agreed standards for Arctic offshore oil and gas activity, convened through the Arctic Council (AOR 2012). Generally, Arctic states should work to harmonize and strengthen drilling standards, with involvement from industry and non-Arctic states.

In addition to the Arctic Offshore Oil and Gas Guidelines, the Arctic Council has led other work of significance on Arctic hydrocarbon development and environmental protection. The AMAP (Arctic Monitoring and Assessment Programme) working group provides assessments on petroleum hydrocarbon pollution and oil and gas activities in the Arctic while the EPPR (Emergency Prevention, Preparedness, and Response) working group supports marine oil pollution preparedness and response. The US and EU can provide continuing support for these initiatives, particularly should the EU be granted Arctic Council observer status in 2013.

The OSPAR Commission has developed a comprehensive database for offshore oil and gas installations and a monitoring system whereby the OSPAR Offshore Industry Committee (OIC) collects data on emission and discharges.[9] Additionally, the OSPAR Commission supports integrated environmental assessments and

---

[8] Commission's 2011 'Proposal for a Regulation on safety of offshore oil and gas prospection, exploration and production activities' <http://eur-lex.europa.eu/LexUriServ/LexUriServ.do?uri=CELEX:52011PC0688:EN:NOT>.

[9] Including use and discharge of drilling fluids and cuttings, discharges of oil in produced water, chemicals used and discharged offshore, and accidental spills of oil and chemicals and emissions to air.

implementation of the EU MSFD within its Joint Assessment Monitoring Programme (JAMP), which covers the implementation of OSPAR monitoring and information collection programmes – resulting in a biannual update of the offshore database as well as annual assessment sheets on discharges, emissions, and spills of oil and hazardous substances. To promote monitoring, transparency, and data exchange for Arctic oil and gas operators, such a monitoring system for discharges, emissions, and spills could be replicated more widely in other forums.

A possible way towards this direction can be already seen in the PAME work program for 2011–2013: In a comparison of the existing regulatory structures of the Arctic states, it aims to include aspects such as organisational structures, planning, implementation, and monitoring. The AOR has suggested "[i]dentify[ing] ways for Arctic Ocean coastal states not party to OSPAR to coordinate further with JAMP and OIS, notwithstanding that OSPAR Region I covers only a portion of the Arctic Ocean and excludes the Canadian, Russian, and US marine Arctic" (AOR 2012).

With regard to accidental oil spills, the new legally binding Arctic MOPPR Agreement is being negotiated under the auspices of the Arctic Council. A task force was created to prepare the international instrument on oil pollution preparedness and response, as mandated at the Arctic Council Ministerial Meeting in Nuuk in May 2011 (Arctic Council 2011). The agreement is currently ready for signature and will be presented at the Arctic Council Ministerial Meeting in May 2013. It builds to the OPRC 90 Convention on marine oil pollution and adds a regional approach as foreseen in Article 10. The Arctic MOPPR Agreement sets forth procedures in advance of a spill for clean-up and coordination, including an obligation of mutual assistance, enhancing collaboration, and promoting training and information sharing between response parties in the region.

The (draft[10]) agreement takes significant steps towards preparedness for an Arctic spill incident, yet still leaves certain areas uncovered and does not ensure adequate investments in infrastructure by the parties or appear to provide minimal standards in this regard. Although the binding framework is to be supplemented with non-binding operational guidelines, these are not expected to add substantial standards beyond formal cooperative efforts.

However, there will be opportunities to influence the success of the agreement upon implementation. For example, parties will be required to participate in training exercises, secure equipment, establish programmes to respond to spills, and develop detailed plans and strategies for response. Here, support could be provided in terms of both financial and technical resources. Additionally, private industry owns many of the emergency response resources available in the region, yet the agreement does not harness these, relying solely on governments. Steps should be taken generally to cooperate with private industry in Arctic oil and gas development, building upon cooperative efforts such as the Arctic Oil Spill

[10] At the time of writing. Draft agreement for signature on file with the authors.

Response Technology Joint Industry Programme (JIP),[11] and as an important first step, agreements should be undertaken to effectively extend private resources for oil response.

11.10 Indigenous Voice in Arctic Governance

Indigenous peoples have inhabited the Arctic for thousands of years and are not only stakeholders in the Arctic, but also rights holders, and deserve a special status in decision making processes. However, their interests can easily be marginalised or neglected in governance institutions.

Traditionally resilient in the face of change, today Arctic indigenous peoples find themselves unable to fully adapt to the rapid rate of change and the range of external stressors, including climate change. Indigenous communities are extremely vulnerable to climate change due to the dependence of their livelihoods on Arctic ecosystems. To support resilience and adaptation to the impacts of climate change, communities must be empowered and have clear rights and access to participation in resource decision making, including through co-management arrangements. To further ensure that efforts are truly participatory, rather than merely procedural, resources must be available to support participation in meetings and planning processes that are often time and money intensive. For EBM processes, stakeholder involvement, including Arctic indigenous peoples, is critical for achieving and understanding social, economic, and environmental goals. EBM can also utilize traditional, as well as scientific, knowledge.

The EU and US have both recognised the particular vulnerability of indigenous communities as central issues within their Arctic policy statements. To support the adaptation of indigenous communities in the marine Arctic, the best forum for transatlantic efforts is the Arctic Council. The Arctic Council affords indigenous groups special status as permanent participants, empowering them to influence the debate on climate change-related issues and include their perspectives in assessments, such as in the Arctic Climate Impact Assessment (ACIA 2005). The Arctic Council's Arctic Resilience Report (ARR) promises to be another important Arctic Council initiative. Initiated at the SAO's meeting in November 2011,[12] the science-based assessment will analyze changes and drivers across multiple scales, engage stakeholders, and identify policy and management options for strengthening resilience and for adaptation. The EU and US should ensure indigenous participation in the ARR and related future assessments, e.g., by providing adequate financial support to allow experts to attend workshops and meetings and supporting on-site monitoring in local communities.

It is important that the status of Arctic indigenous peoples remains strong under the Arctic Council as it continues to evolve. If this status were lost, it would

[11] See <http://www.arcticresponsetechnology.org/>.

[12] See <http://www.arctic-council.org/arr/about/>.

result in significantly less visibility for indigenous peoples' interests. The ICC has emphasized the role of the Arctic Council in exercising rights of self-determination (ICC 2009). The EU and the US should recognise and promote the importance and high-level status of indigenous participation in the Arctic Council as it continues to evolve and in future mechanisms.

The EU and US could also jointly support the creation of an assessment on adaptation in the Arctic to cover issues and challenges associated with the implementation of adaptation policies. Furthermore, across the Arctic, a number of national and sub-national climate change adaptation strategies have been developed. Such strategies have been launched by Canada, the (US) state of Alaska, and Greenland among others. Evaluating existing adaptation strategies and their effectiveness can provide valuable information and best practices for wider use.

11.11 Tackling Climate Change

As highlighted throughout this book, global climate change is driving Arctic change and underlies many of the critical challenges for environmental and cultural sustainability in the region today. While climate change and retreating sea ice may open up new economic opportunities and bring benefits to the region, these will not be without risks to Arctic ecosystems, indigenous communities, and traditional livelihoods. Impacts in the Arctic are predicted to increase faster than elsewhere in the world. For instance, Arctic temperature rise is estimated by models as ranging between 3 and 6 °C by 2080, a greater rate than any other region of the world (AMAP 2011).The rapid rate and extreme level of change is outpacing the ability of management regimes to adapt. Thus, while this book focuses primarily on governance in response to a changed and changing Arctic marine environment, it cannot ignore the forces driving those changes.

Climate change is a global problem in both source and effects, stretching beyond the Arctic region. While climate change has severe impacts within the Arctic, including ocean acidification, permafrost melt, and impacts on biodiversity, these regional impacts also connect to global trends, such as sea level rise, impacts on ocean circulation, and climate feedback loops (see Chap. 2). Approaches in governance have to take both aspects into account—the local and the global drivers.

A required two-fold approach to climate change mitigation should address global warming forcing agents, such as carbon dioxide and methane, as well as mitigation of pollutants that increase the local radiative force within the Arctic, such as black carbon and tropospheric ozone (including its predecessor emissions nitrous oxides and methane). Historically, international efforts have largely focused on creating frameworks for binding commitments to greenhouse gas reductions. Although some progress has been made—particularly by the EU, which has been a climate change leader in international negotiations under the UN

Framework Convention on Climate Change (UNFCCC 1992) and its Kyoto Protocol (1996), as well as internally with the EU's Emissions Trading System and '20–20–20' targets[13]—the potential for a comprehensive global agreement continues to look bleaker and incremental negotiations fail to move at sufficient speed to address the problem. However, new efforts to tackle short-lived climate pollutants (SLCPs) show promise not only for helping to mitigate global climate change, but for the Arctic region in particular. Former US Secretary of State Hillary Clinton launched the Climate and Clean Air Coalition to Reduce Short-Lived Climate Pollutants (CCAC) in February 2012 along with Bangladesh, Canada, Ghana, Mexico, Sweden, and the UN Environment Programme. As of its first anniversary, the CCAC has grown to more than 50 partners, including 27 nations and the European Commission, and has a set of sector-based and cross-cutting initiatives underway aimed at reducing black carbon, methane, and some HFCs (hydrofluorocarbons). A primary focus of the CCAC has been to highlight the multiple benefits of reducing emissions of SLCPs, especially related to public health and agricultural crop production, thus bringing to the table a wide range of partners motivated by a range of issues for a common goal.

Of the major SLCPs targeted by the CCAC, black carbon is most critical in the Arctic region. The largest black carbon sources have been found in land-based transportation (particularly diesel engines), open biomass burning (including agricultural burning, prescribed forest burning, and wildfires), and residential heating (for instance via wood combustion in stoves and boilers) (SLCF Task Force 2011). Black carbon emissions are also expected to increase with rising shipping activities via Arctic routes and from contributions by gas flaring, for instance from offshore oil and gas development.

Following the recommendations of the Arctic Council's Short-Lived Climate Forcers Task Force, the Arctic Council Meeting of the Environmental Ministers in February 2013 has (as mentioned in Sect. 11.2) suggested implementing a common monitoring system for Arctic states' black carbon emissions, and possibly negotiating a high-level agreement on reductions of black carbon emissions (Ministry of the Environment 2013). For the future, such an Arctic Council agreement could tackle an important part of the problem—the emissions of local Arctic sources of black carbon have been identified to contribute more radiative forcing per emission unit—while presently, the CCAC demonstrates successful (and necessary) international cooperation. Lessons learned from the willing and positive participation of a complex array of actors should be taken to move toward stronger cooperation for greenhouse gas emissions reduction.

However, SLCP-mitigation actions must the greater challenge of global greenhouse gas emissions reductions. Here, there remains potential for improvement in transatlantic cooperation and efforts. Beyond traditionally taking different approaches towards the UNFCCC regime, the EU and US have differed in other

[13] Committing to achieve by 2020 a unilateral reduction in greenhouse gas emissions of at least 20 % from 1990 levels, a 20 % increase in renewable energy share in the EU's energy supply, and a 20 % savings in energy consumption.

mitigation approaches, too, such as, for instance, aviation emissions. The EU has sought to cover aviation emissions—including from foreign carriers—through its ETS, a measure strongly opposed by US and other international airlines. In November 2012, US President Barack Obama signed a bill sheltering US airlines from participating in the programme. Although the EU ETS started coverage of emissions from the aviation sector in the beginning of 2012, the EU Commission proposed a derogation from relevant ETS rules in November 2012—as a "gesture of goodwill"[14]—to prevent action against international aircraft operators for non-compliance with ETS requirements and 'stop the clock' until January 2014 (European Commission 2012). However, this conflict could in fact help to revive cooperative efforts to pursue multilateral solutions through other fora, such as the International Civil Aviation Organization (ICAO) and its upcoming Assembly meetings.

To truly manage the impacts of climate change in the Arctic, as well as the potentially catastrophic global implications, stringent mitigation efforts must be undertaken. The EU and US must, as global leaders and major emitters, find ways to promote cooperation and tackle all sides of the problem.

11.12 Conclusion

Global climate change, transboundary pollution, and global markets all drive Arctic change, underlining the imperative for global action and inclusive governance, with cooperation between Arctic states, non-Arctic states, indigenous communities, industry, and NGOs. Already, important steps have been taken to address marine Arctic governance gaps, but further action is needed to adapt the existing legal and policy framework to meet emerging governance challenges in the marine Arctic. Integrated, flexible management structures are needed in order to effectively manage fast-moving, complex challenges.

While Arctic governance clearly extends beyond the jurisdiction of either the EU or US, transatlantic cooperation is needed. Throughout modern environmental policy, both the US and European countries have often been front-runners in developing environmental policies and instruments, offering political leadership that can assist in addressing the challenges facing the marine Arctic.

On the whole, the Arctic Council continues to be the best forum for transatlantic cooperation as it has a strong record on Arctic environmental issues, an established network of stakeholders, and has successfully brought players together in a flexible forum including now, in its evolving role, helping to facilitate the development of legally binding measures. The Arctic Council has promoted coordination not only between states and indigenous groups, but between governance bodies and regimes (e.g., OSPAR, CBD), and is leading the way on many key policy

[14] See <http://ec.europa.eu/clima/policies/transport/aviation/index_en.htm>.

areas such as, for instance, with task forces on EBM and SLCPs. The EU and US should support the continued momentum and success of these efforts, such as by providing resources and sharing knowledge and best practices as models for international action and rule making. A decision on whether the EU will be granted permanent observer status will help determine the potential for transatlantic dialogue and cooperation on the Arctic within this forum.

Enhanced transatlantic cooperation can help leverage emerging opportunities to improve protection of the Arctic marine environment. Through both international and domestic action, the EU and US can promote environmental leadership and provide resources in shaping agendas and policy making. Working together, the EU and US can help ensure that environmental protection measures are on the agenda in international fora, thereby promoting multilateral cooperation on issues of global significance. Working separately, although perhaps in coordination, the two can lead through their own policy measures, setting examples and best practices and serving as a model for international rules and the evolution of future governance frameworks.

References

ACIA (2005) Arctic Climate Impact Assessment. Cambridge University Press, New York

AMAP (1998) AMAP Assessment Report: Arctic Pollution Issues. Arctic Monitoring and Assessment Programme, Oslo

AMAP (2011) Snow, Water, Ice and Permafrost in the Arctic (SWIPA): Climate Change and the Cryosphere. Arctic Monitoring and Assessment Programme, Oslo

AMAP (2012) Arctic Climate Issues 2011: Changes in Arctic Snow, Water, Ice and Permafrost. SWIPA 2011 Overview Report. Oslo: Arctic Monitoring and Assessment Programme

AMSA (2009) Arctic Marine Shipping Assessment 2009 Report. Arctic Council, US

AOR (2012) Draft of 16 Oct 2012 of the Phase II Report of the Arctic Ocean Review Project (doc. AC-SAO-NOV12-4.4a)

Arctic MOPPR Agreement (2013) Agreement on Cooperation on Marine Oil Pollution Preparedness and Response in the Arctic, scheduled to be signed at the Arctic Council's Ministerial Meeting in Kiruna

Arctic SAR Agreement (2011) Agreement on Cooperation on Aeronautical and Maritime Search and Rescue in the Arctic, 12 May 2011, 50 I.L.M. 1119 (2011). Entered into force on 19 Jan 2013

ARPA (1984) Arctic Research and Policy Act of 1984, Pub. L. No. 98–373, amended by Pub. L. No 101–609 (1990)

Best A, Cavalieri S, Jariabka M, Koivurova T, Mehling M, Molenaar EJ (2009a) Transatlantic Policy Options for Supporting Adaptations in the Marine Arctic. Arctic TRANSFORM

Best A, Czarnecki R, Koivurova T, Molenaar EJ (2009b) Comparative policy analyses: U.S., EU and transatlantic Arctic policy. Arctic TRANSFORM

Bird KJ, Charpentier RR, Gautier DL, Houseknecht DW, Klett TR, Pitman JK, Moore TE, Schenk CJ, Tennyson ME, Wandrey CJ (2008) Circum-Arctic Resource Appraisal: Estimates of Undiscovered Oil and Gas North of the Arctic Circle. U.S. Geological Survey, Fact Sheet 2008–3049

Conference of Parliamentarians (2008) Conference Report. The 8th Conference of Parliamentarians of the Arctic Region, Fairbanks, 12–14 Aug 2008

Convention on Biological Diversity (1992) Convention on Biological Diversity, 5 Jun 1992, 1760 U.N.T.S. 79. Entered into force 29 Dec 1991

Council of the European Union (2009) Council conclusions on Arctic issues, 2985th Foreign Affairs Council meeting, 8 Dec 2009
EEA (2012) Report on the future of the EU Energy policy and its implications for the EEA. European Economic Area Joint Parliamentary Committee, Ref. 1118389, 27 Nov 2012
Ehler C (2011) Marine Spatial Planning in the Arctic: A first step toward ecosystem-based management. Aspen Institute
Espoo Convention (1991) Convention on Environmental Impact Assessment in a Transboundary Context, 25 Feb 1989, 1989 U.N.T.S. 309. Entered into force 10 Sept 1997
European Commission (2007) Communication from the Commission to the European Parliament, The Council, the European Economic and Social Committee and the Committee of the Regions: An Integrated Maritime Policy for the European Union, 10 Oct 2007, COM (2007) 575 final
European Commission (2008) Communication from the Commission to the European Parliament and the Council: the European Union and the Arctic Region, 20 Nov 2008, COM 763
European Commission (2012) Decision of the European Parliament and of the Council derogating temporarily from Directive 2003/87/EC of the European Parliament and of the Council establishing a scheme for greenhouse gas emission allowance trading within the Community, COM (2012) 697. <http://ec.europa.eu/clima/policies/transport/aviation/docs/com_2012_697_en.pdf/>
European Commission and High Representative (2012) Joint Communication to the European Parliament and the Council. Developing a European Union Policy towards the Arctic region: progress since 2008 and next steps. 26 June 2012, JOIN(2012) 19 final
Exec. Order No. 13547 (2010) Stewardship of the Ocean, Our coasts, and the Great Lakes. 10 July 2010, 75 Fed. Reg. 43,023
Fish Stocks Agreement (1995) Agreement for the Implementation of the Provisions of the United Nations Convention on the Law of the Sea of 10 Dec 1982 relating to the Conservation and Management of Straddling Fish Stocks and Highly Migratory Fish Stocks, 4 Aug 1995, 2167 U.N.T.S. 3. Entered into force 11 Dec 2001
Hoel AH (2009) Best practices in ecosystem-based oceans management in the Arctic. report no. 129 of the Norwegian Polar Institute, Polar Environmental Centre
ICC (2008) Towards an Inuit Declaration on Arctic sovereignty. Inuit Leaders' Summit on Arctic sovereignty, Kuujjuaq, 6–7 Nov 2008
ICC (2009) A Circumpolar Inuit Declaration on Sovereignty in the Arctic. Inuit Circumpolar Council, US
Ilulissat Declaration (2008) Arctic Ocean Conference. Ilulissat, Greenland. 27 May 2008, 48 I.L.M. 382 (2009)
IUCN/NRDC (2010) IUCN/NRDC Workshop on Ecosystem-based Management in the Arctic Marine Environment. Workshop Report, Washington, D.C. International Union for the Conservation of Nature and Natural Resources Defense Council
Juda L, Hennessey T (2001) Governance profiles and the management of the uses of large marine ecosystems. Ocean Dev Int Law 32:43–69
Kyoto Protocol to the United Nations Framework Convention on Climate Change (1996) 11 Dec 1997, 2303 U.N.T.S. 148. Entered into force 16 Feb 2005
LOS Convention (1982) United Nations Convention on the Law of the Sea, 10 Dec 1982, 1833 U.N.T.S. 396. Entered into force 16 Nov 1994
Marine Strategy Framework Directive (2008) Directive 2008/56/EC of the European Parliament and the Council of 17 June 2009 establishing a framework for community action in the field of marine environmental policy, June 17 2009, 2008 O.J. (L 164)
MARPOL 73/78 (1973/1978) International Convention for the Prevention of Pollution from Ships, 2 Nov 1973, 2 I.L.M. 1319 (1973), as modified by the 1978 Protocol Relating to the International Convention for the Prevention of Pollution from Ships (17 Feb 1978, 17 I.L.M. 546 (1978)) and the 1997 Protocol to Amend the International Convention for the Prevention of Pollution from Ships (26 Sept 1997) and as regularly amended
Ministry of the Environment (2013) Arctic Environment ministers agreed to strengthen cooperation to protect the Arctic environment. Ministery of the Environment, Government Offices of Sweden. <http://www.government.se/sb/d/17129/a/208655>. Accessed 10 Feb 2013

National Ocean Council (2012) Draft National Ocean Policy Implementation Plan, 12 Jan 2012. <http://www.whitehouse.gov/sites/default/files/microsites/ceq/national_ocean_policy_draft_implementation_plan_01-12-12.pdf>. Accessed 5 Feb 2013

NSPD-66 (2009) National Security Presidential Directive and Homeland Security Presidential Directive. Arctic Region Policy, 9 Jan 2009, NSPD-66/HSPD-25

OPRC 90 (1990) International Convention on Oil Pollution Preparedness, Response, and Cooperation, 30 Nov 1990. 30 I.L.M. 733 (1991). Entered into force 13 May 1995

OSPAR Convention (1992) Convention for the Protection of the Marine Environment of the North-East Atlantic, 22 Sept 1992, 32 I.L.M. 1072 (1993). Entered into force 25 Mar 1998

Ottawa Declaration (1996) Declaration on the Establishment of the Arctic Council, 19 Sept 1996, 35 I.L.M. 1382

PAME (2009) Arctic Offshore Oil and Gas Guidelines. Last updated 29 Apr 2009

Pettersen T (2012) 46 vessels through Northern Sea Route. Barents Observer. <http://barentsobserver.com/en/arctic/2012/11/46-vessels-through-northern-sea-route-23-11>. Accessed 10 Feb 2013

Polar Shipping Guidelines (2009) Guidelines for ships operating in polar waters, IMO assembly resolution A.1024(26), 2 Dec 2009

SAO (2007) Meeting of Senior Arctic officials: Final Report. Narvik, 28–29 Nov 2007

SAO (2012a) Arctic Council Ecosystem-Based Management Experts Group Intersessional Report: Knowledge and Process Needs for Arctic EBM. Doc 3.7b. Haparanda, Nov 2012

SAO (2012b) Experts Group on Arctic EBM: Draft Outline of Report to SAOs. Doc 3.7c. Haparanda, Nov 2012

Sherman K (1994) Sustainability, biomass yields, and health of coastal ecosystems: An ecological perspective. Mar Ecol Prog Ser 112:277–301

S.J. Res. 17 (2007) A joint resolution directing the United States to initiate international discussions and take necessary steps with other Nations to negotiate an agreement for managing migratory and transboundary fish stocks in the Arctic Ocean, 110th Cong

SLCF Task Force (2011) Progress Report and Recommendations for Ministers. Arctic Council Task Force on Short-Lived Climate Forcers

Speer L, Laughlin T (2010) IUCN/NRDC Workshop to Identify Areas of Ecological and Biological Significance or Vulnerability in the Arctic Marine Environment. Workshop report, La Jolla

TFEU (2008) Treaty on the Functioning of the European Union. Consolidated version. O.J. C 115/47

United Nations Framework Convention on Climate Change (1992) 1771 U.N.T.S. 107. Entered into force 21 Mar 1994

Annex: List of Relevant Treaties, Instruments, and Agreements

Short title	Full title and citation
AEPS	Arctic Environmental Protection Strategy, 14 January 1991, 30 I.L.M. 1624 (1991)
Antarctic Treaty	The Antarctic Treaty, 1 December 1959, 402 U.N.T.S. 71. Entered into force 23 June 1961
Anti-fouling Convention	International Convention on the Control of Harmful Anti-fouling Systems on Ships, 5 October 2001, IMO Doc. AFS/CONF/26, of 18 October 2001. Entered into force 17 September 2008
Arctic SAR Agreement	Agreement on Cooperation on Aeronautical and Maritime Search and Rescue in the Arctic, 12 May 2011, 50 I.L.M. 1119 (2011). Entered into force on 19 January 2013
Arctic Shipping Guidelines	Guidelines for Ships Operating in Arctic Ice-Covered Waters, IMO MSC/Circ. 1056, MEPC/Circ. 399, 23 December 2002
Basel Convention	Basel Convention on the Control of Transboundary Movements of Hazardous Wastes and their Disposal. 22 March 1989, 1673 U.N.T.S. 57. Entered into force 5 May 1992
BWM Convention	International Convention for the Control and Management of Ships' Ballast Water and Sediments, 13 February 2004, 30 I.L.M. 1455 (1991). Not in force, IMO Doc. BWM/CONF/36, of 16 February 2004
Bunker Oil Convention	International Convention on Civil Liability for Bunker Oil Pollution Damage, 23 March 2001, 402 U.N.T.S. 71. Entered into force 21 November 2008
CBD	Convention on Biological Diversity, 5 June 1992, 1760 U.N.T.S. 79. Entered into force 29 December 1991
CBS Convention	Convention on the Conservation and Management of Pollock Resources in the Central Bering Sea, 16 June 1994, 34 I.L.M. 67 (1995). Entered into force 8 December 1995
CITES	Convention on International Trade in Endangered Species of Wild Fauna and Flora, 3 March 1973, 993 U.N.T.S. 243. Entered into force 1 July 1975
Civil Liability Convention	International Convention on Civil Liability for Oil Pollution Damage, Brussels, 29 November 1969, 9 I.L.M. 45 (1970). Entered into force 19 June 1975. Replaced and entered into force 30 May 1996

(continued)

E. Tedsen et al. (eds.), *Arctic Marine Governance*,
DOI: 10.1007/978-3-642-38595-7, © Springer-Verlag Berlin Heidelberg 2014

(continued)

Short title	Full title and citation
CMS / Bonn Convention	Convention on the Conservation of Migratory Species of Wild Animals, 23 June 1979, 1651 U.N.T.S. 33. Entered into force 1 November 1983
COLREG 72	Convention on the International Regulations for Preventing Collisions at Sea, 20 October 1972, 1050 U.N.T.S. 16. Entered into force 15 July 1977
Compliance Agreement	Agreement to Promote Compliance with International Conservation and Management Measures by Fishing Vessels on the High Seas, 24 November 1993, 33 I.L.M. 969 (1994). Entered into force 24 April 2003
Espoo Convention	Convention on Environmental Impact Assessment in a Transboundary Context, 25 February 1989, 1989 U.N.T.S. 309. Entered into force 10 September 1997
Fish Stocks Agreement	Agreement for the Implementation of the Provisions of the United Nations Convention on the Law of the Sea of 10 December 1982 relating to the Conservation and Management of Straddling Fish Stocks and Highly Migratory Fish Stocks, 4 August 1995, 2167 U.N.T.S. 3. Entered into force 11 December 2001
Fund Convention	International Convention on the Establishment of an International Fund for Compensation for Oil Pollution Damage, 18 December 1971, 11 I.L.M. 284 (1972). Entered into force 16 October 1978
Gothenburg Protocol	Protocol to the 1979 Convention on Long-range Transboundary Air Pollution to Abate Acidification, Eutrophication and Ground-level Ozone, 30 November 1999, EB.AIR/1999/1. Entered into force 17 May 2005
Helsinki Protocol	Protocol to the 1979 Convention on Long-Range Transboundary Air pollution on the Reduction of Sulphur Emissions or their Transboundary Fluxes by at least 30 percent, 14 June 1985, 1480 U.N.T.S. 215. Entered into force 2 September 1987
HNS Convention	International Convention on Liability and Compensation for Damage in Connection with the Carriage of Hazardous and Noxious Substances by Sea, 3 May 1996, 35 I.L.M. 1406 (1996). Not in force
HNS Protocol	Protocol on Preparedness, Response and Co-operation to Pollution Incidents by Hazardous and Noxious Substances, 15 March 2000, IMO Doc. HNS-OPRC/CONF/11/Rev.1, of 15 March 2000. Entered into force 14 June 2007
ICCPR	International Covenant on Civil and Political Rights, 16 December 1966, 999 U.N.T.S. 17. Entered into force 23 March 1976
ICRW	International Convention for the Regulation of Whaling, 2 December 1946, 161 U.N.T.S. 72. Entered into force 10 November 1948
ILO Convention No. 169	Convention concerning Indigenous and Tribal Peoples in Independent Countries, 27 June 1989, 28 I.L.M. 1382 (1989). Entered into force 5 September 1991
Ilulissat Declaration	Ilulissat Declaration, Arctic Ocean Conference. Ilulissat, Greenland. 27 May 2008, 48 I.L.M. 382 (2009)

(continued)

(continued)

Short title	Full title and citation
Kyoto Protocol	Kyoto Protocol to the United Nations Framework Convention on Climate Change, 11 December 1997, 2303 U.N.T.S. 148. Entered into force 16 February 2005
London Convention/LC	Convention on the Prevention of Marine Pollution by Dumping of Wastes and Other Matter. 29 December 1972, 1046 U.N.T.S. 120. Entered into force 30 August 1957
London Protocol/LP	Protocol to the Convention on the Prevention of Marine Pollution by Dumping of Wastes and Other Matter, 7 November 1996, 36 I.L.M. 7 (1997). Entered into force 24 March 2006
LOS Convention	United Nations Convention on the Law of the Sea, 10 December 1982, 1833 U.N.T.S. 396. Entered into force 16 November 1994
LRTAP Convention	Convention on Long-range Transboundary Air Pollution, 13 November 1979, 1302 U.N.T.S. 217. Entered into force 16 March 1983
MARPOL 73/78	International Convention for the Prevention of Pollution from Ships, 2 November 1973, 2 I.L.M. 1319 (1973). Entered into force 2 October 1983
Montreal Protocol	Montreal Protocol on Substances that Deplete the Ozone Layer, 16 September 1987, 1522 U.N.T.S. 3. Entered into force 1 January 1989
Murmansk Treaty	Treaty between the Kingdom of Norway and the Russian Federation concerning Maritime Delimitation and Cooperation in the Barents Sea and the Arctic Ocean, 15 September 2010, U.N.T.S. Reg. No. 49095. Entered into force 7 July 2011
NAFO Convention	Convention on Future Multilateral Cooperation in the Northwest Atlantic Fisheries, 24 October 1978, 1135 U.N.T.S. 369. Entered into force 1 January 1979
NASCO Convention	Convention for the Conservation of Salmon in the North Atlantic Ocean, 2 March 1982, 1338 U.N.T.S. 33 (1983). Entered into force 1 October 1983
NEAFC Convention	Convention on Future Multilateral Cooperation in the North-East Atlantic Fisheries, 18 November 1980, 1285 U.N.T.S. 129. Entered into force 17 March 1982
NPAFC	Convention for the Conservation of Anadromous Stocks in the North Pacific Ocean, 11 February 1992. 22 Law of the Sea Bulletin 21 (1993). Entered into force 16 February 1993
OPRC Convention/OPRC 90	International Convention on Oil Pollution Preparedness, Response, and Cooperation, 30 November 1990. 30 I.L.M. 733 (1991). Entered into force 13 May 1995
OSPAR Convention	Convention for the Protection of the Marine Environment of the North-East Atlantic, 22 September 1992, 32 I.L.M. 1072 (1993). Entered into force 25 March 1998
Ottawa Declaration	Declaration on the Establishment of the Arctic Council, 19 September 1996, 35 I.L.M. 1382 (1996)
Part XI Deep-Sea Mining Agreement	Agreement relating to the Implementation of Part XI of the United Nations Convention on the Law of the Sea of 10 December 1982, 28 July 1994. 1836 U.N.T.S. 3. Entered into force 28 July 1996

(continued)

(continued)

Short title	Full title and citation
Polar Bear Agreement	International Agreement on the Conservation of Polar Bears, 15 November 1973, 13 I.L.M. 3 (1974). Entered into force 26 May 1976
Polar Shipping Guidelines	Guidelines for ships operating in polar waters, IMO Assembly Resolution A.1024(26), 2 December 2009
POPs Protocol	Protocol to the 1979 Convention on LRTAP on Persistent Organic Pollutants (POPs), 24 June 1998, 2230 U.N.T.S. 79. Entered into force 23 October 2003
Port State Measures Agreement	Agreement on Port State Measures to Prevent, Deter and Eliminate Illegal, Unreported and Unregulated Fishing, 22 November 2009. Not in force (at 27 November 2012)
Ramsar Convention	Convention on Wetlands of International Importance Especially as Waterfowl Habitat, 2 February 1971, 996 U.N.T.S. 245. Entered into force 21 December 1975
Rovaniemi Declaration	Rovaniemi Declaration (1991). Declaration on the Protection of the Arctic Environment, 14 June 1991, 30 I.L.M. 1624 (1991)
SAR Convention	International Convention on Maritime Search and Rescue, 27 April 1979. 1405 U.N.T.S. 118. Entered into force 22 June 1985
Stockholm Convention	Stockholm Convention on Persistent Organic Pollutants, 22 May 2001, 2256 U.N.T.S. 119. Entered into force 17 May 2004
SEA Protocol	Protocol on Strategic Environmental Assessment to the Convention on Environmental Impact Assessment in a Transboundary Context, 21 May 2001, UNECE Document ECE/MP.EIA/2003/2. Entered into force 11 July 2010
Sofia Protocol	Protocol to the 1979 Convention on Long-Range Transboundary Air Pollution Concerning the Control of Emissions of Nitrogen Oxides or their Transboundary Fluxes, 31 October 1988, 28 I.L.M. 212 (1989). Entered into force 14 February 1991
SOLAS	International Convention for the Safety of Life at Sea, 1 November 1974, 1184 U.N.T.S. 278. Entered into force 25 May 1980
Spitsbergen Treaty	Treaty Concerning the Archipelago of Spitsbergen, 9 February 1920, 2 L.N.T.S. 7. Entered into force 14 August 1925
STCW Convention	International Convention on Standards of Training, Certification and Watchkeeping for Seafarers, 1 December 1978, 1361 U.N.T.S. 2 Entered into force 28 Apr l 1984. As amended and modified by the 1995 Protocol
UNDRIP	United Nations Declaration on the Rights of Indigenous Peoples. General Assembly Resolution. New York, 13 September 2007. A/RES/61/295
UNESCO World Heritage Convention	Convention Concerning the Protection of the World Cultural and Natural Heritage, 16 November 1972, 1037 U.N.T.S. 151. Entered into force 17 December 1975
UNFCCC	United Nations Framework Convention on Climate Change, 9 May 1992, 1771 U.N.T.S. 107. Entered into force 21 March 1994

(continued)

(continued)

Short title	Full title and citation
Vienna Convention	Vienna Convention for the Protection of the Ozone Layer, 22 March 1985, 1513 U.N.T.S. 293. Entered into force 22 September 1988
VOC Protocol	Protocol to the 1979 Convention on LRTAP concerning the Control of Emissions from Volatile Organic Compounds or their Transboundary Fluxes, 18 November 1991, 2001 U.N.T.S. 187. Entered into force 29 September 1997

MIX
Papier aus verantwortungsvollen Quellen
Paper from responsible sources
FSC® C105338

If you have any concerns about our products,
you can contact us on
ProductSafety@springernature.com

In case Publisher is established outside the EU,
the EU authorized representative is:
Springer Nature Customer Service Center GmbH
Europaplatz 3, 69115 Heidelberg, Germany

Printed by Libri Plureos GmbH
in Hamburg, Germany